Optional Processes

CHAPMAN & HALL/CRC
Financial Mathematics Series

Aims and scope:
The field of financial mathematics forms an ever-expanding slice of the financial sector. This series aims to capture new developments and summarize what is known over the whole spectrum of this field. It will include a broad range of textbooks, reference works and handbooks that are meant to appeal to both academics and practitioners. The inclusion of numerical code and concrete real-world examples is highly encouraged.

Series Editors

M.A.H. Dempster
Centre for Financial Research
Department of Pure Mathematics and Statistics
University of Cambridge

Dilip B. Madan
Robert H. Smith School of Business
University of Maryland

Rama Cont
Department of Mathematics
Imperial College

Interest Rate Modeling

Theory and Practice, 2nd Edition

Lixin Wu

Metamodeling for Variable Annuities

Guojun Gan and Emiliano A. Valdez

Modeling Fixed Income Securities and Interest Rate Options

Robert A. Jarrow

Financial Modelling in Commodity Markets

Viviana Fanelli

Introductory Mathematical Analysis for Quantitative Finance

Daniele Ritelli, Giulia Spaletta

Handbook of Financial Risk Management

Thierry Roncalli

Optional Processes: Theory and Applications

Mohamed Abdelghani, Alexander Melnikov

For more information about this series please visit: *https://www.crcpress.com/Chapman-and-HallCRC-Financial-Mathematics-Series/book-series/CHFINANCMTH*

Optional Processes
Theory and Applications

Mohamed Abdelghani
Morgan Stanley, New York, USA

Alexander Melnikov
University of Alberta

CRC Press
Taylor & Francis Group
Boca Raton London New York

CRC Press is an imprint of the
Taylor & Francis Group, an **informa** business

A CHAPMAN & HALL BOOK

First edition published 2020
by CRC Press
6000 Broken Sound Parkway NW, Suite 300, Boca Raton, FL 33487-2742

and by CRC Press
2 Park Square, Milton Park, Abingdon, Oxon, OX14 4RN

First issued in paperback 2022

© 2020 Taylor & Francis Group, LLC
CRC Press is an imprint of Taylor & Francis Group, an Informa business

Visit the Taylor & Francis Web site at
http://www.taylorandfrancis.com

and the CRC Press Web site at
http://www.crcpress.com

ISBN 13: 978-0-367-50851-7 (pbk)
ISBN 13: 978-1-138-33726-8 (hbk)
ISBN 13: 978-0-429-44249-0 (ebk)

DOI: 10.1201/9780429442490

Typeset in CMR
by Nova Techset Private Limited, Bengaluru & Chennai, India

Contents

Preface

This book deals with the subject of stochastic processes on *un*usual probability spaces and their applications to mathematical finance, stochastic differential equations and filtering theory. *Un*usual probability spaces are probability spaces where the information σ-algebras are neither right- nor left, neither continuous nor complete. The term *un*usual belongs to the famous expert in the general theory of stochastic processes, C. Dellacherie, who initiated the study of *un*usual probability spaces about 50 years ago. These spaces allow for the existence of a richer class of stochastic processes, such as *optional* martingales and semimartingales.

Further developments at that period were done by Lepingle, Horowitz, Lenglart and Galtchouk. In these publications, a modern version of stochastic analysis was created under the "*un*usual conditions". In the usual theory, right-continuous and left limits (r.c.l.l.) semimartingales are measurable with respect to an optional σ-algebra on the product of sample space and time, generated by all right-continuous processes adapted to a right-continuous and complete filtration. However, in the *un*usual case they are not, and it is necessary to assume that they are optional processes so as to provide the existence of their regular modifications which admit finite right limits and left limits (r.l.l.l.). The existence of such initial theory calls for its further development as well as for further applications in today's times with new research challenges.

Besides the pure theoretical motivation for studying optional processes, we should note that optional processes can be used in different areas of applications. In particular, one can provide a strong motivation for this research that comes from the modeling of electricity spot prices, the complexity of which lies in the modeling of severe and frequent price spikes. In this case, optional processes allow for a more accurate representation of the spot price dynamics compare to cadlag processes.

Therefore, in this book we present a comprehensive treatment of the stochastic calculus of optional processes on *un*usual probability spaces and its applications. We begin with a foundation chapter on the analytic basis of optional processes, then there are several chapters on the stochastic calculus of martingales and semimartingales. We will cover many topics such as linear stochastic differential equations with respect to optional semimartingales; solutions to the nonhomogeneous linear stochastic differential equation; Gronwall lemma; existence and uniqueness of solutions of stochastic equations of optional semimartingales under monotonicity conditions; and theorems on comparison of solutions of stochastic equations of optional semimartingale under Yamada conditions.

Furthermore, a financial market model based on optional semimartingales is presented and methods for finding local martingale deflators are also given. Arbitrage pricing and hedging of contingent claims in these markets are treated. A new theory of defaultable markets on *un*usual probability spaces is presented. Also, several examples of financial applications are given: a laglad jump diffusion model and a portfolio of a defaultable bond and a stock.

The book also contains a version of the uniform Doob-Meyer decomposition of optional semimartingales. Its fundamental role in mathematical finance is well established, due to its application in superhedging problems in incomplete markets. It is shown here that this decomposition may be very helpful in the construction of optimal filters in the filtering problem for optional semimartingales which covers a variety of well-known models.

Our work is and has been supported by Natural Sciences and Engineering Research Council of Canada discovery grants #5901 and #RES0043487, University of Alberta, Canada, which we are grateful for. The authors would like to thank Professor Charles Tapiero (New York University), Professor Robert Elliott (University of Calgary) and Professor Svetlozar Rachev (Texas Tech University) for their enthusiastic support of such a challenging book project. We are also grateful to Andrey Pak, a PhD candidate in the mathematical finance program of the University of Alberta, for his help. We are very grateful to Mansi Kabra and editorial assistants of CRC Press.

We would like to dedicate this work to our families and hope it will better the future prospects of humanity. Finally, gratias ago ∞.

Mohamed Abdelghani
Morgan Stanley, New York City, USA

Alexander Melnikov
University of Alberta, Edmonton, Canada

Introduction

The stochastic basis: $(\Omega, \mathcal{F}, \mathbf{F} = (\mathcal{F}_t)_{t \geq 0}, \mathbf{P})$ – is a probability space $(\Omega, \mathcal{F}, \mathbf{P})$ equipped with a non-decreasing family, $\mathbf{F}$, of σ-algebras also known as a filtration or information flow, such that $\mathcal{F}_t \in \mathbf{F}$, $\mathcal{F}_s \subseteq \mathcal{F}_t$, for all $s \leq t$. It is a key notion of the general theory of stochastic processes. The theory of stochastic processes is well developed under the so-called "usual conditions": $\mathcal{F}_t$ is complete for all time t, that is $\mathcal{F}_0$ is augmented with $\mathbf{P}$ null sets from $\mathcal{F}$, and $\mathcal{F}_t$ is right-continuous, $\mathcal{F}_t = \mathcal{F}_{t+} = \cap_{u>t} \mathcal{F}_u$. Under these convenient conditions, adapted processes from a very large class, known as semimartingales, can be seen as processes with right-continuous and left limits paths (r.c.l.l.).

The stochastic calculus of r.c.l.l. semimartingales on usual spaces is also a well-developed part of the theory of stochastic processes. This calculus has a number of excellent applications in different areas of modern probability theory, mathematical statistics and other fields, especially mathematical finance. Many fundamental results and constructs of modern mathematical finance were also proved with the help of the general theory of stochastic processes under the usual conditions. It is difficult to imagine how to get these results using other techniques and approaches. Moreover, these areas of research and applications are interconnected; namely, not only is the general theory of stochastic processes, often called stochastic analysis, important for mathematical finance, but also the needs of mathematical finance sometimes lead to fundamental results for stochastic analysis. An excellent example of such influence is the so-called optional or uniform Doob-Meyer decomposition of positive supermartingale.

In spite of the pervasiveness of the "usual conditions" in stochastic analysis, there are examples showing the existence of a stochastic basis without the "usual conditions" (see for instance Fleming & Harrington (2011) [1] p.24). If we suppose

$$X_t = \mathbf{1}_{t>t_0} \mathbf{1}_A,$$

where A is $\mathcal{F}$-measurable with $0 < \mathbf{P}(A) < 1$ then filtration $\mathcal{F}_t$ generated by the history of X, $\mathcal{F}_t = \sigma(X_s, s \leq t)$ is not right continuous at t_0, i.e., $A \notin \mathcal{F}_{t_0}$, but $A \in \mathcal{F}_{t_0+}$, and it is not possible to make it right-continuous in a useful way. Furthermore, is it "*un*natural" to assume the usual conditions are true? The completion and right continuity are an arbitrary construct to make it easy to carry on with analysis and prove certain results that would have been rather difficult to prove otherwise. Let us consider the notion of completion; completion requires that we know a priori all the null sets of $\mathcal{F}$ and augment the initial σ-algebra $\mathcal{F}_0$ with these null sets. In other words, it is an initial completion of the probability space by future null sets. Moreover, assuming that the σ-algebras $(\mathcal{F}_t)$ are right-continuous is also rather *un*natural; it means that the immediate future is equivalent to the present which is different from the past! Also, with right continuity of filtration, events like "$(\tau(\omega) = t)$" for all t must have a null total probability. All this leads us to believe that the usual conditions are too restrictive and rather not natural to assume.

Moreover, as we have noted earlier stochastic processes on usual probability spaces lead to a calculus of r.c.l.l. semimartingales only. However, as we will see later in this work there are many processes that are not r.c.l.l., consider, for example, the sum or product of a left continuous and a right continuous semimartingale. Does a calculus of such processes exist? And on what types of probability spaces can it be defined? We know that such processes must be mostly excluded from the framework of stochastic processes on a usual basis, and, if they were to be considered their use must be loaded with assumptions and restrictions. Consequently, famous experts of stochastic analysis, Doob, Meyer and Dellacherie, initiated studies of stochastic processes without the assumptions of the usual conditions and cadlag paths; one may also note that this problem is emphasized in the book of Kallianpur (2013) [2] in context of filtering theory. Dellacherie called this case the "*un*usual conditions", and the stochastic basis became known as the *un*usual stochastic basis or the *un*usual probability space. We will follow this terminology in this work. Dellacherie and Meyer began their studies with the process,

$$\text{"} \mathbf{E}\left[X|\mathcal{F}_t\right] \text{"} \tag{0.1}$$

where X is some bounded random variable in $\mathcal{F}$, and $\mathcal{F}_t$ is not complete, or right- or left-continuous. Their goal was to find out if there exists a reasonable adapted modification of the conditional expectation (0.1). They proved the following projection theorem:

Theorem 0.0.1 Let X be a bounded random variable, and then there is a version X_t of the martingale $\mathbf{E}\left[X|\mathcal{F}_t\right]$ possessing the following properties: X_t is an optional process and for every stopping time T, $X_T \mathbf{1}_{T<\infty} = \mathbf{E}\left[X\mathbf{1}_{T<\infty}|\mathcal{F}_T\right]$ a.s.

It turns out that optional processes on *un*usual stochastic basis have right limits and left limits (r.l.l.l.) but are not necessarily right- or left-continuous. Actually, stochastic processes that are r.l.l.l. appear to exist even when the usual conditions are satisfied. Mertens (1972) [3] showed that under the usual conditions for X, a positive optional strong supermartingale of class D with $X_{0-} = X_0$ and $X_\infty = 0$, there exists an integrable, i.e., $\mathbf{E}[A] < \infty$, increasing, predictable, r.l.l.l. process A such that,

$$X_T = \mathbf{E}\left[A_\infty|\mathcal{F}_T\right] - A_T, \tag{0.2}$$

for every stopping time T. Moreover, A is unique and the following equalities between processes hold: $\triangle^+ A = A_+ - A = X - {}^p X$, $\triangle A = A - A_- = X - {}^o\left(X_+\right)$ where ${}^p X$ and ${}^o X$ are the predictable and optional projections of the process X, respectively. In particular, A is right-continuous if and only if X is right-continuous and left-continuous if and only if $X_- = {}^p X$. An optional process X is an optional strong martingale (resp. supermartingale) if for every bounded stopping time T, X_T is integrable and for every pair of bounded stopping times S, T such that $S < T$, $X_S = \mathbf{E}\left[X_T|\mathcal{F}_S\right]$ (resp. $X_S > \mathbf{E}\left[X_T|\mathcal{F}_S\right]$ a.s.). Mertens decomposition (0.2) was later generalized by Dellacherie and Meyer (1975) [4] under the *un*usual conditions.

Many mathematicians have contributed directly or indirectly to the theory of optional processes on *un*usual probability spaces such as Doob (1975) [5] and Lepingle (1977) [6], Horowitz (1978) [7], Lenglart (1980) [8]. However, much of the theoretical foundation and stochastic calculus of the theory of stochastic processes on *un*usual probability spaces were formulated mostly by Galchuk in several papers published between 1975 and 1985 [9, 10, 11]. In these publications, a parallel theory of stochastic analysis was constructed for optional processes on *un*usual probability spaces. The existence of such theory calls for a new initiative for its further developments as well as for further applications in a very well-developing area of mathematical finance as a natural and promising reserve for further studies (see, for example, Gasparyan [12], Khun and Stroh [13] and recent papers by Abdelghani and Melnikov [14, 15, 16, 17, 18]).

This book is organized as follows. The first three chapters are foundational knowledge on probability theory, stochastic processes and martingale theory, respectively. Chapter 4 is the first work that we know of on optional strong supermartingales with the *un*usual conditions; in this chapter we cover the theory of strong supermartingales, Mertenz decomposition and Snell's envelope. In Chapter 5 we introduce optional martingales decomposition and integration theory. In Chapter 6 we cover optional supermartingale Doob-Meyer-Glachuk decomposition in detail. Chapter 7 is on the calculus of optional semimartingale: formula for change of variables, integration with respect to random optional measures and uniform Doob-Meyer decomposition of optional supermartingales. At this point we come to the second part of the book where we present applications to differential equations, financial markets and filtering theory. In Chapter 8 of the second part of the book, we study optional linear stochastic equations: stochastic exponential, logarithms and nonhomogeneous linear equations. Furthermore, we explore comparison of solutions and existence and uniquess of nonlinear optional stochastic integral equations under monotonicity conditions. Chapter 9 is where we bring in some financial market applications of the theory of optional processes; we introduce an optional portfolio theory and discuss pricing and hedging and non-arbitrage conditions. We also present optional defaultable market and some results on the probability of default. Finally in the last chapter we present the application of uniform Doob-Meyer decomposition to the theory of optimal filtering of optional supermartingales and conclude the book with a financial application of filtering theory.

We recommend that the book be read in the following order. One should begin with Chapter 5 on optional martingales followed by Chapters 6 and 7, using the first four chapters of the book as references. Then, read Chapter 8, specifically the sections on linear stochastic equations. After that, the reader can proceed to the chapter on financial applications, Chapter 9. Chapter 11 on filtering theory can be read after that, followed by the sections on nonlinear stochastic equations. Finally, if the readers wish, they can read the first four chapters in some detail.

Chapter 1

Spaces, Laws and Limits

In this chapter and the next three, we bring to readers parts of the wonderful work carried out by Dellacherie and Meyer in their books, Probability and Potential A [19] and B [20], especially chapters IV on stochastic processes and VI on theory of martingales as well as the appendix of book B on strong supermartingales. Excerpts and summary of concepts, definitions and theorems stated there, that are requisite for the development of the calculus of optional processes on *un*usual probability spaces, are therefore brought here to lay down the basic foundation of the theory of optional processes.

1.1 Foundation

Intuitively, a set is a collection of objects described by simply listing its elements separated by commas "," within braces "{}". A set can also be characterized by a property of its elements. Examples of sets are, the natural numbers $\mathbb{N}$ and the real numbers $\mathbb{R}$. Interestingly, the natural numbers can be characterized entirely by sequences of ternary abjad: comma, left and right braces, as follows: let 0 be the empty set {}, 1 be {{}}, $2 := \{\{\}, \{\{\}\}\}$ and so on.

Sets are usually endowed with few simple operations. A union $\cup$ of two sets A and B is $A \cup B$ which means, to simply combine their elements. The intersection $\cap$ of two sets, $A \cap B$, means to find the common elements between the two sets. The complement of a set A is denoted by $\complement A$ or by A^c, means to find all those elements that are not in A. The difference between two sets is $A \backslash B = A \cap B^c$ and the symmetric difference is $A \triangle B = (A \backslash B) \cup (B \backslash A)$. The set of all elements x in E ($x \in E$) with some property P is denoted by $\{x \in E : P(x)\}$ or by $\{x : P(x)\}$, or simply by $\{P\}$ if there is no ambiguity. A subset B of E is denoted by $B \subseteq E$ if all elements $x \in B$ then $x \in E$.

Given a family of subsets $\mathcal{E}$ of E, the restriction to A is denoted by $\mathcal{E}|_A$ – is the set of traces on A of elements of $\mathcal{E}$: $\mathcal{E}|_A = \{B \cap A, B \in \mathcal{E}\}$. A family $\mathcal{E}$ of sets is closed under (...), where the brackets contain set theoretic operation symbols, sometimes followed by the letters f, c, a, m, which abbreviate respectively: finite, countable, arbitrary and monotone. For example, $\mathcal{E}$ is closed under

($\cup$f,$\cap$a) means finite unions of elements of $\mathcal{E}$ and arbitrary intersections of elements of $\mathcal{E}$ still belong to $\mathcal{E}$; $\mathcal{E}$ is closed under ($\cup$mc,^{c}) means that monotone countable unions of elements of $\mathcal{E}$ (i.e., unions of increasing sequences in $\mathcal{E}$) still belong to $\mathcal{E}$ and complements of elements of $\mathcal{E}$ still belong to $\mathcal{E}$. The closure of a family of subsets under ($\cup$c) (resp. ($\cap$c)) is denoted by $\mathcal{E}_\sigma$ (resp. $\mathcal{E}_\delta$). We write $((\mathcal{E})_\sigma)_\delta = \mathcal{E}_{\sigma\delta}$.

A function f is a special relation that assigns to each input element of a set, its domain, a single output element from another set, its range. Suppose the function f is defined on E, and A is a subset of E ($A \subset E$) then the restriction of f to the set A is denoted by $f|_A$. The limit $s \uparrow t$ means $s \to t$ and $s \le t$, that is, s can be arbitrarily close to t, within some $\epsilon > 0$, and possibly equal to it. Whereas, the limit $s \uparrow\uparrow t$ means $s \to t$ and $s < t$. Similarly, for sequences $s_n \uparrow t$ and $s_n \uparrow\uparrow t$ with the additional meaning that (s_n) is increasing. Obvious changes are required if $\downarrow$ and $\downarrow\downarrow$ appear instead.

A topological space is a set E endowed with a topology $\mathcal{E}$. The topology $\mathcal{E}$ is a set of subsets of E that is closed under ($\cup$a,$\cap$f). The sets of $\mathcal{E}$ are called open sets. A topological space E is *separable* if it contains a countable dense set. It is compact if each of its open covers has a finite subcover. An open cover if each of its members is an open set whose union contains a given set, for example E. A topological space is *Hausdorff* space, separated space or T2 space if each pair of distinct points can be separated by a disjoint open set.

A filter $\mathfrak{F}$ on a set E is a family of non-empty subsets of E such that it is closed under ($\cap$f) and if $A \in \mathfrak{F}$ and $A \subset B \subset E$ then $B \in \mathfrak{F}$. An ultrafilter is a filter of E such that it is properly contained in no filter in E.

A metric space is a set E with a metric d – a function from $E \times E$ to $\mathbb{R}$ such that for any $x, y, z \in E$, $\mathrm{d}(x, y) = 0$ if and only if $x = y$, $\mathrm{d}(x, y) = \mathrm{d}(y, x)$ and $\mathrm{d}(x, z) \le \mathrm{d}(x, y) + \mathrm{d}(y, z)$. A complete metric space (or a *Cauchy* space), if every Cauchy sequence of points in the metric space has a limit in the metric space. A completely metrizable topological space is a topological space $(E, \mathcal{E})$ for which there exists at least one metric d on E such that (E, d) is a complete metric space and d induces the topology $\mathcal{E}$.

A topological space is *Polish* if it is a separable and completely metrizable topological space, or, in other words, if there exists a distance compatible with its topology under which E is complete and separable. A space is locally compact if every point of the space has a compact neighborhood. Locally compact spaces with countable base are called LCC spaces.

A function f is continuous if pre-image of every open set is an open set or the limit exists for every point in the domain. A function f is lower semi-continuous (l.s.c.) if $f(x) \le \liminf_{y \to x} f(y)$ and upper semi-continuous (u.s.c.) if $f(x) \ge \liminf_{y \to x} f(y)$ for any x and y in the domain.

A *measurable space* is a set E together with an algebra of subsets of E including E, and is closed under any countable unions of subsets of E and complement – ($\cup$c,$\complement$). A *measure space* is a measurable space endowed with a measure. A *measure* without qualification always means positive countably additive set function on a measurable space. Some of the measures that we will consider in this book are σ-finite where the space E is the union of a countable family of sets of finite measure.

For a *signed-measure* μ, μ^+ and μ^- are positive and negative variations, are both positive, and the total variation is $|\mu| = \mu^+ + \mu^-$. $|\mu|$ also denotes the total mass $< \mu, 1 >$ of μ which is sometimes infinite. The sup and inf of two measures are denoted by $\lambda \vee \mu = \sup(\lambda, \mu)$ and $\lambda \wedge \mu = \inf(\lambda, \mu)$. The integral of a function f with respect to a measure μ is $\int f(x)\mu(dx)$ also denoted by $\mu(f)$ or $< \mu, f >$ and often abridged to $\int f\mu$. The "ratio" $\frac{\mu}{\nu}$ denotes a Radon-Nikodym density, without "d"; when the measure μ on $\mathbb{R}$ appears as the derivative of an increasing function F, we use the standard notation with "d" for Stieltjes integral $\int f(x)dF(x)$ instead of $F(dx)$. If μ is a probability law, we often write $\mathbf{E}[f]$ for $\int f\mu$ and $\mathbf{E}[f, A]$ for $\int_A f\mu$.

If E is a topological space, $\mathcal{C}(E), \mathcal{C}_{\mathrm{b}}(E), \mathcal{C}_{\mathrm{c}}(E), \mathcal{C}_0(E)$ denote the spaces of real-valued functions which are, respectively, continuous, bounded and continuous, continuous with compact support, continuous and tending to 0 at infinity. Adjoining a "$+$" to this notation, for example $\mathcal{C}^+(E)$ and so on, enables us to denote the corresponding cones of positive functions. As usual $\mathcal{C}_{\mathrm{c}}^\infty(E)$ denotes the space of infinitely differentiable functions with compact support.

If $(E, \mathcal{E})$ is a measurable space, the notation $\mathfrak{M}(\mathcal{E})$ (resp. $\mathcal{B}(\mathcal{E})$) denotes the space of $\mathcal{E}$-measurable (resp. bounded $\mathcal{E}$-measurable) real functions. Spaces of measures are used only on a Hausdorff topological space E: $\mathfrak{M}_{\mathrm{b}}^+(E), \mathfrak{M}^+(E)$ then are the cones of bounded (resp. arbitrary) Radon measures on E and $\mathfrak{M}_{\mathrm{b}}(E), \mathfrak{M}(E)$ are the vector spaces generated by these cones.

Other relevant definitions and theorems will be introduced when needed.

1.2   Measurable Spaces, Random Variables and Laws

1.2.1   Measurable spaces

The heart of probability theory are algebras of possible events known as the σ-fields.

Definition 1.2.1 Let Ω be a set; a σ-field $\mathcal{F}$ on Ω is a family of subsets which contains the empty set and is closed under the operations ($\cup$c,$\cap$c,^{c}). The pair $(\Omega, \mathcal{F})$ is called a *measurable space* and the elements of $\mathcal{F}$ are called measurable or $\mathcal{F}$-measurable sets.

The word σ-field is synonymous with σ-algebra; for that, there is an algebra of sets defined by ($\cup$c,$\cap$c,^{c}).

In probability theory, measurable sets are called events, where Ω is called the *sure event* and the empty set is called the *impossible event*. The operation of taking complements is called passing to the *opposite event*. Sometimes, we may use the phrases: "the event A occurs", "the events A and B occur simultaneously" and "the events A and B are incompatible", to express the set theoretic relations: $\omega \in A$, $\omega \in A \cap B$ and $A \cap B = \emptyset$, respectively.

1.2.2   Measurable functions

Definition 1.2.2 Let $(\Omega, \mathcal{F})$ and $(E, \mathcal{E})$ be two measurable spaces. The mapping f of Ω into E is measurable if

$$f^{-1}(A) \in \mathcal{F} \text{ for all } A \in \mathcal{E}.$$

In the language of probability theory, f is also called a random variable (r.v.). The sum, inverse and composition of random variables is a random variable – measurable function in analysis.

Definition 1.2.3 (a) Let Ω be a set, and $\mathcal{G}$ a family of subsets of Ω. The σ-field generated by $\mathcal{G}$, denoted by $\sigma(\mathcal{G})$, is the smallest σ-field of subsets of Ω containing $\mathcal{G}$.
(b) Let $(f_i)_{i \in I}$ be a family of mappings of Ω into measurable spaces $(E_i, \mathcal{E}_i)_{i \in I}$. The σ-field generated by the mappings f_i denoted by $\sigma(f_i, i \in I)$, is the smallest σ-field of subsets of Ω with respect to which all the mappings f_i are measurable.

In Definition 1.2.3 (a) and (b) are related, since, the σ-field generated by a set of subsets is also generated by indicator functions of these subsets; moreover, the σ-field generated by the mappings f_i is also generated by the family of subsets $f_i^{-1}(A_i)$ where $A_i \in \mathcal{E}_i$ for all i.

Suppose f is a mapping from the measurable space $(E, \mathcal{E})$ into Ω. Then, f is measurable with respect to the σ-field $\sigma(\mathcal{A})$ if and only if $f^{-1}(A) \in \mathcal{E}$ for all $A \in \mathcal{A} \subseteq \Omega$. Similarly, f is measurable with respect to the σ-field $\sigma(f_i, i \in I)$ on Ω if and only if each mapping $f_i \circ f$ is measurable. The σ-field generated by a family of functions $(f_i)_{i \in I}$ is identical to the union in Boolean Algebra on Ω ($\mathfrak{B}(\Omega)$) of all the σ-fields $\sigma(f_i, i \in J)$ with J running through the family of all countable subsets of I.

Definition 1.2.4 Let $(E_i, \mathcal{E}_i)_{i \in I}$ be a family of measurable spaces and the product set $E = \prod_{i \in I} E_i$. Let $\pi_i : E \to E_i$ be the coordinate mappings for $i \in I$. The σ-field $\sigma(\pi_i, i \in I)$ is called the product σ-field of the σ-fields $\mathcal{E}_i$ and is denoted by $\prod_{i \in I} \mathcal{E}_i$.

Notation 1.2.5 We denote the product of two σ-fields by $\mathcal{E}_1 \times \mathcal{E}_2$ and sometimes we may denote the product of many σ-fields as $\times_{i \in I} \mathcal{E}_i$.

1.2.3 Atoms and separable fields

Let $(\Omega, \mathcal{F})$ be a measurable space. The *atoms* of $\mathcal{F}$ are the *equivalence* classes in Ω for the relation

$$\mathbf{1}_A(\omega) = \mathbf{1}_A(\omega'), \text{ for all } A \in \mathcal{F} \tag{1.1}$$

and ω, $\omega' \in \Omega$. Every measurable mapping on Ω with values in a separable metrizable space (e.g., $\mathbb{R}$) – being a limit of elementary functions – is constant on atoms.

The measurable space $(\Omega, \mathcal{F})$ is called *Hausdorff* if the atoms of $\mathcal{F}$ are the points of Ω. If otherwise the space $(\Omega, \mathcal{F})$ is not Hausdorff, then we can define an associated Hausdorff space by defining the quotient space $\mathring{\Omega}$ of Ω by using the equivalence relation (1.1) and $\mathring{\mathcal{F}}$ is the σ-field consisting of the images of elements of $\mathcal{F}$ into $\mathring{\Omega}$ under the canonical mapping of Ω onto $\mathring{\Omega}$.

A measurable space $(\Omega, \mathcal{F})$ is *separable* if there exists a sequence of elements of $\mathcal{F}$ which generates $\mathcal{F}$. A Hausdorff space associated with a separable space is also separable.

Two measurable spaces are *isomorphic* if there is a measurable bijection between them with a measurable inverse. A measurable bijection between topological spaces with Borel fields is a *Borel isomorphism*. A measurable space isomorphic to a separable metrizable space with a Borel σ-field is a *separable Hausdorff space*. Let $(E, \mathcal{E})$ be a measurable separable Hausdorff space, then it is isomorphic to, a not necessarily Borel subspace of $\mathbb{R}$ with it Borel σ-field $\mathcal{B}(\mathbb{R})$.

Let $(\Omega, \mathcal{F})$ be a measurable space and $(E, \mathcal{E})$ be a separable Hausdorff measurable space, then examples of useful measurable sets are: the diagonal of $E \times E$ belongs to the product σ-field $\mathcal{E} \times \mathcal{E}$; if f is a measurable mapping of Ω into E, the graph of f in $\Omega \times E$ belongs to the product σ-field $\mathcal{F} \times \mathcal{E}$; if f and g are measurable mappings of into E, the set $\{f = g\}$ belongs to $\mathcal{F}$.

1.2.4 The case of real-valued random variables

Elementary functions describe random variables that take countably or finitely many values. Functions that take values in $\bar{\mathbb{R}} = \mathbb{R} \cup \{+\infty, -\infty\}$ are called *extended real-valued* functions, whereas real-valued functions aren't allowed the values $\pm\infty$.

Let f and g be two extended real-valued random variables defined on $(\Omega, \mathcal{F})$, then the functions $f \wedge g$, $f \vee g$, $f + g$ and fg, if defined everywhere, are random variables.

If a sequence of extended real-valued r.v. $(f_n)_{n \in \mathbb{N}}$ converges pointwise to f, then f is a random variable. Pointwise convergence means that for any $\omega \in \Omega$ and $\epsilon > 0$ there is $N_\epsilon(\omega) \in \mathbb{N}$ such that f_n is to within an ϵ closeness to f for all $n \geq N_\epsilon(\omega)$.

Theorem 1.2.6 An extended real-valued function f is measurable if and only if there exists a sequence (f_n) of measurable elementary functions which increases to f.

The sequence,

$$f_n = \sum_{k \in \mathbb{Z}} k2^{-k} \mathbf{1}_{\{k2^{-k} < f \leq (k+1)2^{-k}\}} + (-\infty) \mathbf{1}_{\{f = -\infty\}} \tag{1.2}$$

is useful and is known as the Lebesgue's approximation of f which converges uniformly to f.

Uniform convergence is stronger than pointwise and it means that for any $\epsilon > 0$ there is $N_\epsilon \in \mathbb{N}$ for all such that f_n is to within an ϵ nearness to f for all $n \geq N_\epsilon$ where N_ϵ is independent of any $\omega \in \Omega$.

The next theorem shows that the notion of a $\sigma(f)$ measurable r.v. may be replaced by the notion of measurable functions of f.

Theorem 1.2.7 Let f be a random variable defined on $(\Omega, \mathcal{F})$ with values in $(E, \mathcal{E})$ and g a real-valued function defined on Ω. Then, g is $\sigma(f)$-measurable if and only if there exists a real-valued random variable h on E such that $g = h \circ f$.

1.2.5   Monotone class theorem

We come to an extremely useful and fundamental result of probability theory known as the *monotone class theorem*:

Theorem 1.2.8 Let $\mathcal{C}$ be a family of subsets of Ω containing $\emptyset$ and closed under $(\cup f, \cap f)$. Let $\mathfrak{M}$ be a family of subsets of Ω containing $\mathcal{C}$, and closed under $(\cup mc, \cap mc)$; $\mathfrak{M}$ is known as a *monotone class*.

Then, $\mathfrak{M}$ contains the closure $\mathcal{K}$ of $\mathcal{C}$ under $(\cup c, \cap c)$. Furthermore, If $\mathcal{C}$ is closed under the complement operation, $\complement$, then $\mathfrak{M}$ also contains the σ-field generated by $\mathcal{C}$.

Proof. First, we call a set of subsets closed under $(\cup f, \cap f)$ a *horde*. Let $\mathfrak{H}$ be a maximal horde among the hordes contained in $\mathfrak{M}$ and containing $\mathcal{C}$ (Zorn's Lemma). We will show that $\mathfrak{H}$ is closed under $(\cup c, \cap c)$.

Let (A_n) be a decreasing sequence of elements of $\mathfrak{H}$ and let $A = \cap A_n$. The family of all subsets of the form: $(H \cap A) \cup H'$ with $H \in \mathfrak{H} \cup \{\Omega\}$ and $H' \in \mathfrak{H}$, is a horde containing $\mathfrak{H}$ (by taking $H = \emptyset$) and A (by taking $H = \Omega$ and $H' = \emptyset$) and contained in $\mathfrak{M}$. But since $\mathfrak{H}$ is maximal, this horde must be identical to $\mathfrak{H}$, therefore, $A \in \mathfrak{H}$. In other words, $\mathfrak{H}$ is closed under $(\cap c)$, so that $\mathfrak{H}$ contains the closure of $\mathcal{C}$ under $(\cap c)$. The argument is similar for $(\cup c)$.

Let $\mathfrak{J}$ be the set of all $A \in \mathcal{K}$ such that $A^c \in \mathcal{K}$. If the complement of every element of $\mathcal{C}$ belongs to $\mathcal{C}$, or more generally to $\mathcal{K}$, then $\mathfrak{J}$ contains $\mathcal{C}$, in particular, $\emptyset \in \mathfrak{J}$. Obviously then $\mathfrak{J}$ is a σ-field contained in $\mathfrak{M}$, hence, the last sentence of the theorem is true. ∎

The following lemma illustrates an application of the monotone class Theorem 1.2.8.

Lemma 1.2.9 Let $\mathcal{F}_0$ be a set of subsets of Ω closed under $(\cup f,^c)$. Let $\mathbf{P}$ and $\mathbf{P}'$ be two measures on $\mathcal{F} = \sigma(\mathcal{F}_0)$ such that $\mathbf{P}(A) = \mathbf{P}'(A)$ for all $A \in \mathcal{F}_0$. Then, $\mathbf{P}$ and $\mathbf{P}'$ are equal on $\mathcal{F}$.

Another most often used statement of the monotone class theorem is a functional form of it, stated as follows:

Theorem 1.2.10 Let $\mathcal{K}$ be a vector space of bounded real-valued functions on Ω which contains the constants. Furthermore, let $\mathcal{K}$ be closed under uniform convergence and has the following property: for every uniformly bounded increasing sequence of positive functions $f_n \in \mathcal{K}$, the function $f = \lim f_n$ belongs to $\mathcal{K}$. Let $\mathcal{C}$ be a subset of $\mathcal{K}$ which is closed under multiplication. The space $\mathcal{K}$ then contains all bounded functions measurable with respect to the σ-field $\sigma(\mathcal{C})$.

A family of uniformly bounded functions is a family of functions that are all bounded by the same constant.

1.2.6 Probability and expectation

Definition 1.2.11 A probability law $\mathbf{P}$ on a measurable space $(\Omega, \mathcal{F})$ is a measure defined on $\mathcal{F}$ such that $\mathbf{P}$ is a *positive function* on $\mathcal{F}$ with $\mathbf{P}(\Omega) = 1$, which satisfies the *countable additivity* property: $\mathbf{P}(\cup_n A_n) = \sum_n \mathbf{P}(A_n)$ for every sequence $(A_n)_{n \in \mathbb{N}}$ of *disjoint events*. The triple $(\Omega, \mathcal{F}, \mathbf{P})$ is called a probability space.

The number $\mathbf{P}(A)$ is called the probability of the event A. An event whose probability is equal to 1 is said to be almost sure. Let f and g be two random variables defined on $(\Omega, \mathcal{F})$ with values in the same measurable space $(E, \mathcal{E})$. If the set $\{\omega : f(\omega) = g(\omega)\}$ is an event of probability 1, we write $f = g$ a.s. where "a.s." is an abbreviation of almost surely. Similarly, we shall write $A = B$ a.s. to express that two events A and B differ only by a set of zero probability. Generally, we use the expression "almost surely" in the same way "almost everywhere" is used in measure theory.

Definition 1.2.12 A probability space $(\Omega, \mathcal{F}, \mathbf{P})$ is called complete if every subset A of, which is contained in a $\mathbf{P}$-negligible set belongs to the σ-field (then necessarily $\mathbf{P}(A) = 0$).

Note that any probability space can be completed, but we will not prove this here.

Definition 1.2.13 Let $(\Omega, \mathcal{F}, \mathbf{P})$ be a probability space and f be a random variable. The integral $\int_\Omega f(\omega)\mathbf{P}(d\omega)$ is called the expectation of the random variable f and is denoted by the symbol $\mathbf{E}[f]$.

A random variable f is integrable if $\mathbf{E}|f| < \infty$. Let f be an integrable random variable which is measurable with respect to a sub-σ-field $\mathcal{G}$ of $\mathcal{F}$. Then, f is a.s. positive, if and only if, $\int_A f(\omega)\mathbf{P}(d\omega) \geq 0$ for all $A \in \mathcal{G}$; to show this, take A to be the event $\{f < 0\}$.

It follows, in particular, that two integrable random variables f and g which are both $\mathcal{G}$-measurable and have the same integral on every set A of $\mathcal{G}$ are a.s. equal.

Next, we state two of the most used results on measurable functions: the dominated convergence theorem and Fatou's lemma.

Theorem 1.2.14 Dominated (Lebesgue's) Convergence. Let $(f_n)_{n\in\mathbb{N}}$ be a sequence or real-valued random variables that converge almost surely to the random variable f, $\lim_n f_n = f$ a.s. If f_n are bounded in absolute value by some integrable function g, then, f is integrable and $\mathbf{E}[f] = \lim_n \mathbf{E}[f_n]$.

The $\lim_n f_n = f$ a.s. means $\mathbf{P}\{\omega : \lim_n f_n(\omega) = f(\omega)\} = 1$.

Suppose f is a positive random variable which may not be finite but is not integrable, then, we use the convention $\mathbf{E}[f] = +\infty$. In such case, the following lemma hold:

Lemma 1.2.15 Fatou's. Let $(f_n)_{n\in\mathbb{N}}$ be a sequence of positive random variables, then we have

$$\mathbf{E}[\liminf_n f_n] \leq \liminf_n \mathbf{E}[f_n].$$

Remark 1.2.16 We can obtain the *Lebesgue's monotone convergence* theorem if the inequality in Fatou's Lemma is replaced by equality when the sequence of f_n is increasing, whether the expectations are finite or not.

Let's denote by $\mathrm{L}^p(\Omega, \mathcal{F}, \mathbf{P})$ (L^p for short) the vector space of real-valued random variables whose p-th power is integrable ($1 \leq p < \infty$). And, by L^p, the quotient space of L^p defined by the equivalence relation of almost sure equality. For every real-valued measurable function f, we set

$$||f||_p = (\mathbf{E}[|f|^p])^{\frac{1}{p}} \quad (\text{possibly } +\infty).$$

Similarly, we denote by $\mathrm{L}^\infty(\Omega, \mathcal{F})$ the space (note the independence from $\mathbf{P}$) of bounded random variables, with the norm of uniform convergence, $|f|_\infty = \sup_\Omega |f|$. Denote by $\mathsf{L}^\infty(\Omega, \mathcal{F}, \mathbf{P})$ the quotient space of $\mathrm{L}^\infty(\Omega, \mathcal{F})$ by the equivalence relation of a.s. equality. The norm of an element f of L^∞ is the essential supremum of $|f|$ denoted by $||f||_\infty$.

Some useful results to keep in mind about L^p spaces are: L^p is a Banach space, the dual of L^1 is L^∞ and Holder's inequality and Radon-Nikodym theorem applies.

Furthermore, let f and g be two integrable random variables; we say that f and g are orthogonal if the product fg is integrable and has zero expectation. Let $\mathcal{G}$ denote a sub-σ-field of $\mathcal{F}$, U be the closed subspace of L^1 consisting of all classes of $\mathcal{G}$ measurable random variables, and V be the subspace of L^∞ consisting of all classes of bounded random variables orthogonal to every element of U. It follows from the Hahn-Banach theorem that every random variable $f \in L^1$ orthogonal to every element of V is a.s. equal to $\mathcal{G}$-measurable function.

1.2.6.1 Convergence of random variables

We now recall convergence of real-valued random variables restricting ourselves to the case of sequences of random variables. Let (f_n), $n \in \mathbb{N}$, be a sequence of random variables defined on $(\Omega, \mathcal{F}, \mathbf{P})$. We say that the sequence (f_n) converges to a random variable f:

- almost surely if $\mathbf{P}\{\omega : f_n(\omega) \to f(\omega)\} = 1$;

- in probability if $\lim_n \mathbf{P}\{(u : |f_n(\omega) - f(\omega)| > \epsilon\} = 0$ for all $\epsilon > 0$;

- in the strong sense in L^p if the f_n and f belong to L^p and $\lim_n \mathbf{E}[|f_n - f|^p] = 0$;

- in the weak sense in L^1 if f_n and f belong to L^1 and, for every random variable $g \in L^\infty$, $\lim_n \mathbf{E}[fg] = \mathbf{E}[fg]$;

- in the weak sense in L^2 if f_n and f belong to L^2 and, for every random variable $g \in L^2$, $\lim_n \mathbf{E}[f_n g] = \mathbf{E}[fg]$.

Recall, also, the following facts: almost sure convergence and strong convergence in L^p imply convergence in probability and that every sequence which converges in probability contains a subsequence which converges almost surely.

1.2.6.2 Fubini's Theorem

Definition 1.2.17 Let $(\Omega, \mathcal{F})$ and $(E, \mathcal{E})$ be two measurable spaces. A family $(\mathbf{P}_x)_{x \in E}$ of probability laws on $(\Omega, \mathcal{F})$ is said to be $\mathcal{E}$-measurable if the function $x \longmapsto \mathbf{P}_x(A)$ is $\mathcal{E}$-measurable for all $A \in \mathcal{F}$.

Given the family $(\mathbf{P}_x)_{x \in E}$ we state the Fubini theorem:

Theorem 1.2.18 Fubini's. Let $\mathbf{Q}$ be a probability laws on $(E, \mathcal{E})$ and $(U, \mathcal{U})$ be the measurable space $(E \times \Omega, \mathcal{E} \times \mathcal{F})$. Then,
(1) let f be a real-valued r.v. defined on $(U, \mathcal{U})$; each one of the partial mappings $x \mapsto f(x, \omega)$ and $\omega \mapsto f(x, \omega)$ is measurable on the corresponding factor space;
(2) there exists one and only one probability law ξ on $(U, \mathcal{U})$ such that, for all $A \in \mathcal{E}$ and $B \in \mathcal{F}$,

$$\xi(A \times B) = \int_A \mathbf{P}_X(B)\mathbf{Q}(dx). \tag{1.3}$$

(3) if f is a positive r.v. on $(U, \mathcal{U})$, then, the function

$$x \mapsto \int_\Omega f(x, \omega) \mathbf{P}_x(d\omega).$$

is $\mathcal{E}$-measurable and

$$\int_U f(x, \omega) \xi(dx, d\omega) = \int_E \mathbf{Q}(dx) \int_\Omega f(x, \omega) \mathbf{P}(d\omega). \tag{1.4}$$

When all $\mathbf{P}_x$ are equal to the same law $\mathbf{P}$, the law $\mathbf{D}$ is called the product law of $\mathbf{Q}$ and $\mathbf{P}$, denoted by $\mathbf{Q} \otimes \mathbf{P}$. The probability space $(U, \mathcal{U}, \mathbf{Q} \otimes \mathbf{P})$ may not in general be complete.

1.2.6.3 Uniform integrability

Let H be a subset of the space $L^1(\Omega, \mathcal{F}, \mathbf{P})$;

Definition 1.2.19 H is called a uniformly integrable set if the integrals

$$\int_{\{|f| \geq c\}} |f(\omega)| \mathbf{P}(d\omega) \quad (f \in H) \tag{1.5}$$

tend uniformly to 0 as the positive number c tends to $+\infty$.

Conditions whereby one can assess that a set of functions are uniformly integrable follows from the following theorem:

Theorem 1.2.20 For H to be uniformly integrable, it is necessary and sufficient that the following conditions hold: (a) the expectations $\mathbf{E}|f|$ for all f in H, are uniformly bounded; (b) for every $\epsilon > 0$, there exists a number $\delta > 0$ such that the conditions $A \in \mathcal{F}$, $\mathbf{P}(A) \leq \delta$ imply the inequality

$$\int_A |f(\omega)| \mathbf{P}(d\omega) \leq \epsilon \quad f \in H \tag{1.6}$$

The next theorem helps us understand the significance of uniform integrability.

Theorem 1.2.21 The following properties are equivalent:
(1) H is uniformly integrable.
(2) There exists a positive function $G(t)$ defined on $\mathbb{R}_+$ such that $\lim \frac{1}{t} G(t) = \infty$ and

$$\sup_{f \in H} \mathbf{E}\left[G \circ |f|\right] < \infty. \tag{1.7}$$

The next result generalizes dominated convergence theorem where uniform integrability is used.

Theorem 1.2.22 Let (f_n) be a sequence of integrable random variables which converges almost everywhere to a random variable f. Then, f is integrable and f_n converges to f in the strong sense in L^1, if and only if the f_n are uniformly integrable. Furthermore, If f_n's are positive, it is also necessary and sufficient that:

$$\lim_n \mathbf{E}[f_n] = \mathbf{E}[f] < \infty. \tag{1.8}$$

1.2.6.4   Completion of probability spaces

Definition 1.2.23 Let $(\Omega, \mathcal{F}, \mathbf{P})$ be a probability space. A set $A \subset \Omega$ is called internally $\mathbf{P}$-negligible if $\mathbf{P}(B) = 0$ for every $B \in \mathcal{F}$ contained in A.

Theorem 1.2.24 Let $\mathcal{N}$ be a family of subsets of Ω which satisfies the following conditions: (1) Every element of $\mathcal{N}$ is internally $\mathbf{P}$-negligible. (2) $\mathcal{N}$ is closed under ($\cup c$). Let $\mathcal{F}^*$ be the σ-field generated by $\mathcal{F}$ and $\mathcal{N}$. The law then can be extended uniquely to a law $\mathbf{P}^*$ on $\mathcal{F}^*$ such that every element of $\mathcal{N}$ is $\mathbf{P}^*$- negligible.

Theorem 1.2.24 implies the possibility of completing any probability space $(\Omega, \mathcal{F}, \mathbf{P})$.

Remark 1.2.25 Let $\mathcal{F}^{\mathbf{P}}$ be a completed σ-field, we can characterize the elements of $\mathcal{F}^{\mathbf{P}}$ as follows: every element of $\mathcal{F}^{\mathbf{P}}$ can be expressed as $F \Delta M$, where F belongs to $\mathcal{F}$ and M is contained in some $\mathbf{P}$-negligible set $N \in \mathcal{F}$. Then, $F \Delta M$ lies between the two sets $F \backslash N$ and $F \cup N$, which belong to $\mathcal{F}$ and differ only by a negligible set.

Remark 1.2.26 A real-valued function f is measurable relative to the completed σ-field $\mathcal{F}^{\mathbf{P}}$, if and only if there exist two $\mathcal{F}$-measurable real-valued functions g and h such that $g \le f \le h$ and $\mathbf{P}\{g \ne h\} = 0$.

An interesting concept is that of universal completion, where by which every law on $\mathbf{P}$ on $(\Omega, \mathcal{F})$ can be extended uniquely to a law $\hat{\mathbf{P}}$ on $\hat{\mathcal{F}}$, as defined next.

Definition 1.2.27 Universal Completion. Let $(\Omega, \mathcal{F})$ be a measurable space; for each law $\mathbf{P}$ on $(\Omega, \mathcal{F})$ consider the completed σ-field $\mathcal{F}^{\mathbf{P}}$ and denote by $\hat{\mathcal{F}}$ the intersection of all the σ-fields $\mathcal{F}^{\mathbf{P}}$: the measurable space $(\Omega, \hat{\mathcal{F}})$ is called the universal completion of $(\Omega, \mathcal{F})$.

1.2.6.5   Independence

Where would we be without the concept of independence of random variables?

Definition 1.2.28 Suppose $(X_i)_{i \in I}$ is a *finite* family of random variables from a probability space $(\Omega, \mathcal{F}, \mathbf{P})$ to measurable spaces $(E_i, \mathcal{E}_i)_{i \in I}$. And, let X be the random variable $(X_i)_{i \in I}$ with values in the product space $(\prod_i E_i, \prod_i \mathcal{E}_i)_{i \in I}$. The random variables X_i (or the family (X_i)) are said to be independent if the law of X is the product of the laws of the X_i.

Furthermore, if $(X_i)_{i \in I}$ is an *arbitrary* family of random variables, then the family (X_i) is said to be independent if every finite subfamily is independent. In other words, the random variables $(X_i)_{i \in I}$ are independent if and only if

$$\mathbf{P}\{\forall i \subset J, X_i \in A_i\} = \prod_{i \in J} \mathbf{P}\{X_i \in A_i\}$$

for every finite subset $J \subset I$ and every family $(A_i)_{i \in J}$ such that $A_i \in \mathcal{E}_i$ for all $i \in J$.

Independence as regards σ-fields is defined as follows:

Definition 1.2.29 Let $(\Omega, \mathcal{F}, \mathbf{P})$ be a probability space and let $(\mathcal{F}_i)_{i \in I}$ be a family of sub-σ-fields of $\mathcal{F}$. The σ-fields are called independent if $\mathbf{P}(\cap_{i \in J} A_i) = \prod_{i \in J} \mathbf{P}(A_i)$ for every finite subset $J \subset I$ and every family of sets $(A_i)_{i \in J}$ such that $A_i \in \mathcal{F}_i$ for all $i \in J$.

Theorem 1.2.30 Let $\mathcal{F}_1, \mathcal{F}_2, ..., \mathcal{F}_n$ be independent σ-fields and let $f_1, f_2, ..., f_n$ be integrable real-valued random variables measurable relative to the corresponding σ-fields. Then, the product $f_1 f_2 ... f_n$ is integrable and

$$\mathbf{E}[f_1 f_2 ... f_n] = \mathbf{E}[f_1]\mathbf{E}[f_2]...\mathbf{E}[f_n].$$

1.2.6.6   Conditional expectation

The notion of conditional probability and expectation is essential to probability theory.

Theorem 1.2.31 Let $(\Omega, \mathcal{F}, \mathbf{P})$ be a probability space and f be a r.v. from $(\Omega, \mathcal{F})$ to some measurable space $(E, \mathcal{E})$. Let $\mathbf{Q}$ be the image law of $\mathbf{P}$ under f. Let X be a $\mathbf{P}$-integrable r.v. on $(\Omega, \mathcal{F})$. There exists a $\mathbf{Q}$-integrable random variable Y on $(E, \mathcal{E})$ such that, for every set $A \in E$:

$$\int_A Y(x)\mathbf{Q}(dx) = \int_{f^{-1}(A)} X(\omega)\mathbf{P}(d\omega). \tag{1.9}$$

If Y' is any r.v. satisfying (1.9), then $Y' = Y$ a.s.

The image law of **P** under f, denoted by $f(\mathbf{P})$, is the law **Q** on $(E, \mathcal{E})$ defined by: $Q(A) = \mathbf{P}(f^{-1}(A))$ $(A \in \mathcal{E})$. This law is also called the law of, or the distribution of, f.

Definition 1.2.32 Let Y be $\mathcal{F}$-measurable and **Q**-integrable r.v. satisfying relation (1.9). We call Y *a version of* the conditional expectation of X given f, denoted by $\mathbf{E}[X|f]$.

If $X = \mathbf{1}_B$ is the indicator of an event B, then $\mathbf{E}[\mathbf{1}_B|f] = \mathbf{P}(B|f)$ is the conditional probability of B given f. Keep in mind that this "probability" is a random variable, up to equivalence, not a number.

Let's make few more remarks before moving on to define conditioning on σ-algebras.

Remark 1.2.33 (a) Partition Ω into a sequence of measurable sets A_n and let f be the mapping of Ω into $\mathbb{N}$, equal to n on A_n nil otherwise. Define a measure **Q** on $\mathbb{N}$ by $\mathbf{Q}(\{n\}) = \mathbf{P}(A_n)$ and let X be an integrable r.v. on Ω; we can compute $Y = \mathbf{E}[X|f]$ by

$$Y(n) = \mathbf{P}(A_n)^{-1} \int_{A_n} X\mathbf{P} \text{ for all such that } \mathbf{P}(A_n) \neq 0.$$

If $\mathbf{P}(A_n)$ is zero, $Y(n)$ can be chosen arbitrarily. Suppose in particular that X is the indicator of an event B then

$$Y(n) = \frac{\mathbf{P}(B \cap A_n)}{\mathbf{P}(A_n)} \text{ if } \mathbf{P}(A_n) \text{ is non-zero;}$$

from this we recognize the number which is called the conditional probability of B given that A_n, in elementary probability theory.

(b) Given an arbitrary r.v. X; X has a generalized conditional expectation if $\mathbf{E}[X^+|f]$ and $\mathbf{E}[X^-|f]$ are finite a.s. and we set $\mathbf{E}[X|f] = \mathbf{E}[X^+|f] - \mathbf{E}[X^-|f]$.

A far more important variant of Definition 1.2.32 of conditional expectations and most commonly used, we get by taking E to be Ω, $\mathcal{E}$ to be a sub-σ-field of $\mathcal{F}$ and f to be the identity mapping. The image measure **Q** is then the restriction of **P** to $\mathcal{E}$ and we get the following definition:

Definition 1.2.34 Let $(\Omega, \mathcal{F}, \mathbf{P})$ be a probability space, $\mathcal{E}$ be a sub-σ-field of $\mathcal{F}$ and X be an integrable r.v. A version of the conditional expectation of X given $\mathcal{E}$ is any $\mathcal{E}$-measurable integrable random variable Y such that

$$\int_A X(\omega)\mathbf{P}(d\omega) = \int_A Y(\omega)\mathbf{P}(d\omega) \text{ for all } A \in \mathcal{E}. \tag{1.10}$$

Often we omit the word version and simply denote Y by the notation $\mathbf{E}[X|\mathcal{E}]$ or $\mathcal{E}X$ by some authors. A random variable X when not assumed to be positive or integrable has a generalized conditional expectation given $\mathcal{E}$ if and only if the measure $|X|\mathbf{P}$ is σ-finite on $\mathcal{E}$.

Remark 1.2.35 (a) If $\mathcal{E}$ is the field $\sigma(f_i, i \in I)$ generated by a family of random variables, then the conditional expectation of X given the f_i's is written as $\mathbf{E}[X|f_i, i \in I]$. In particular, if X is the indicator of an event A, then we may speak of the conditional probability of A given $\mathcal{E}$ and write $\mathbf{P}(A|\mathcal{E})$ or $\mathbf{P}(A|f_i, i \in I)$.

(b) If $\mathcal{E}$ is the σ-field $\sigma(f)$, then we have that $\mathbf{E}[X|\mathcal{E}] = Y \circ f$ a.s. – the composition of Y with f.

(c) Often, it happens that conditional expectations are iterated as in $\mathbf{E}[\mathbf{E}[X|\mathcal{F}_1]|\mathcal{F}_2]$, where $\mathcal{F}_1$ and $\mathcal{F}_2$ are sub-$\mathcal{F}$-σ-fields which we may write in the simpler notation $\mathbf{E}[X|\mathcal{F}_1|\mathcal{F}_2]$.

Next we list fundamental properties of conditional expectations:

Lemma 1.2.36 Let all random variables here be defined on $(\Omega, \mathcal{F}, \mathbf{P})$. Then, (1) Let X and Y be integrable random variables and a, b, c be constants. Then, for every σ-field $\mathcal{E} \subset \mathcal{F}$,

$$\mathbf{E}[aX + bY + c|\mathcal{E}] = a\mathbf{E}[X|\mathcal{E}] + b\mathbf{E}[Y|\mathcal{E}] + c \quad \text{a.s.} \tag{1.11}$$

(2) Let X and Y be integrable random variables such that $X \leq Y$ a.s. Then $E[X|\mathcal{E}] \leq E[Y|\mathcal{E}]$ a.s.
(3) Let X_n, $n \in \mathbb{N}$ being integrable random variables which increase to an integrable r.v. X. Then

$$\mathbf{E}[X|\mathcal{E}] = \lim_n \mathbf{E}[X_n|\mathcal{E}] \quad \text{a.s.} \tag{1.12}$$

(4) Jensen's inequality. Let c be a convex mapping of into and let X be an integrable r.v. such that $c \circ X$ is integrable. We then have

$$c \circ \mathbf{E}[X|\mathcal{E}] \leq \mathbf{E}[c \circ X|\mathcal{E}] \quad \text{a.s.} \tag{1.13}$$

(5) Let X be an integrable r.v.; then $\mathbf{E}[X|\mathcal{E}]$ is $\mathcal{E}$-measurable if X $\mathcal{E}$-measurable, then $X = \mathbf{E}[X|\mathcal{E}]$ a.s.
(6) Let $\mathcal{D}$, $\mathcal{E}$, be two sub-σ-fields of $\mathcal{F}$ such that $\mathcal{D} \subset \mathcal{E}$. Then for every integrable r.v. X

$$\mathbf{E}[X|\mathcal{E}|\mathcal{D}] = \mathbf{E}[X|\mathcal{D}] \quad \text{a.s.} \tag{1.14}$$

And in particular

$$\mathbf{E}[\mathbf{E}[X|\mathcal{E}]] = E[X]. \tag{1.15}$$

(7) Let X be an integrable r.v. and Y be an $\mathcal{E}$-measurable r.v. such that XY is integrable. Then

$$\mathbf{E}[XY|\mathcal{E}] = Y\mathbf{E}[X|\mathcal{E}] \quad \text{a.s.} \tag{1.16}$$

It is well known that a continuous linear operator on a Banach space B is still continuous when B is given its weak topology $\sigma(\mathrm{B}, \mathrm{B}^\star)$, see Bourbaki [21]. Consider the function $|x|^p$ for $(1 \leq p \leq \infty)$. Then,

$$||\mathbf{E}[X|\mathcal{E}]||_p \leq ||X||_p \tag{1.17}$$

by Jensen's inequality; the same inequality is obvious for $p = \infty$. So, the mapping $X \mapsto \mathbf{E}[X|\mathcal{E}]$ is an operator of norm ≤ 1 on L^p $(1 \leq p \leq \infty)$. As a result we have the following theorem:

Theorem 1.2.37 Continuity of Expectation. The conditional expectation operators are continuous for the weak topologies $\sigma(L^1, L^\infty)$ and $\sigma(L^2, L^2)$.

Definition 1.2.38 Conditional Independence. Let $(\Omega, \mathcal{F}, \mathbf{P})$ be a probability space and $\mathcal{F}_1$, $\mathcal{F}_2$, $\mathcal{F}_3$ be three sub-σ-fields of $\mathcal{F}$. $\mathcal{F}_1$ and $\mathcal{F}_3$ are called conditionally independent given $\mathcal{F}_2$ if

$$\mathbf{E}[Y_1 Y_3 | \mathcal{F}_2] = \mathbf{E}[Y_1 | \mathcal{F}_2] \mathbf{E}[Y_3 | \mathcal{F}_2] \quad \text{a.s.} \tag{1.18}$$

where Y_1, Y_3 denote positive random variables measurable with respect to the corresponding σ-fields $\mathcal{F}_1$, $\mathcal{F}_3$.

We stop here with basic probability theory and direct the reader back to the book of Dellacherie and Meyer [19] for further details and proofs and move on to review some elements of analytic set theory.

1.3 Analytic Set Theory

Here we cover elements of analytic set theory necessary to prove Choquet's Theorem on capacitability and other useful results such as Blackwell's Theorem. These results are pertinent to the development of the calculus of optional processes.

1.3.1 Paving and analytic sets

Definition 1.3.1 A paving on a set E is a family, $\mathcal{E}$ of subsets of E, which contains the empty set. The pair $(E, \mathcal{E})$ is called a paved space.

Let $(E_i, \mathcal{E}_i)$ be a family of paved sets. The product (resp. sum) pavings of $\mathcal{E}_i$ is the paving on the set $\prod_i E_i$. (resp. $\sum_i E_i$) consisting of the subsets of the form $\prod_i A_i$ (resp. $\sum_i A_i$) where $A_i \subset E_i$ belongs to $\mathcal{E}_i$ for all i – the $\sum_i E_i$ means $\bigcup_i E_i \times \{i\}$.

Definition 1.3.2 Let $(E, \mathcal{E})$ be a paved set and $(K_i)_{i \in I}$ be a family of elements of $\mathcal{E}$. We say that this family has the finite intersection property if $\bigcap_{i \in I_0} K_i \neq \emptyset$ for every finite subset $I_0 \subset I$.

Definition 1.3.2 amounts to saying that the sets K_i belong to a *filter*, and by Bourbaki *ultrafilter theorem* ([21] (3rd edition), chapter 1 section 6.4, theorem 1) we can say that there exists an ultrafilter $\mathfrak{U}$ such that $K_i \in \mathfrak{U}$ for all $i \in I$.

Definition 1.3.3 Let $(E, \mathcal{E})$ be a paved set. The paving $\mathcal{E}$ is said to be compact (resp. semi-compact) if every family (resp. every countable family) of elements of $\mathcal{E}$, which has the finite intersection property, has a non-empty intersection.

For example, if E is a Hausdorff topological space (i.e., any two distinct points belong to two disjoint open sets) the paving on E consisting of the compact subsets of E (henceforth denoted by $\mathcal{K}(E)$) is compact.

Theorem 1.3.4 Let E be a set with a compact (resp. semi-compact) paving $\mathcal{E}$ and let $\mathcal{E}'$ be the closure of $\mathcal{E}$ under $(\cup f, \cap a)$ (resp. $(\cup f, \cap c)$). Then, the paving $\mathcal{E}'$ is compact (resp. semi-compact).

Theorem 1.3.5 Let $(E_i, \mathcal{E}_i)_{i \in I}$ be a family of paved sets. If each of the pavings $\mathcal{E}_i$ is compact (resp. semi-compact) so are the product paving $\prod_{i \in I} \mathcal{E}_i$ and the sum paving $\sum_{i \in I} \mathcal{E}_i$.

Theorem 1.3.6 Let $(E, \mathcal{E})$ be a paved set and let f be a mapping of E into a set F. Suppose that, for all $x \in F$, the paving consisting of the sets $f^{-1}(\{x\}) \cap A$, $A \in \mathcal{E}$, is semi-compact. Then, for every decreasing sequence $(A_n)_{n \in N}$ of elements of $\mathcal{E}$,

$$f \left(\bigcap_{n \in \mathbb{N}} A_n \right) = \bigcap_{n \in \mathbb{N}} f(A_n). \tag{1.19}$$

Next, we present the complicated definition of analytic sets involving paving, compactness, metrizability and projection:

Definition 1.3.7 Let $(F, \mathcal{F})$ be a paved set. A subset A of F is called $\mathcal{F}$-analytic if there exists an auxiliary compact metrizable space E and a subset $B \subset E \times F$ belonging to $(\mathcal{K}(E) \times \mathcal{F})_{\sigma\delta}$ such that A is the projection of B onto F ($\pi_F B$ or B_F for short). The paving on F consisting of all $\mathcal{F}$-analytic sets is denoted by $\mathfrak{a}(\mathcal{F})$.

It follows from the definition that every $A \in \mathfrak{a}(\mathcal{F})$ is contained in some element of $\mathcal{F}_\sigma$. Also, the space F is $\mathcal{F}$-analytic if and only if it belongs to $\mathcal{F}_\sigma$.

Theorem 1.3.8 $\mathcal{F} \subset \mathfrak{a}(\mathcal{F})$ and the paving $\mathfrak{a}(\mathcal{F})$ is closed under $(\cup c, \cap c)$.

Theorem 1.3.9 (a) Let $(E, \mathcal{E})$ and $(F, \mathcal{F})$ be two paved sets; we have $a(\mathcal{E}) \times a(\mathcal{F}) \subset a(\mathcal{E} \times \mathcal{F})$.

(b) Suppose that E is a compact metrizable space and $\mathcal{E} = \mathcal{K}(E)$ and let A' be an element of $a(\mathcal{E} \times \mathcal{F})$. The projection A of A' onto F is $\mathcal{F}$-analytic.

Theorem 1.3.10 Let $(F, \mathcal{F})$ be a paved set and $\mathcal{G}$ be a paving such that $\mathcal{F} \subset \mathcal{G} \subset a(\mathcal{F})$. Then, $a(a(\mathcal{F})) = a(\mathcal{G}) = a(\mathcal{F})$. In particular, $a(\mathcal{G}) = a(\mathcal{F})$ if $\mathcal{G}$ is the closure of $\mathcal{F}$ under $(\cup c, \cap c)$.

Theorem 1.3.11 Let $(F, \mathcal{F})$ and $(G, \mathcal{G})$ be two paved sets and f be a mapping of F into G such that $f^{-1}(\mathcal{G}) \subset a(\mathcal{G})$. Then, $f^{-1}(a(\mathcal{G})) \subset a(\mathcal{F})$.

Theorem 1.3.12 $a(\mathcal{G})$ contains the σ-field $a(\mathcal{F})$ generated by $\mathcal{F}$ if and only if the complement of every element of $\mathcal{F}$ is $\mathcal{F}$-analytic.

Let $\mathcal{B}$ be the Borel σ-field of $\mathbb{R}$, and $\mathcal{K}$ a paving consisting of all compact subsets of $\mathbb{R}$. We could also replace $\mathbb{R}$ by any locally compact spaces with countable base (LCC) space or, in particular, any metrizable compact space. Then,

Theorem 1.3.13 (a) $\mathcal{B} \subset a(\mathcal{H})$, $a(\mathcal{B}) = a(\mathcal{H})$.

(b) Let $(\Omega, \mathcal{F})$ be a measurable space. The product σ-field $\mathcal{G} = \mathcal{B} \times \mathcal{F}$ on $\mathbb{R} \times \Omega$ is contained in $a(\mathcal{H} \times \mathcal{F})$.

(c) The projection onto Ω of an element of $\mathcal{G}$ (or, more generally, of $a(\mathcal{G})$) is $\mathcal{F}$-analytic.

Next, we present the separation theorem for analytic sets, then study analytic sets in special spaces where the separation theorem leads us to simple but useful results such as the Souslin-Lusin and Blackwell's Theorems.

1.3.2 Separable sets

Let $(F, \mathcal{F})$ be a paved space and by $\mathcal{C}(\mathcal{F})$ the closure of $\mathcal{F}$ under $(\cup c, \cap c)$. For example, if F is a LCC space and $\mathcal{F}$ is the paving of compact subsets of F, $\mathcal{C}(\mathcal{F})$ is the Borel σ-field.

Two subsets A and A' of F are called separable by elements of $\mathcal{C}(\mathcal{F})$ if there exist two disjoint elements of $\mathcal{C}(\mathcal{F})$ containing, respectively, A and A'. Then, the *separation theorem* is as follows:

Theorem 1.3.14 Suppose that the paving $\mathcal{F}$ is semi-compact and let A and A' be two disjoint $\mathcal{F}$-analytic sets; then, A and A' can be separated by elements of $\mathcal{C}(\mathcal{F})$.

Proof. For convenience, suppose that $F \in \mathcal{F}$ and note that this condition does not weaken the statement of the theorem. We begin with an auxiliary result:

Let (C_n) and (D_n) be two sequences of subsets of F such that C_n and D_m are separable by elements of $\mathcal{C}(\mathcal{F})$ for every pair (n, m) then the sets $\cup_n C_n$ and $\cup_m D_m$ are separable.

Indeed, we choose for each pair (n, m) two elements E_{nm}, F_{nm} of $\mathcal{C}(\mathcal{F})$ such that $C_n \subset E_{nm}$, $D_m \subset F_{nm}$, $E_{nm} \cap F_{nm} = \emptyset$ and set

$$E' = \cup_n \cap_m E_{nm}, \quad F' = \cup_p \cap_n F_{np};$$

these sets belong to $\mathcal{C}(\mathcal{F})$ and $\cup C_n \subset E'$, $\cup D_m \subseteq F'$ and $E' \cap F' = \emptyset$.

Having established this result, let's consider two disjoint $\mathcal{F}$-analytic sets A and A'; by a preliminary construction, that of a product, we can assume that there exists one single compact metrizable space E with its paving $\mathcal{K}(E) = \mathcal{E}$, such that, A and A' are, respectively, the projections of sets:

$$J = \cup_n \cap_m J_{nm}, \quad J' = \cap_n \cup_m J'_{nm}$$

where the sets $J_{nm} = E_{nm} \times F_{nm}$ and $J'_{nm} = E'_{nm} \times F'_{nm}$ belonging to $\mathcal{E} \times \mathcal{F}$. To abbreviate, let us say that two subsets of $E \times F$ are separable if their projections onto F are separable by elements of $\mathcal{C}(\mathcal{F})$. We then assume that J and J' are not separable and deduce that A and A' are not disjoint, contrary to the hypothesis.

Let $m_1, m_2, \ldots, m_i$ be integers and set

$$L_{m_1 m_2 \ldots m_i} = J_{1m_1} \cap J_{2m_2} \cap \ldots \cap J_{im_i} \cap (\cap_{n > i} \cup_m J_{nm}),$$

and $L'_{m_1 m_2 \ldots m_i}$ is defined similarly. Since $J = \cup_{m_1} L_{m_1}$, $J' = \cup_{m'_1} L'_{m'_1}$ and J and J' are not separable, the above lemma implies the existence of two integers m_1, m'_1 such that L_{m_1} and $L'_{m'_1}$ are not separable but $L_{m_1} = \cup_{m_i} L_{m_1 m_i}$ and $L'_{m'_1} = \cup_{m'_i} L'_{m'_1 m'_i}$. Hence, there exist two integers m_2 and m'_2 such that $L_{m_1 m_2}$ and $L_{m'_1 m'_2}$ are not separable. Thus, we construct inductively two infinite sequences $m_1, m_2, \ldots, m'_1, m'_2 \ldots$ such that $L_{m_1 m_2 \ldots m_i}$ and $L'_{m'_1 m'_2 \ldots m'_i}$ aren't separable.

These sets cannot be empty, since every subset of $E \times F$ can be separated from the empty set (here we use the assumption that $F \in \mathcal{C}(\mathcal{F})$). Hence

$$E_{1m_1} \cap E_{2m_2} \cap \ldots \cap E_{im_i} \neq \emptyset, \quad E'_{1m'_1} \cap E'_{2m'_2} \cap \ldots \cap E'_{im'_i} \neq \emptyset$$

Similarly,

$$(F_{1m_1} \cap F_{2m_2} \cap \ldots \cap F_{im_i}) \cap \left(F'_{1m'_1} \cap F'_{2m'_2} \cap \ldots \cap F'_{im'_i} \right) \neq \emptyset$$

because the two sets in brackets belong to $\mathcal{C}(\mathcal{F})$ and $L_{m_1 m_2 \ldots m_i}$ and $L'_{m'_1 m'_2 \ldots m'_i}$ are not separable. The pavings $\mathcal{E}$ and $\mathcal{F}$ are semi-compact, so there exist $x \in \cap_i E_{im_i}$, $x' \in \cap_i E'_{im'_i}$ and $y \in \cap_i \left(F_{im_i} \cap F'_{im'_i} \right)$. Then, $(x, y) \in J$, $(x', y) \in J'$ and finally $y \in A \cap A'$ which leads to the desired contradiction. ∎

1.3.3 Lusin and Souslin spaces

In the previous section, we were concerned with $\mathcal{F}$-analytic subsets of a paved space $(F, \mathcal{F})$ where a statement of the type "A is $\mathcal{F}$-analytic" expresses something about its position within a larger set but does not express an intrinsic property of the set A.

Here, we are going to change our point of view where $(F, \mathcal{F})$ will be a measurable space and we will study subsets A of F characterized by intrinsic properties of the measurable space $(A, \mathcal{F}|_A)$. We start by recalling a few definitions:

Definition 1.3.15 Two measurable spaces $(E, \mathcal{E})$ and $(F, \mathcal{F})$ are said to be isomorphic if there exists a bijection between E and F which is measurable and has a measurable inverse.

A space $(F, \mathcal{F})$ is called Hausdorff if the atoms of $\mathcal{F}$ are the points of F.

A Hausdorff separable measurable space $(F, \mathcal{F})$ is isomorphic to a space $(U, \mathcal{B}(U))$, where U is a (not necessarily Borel) subset of $\mathbb{R}$.

Lemma 1.3.16 Let $(F, \mathcal{F})$ be a paved set, E be a subset of F and $\mathcal{E}$ be the paving $\mathcal{F}|_E$, i.e., the trace of F on E. Then $\mathrm{a}(\mathcal{E}) = \mathrm{a}(\mathcal{F})|_E$.

Given a topological space E, we denote by $\mathcal{K}(E)$, $\mathcal{G}(E)$, $\mathcal{B}(E)$ simply as $\mathcal{K}$, $\mathcal{G}$, $\mathcal{B}$ the pavings consisting of the compact, the open and the Borel subsets of E, respectively. If E is metrizable, the complement of an open set is in $\mathcal{G}_\delta$, hence $\mathcal{G} \subset \mathcal{B} \subset \mathrm{a}(\mathcal{G})$ and $\mathrm{a}(\mathcal{B}) = \mathrm{a}(\mathcal{G})$. The latter paving will be denoted by $\mathrm{a}(E)$ and its elements will be called analytic in E. In the metrizable case, these are the same as the $\mathcal{C}$-analytic sets, where $\mathcal{C}$ is the paving of closed subsets of E, and the same as the $\mathcal{K}$-analytic sets if E is compact or LCC.

A topological space is a Lusin space if it is homeomorphic to a Borel subset of a compact metric space. Some stronger topology makes a Lusin into a Polish space.

Definition 1.3.17 (a) A metrizable topological space is said to be Lusin (resp. Souslin, cosouslin) if it is homeomorphic to a Borel subset (resp. an analytic subset, a complement of an analytic subset) of a compact metrizable space.

(b) A measurable space $(F, \mathcal{F})$ is said to be Lusin (resp. Souslin, cosouslin) if it is isomorphic to a measurable space $(H, \mathcal{B}(H))$, where H is a Lusin (resp. Souslin, cosouslin) metrizable space.

(c) In a Hausdorff measurable space $(F, \mathcal{F})$, a set E is said to be Lusin (resp. Souslin, cosouslin) if the measurable space $(E, \mathcal{F}|_E)$ is Lusin (resp . Souslin, cosouslin). We denote by $\mathcal{L}(\mathcal{F})$, $\mathcal{S}(\mathcal{F})$, $\mathcal{S}'(\mathcal{F})$ the pavings consisting of the Lusin, Souslin, cosouslin sets in $(F, \mathcal{F})$.

A Lusin metrizable (resp. measurable) space is both Souslin and cosouslin and the converse is also true. Also, every Lusin, Souslin or cosouslin metrizable (resp. measurable) space is separable and Hausdorff. A general definition of Lusin (resp. Souslin) topological spaces are still Hausdorff but not necessarily metrizable; however, their Borel σ-field is a Lusin (resp. Souslin) σ-field in the sense of Definition 1.3.17(b).

The introduction of Lusin and cosouslin spaces is not a mere luxury but it is done, because the spaces of paths of processes is either Lusin or cosouslin.

Next, we will first give the means to construct Lusin measurable spaces and the usual example of a non-compact Lusin metrizable space.

Theorem 1.3.18 (a) Let $(F, \mathcal{F})$ be a measurable space. If F is Lusin, then $\mathcal{F} \subset \mathcal{L}(\mathcal{F})$. If F is Souslin, then $a(\mathcal{F}) \subset \mathcal{S}(\mathcal{F})$. If F is cosouslin, the complement of every element of $a(\mathcal{F})$ belongs to $\mathcal{S}(\mathcal{F})$.

(b) Furthermore, every Polish space is Lusin, and hence, so is every Borel subspace of a Polish space.

The following theorem on mappings between two Hausdorff separable measurable spaces is a useful results.

Theorem 1.3.19 Let $(F, \mathcal{F})$ and $(F', \mathcal{F}')$ be two Hausdorff separable measurable spaces. Assume that F and F' are embedded in compact metric spaces C and C' and that $\mathcal{F} = \mathcal{B}(C)|_F$, $\mathcal{F}' = \mathcal{B}(C')|_{F'}$. Let f be a measurable mapping of F into F'. Then,
(a) f can be extended to a Borel mapping g of C into C';
(b) If f is also an isomorphism between F and F', then there exist two Borel subsets $B \supset F$, $B' \supset F'$ of C and C', respectively, such that g induces an isomorphism between B and B';
(c) for all $A \in \mathcal{S}(\mathcal{F})$, we have $f(A) \in \mathcal{S}(\mathcal{F}')$;
(d) $\mathcal{S}(\mathcal{F}) \subset a(F)$, and $\mathcal{S}(\mathcal{F}) = a(F)$ if and only if F is Souslin. In particular, if F is Souslin, then $f(a(\mathcal{F})) \subset a(\mathcal{F}')$.

Now we give analogous results to part (d) of Theorem 1.3.19 for the families $\mathcal{L}(\mathcal{E})$ and $\mathcal{S}'(\mathcal{E})$. Denote by $a'(\mathcal{E})$ the family of subsets of E whose complements are $\mathcal{E}$-analytic.

Theorem 1.3.20 Let $(E, \mathcal{E})$ be a Hausdorff separable measurable space. Then,
(a) $\mathcal{L}(\mathcal{E}) \subset \mathcal{E}$, with equality if and only if E is Lusin;
(b) $\mathcal{S}(\mathcal{E}) \subset a(\mathcal{E})$, with equality if and only if E is Souslin;
(c) $\mathcal{S}'(\mathcal{E}) \subset a'(\mathcal{E})$, with equality if and only if E is cosouslin.

The next theorem is interesting; it tells us that Lusin spaces are the same as the space of Borel subsets of the unit interval.

Theorem 1.3.21 (1) Every Lusin (Souslin) measurable space is isomorphic to an (analytic) Borel subset of $[0, 1]$. (2) Every Lusin (Souslin) metrizable space is homeomorphic to an analytic Borel subspace of the cube $[0, 1]^{\mathbb{N}}$.

1.3.3.1 Souslin-Lusin Theorem

To apply Theorem 1.3.20, it is necessary to know whether a given subset A of E belongs for example to $\mathcal{L}(\mathcal{E})$: this means that we are able to construct between A and some known Lusin space L a measurable bijection f with a measurable inverse. The Souslin-Lusin theorem will spare us worrying about f. It is one of the greatest tools of measure theory and we shall use it whenever possible.

Theorem 1.3.22 Let $(F, \mathcal{F})$ and $(F', \mathcal{F}')$ be two Hausdorff separable measurable spaces and h be an injective measurable mapping of F into F'; (1) If F is Souslin, h is an isomorphism of F into $h(F)$. (2) Further, if F is Lusin, then $h(\mathcal{F}) \subset L(\mathcal{F}') \subset \mathcal{F}'$.

Proof. By the definitions stated in 1.3.15, there is no loss in generality in supposing that $(F', \mathcal{F}')$ is the interval $[0, 1]$ with its Borel σ-field. Let $A \in \mathcal{F}$. The sets $h(A)$ and $h(A^c)$ are Souslin (1.3.19) and disjoint since h is injective. By the separation Theorem 1.3.14, since the Souslin subsets of $[0, 1]$ are $\mathcal{K}$-analytic, $h(A)$ and $h(A^c)$ can be separated by two Borel subsets B and B' of $[0, 1]$. Then $h(B)$ and $h(B')$ are elements of 5 which separate A and A^c and hence $A = h^{-1}(B)$, $A^c = h^{-1}(B')$ and $h(A) = B \cap h(F)$. Thus $h(A) \in \mathcal{F}'$, which means that h is a measurable isomorphism. The rest of the statement follows from Theorem 1.3.20. ■

Other extension of this theorem can be found in [19], chapter III.

1.3.4 Capacities and Choquet's Theorem

Choquet's Theorem on capacitability has become one of the fundamental tools of probability theory. Here we will define the Choquet Capacity and prove the Chouquet Theorem. This section can be understood with only the definition of analytic sets and some elementary properties of compact pavings.

Definition 1.3.23 Choquet's Capacity. Let F be a set with a paving $\mathcal{F}$ closed under $(\cup f, \cap f)$. Choquet capacity on $\mathcal{F}$ ($\mathcal{F}$-capacity) is an extended real-valued set function λ defined for all subsets of $\mathcal{F}$ with the following properties
(a) λ is increasing $(A \subset B \Rightarrow \lambda(A) \leq \lambda(B))$;
(b) For every increasing sequence $(A_n)_n$ of subsets of $\mathcal{F}$,

$$\lambda(\cup A_n) = \sup_n \lambda(A_n) \tag{1.20}$$

(c) For every decreasing sequence $(A_n)_n$ of elements of $\mathcal{F}$

$$\lambda(\cap A_n) = \inf_n \lambda(A_n). \tag{1.21}$$

A subset A of $\mathcal{F}$ is called capacitable if

$$\lambda(A) = \sup\left\{\lambda(B) : B \in \mathcal{F}_{\sigma\delta}, B \subset A\right\}. \tag{1.22}$$

Theorem 1.3.24 Choquet's Theorem. Let λ be an $\mathcal{F}$-capacity. Then, every $\mathcal{F}$-analytic set is capacitable relative to λ.

The proof of the theorem involves the following two lemmas:

Lemma 1.3.25 Relative to λ, every element of $\mathcal{F}_{\sigma\delta}$ is capacitable.

Proof. Let A be an element of $\mathcal{F}_{\sigma\delta}$ such that $\lambda(A) > -\infty$. A is the intersection of a sequence $(A_n)_{n\geq1}$ of elements of $\mathcal{F}_\sigma$ and each A_n is the union of an increasing sequence $(A_{nm})_{m\geq1}$ of elements of $\mathcal{F}$. We will show that there exists, for every number $a < \lambda(A)$, an element B of $\mathcal{F}_\delta$ such that $B \subset A$ and $\lambda(B) \geq a$.

First, we prove that there is a sequence $(B_n)_{n\geq1}$ of elements of $\mathcal{F}$ such that $B_n \subset A_n$ and $\lambda(C_n) > a$, where $C_n = A \cap B_1 \cap B_2 \cap \ldots \cap B_n$. To construct B_1, we have by (1.20)

$$\lambda(A) = \lambda(A \cap A_1) = \sup_m \lambda(A \cap A_m).$$

Then, we choose B_1 to be one of the sets A_{1m}, where m is chosen sufficiently large as to have $\lambda(A \cap A_{1m}) > a$. Then, we suppose that the construction has been made up to the $(n-1)^{th}$ term where by hypothesis $C_{n-1} \subset A$ and $\lambda(C_{n-1}) > a$. Consequently:

$$\lambda(C_{n-1}) = \lambda(C_{n-1} \cap A_n) = \sup_m \lambda(C_{n-1} \cap A_{nm}).$$

Then we take B_n to be one of the sets A_{nm}, where m is sufficiently large, so that $\lambda(C_{n-1} \cap A_{nm}) = \lambda(C_n) > a$.

Having constructed the sequence $(B_n)_{n\geq1}$, we set $B'_n = B_1 \cap B_2 \cap \ldots \cap B_n$ and $B = \cap_n B_n = \cap_n B'_n$. The sets B'_n belong to $\mathcal{F}$ and are decreasing in n, and we have $C_n \subset B'_n$. Hence, $\lambda(B'_n) > a$ and $\lambda(B) \geq a$ by (1.21). We have $B_n \subset A_n$, hence, $B \subset A$. The set B satisfies the required conditions and the lemma is established. ∎

Now let A be $\mathcal{F}$-analytic, then there exists a compact metric space E with its compact paving $\mathcal{K}(E) = \mathcal{E}$ and an element B of $(\mathcal{E} \times \mathcal{F})_{\sigma\delta}$ such that the projection of B onto F is equal to A. Let π denote the projection of $E \times F$ onto F and $\mathcal{H}$ denote the paving consisting of all finite unions of elements of $\mathcal{E} \times \mathcal{F}$. By Theorem 1.3.4, there will be no loss of generality in supposing that $\mathcal{E}$ is closed under $(\cup f, \cap f)$ and then $\mathcal{H}$ is closed under $(\cup f, \cap f)$.

Lemma 1.3.26 The set function J defined for all $H \subset E \times F$ by:

$$J(H) = \lambda(\pi(H))$$

is an $\mathcal{H}$-capacity on $E \times F$.

Proof. The function J is obviously increasing and satisfies (1.20). Property (1.21) follows immediately from the relation: $\cap_n \pi(B_n) = \pi(\cap_n B_n)$ which holds, according to Theorem 1.3.6, for every decreasing sequence $(B_n)_{n \in \mathbb{N}}$ of elements of $\mathcal{H}$.

Since B is capacitable relative to J by Lemma 1.3.25, there exists an element D of $\mathcal{H}_\delta$ such that $D \subset B$, $J(D) \geq J(B) - \epsilon$ ($\epsilon > 0$). Let C be the set $\pi(D)$: the above equality shows that C is an element of $\mathcal{F}_\delta$ and we have $C \subset A$, $\lambda(C) \geq \lambda(A) - \epsilon$. ∎

Remark 1.3.27 Let $\mathcal{C}$ be the class of all sets A such that $\lambda(A) > a$: $\mathcal{C}$ has the properties:

$$A \in \mathcal{C}, \quad A \subset B \Rightarrow B \in \mathcal{C}, \tag{1.23}$$

if (A_n) is an increasing sequence of subsets of F, whose union belongs to $\mathcal{C}$, $\qquad$ (1.24)
then some A_n belongs to $\mathcal{C}$.

On the other hand, the property we established can be stated as follows:

if a $\mathcal{F}$-analytic set belongs to $\mathcal{C}$, it contains the intersection of $\qquad$ (1.25)
a decreasing sequence of elements of $\mathcal{F} \cap \mathcal{C}$.

Remark 1.3.28 Lemma 1.3.25 is basically saying that, any $\mathcal{F}_{\sigma\delta}$ belonging to $\mathcal{C}$ satisfies (1.25) while Lemma 1.3.26 explains the fact that the class $\mathcal{C}'$ in $E \times F$ consisting of the sets whose projection on F belongs to $\mathcal{C}$ still satisfies (1.23) and (1.24). Then, Lemma 1.3.25 is applied in $E \times F$, and because of the compactness of the paving $\mathcal{E}$ (Theorem 1.3.6) the projection and intersection commute.

Sion calls the class $\mathcal{C}$ satisfying (1.23) and (1.24) a capacitance. The validity of (1.25) then is "Sion's Capacitability Theorem", which is a little bit more general than that of Choquet, see Sion [22].

1.3.4.1 Constructing capacities

It is not trivial to find a Choquet's set function which is from the start defined for all subsets of a set F. But, it is otherwise natural to consider a set function defined on a paving and to determine whether one can extend it to the whole of the Boolean algebra, $\mathfrak{B}(F)$, as a Choquet capacity. We now describe such an extension procedure for strongly subadditive set functions but we limit ourselves to the case of positive set functions.

Definition 1.3.29 Let $\mathcal{F}$ be paving on the set F, closed under $(\cup f, \cap f)$. Let λ be a positive and increasing set function defined on $\mathcal{F}$. We say that λ is strongly subadditive if for every pair (A, B) of elements of $\mathcal{F}$

$$\lambda(A \cup B) + \lambda(A \cap B) \leq \lambda(A) + \lambda(B). \tag{1.26}$$

If the symbol $\leq$ is replaced by $=$, we get the definition of an additive function on $\mathcal{F}$.

Theorem 1.3.30 Let $\mathcal{F}$ be a paving on F which is closed under $(\cup f, \cap f)$ and let λ be an increasing and positive set function on $\mathcal{F}$. The following properties are equivalent:
(a) λ is strongly subadditive;
(b) $\lambda(P \cup Q \cup R) + \lambda(R) \leq \lambda(P \cup R) + \lambda(Q \cup R)$ for all $P, Q, R \in \mathcal{F}$;
(c) $\lambda(Y \cup Y') + \lambda(X) + \lambda(X') \leq \lambda(X \cup X') + \lambda(Y) + \lambda(Y')$ for all pairs (X, Y), (X', Y') of elements such that $X \subset Y$, $X' \subset Y'$.

With the next theorem, we associate an outer capacity to every strongly subadditive and increasing set function and show that this procedure yields a Choquet capacity.

Theorem 1.3.31 Let F be a set with a paving $\mathcal{F}$ closed under $(\cup f, \cap f)$. Let λ be a set function $\mathcal{F}$ defined on $\mathcal{F}$ positive, increasing and strongly subadditive, which satisfies the following property:

for every increasing sequence $(A_n)_{n \geq 1}$ of elements of $\mathcal{F}$ whose union A belongs to $\mathcal{F}$, $\quad$ (1.27)
$$\lambda(A) = \sup_n \lambda(A_n).$$

For every set $A \in \mathcal{F}_\sigma$ we define

$$\lambda(A) = \sup_{B \in \mathcal{F}, B \subset A} \lambda(B). \tag{1.28}$$

and, for every subset C of F:

$$\lambda^\star(C) = \inf_{A \in \mathcal{F}_\sigma, A \supset C} \lambda^\star(A) \quad (\inf \emptyset = +\infty). \tag{1.29}$$

$\lambda^\star$ is called the outer capacity associated with λ.

Then the function λ^* is increasing and has the following properties: (a) for every increasing sequence $(X_n)_{n \geq 1}$ of subsets of F,

$$\lambda^\star(\cup_n X_n) = \sup_n \lambda^\star(X_n). \tag{1.30}$$

(b) Let (X_n), (Y_n) be two sequences of subsets of F such that $X_n \subset Y_n$ for all n. Then:

$$\lambda^\star(\cup_n Y_n) + \sum_n \lambda^\star(X_n) \leq \lambda^\star(\cup_n X_n) + \sum_n \lambda^\star(Y_n). \qquad (1.31)$$

(c) The function $\lambda^\star$ is an $\mathcal{F}$-capacity, if and only if

$$\lambda^\star(\cap_n A_n) = \inf_n \lambda(A_n) \qquad (1.32)$$

for every decreasing sequence $(A_n)_{n \geq 1}$ of elements of $\mathcal{F}$.

Finally, it is immediate that condition (32.6) is necessary and sufficient for $\lambda^\star$ to be a Choquet $\mathcal{F}$-capacity.

Next, we present an application of Choquet Capacity Theorem 1.3.24 and Outer Capacity Theorem 1.3.31 to measure theory.

Applications to measure theory

Let $(\Omega, \mathfrak{A}, \mathbf{P})$ be a complete probability space and let $\mathcal{F}$ be a family of subsets of Ω, contained in $\mathfrak{A}$ and closed under $(\cup f, \cap f)$. Let λ be the restriction of $\mathbf{P}$ to $\mathcal{F}$. Obviously $\lambda^*(A) = \mathbf{P}(A)$ for every element A of $\mathcal{F}_\sigma$ and consequently also $\lambda^*(A) = \mathbf{P}(A)$ for every element A of $\mathcal{F}_\delta$ by (1.29). Conditions (1.27) and (1.32) are obviously satisfied.

Let A be an $\mathcal{F}$-analytic subset of Ω; Choquet's Theorem implies that

$$\sup_{B \in \mathcal{F}_\delta, B \subset A} \mathbf{P}(B) = \inf_{C \in \mathcal{F}_\sigma, C \supset A} \mathbf{P}(C).$$

So there exist an element B' of $\mathcal{F}_{\delta\sigma}$ and an element C' of $\mathcal{F}_{\sigma\delta}$ such that $B' \subset A \subset C'$ and $\mathbf{P}(B') = \mathbf{P}(C')$. This implies in particular that $A \in \mathfrak{A}$.

Next, we consider Carathéodory's extension theorem; let's go back to Theorem 1.3.31 and suppose that λ is additive on $\mathcal{F}$ and that (1.32) holds. Let $(A_n)_{n \in \mathbb{N}}$ and $(B_n)_{n \in \mathbb{N}}$ be two decreasing sequences of elements of $\mathcal{F}$. Passing to the limit (according to (1.32)) in the formula

$$\lambda(A_n \cup B_n) + \lambda(A_n \cap B_n) = \lambda(A_n) + \lambda(B_n)$$

we see that $\lambda^\star$ is additive on $\mathcal{F}_\delta$. Then let A and B be two elements of $\mathfrak{A}(\mathcal{F})$ and ϵ a number > 0; we choose two sets A' and B', belonging to $\mathcal{F}_\delta$, contained respectively in A and B and such that:

$$\lambda^\star(A') \geq \lambda^\star(A) - \epsilon; \quad \lambda^\star(B') \geq \lambda^\star(B) - \epsilon.$$

Then we have:

$$\begin{aligned}
\lambda^\star(A \cup B) + \lambda^\star(A \cap B) &\geq \lambda^\star(A' \cup B') + \lambda^\star(A' \cap B') \\
&= \lambda^\star(A') + \lambda^\star(B') \\
&\geq \lambda^\star(A) + \lambda^\star(B) - 2\epsilon.
\end{aligned}$$

Since the function $\lambda^\star$ is strongly subadditive and ϵ is arbitrary, we see that $\lambda^\star$ is additive on $\mathfrak{A}(\mathcal{F})$.

Having established this, we consider a Boolean algebra $\mathcal{F}$ and on it an additive set function λ, which is positive and finite and satisfies Carathéodory's condition: If $A_n \in \mathcal{F}$ are decreasing and $\cap A_n = \emptyset$, then $\lim_n \lambda(A_n) = 0$. Then obviously 1.27 is satisfied. We show that (1.32) is also satisfied. This condition can be stated as follows: If (G_n) is an increasing sequence and (F_n) a decreasing sequence of elements of $\mathcal{F}$ and $\cup G_n \supset \cap F_n$, then $\sup_n \lambda(G_n) \geq \inf_n \lambda(F_n)$. Now let $H_n = F_0 \backslash F_n \in \mathcal{F}$; then H_n is increasing and $F_0 \subset \cup(G_n \cup H_n)$. By (1.27), $\sup_n \lambda(G_n \cup H_n) \geq \lambda(F_0)$ and for a fortiori $\sup_n(\lambda(G_n) + \lambda(H_n)) \geq \lambda(F_0)$, whence subtracting

$$\sup_n \lambda(G_n) \geq \inf_n(\lambda(F_0) - \lambda(H_n)) = \inf_n \lambda(F_n).$$

Then, by Theorem 1.3.31 and remark 1.3.27 we see that λ^* is additive on $\mathfrak{A}(\mathcal{F})$. Hence, it is also additive on $\sigma(\mathcal{F}) \subset \mathfrak{A}(\mathcal{F})$. Since λ^* passes through the limit along increasing sequences, λ^* is a measure on $\sigma(\mathcal{F})$ which extends λ and we have established the classical Carathéodory extension theorem from probability theory.

Similarly, other application of Choquet's Theorem is to the representation theorem of Riesz and Daniell's, see [19].

1.3.5 Theorem of cross-section

Definition 1.3.32 Debut of Set. Let $(\Omega, \mathcal{F})$ be a measurable space and A a subset of $\mathbb{R}_+ \times \Omega$. We write, for all $\omega \in \Omega$,

$$\mathsf{D}_A(\omega) = \inf\{t \in \mathbb{R}_+ : (t, \omega) \in A\} \tag{1.33}$$

with the usual convention that $\inf \emptyset = +\infty$. The function D_A is called the *debut of* A.

Theorem 1.3.33 Cross-Section. Suppose that A belongs to the σ-field $\mathcal{B}(\mathbb{R}_+) \times \mathcal{F}$, or, more generally, that A is $(\mathcal{B}(\mathbb{R}_+) \times \mathcal{F})$-analytic. Then,
(a) The debut D_A is measurable relative to the σ-field $\hat{\mathcal{F}}$ – the universal completion (Definition 1.2.27) of $\mathcal{F}$.
(b) Let $\mathbf{P}$ be a probability law on $(\Omega, \mathcal{F})$. There exists an $\mathcal{F}$-measurable r.v. T with values in $[0, \infty]$ such that

$$T(\omega) < \infty \Rightarrow (T(\omega), \omega) \in A \quad (\text{``}T \text{ is a cross-section of } A\text{''}) \tag{1.34}$$

$$\mathbf{P}(T < \infty) = \mathbf{P}(\mathsf{D}_A < \infty). \tag{1.35}$$

In other words, T is an almost-complete cross-section of A.

Proof. Let $r > 0$. The set $\{\mathsf{D}_A < r\}$ is the projection on Ω of $\{(t, \omega) : t < r(t, \omega) \in A\}$. By Theorem 1.3.13, $\{\mathsf{D}_A < r\}$ is $\mathcal{F}$-analytic and by Section 1.3.4.1, it belongs to every completed σ-field of $\mathcal{F}$, whence assertion (a).

Associate with $\mathbf{P}$ the set function $\mathbf{P}^*$ as in Theorem 1.3.31 ($\mathbf{P}^*$ is the classical "outer" probability of Carathéodory). $\mathbf{P}^*$ is an $\mathcal{F}$-capacity equal to $\mathbf{P}$ on $\mathcal{F}$ and on the completed σ-field of $\mathcal{F}$ (Definition 1.2.27(b)). Let π be the projection of $\mathbb{R}_+ \times \Omega$ onto Ω and let λ be the set function $A \mapsto \mathbf{P}^*(\pi(A))$. λ is a capacity relative to the paving $\mathfrak{P}$, the closure of $\mathcal{K}(\mathbb{R}_+) \times \mathcal{F}$ under $(\cup c, \cap c)$ (Lemma 1.3.26). By Theorem 1.3.13, every element of the product σ-field $\mathcal{B}(\mathbb{R}_+) \times \mathcal{F}$, or, more generally of $\mathfrak{A}(\mathcal{B}(\mathbb{R}_+) \times \mathcal{F})$, is $\mathfrak{P}$-analytic and the capacitability Theorem 1.3.24 implies the existence of an element B of $\mathfrak{P}_\delta = \mathfrak{P}$ contained in A such that, for all $\epsilon > 0$, $\lambda(B) > \lambda(A) - \epsilon$ which can also be written as

$$\mathbf{P}(\mathsf{D}_B < \infty) > \mathbf{P}(\mathsf{D}_A < \infty) - \epsilon,$$

since for all $\omega \in \Omega$ the set $B(\omega) = \{t : (t, \omega) \in B\}$ is compact, the graph of D_B in $\mathbb{R}_+ \times \Omega$ is contained in A. Then let S_ϵ be an $\mathcal{F}$-measurable positive r.v., equal almost everywhere to D_B; we write

$$T_\epsilon(\omega) = \begin{cases} S_\epsilon(\omega) & \text{if} \quad (S_\epsilon(\omega), \omega) \in A \\ +\infty & \text{otherwise} \end{cases}$$

Then T_ϵ satisfies (1.34) and a weaker condition than (1.35) $\mathbf{P}(T_\epsilon < \infty) > \mathbf{P}(\mathsf{D}_A > \infty) - \epsilon$. Let us say (in this proof only) that, given $C \in \mathcal{B}(\mathbb{R}_+) \times \mathcal{F}$, a positive $\mathcal{F}$-measurable function S such that $(S(\omega), \omega) \in C$ for all $\omega \in \{S < \infty\}$ is a section of C with remainder $\mathbf{P}(S = \infty, \mathsf{D}_C < \infty)$.

By the above, C has a section with remainder $< \epsilon$ for all $\epsilon > 0$. We construct sections of A inductively as follows. $T_0 = +\infty$ identically. If T_n has been defined; Next construct a section S_n of $A_n = A \cap \{(t, \omega) : T_n(\omega) = \infty\}$ such that $\mathbf{P}(S_n < \infty) \geq \frac{1}{2}\mathbf{P}(\mathsf{D}_{A_n} < \infty)$, and we set $T_{n+1} = T_n \wedge S_n$, a section of A which "extends" T_n. At each step, the remainder is at most half of the proceeding one. So $T = \inf_n T_n$ is a section with remainder zero, which therefore satisfies (1.34) and (1.35). ∎

In part (b), the cross-section T of A is said to be complete if $T(\omega) < \infty$ for every ω such that $\mathsf{D}_A(\omega) < \infty$. At the cost of minor modifications, the cross-section theorem is still valid if $(\mathbb{R}_+, \mathcal{B}(\mathbb{R}_+))$ is replaced by a Souslin measurable space $(S, \mathcal{S})$, which, from a measure theoretic point of view is not distinguishable from an analytic subset of $\mathbb{R}$: (a) is no longer meaningful but the projection $\pi(A)$ of A onto Ω still belongs to $\mathcal{F}$; (b) remains true provided $\mathbf{P}(\mathsf{D}_A < \infty)$ is replaced by $\mathbf{P}(\pi(A))$ and $[0, \infty]$ by $S \cup \{\infty\}$ where "∞" is a point added to S.

Chapter 2

Stochastic Processes

In the first two sections of this chapter we study stochastic processes and methods leading to the construction of suitable versions of them. In the latter sections of the chapter we study fundamental structures of a probability space provided with an increasing family of σ-fields. Then, we define the notions of adapted, progressive optional and predictable processes and study the nature and classification of random times.

2.1   Construction

Definition 2.1.1 Let $(\Omega, \mathcal{F}, \mathbf{P})$ be a probability space, $\mathbb{T}$ be any set and $(E, \mathcal{E})$ be a measurable space. A stochastic process (or simply a process) defined on Ω, with time set $\mathbb{T}$ and state space E, is any family $(X_t)_{t \in \mathbb{T}}$ of E valued random variables indexed by $\mathbb{T}$.

The space Ω is often called the sample space of the process, and the r.v. X_t is called the state at time t. For every $\omega \in \Omega$ the mapping $t \mapsto X_t(\omega)$ from $\mathbb{T}$ into E is called the sample path of ω. Usually, certainly in this book also, $\mathbb{T}$ will always be a subset of the extended real line $\overline{\mathbb{R}}$: usually an interval of $\overline{\mathbb{R}}$ in the continuous case or of $\overline{\mathbb{Z}}$ in the discrete case and sometimes a dense countable set of $\mathbb{Q}$, for example. This is where the terminology of time, instants, and paths originated.

However, there exist parts of the theory of processes where $\mathbb{T}$ is only a partially ordered set. For example, in statistical mechanics $\mathbb{T}$ may be the family of subsets of a finite or countable set, partially ordered by inclusion or even has no order structure at all; also in some problems of ergodic theory $\mathbb{T}$ may be a group; and in problems concerning regularity of paths of Gaussian processes $\mathbb{T}$ is just a metric space.

Remark 2.1.2 (a) The notion of a random variable was related to a measurable space $(\Omega, \mathcal{F})$ not to a probability space $(\Omega, \mathcal{F}, \mathbf{P})$, as well, the notion of a process does not really require a law $\mathbf{P}$, and from time to time we may speak of a process on some space without emphasis on any particular law on it.

(b) We have defined a process as a family $(X_t)_{t \in \mathbb{T}}$ of r.v.; that is a mapping of $\mathbb{T}$ into the set of E-valued random variables. A process can also be considered as a mapping $(t, \omega) \mapsto X_t(\omega)$ of $\mathbb{T} \times \Omega$ into E or as a mapping $\omega \mapsto (t \to X_t(\omega))$ of Ω into the set of all possible paths. In the latter interpretation, the process appears as a r.v. with values in the set of paths (a random function), but this notion is not complete from a mathematical point of view because it lacks a σ-field given on the set of all paths.

The point of view where a process is a function on $\mathbb{T} \times \Omega$ will be the most useful. We illustrate it by a specific definition:

Definition 2.1.3 Suppose that $\mathbb{T}$ is given a σ-field $\mathcal{T}$. The process $(X_t)_{t \in \mathbb{T}}$ is said to be *measurable* if the mapping $(t, \omega) \mapsto X_t(\omega)$ is measurable on $\mathbb{T} \times \Omega$ with respect to $\mathcal{T} \times \mathcal{F}$.

In the discrete case $\mathbb{T} \subset \overline{\mathbb{Z}}$, the σ-field $\mathcal{T}$ is that of all subsets of $\mathbb{T}$ and the notion of measurability is trivial: every process is measurable.

We think of a continuous time stochastic process as a mathematical model we use to describe a natural phenomenon whose evolution is governed by chance. It is, then, natural to ask under what conditions two processes describe the same phenomenon and how observations of the phenomenon can be used to construct a process which describes it?

To give answers to these questions, it may be intuitive to assume that at any finite set of time instants $t_1, t_2, ..., t_n$ we can determine with arbitrary "precision" the state of the process by performing a large number of independent experiments; it is then possible to estimate with arbitrary precision probabilities of the type

$$\mathbf{P}(X_{t_1} \in A_1, \ldots, X_{t_n} \in A_n) \quad (A_1, \ldots, A_n) \in \mathcal{E}) \tag{2.1}$$

and in general observation can give nothing more. Note that, we required the notion of a law $\mathbf{P}$ to assess the suitability of the process (X_t) to describe a natural phenomenon. Equation 2.1 is known as the *time law* or the *finite dimensional distribution* of the stochastic process (X_t).

As a consequence of 2.1, we are able to give the following definition that expresses, reasonably, the fact that two processes, (X_t) and (X_t'), represent the same natural phenomenon.

Definition 2.1.4 We consider two stochastic processes with the same time set $\mathbb{T}$ and state space $(E, \mathcal{E})$: $(\Omega, \mathcal{F}, \mathbf{P}, (X_t)_{t \in \mathbb{T}})$ and $(\Omega', \mathcal{F}', \mathbf{P}', (X_t')_{t \in \mathbb{T}})$. The processes (X_t) and (X_t') are called *equivalent* if:

$$\mathbf{P}(X_{t_1} \in A_1, \ldots, X_{t_n} \in A_n) = \mathbf{P}'(X_{t_1}' \in A_1, \ldots, X_{t_n}' \in A_n)$$

for every finite system of instants $t_1, t_2, \ldots, t_n$ and elements $A_1, A_2, ..., A_n$ of $\mathcal{E}$.

2.1.1 Time law

We often say that (X_t) and (X'_t) have the *same time law*, or simply the same law, or that they are versions of each other.

Remark 2.1.5 (a) However, the notion of a time law leads to criticism. On the one hand, it is too precise. For that it is impossible in practice to determine a precise measurement value at any given time instant t. All that instruments can give are average results over small time intervals. In other words, we have no direct access to the r.v. X_t themselves, but only to r.v. of the form

$$\frac{1}{b-a} \int_a^b f(X_u) du$$

where f is a function on the state space E, considering such integrals of course requires some metastability from the process. This difficult lead to a notion of almost-equivalence which we will describe later in nos. 2.3.24-2.3.41.

(b) On the other hand, the time law notion is insufficiently precise, because it concerns only finite subsets of a set $\mathbb{T}$ which in general is uncountable. Consider as an example the probability space $\Omega = [0,1]$ with the Borel σ-field $\mathcal{F} = \mathcal{B}([0,1])$ and Lebesgue measure $\mathbf{P}$ and $\mathbb{T} = [0,1]$; let there be two real-valued processes (X_t) and (Y_t) defined as follows:

$$\begin{aligned} X_t(\omega) &= \; 0 \text{ for all } \omega \text{ and all } t, \\ Y_t(\omega) &= \; \begin{cases} 0 & \text{for all } \omega \text{ and all } t \neq \omega, \\ 1 & \text{otherwise} \end{cases} \end{aligned} \qquad (2.2)$$

For each t, $Y_t = X_t$ a.s. but the set of ω such that $X.(\omega) = Y.(\omega)$ is empty. The two processes have the same time law, but the first one has all its paths continuous while the paths of the second one are almost all discontinuous.

We give formal definitions of several different notions of "equalities" of stochastic processes in light of the remarks we have made above (Remark 2.1.5) about their equivalence, Definition 2.1.4. The first one is a little more precise than equivalence:

Definition 2.1.6 Let $(X_t)_{t \in \mathbb{T}}$ and $(Y_t)_{t \in \mathbb{T}}$ be two stochastic processes defined on the same probability space $(\Omega, \mathcal{F}, \mathbf{P})$ with values in the same state space $(E, \mathcal{E})$. We say that $(Y_t)_{t \in \mathbb{T}}$ is a *standard modification* of $(X_t)_{t \in \mathbb{T}}$ if $X_t = Y_t$ a.s. for each $t \in \mathbb{T}$.

The second definition expresses the greatest possible precision from the probabilistic point of view of two indistinguishable processes.

Definition 2.1.7 Let the processes $(X_t)_{t \in \mathbb{T}}$ and $(Y_t)_{t \in \mathbb{T}}$ be two stochastic processes defined on the same probability space $(\Omega, \mathcal{F}, \mathbf{P})$ with values in the same state space $(E, \mathcal{E})$ are said to be $\mathbf{P}$ *indistinguishable* (or simply indistinguishable) if for almost all $\omega \in \Omega$

$$X_t(\omega) = Y_t(\omega) \text{ for all } t.$$

For example, if two real-valued processes (X_t) and (Y_t) have right-continuous (or left-continuous) paths on $\mathbb{T} = \mathbb{R}$ and if, for each rational t, $X_t = Y_t$ a.s., then they are indistinguishable: the paths $X.(\omega)$ and $Y.(\omega)$ are indeed a.s. equal on the rationals and hence everywhere on $\mathbb{R}$.

Remark 2.1.8 The definition of indistinguishable processes can be expressed differently: A random set is a subset A of $\mathbb{T} \times \Omega$ whose indicator $\mathbf{1}_A$. The indicator of A as a function of (t, ω) is a stochastic process (i.e., $\omega \mapsto \mathbf{1}_A(t, \omega)$ is a r.v. for all t). The set A is said to be evanescent if the process $\mathbf{1}_A$ is indistinguishable from 0, which also means that the projection of A on Ω is contained in a **P**-negligible set. Two processes, (X_t) and (Y_t), then are indistinguishable if and only if the set $\{(t, \omega) : X_t(\omega) \neq Y_t(\omega)\}$ is evanescent.

2.1.2   Canonical process

Of all the processes with a given time law we want to distinguish a process defined naturally and unambiguously using no more information about the process other than its time law. A process that we will define this way is called *canonical*. Here is how it is done.

Consider a stochastic process $(\Omega, \mathcal{F}, \mathbf{P}, (X_t)_{t \in \mathbb{T}})$ with values in $(E, \mathcal{E})$. Denote by τ the mapping of Ω into $E^{\mathbb{T}}$ which associates with $\omega \in \Omega$ the point $(X_t(\omega))_{t \in \mathbb{T}}$ of $E^{\mathbb{T}}$ – the path of ω.

The mapping τ is measurable given the product σ-field $\mathcal{E}^{\mathbb{T}}$ on $E^{\mathbb{T}}$; hence we can consider the image law $\tau(\mathbf{P})$ on the space $(E^{\mathbb{T}}, \mathcal{E}^{\mathbb{T}})$. We denote by Y_t the coordinate mapping of index t on $E^{\mathbb{T}}$. The processes $(\Omega, \mathcal{F}, \mathbf{P}, (X_t)_{t \in \mathbb{T}})$ and $(E, \mathcal{E}, \tau(\mathbf{P}), (Y_t)_{t \in \mathbb{T}})$ are then equivalent by the very definition of image laws and we can state the definition:

Definition 2.1.9 With the above notation, the process $(E^{\mathbb{T}}, \mathcal{E}^{\mathbb{T}}, \tau(\mathbf{P}), (Y_t)_{t \in \mathbb{T}})$ is called the *canonical process* associated with or equivalent to the process $(X_t)_{t \in \mathbb{T}}$.

Consequently, two processes (X_t) and (X'_t) are equivalent if and only if they are associated with the same canonical process.

The canonical process is rarely used directly for uncountable time set $\mathbb{T}$. The σ-fields $\mathcal{E}^{\mathbb{T}}$ contains just events which depend only on countably many variables Y; whereas, the most interesting properties of the process, such as, continuity of paths, involve all these random variables. A canonical process is useful mainly as a step in constructing more complicated processes.

Remark 2.1.10 We point out the fact that the canonical character of the process (X_t) depends on the available information about it. In the absence of information other than the time law, then, we don't have a choice but to be satisfied with the canonical process (Y_t). However, if it is also known that the process (X_t) has a version with continuous paths under some topology on $\mathbb{T}$, then, it would be silly to use this information. The set $E^{\mathbb{T}}$ of all mappings of $\mathbb{T}$ into E will be replaced by that of all continuous mappings of $\mathbb{T}$ into E, onto which the measure will be carried by the same procedure as above, thus defining a *canonical continuous process*.

The notion of a canonical process leads to a simple but hardly satisfying solution to the problem of constructing stochastic processes by itself; additional tooling is required.

For the following discussion, we start by recalling the general theorem on the construction of stochastic processes due to Kolmogorov:

Theorem 2.1.11 Let E be a separable (metrizable) space, $\mathbb{T}$ be any index set and F be the product set $E^{\mathbb{T}}$ with the product σ-field $\mathcal{F} = (\mathcal{B}(E))^{\mathbb{T}}$. For every *finite subset* U of $\mathbb{T}$, let F_U denote the (metrizable) space E^U and q_U the projection of F onto F_U and let μ_U be a *tight* probability law on F_U.

There *exists a probability law* μ on $(F, \mathcal{F})$ such that $q_U(\mu) = \mu_U$ for every finite $U \subset T$, if and only if the following condition is satisfied:

$$\text{For every pair } (U, V) \text{ of finite subsets such that} U \subset V, \ \mu_U \text{ is the image of} \tag{2.3}$$

$$\mu_V \text{ under the projection of } F_V \text{ onto } F_U. \tag{2.4}$$

The measure μ then is *unique*.

Remark 2.1.12 A finite Borel measure μ on E is called tight if for every $\epsilon > 0$ there exists a compact set $K_\epsilon \subset E$ such that $\mu(E \backslash K_\epsilon) < \epsilon$. A tight finite Borel measure is also called a Radon measure (see also [19] III.46).

A more of a general definition of a tight collection of measures is as follows: let $(E, \mathcal{E})$ be a topological space, and let $\mathcal{F}$ be a σ-algebra on E that contains the topology $\mathcal{E}$. So, every open subset of E is a measurable set, and $\mathcal{F}$ is at least as fine as the Borel σ-algebra on E. Let $\mathfrak{M}$ be a collection of possibly signed or complex measures defined on $\mathcal{F}$. The collection $\mathfrak{M}$ is called tight or sometimes uniformly tight if, for any $\epsilon > 0$, there is a compact subset K_ϵ of E such that, for all measures $\mu \in \mathfrak{M}$, $|\mu|(E \backslash K_\epsilon) < \epsilon$ where $|\mu|$ is the total variation measure of μ.

If the tight collection $\mathfrak{M}$ consists of a single measure μ, then some authors call μ a tight measure, while others call it an inner regular measure.

In probability theory, if X is an E valued random variable whose probability distribution on E is a tight measure, then X is said to be a separable random variable or a Radon random variable.

Let us return to the situation described in Definition 2.1.4: we have observed some "random phenomenon" which we wish to represent by means of a process. Since it can only be defined to within an equivalence, the choice that comes to mind is that of the canonical process. Hence we use the measurable space $(E^{\mathbb{T}}, \mathcal{E}^{\mathbb{T}})$ and the coordinate mappings $(Y_t)_{t \in \mathbb{T}}$. It remains to construct a probability law $\mathbf{P}$ on this space such that

$$\mathbf{P}(Y_{t_1} \in A_1, \dots, Y_{t_n} \in A_n) = \Phi(t_1, \dots, t_n; A_1, \dots, A_n)$$

for every finite subset $u = \{t_1, t_2, \dots, t_n\}$ of $\mathbb{T}$ and every finite family $A_1, A_2, ..., A_n$ of measurable subsets of E, the functions Φ are given by observation. For the construction to be possible, it is necessary that the set function

$$A_1 \times A_2 \times ... \times A_n \mapsto \Phi(t_1, t_2, \dots, t_n; A_1, A_2, \dots, A_n)$$

be extendable to a probability law $\mathbf{P}_u$ on $(E^u, \mathcal{E}^u)$, a probability law which moreover is uniquely determined by Φ (by Theorem 1.2.9), applied to the set of finite unions of subsets of E^u of the form $A_1 \times A_2 \times ... \times A_n$. On the other hand, it is necessary that

$$\pi_{uv}(\mathbf{P}_v) = \mathbf{P}_u$$

for every pair of finite subsets u, v of $\mathbb{T}$ such that $u \subset v$, where π_{uv} denotes the projection of E^v onto E^u. We recognize here the definition of an inverse system of probability laws (2.1.11) and the possibility of constructing the law $\mathbf{P}$ appears to be equivalent to the existence of an inverse limit for the inverse system $(\mathbf{P}_u)$. Theorem 2.1.11 then gives a simple condition that implies the existence of $\mathbf{P}$.

The above-described procedure is known as the Kolmogorov extension (consistency) theorem or the Daniell-Kolmogorov theorem. It guarantees that a suitable collection of time laws can define a stochastic process.

2.2   Processes on Filtrations

Henceforth, we assume that the time set $\mathbb{T}$ is the closed positive half-line $\mathbb{R}_+$. And, in what follows, we will introduce some terminology which will be used throughout this book.

Let $(\Omega, \mathcal{F})$ be a measurable space and let $(\mathcal{F}_t)_{t \in \mathbb{R}}$ be a family of sub-σ-fields of $\mathcal{F}$, such that $\mathcal{F}_s \subset \mathcal{F}_t$ for any $s \leq t$ and we say that $(\mathcal{F}_t)$ is an increasing family of σ-fields on $(\Omega, \mathcal{F})$ or a filtration of $(\Omega, \mathcal{F})$; $\mathcal{F}_t$ is called the σ-field of events prior to t and we define

$$\mathcal{F}_{t+} = \bigcap\nolimits_{s>t} \mathcal{F}_s, \quad \mathcal{F}_{t-} = \bigvee\nolimits_{s<t} \mathcal{F}_s \quad (t > 0). \tag{2.5}$$

The family $(\mathcal{F}_t)$ is said to be right-continuous if $\mathcal{F}_t = \mathcal{F}_{t+}$ for all t. The family $(\mathcal{F}_{t+})_{t \in \mathbb{R}}$ is right-continuous for every family $(\mathcal{F}_t)$.

When the time set is $\mathbb{N}$, definitions (2.5) still have a meaning: $\mathcal{F}_{n+}$ and $\mathcal{F}_{n-}$ must be interpreted as $\mathcal{F}_{n+1}$ and $\mathcal{F}_{n-1}$. It turns out that the latter analogy between $\mathcal{F}_{n-1}$ and $\mathcal{F}_{t-}$ is interesting, while the former isn't.

2.2.1   Adapted processes

Definition 2.2.1 Let $(X_t)_{t \in \mathbb{R}_+}$ be a process defined on a measurable space $(\Omega, \mathcal{F})$ and let $(\mathcal{F}_t)_{t \geq 0}$ be a filtration. The process (X_t) is said to be *adapted* to $(\mathcal{F}_t)$ if X_t is $\mathcal{F}_t$-measurable for every $t \in \mathbb{R}_+$.

Remark 2.2.2 Every process (X_t) is adapted to the family of σ-fields $\mathcal{F}_t = \sigma(X_s, s \leq t)$ which is often called the natural family for this process.

Remark 2.2.3 An intuitive meaning of the above definitions is that, if we interpret the parameter t as time and each event as a physical phenomenon, the sub-σ-field $\mathcal{F}_t$ consists of the events which represent phenomena prior to the instance t. In the same way, random variables that are $\mathcal{F}_t$ measurable are those which depend only on the evolution of the universe prior to t. Alternatively, one can see that, it is really the introduction of a filtration which expresses the parameter t as time and that the future is uncertain, whereas the past is knowable at least for an *ideal observer*. This fundamental idea is due to Joseph Leo Doob.

The presence of a filtration in the formulation of stochastic processes and probability spaces is not a restriction, for it is permissible to take $\mathcal{F}_t = \mathcal{F}$ for all t. This choice corresponds to the deterministic world view, where the ideal observer may predict at any time, through the integration of a complicated differential system, all the future evolution of the universe. If there has ever been any real intervention of chance, it has taken place at the initial instant, and causality has left no room for it thereafter. However, this does not quite prevent probabilities from occurring in the deterministic description of the universe, because of the imprecise nature of our measurements.

2.2.2 Progressive measurability

On a space $(\Omega, \mathcal{F})$ filtered by a family $(\mathcal{F}_t)_{t \geq 0}$ the notion of a measurable process may be made more precise by introducing the notion of progressive measurability, as follows:

Definition 2.2.4 Let $(\Omega, \mathcal{F})$ be a measurable space and let $(\mathcal{F}_t)_{t \geq 0}$ be a filtration on it. Let $(X_t)_{t \geq 0}$ be a process defined on this space with values in $(E, \mathcal{E})$; we say that every (X_t) is *progressively measurable* or progressive with respect to the family $(\mathcal{F}_t)$ if for every $t \in \mathbb{R}_+$ the mapping $(s, \omega) \mapsto X_s(\omega)$ of $[0, t] \times \Omega$ into $(E, \mathcal{E})$ is measurable with respect to the σ-field $\mathcal{B}([0, t]) \times \mathcal{F}_t$.

Remark 2.2.5 Progressive processes are obviously adapted. Furthermore, (a) if (X_t) is progressive with respect to the family $(\mathcal{F}_{t+\epsilon})$ for every $\epsilon > 0$ but adapted to $(\mathcal{F}_t)$, then it is also progressive with respect to $(\mathcal{F}_t)$. (b) If the adaptation condition is omitted, we can still assert that (X_t) is progressive with respect to the family $(\mathcal{F}_{t+})$.

To check if a process is progressively measurable, we have to study sets of the form $\{(t, \omega) : X_t(\omega) \in A\}$, $A \in \mathcal{E}$, and can hence reduce to the real-valued processes, for $s \leq t$ we have

$$X_s = \lim_{\epsilon \to 0} X_s \mathbf{1}_{[0, t-\epsilon[}(s) + X_t \mathbf{1}_{\{t\}}(s) \tag{2.6}$$

and the right-hand side is, for all $\epsilon > 0$, a measurable function with respect to $\mathcal{B}([0, t]) \times \mathcal{F}_t$ (or $\mathcal{F}_{t+}$ if X_t is only $\mathcal{F}_{t+}$-measurable).

If the intervals $[0, t]$ in Definition 2.2.4 are replaced by intervals $[0, t[$, one gets only progressivity with respect to the family $(\mathcal{F}_{t+})$. Here is the easiest example of a progressive process:

Theorem 2.2.6 Let (X_t) be a process with values in metrizable space E, adapted to $(\mathcal{F}_t)$ and with right-continuous paths. Then, (X_t) is progressive with respect to $(\mathcal{F}_t)$. The same conclusion holds for a process with left-continuous paths.

Proof. For every $n \in \mathbb{N}$ we define

$$X_t^n = X_{(k+1)2^{-n}} \text{ if } t \in [k2^{-n}, (k+1)2^{-n}[.$$

(X_t^n) is obviously progressive with respect to the family $(\mathcal{F}_{t+\epsilon})$ provided $\epsilon > 2^{-n}$. Hence, the process X_t, which is equal to $\lim_n X_t^n$ by right-continuity, is progressive with respect to each family $(\mathcal{F}_{t+\epsilon})$. Since, it is adapted, we conclude by Remark 2.2.5 (a) that it is progressive with respect to $(\mathcal{F}_t)$. A similar argument applies for a process with left-continuous paths. ∎

2.3   Paths Properties

Let's talk paths of real-valued processes and as an example consider the path of a particle getting hit from all sides. Its path is governed by chance and has no reason for it to be regular. Questions that arise naturally about rough paths of such real-valued processes are of the following kind: are the paths measurable? Are the paths locally integrable? Are the paths locally bounded or at least does the process have a modification with these properties?

Furthermore, are quantities of the type $\sup_{t \in I} |X_t|$ random variables when I is an interval of a non-countable set, and does there exist a method for determining their law? Also, are the paths continuous at a point or continuous on an interval? Continuity itself is a strong property, since typical paths of "nice" processes may have jumps with limits on both sides.

This section is divided into three parts. First, the study of paths along a countable dense set (nos. 2.3.1-2.3.9), leading to the theory of separability (nos. 2.3.3-2.3.18). Next, the direct study of a measurable process on the whole of $\mathbb{R}_+$ using the theory of analytic sets (nos. 2.3.20-2.3.23). Finally, the study of paths of processes of lim sup to negligible sets (nos. 2.3.24-2.3.41). It is important that the reader keep this plan in mind, since the same properties are studied three times from three different points of view; see for example, Theorems 2.3.3-2.3.4, then 2.3.23, then 2.3.42. Similarly, Theorems 2.3.1, 2.3.22 and 2.3.28.

Throughout, time is $\mathbb{R}_+$ unless otherwise mentioned and processes are defined on a probability $(\Omega, \mathcal{F}, \mathbf{P})$ provided with a filtration $(\mathcal{F}_t)_{t \geq 0}$. We use the following abbreviations: r.c. for right-continuous, l.c. for left-continuous, r.c.l.l. for right-continuous on $[0, +\infty[$ with finite left-hand limits on $]0, +\infty[$, and r.l.l.l. for finite right and left limits.

2.3.1 Processes on dense sets

Let D be a countable dense subset of $\mathbb{R}_+$, and $(X_s)_{s \in D}$ is an adapted real-valued (i.e., X_s is $\mathcal{F}_s$-measurable for every $s \in D$).

Theorem 2.3.1 For all $t \geq 0$ we set

$$\overline{Y}_t^+(\omega) = \limsup_{s \in D, s \downarrow t} X_s(\omega), \quad \underline{Y}_t^+(\omega) = \liminf_{s \in D, s \downarrow t} X_s(\omega). \tag{2.7}$$

Then the two processes $\overline{Y}_t^+$ and $\underline{Y}_t^+$ are progressive relative to the family $(\mathcal{F}_{t+})$. So are the processes defined on $]0, \infty[$ by

$$\overline{Z}_t^-(\omega) = \limsup_{s \in D, s \uparrow\uparrow t} X_s(\omega), \quad \underline{Z}_t^-(\omega) = \liminf_{s \in D, s \uparrow\uparrow t} X_s(\omega) \tag{2.8}$$

relative to the family $(\mathcal{F}_t)$.

Proof. For every integer n we define a process (Y_t^n) as follows: if $t \in [k2^{-n}, (k+1)2^{-n}[$, $Y_t^n = \sup_{s \in D_t} X_s$, where $D_t = D \cap]t, (k+1)2^{-n}[$. This process is adapted to the family $(\mathcal{F}_{t+\epsilon})$ for all $\epsilon > 2^{-n}$, it is right-continuous and hence progressive relative to the family $(\mathcal{F}_t)$. Therefore, so is the process

$$Y_t = \limsup_{s \in D, s \uparrow\uparrow t} X_s = \lim_n Y_t^n.$$

It follows from Definition 2.2.4 that (Y_t) is progressive with respect to $(\mathcal{F}_{t+})$. To deal with $(\overline{Y}_t^+)$, we simply note that

$$\overline{Y}_t^+ = Y_t \mathbf{1}_{\{t \notin D\}} + (Y_t \vee X_t)\mathbf{1}_{t \in D}.$$

The argument is similar for the other processes (2.7) and (2.8). ∎

Remark 2.3.2 We emphasize that $\lim_{s \to t}$ or $\lim_{s \uparrow t}$ are limits with t included, so that if $\lim_{s \to t} f(s)$ exits, f must be continuous at t. On the other hand, $\lim_{s \to t, s \neq t}$, $\lim_{s \uparrow\uparrow t}$ exclude t.

The following statement is the first in a series of theorems that analyzes the path properties of stochastic processes. The first theorem looks at the measurability of subsets of Ω when the time parameter of a stochastic process is restricted to countably dense subsets of $\mathbb{R}$.

Theorem 2.3.3 (a) The set W (resp. W') of $\omega \in \Omega$ such that the path $X_.(\omega)$ is the restriction to D of right-continuous (resp. r.c.l.l) mapping on $\mathbb{R}_+$ the complement of an $\mathcal{F}$-analytic set; hence, it belongs to the universal completion σ-field of $\mathcal{F}$. This result extends to processes with values in a cosouslin metrizable space E.

(b) In the real case, or, more generally for processes with values in a Polish space, E, it can even be affirmed that W' belongs to $\mathcal{F}$.

Proof. (a) Embed the separable metrizable space E in the cube $I = \bar{\mathbb{R}}^{\mathbb{N}}$ (or in $\bar{\mathbb{R}}$ if $E = \mathbb{R}$) and denote by J the compact metrizable space obtained by adjoining an isolated point α to I. Write

$$
\begin{aligned}
X_{t+}(\omega) &= \lim_{s \in D, s \downarrow t} X_s(\omega) \text{ if this limit exists in I} \\
&= \alpha \text{ otherwise,}
\end{aligned}
$$

similarly

$$
\begin{aligned}
X_{t-}(\omega) &= \lim_{s \in D, s \uparrow\uparrow t} X_s(\omega) \text{ if this limit exists,} \\
&= \alpha \text{ otherwise.}
\end{aligned}
$$

For $t \in D$, the existence of $X_{t+}(\omega)$ implies $X_{t+}(\omega) = X_t(\omega)$, since, X_{t+} is a right limit at t with t included. The mappings $(t, \omega) \leftrightarrow X_{t+}(\omega)$, $X_{t-}(\omega)$ are $\mathcal{B}(\mathbb{R}_+) \times \mathcal{F}$-measurable. Since $I = \bar{\mathbb{R}}^{\mathbb{N}}$, we can immediately reduce to the real case and then we have

$$
X_{t+}(\omega) = \overline{Y}_t^+(\omega) \text{ if } \overline{Y}_t^+(\omega) = \underline{Y}_t^+(\omega), \ X_{t+}(\omega) = \alpha \text{ otherwise}
$$

and we apply Theorem 2.3.1; similarly for X_{t-} using $\overline{Z}^-$, $\underline{Z}^-$. We denote by A the set J\E; since E is cosouslin, A is analytic in J (1.3.20) and the set

$$
H = \{(t, \omega) : X_{t+}(\omega) \in A\} \ (\text{resp. } H' = \{(t, \omega) : X_{t+}(\omega) \in A \text{ or } X_{t-}(\omega) \in A\})
$$

is analytic, being the inverse image of an analytic set under a measurable mapping (Theorem 1.3.11). We conclude by noting that the complement of W (resp. W') is the projection of H (resp. H') onto Ω and applying Theorem 1.3.13.

It remains to show that W' is $\mathcal{F}$-measurable if E is Polish. We shall see later (Theorem 2.3.9) a similar result proved quite differently.

We give E a metric d, under which E is complete, and set $d(\alpha, E) = +\infty$. For $\epsilon > 0$ we define the following functions inductively:

$$
T_0^\epsilon(\omega) = 0
$$

$Z_0^\epsilon(\omega) = \lim_{s \in D, s \downarrow 0} X_t(\omega)$ if this limit exists α otherwise, then

$$
T_{n+1}^\epsilon(\omega) = \inf\{t \in D, t > T_n(\omega), \ d(X_t(\omega), Z_n(\omega)) > \epsilon\} \quad (\inf \emptyset = +\infty)
$$

$Z_{n+1}^\epsilon(\omega) = \lim_{s \in D, s \downarrow T_{n+1}(\omega)} X_s(\omega)$ if $T_{n+1}^\epsilon(\omega) < \infty$ and this limit exists, $= \alpha$ otherwise. It is easy to verify that the functions T_n^ϵ, Z_n^ϵ on E^D are $\mathcal{F}$-measurable. The statement then follows from the following Lemma (2.3.4). ∎

Lemma 2.3.4

$$
W' = \left\{\omega \in \Omega : \forall k \in \mathbb{N}, \lim_n T_n^{2^{-k}}(\omega) = +\infty\right\}.
$$

Proof. (a) If $\omega \in W'$, ω is the restriction to D of a right-continuous mapping on $\mathbb{R}_+$ into E. It follows by right-continuity that the limits in the preceding definition always exist and that for all ϵ and all n such that $T_n^\epsilon(\omega) < \infty$

$$
T_{n+1}^\epsilon(\omega) > T_n^\epsilon(\omega), \quad d(Z_n^\epsilon(\omega), Z_{n+1}^\epsilon(\omega)) \geq \epsilon \text{ if } T_{n+1}^\epsilon(\omega) < \infty.
$$

The oscillation of ω on the interval $[T_n^\epsilon(\omega), Z_{n+1}^\epsilon(\omega)]$ is therefore at least ϵ: if $T_{n+1}^\epsilon(\omega) < \infty$, and the existence of left-hand limits therefore prevents the $T_n^\epsilon(\omega)$ from accumulating at a finite distance. Consequently $\lim_n T_n^\epsilon(\omega) = +\infty$ for all $\epsilon > 0$.

(b) Conversely, suppose that $T_n^\epsilon(\omega) \mapsto +\infty$. Then define r.c.l.l. mapping F_ϵ of D into E by writing $f_\epsilon(t) = Z_n^\epsilon(\omega)$ for $t \in D \cap [T_n^\epsilon(\omega), T_{n+1}^\epsilon(\omega)[$. Then $\mathrm{d}(X_t(\omega), f_\epsilon(t)) \leq 2^{-\epsilon}$; for all $t \in D$. If the above property is satisfied for values of ϵ tending to 0 - for example $\epsilon = 2^{-k}$ - we see that $X.(\omega)$ is the uniform limit on D of a sequence of r.c.l.l. mappings on $\mathbb{R}_+$. It follows immediately that it can be extended to a r.c.l.l. mapping on $\mathbb{R}_+$. ■

Remark 2.3.5 (a) Theorem 2.3.3 can be put into a canonical form. We consider the set W (resp. W′) of all right-continuous (resp. r.c.l.l.) mappings of $\mathbb{R}_+$ into a cosouslin metrizable space E and give it the σ-field generated by the coordinate mappings. The mapping that associates with each w $\in$ W (resp. W′) its restriction to D is a measurable isomorphism of W (resp. W′) into $\Omega = E^D$ with its Borel σ-field; Ω is cosouslin and Polish if E is Polish. Applying Theorem 2.3.3 to the process $(X_t)_{t \in D}$ consisting of the coordinate mappings on Ω, it follows that W is the complement of an analytic set in Ω, and W′ a Borel subset of Ω if E is Polish. Hence the measurable space W is cosouslin and the measurable space W′ is Lusin if E is Polish. The proof also indicates a cosouslin (respo Lusin) topology on W (resp. W′), that of pointwise convergence on 0, but this topology is uninteresting in general, since it involves an arbitrary choice of a countable dense set D, while the measure theoretic statement is intrinsic.

When E is Polish, there exists an interesting topology on W′ under which W′ is Polish: the Skorokhod topology. See for example Maisonneuve [23].

(b) We adjoin to E an isolated point denoted by δ and we denote by Ω the set of all right-continuous mappings ω of $\mathbb{R}_+$ into $E \cup \{\delta\}$, which keep the value δ from the first instant they assume it, so that the set $\{t : \omega(t) = \delta\}$ is a closed half-line (possibly empty) $[\zeta(\omega), +\infty[$. It is easy to see that the lifetime ζ is a measurable function relative to the σ-field $\mathcal{F}^0$ on Ω generated by the coordinate mappings if E is cosouslin, and that Ω is a cosouslin space under the topology of pointwise convergence on D. On the other hand, if E is Polish, the space Ω' of elements ω with a left limit in E at every point of the interval $]0, \zeta(\omega)[$ but not necessarily at the instant $\zeta(\omega)$ itself is Lusin under the topology of pointwise convergence on D. The idea is the same as in the proof of Remark 2.2.3; one just has to write $\lim_n T_n^\epsilon \geq \zeta$ instead of $\lim_n T_n^\epsilon = +\infty$.

2.3.2 Upcrossings and downcrossings

Here we will study the numbers of upcrossings and downcrossings which are important in martingale theory. We will not require right continuity of processes, but rather the existence of right-hand and left-hand limits.

Let f be a mapping of $\mathbb{R}_+$ into a Hausdorff space E. We say that f is *free of oscillatory discontinuities* if the right-hand limit

$$f(t+) = \lim_{s \downarrow \downarrow t} f(s)$$

exists in E at every point t of $\mathbb{R}_+$ and the left-hand limit

$$f(t-) = \lim_{s\uparrow\uparrow t} f(s)$$

also exists in E at every point of $\mathbb{R}_+\backslash\{0\}$, but it does not necessarily exist at infinity.

We start by considering extended real-valued functions and giving a simple criterion of freedom from oscillatory discontinuities in $\overline{\mathbb{R}}$; let f be a mapping of $\mathbb{R}_+$ into $\overline{\mathbb{R}}$ and denote by a, b two finite real numbers such that $a < b$ and by $\mathbb{U}$ a finite subset of $\mathbb{R}_+$ whose elements are $s_1, s_2, \ldots, s_n$ arranged by order of magnitude. Define inductively the instants $t_1, \ldots, t_n \in \mathbb{U}$ as follows:
   - t_1 is the first of the elements s_i of $\mathbb{U}$ such that $f(s_i) < a$, or s_n if no such new element exists;
   - t_k is, for every even (resp. odd) integer lying between 1 and n, the first of the elements s_i of $\mathbb{U}$ such that $s_i > t_{k-1}$ and $f(s_i) > b$ (resp. $f(s_i) < a$). If no such element exists, we write $t_k = s_n$.

We consider the last even integer $2k$ such that

$$f(t_{2k-1}) < a, \quad f(t_{2k}) > b;$$

if no such integer exists, we write $k = 0$. The intervals

$$(t_1, t_2), (t_3, t_4), \ldots, (t_{2k-1}, t_{2k})$$

of $\mathbb{U}$ represent periods of time during which the function f goes upward, from below a to above b, whereas the intermediate intervals represent downward periods. The number k is called the number of *upcrossings* by f (considered on $\mathbb{U}$) of the interval $[a, b]$ and is denoted by

$$U(f; \mathbb{U}; [a, b]). \tag{2.9}$$

We define similarly the number of *downcrossings* of f (considered on $\mathbb{U}$) on the interval $[a, b]$;

$$D(f; \mathbb{U}; [a, b]) = U(-f, \mathbb{U}, [-b, -a]). \tag{2.10}$$

We can also define the upcrossings and downcrossings of an interval of the form $]a, b[$, replacing strict inequalities by choosing inequalities in the definition of the instants t_i. Now let S be any subset of $\mathbb{R}_+$. We write:

$$U(f; S; [a, b]) = \sup_{\mathbb{U} \text{ finite, } \mathbb{U} \subset S} U(f; \mathbb{U}; [a, b]). \tag{2.11}$$

Definition (2.10) can be similarly extended.

The principal interest in the upcrossing and downcrossing numbers arises from the following theorem:

Theorem 2.3.6 Let f be a function on $\mathbb{R}_+$ with values in $\overline{\mathbb{R}}$. For f free of oscillatory discontinuities, it is necessary and sufficient that

$$U(f; \mathrm{I}; [a, b]) < +\infty \tag{2.12}$$

for every pair of rational numbers a, b such that $a < b$ and every compact interval I of $\mathbb{R}_+$.

Proof. Suppose that there exists a point t where the function f has an oscillatory discontinuity, for example, where it has no left-hand limit. Then we can find a sequence of points t_n increasing to t such that

$$\liminf_{n\to\infty,\ n\ \text{odd}} f(t_n) = c > d = \limsup_{n\to\infty,\ n\ \text{even}} f(t_n).$$

we then choose a sufficiently large interval I and two rational numbers a and b such that $d < a < b < c$. It is immediately verified, removing finite subsets from the set of points t that $U(f; I; [a, b]) = +\infty$. The converse follows from a property which the reader can prove easily: if r, s, t are three instants such that $r < s < t$, then:

$$U(f; [r, t]; [a, b]) \le U(f; [r, s]; [a, b]) + U(f; [s, t]; [a, b]) + 1.$$

Let α and β be the end-points of I. Suppose that the function f has no oscillatory discontinuities; then we can associate with each point $t \in I$ an open interval I_t containing t, such that the oscillation of f on each one of the intervals $I_t \cap]t, \beta]$, $[\alpha, t[\cap I_t$ is strictly less than $b - a$. We can cover the interval I with a finite number of intervals $I_{t_1}, I_{t_2}, \dots, I_{t_k}$. We arrange by order of magnitude the points α and β, the points $t_1, t_2, \dots, t_k$ and the end-points of the intervals $I_{t_1}, I_{t_2}, \dots, I_{t_k}$; we thus get a finite set of points: $\alpha = s_0 < s_1 < \dots < s_n = \beta$, such that the oscillation of f on each of the intervals $]s_i, s_{i+1}[$ is no greater than $b - a$. Then we have $U(f;]s_i, s_{i+1}[, [a, b]) = 0$ and consequently also $U(f; [s_i, s_{i+1}], [a, b]) \le 1$. The inequality quoted above then gives

$$U(f; I; [a, b]) \le 2n - 1$$

and the converses established. ∎

Remark 2.3.7 The numbers $U(f; \mathbb{U}; [a, b])$ and $D(f; \mathbb{U}; [a, b])$ have the advantage of defining lower semicontinuous functions of f for pointwise convergence. This property extends to the number of upcrossings or downcrossings on any set S.

Remark 2.3.8 The statement of the theorem above concerns $\overline{\mathbb{R}}$. One can also express, using numbers of upcrossings, whether a finite function f on $\mathbb{R}_+$ has finite right-hand and left-hand limits. A finite function with finite right-hand and left-hand limits is bounded in the neighborhood of every point and hence bounded on every compact interval I, so that for every rational a

$$\lim_n U(f; I; [a, a + n]) = 0 = \lim_n U(f; I; [a - n, a]). \tag{2.13}$$

Whereas conversely, if f is not bounded above for example, we can find some a such that the left-hands side of (2.13) is ≥ 1 for all n.

Here is the application to stochastic processes.

Theorem 2.3.9 Let E be an LCC space and let $(X_t)_{t \in D}$ be a process with values in E, defined on $(\Omega, \mathcal{F}, \mathbf{P})$ with time set a countable dense set D. The set of all $\omega \in \Omega$ such that the path $X_\cdot(\omega)$ on D can be extended to a mapping of $\mathbb{R}_+$ into E without oscillatory discontinuities is measurable $\mathcal{F}$-measurable.

Proof. We may assume that E is the complement of a point x_0 in a compact metric space F, whose distances are denoted by d. Let $(x_n)_n \geq 1$ be a sequence dense in E. We write $h_n(x) = d(x_n, x)$ for $n \geq 1$ (so that the sequence (h_n) of continuous functions separates the points) and $h_0(x) = 1/d(x_0, x)$. We want to express that each one of the real processes $(h_n \circ X_t)_n \geq 1$ has right-hand and left-hand limits along D and that the process $(h_0 \circ X_t)$ has finite right-hand and left-hand limits along D. This follows immediately using the numbers of upcrossings of paths, considered on D. ∎

Remark 2.3.10 (a) The result extends to the case of a Polish space E, since every Polish space E can be considered (1.3.18) as a $\mathcal{G}_\delta$ in some compact metric space and hence as an intersection of LCC spaces E_n. We then write down the preceding conditions for each of the E_n. If E were cosouslin (in particular Lusin), the set in the statement would be the complement of an $\mathcal{F}$-analytic set: we leave this aside.

(b) We have been concerned here with r.c.l.l. or r.l.l.l. mappings, but we might consider continuous mappings analogously. The method would be more classical: To express that a mapping of D into a Polish space E can be extended to a continuous mapping of $\mathbb{R}_+$ into E, one just writes for every integer n the condition for uniform continuity on $D \cap [0, n]$.

2.3.3   Separability

The notion of a separable stochastic process is one where the behaviors of its paths are essentially determined by their behavior on countable subsets. Our problem now, which is intimately connected to the notion of separability, is how can we recognize whether a given process (X_t) admits a modification (Y_t) with nice properties, for example, a modification with r.c.l.l. or r.l.l.l. paths or a modification with bounded paths.

However, it is sometimes not possible to modify a given process without destroying its character, considered as is in Remark (2.1.5) a process

$$X_t = 0 \text{ if } B_t \neq 0, \ X_t = 1 \text{ if } B_t = 0, \tag{2.14}$$

where (B_t) is one-dimensional Brownian motion, with continuous paths. If we are just looking for a modification with regular paths, we may simply take the modification $Y_t = 0$. Here, if one considers a separable modification, we would destroy the structure of the process.

The theory of separability was developed by Doob for continuous time processes $(X_t)_{t \in \mathbb{R}_+}$ and extends without difficulty to processes whose time set is a topological space with countable base. Instead of this, we study processes indexed by $\mathbb{R}_+$, but under the *right-topology* (2.3.11) on $\overline{\mathbb{R}}$, which hasn't a countable base; this extension is due to Chung-Doob [24]. On the other hand, the theory of separability can be extended to processes with values in a compact metrizable space, whereas we will only consider processes with values in $\overline{\mathbb{R}}$ (beware of the distinction between $\overline{\mathbb{R}}$ and $\mathbb{R}$ which is important here).

Remark 2.3.11 The right-topology, also known as the lower-limit-topology, is the one generated by half-open intervals (open on the right). The neighborhoods of $x \in \mathbb{R}_+$ for the right-topology are the sets containing an interval $[x, x + \epsilon[$, $\epsilon > 0$, so that a left-closed interval $[a, b[$ is closed under the right-topology.

Let us now define the notion of a separable mapping.

Definition 2.3.12 Let f be a mapping of a topological space $\mathbb{T}$ into a topological space E and let D be a dense set in $\mathbb{T}$. We saw that f is D-separable if the set of points $(t, f(t))$, $t \in D$, is dense in the graph of f (for the product topology on $\mathbb{T} \times E$).

Henceforth, we take $\mathbb{T} = \mathbb{R}_+$ with the right-topology and $E = \overline{\mathbb{R}}$. On the other hand, D will be countable. We then say that f is right D-separable (D-separable if the ordinary topology of $\mathbb{R}_+$ is used.)

Definition 2.3.13 Let $(X_t)_{t \in \mathbb{R}_+}$ be a process with values in $\overline{\mathbb{R}}$, defined on a probability space $(\Omega, \mathcal{F}, \mathbf{P})$. (X_t) is called *right-separable* if there exists a countable dense set D such that, for almost all $\omega \in \Omega$, the path $X_\cdot(\omega)$ is right D-separable.

The following Theorem is a modification of Doob [25] (Stochastic Processes, pp. 56-57).

Theorem 2.3.14 Let (X_t), $t \in \mathbb{R}_+$ be a process with values in $\overline{\mathbb{R}}$. There exists a countable dense set D with the following property: for every closed set F of $\overline{\mathbb{R}}$ and every open set $I \subset \mathbb{R}_+$ under the right-topology,

$$\mathbf{P}\{X_t \in F \text{ and } X_u \in F \text{ for all } t \in D \cap I\} = 0 \text{ for every } u \in I. \tag{2.15}$$

Also, for every countable set S

$$\mathbf{P}\{X_t \in F \text{ for all } t \in D \cap I\} \leq \mathbf{P}\{X_t \in F \text{ for all } t \in S \cap I\}. \tag{2.16}$$

Proof. Equations (2.15) and (2.16) are equivalent, which is easy to show.

Now, choose a countable set $\mathfrak{H}$ of closed subsets of $\overline{\mathbb{R}}$, such that, every closed set is the intersection of a decreasing sequence of elements of $\mathfrak{H}$ and a countable set $\mathfrak{G}$ of open subsets of $\mathbb{R}_+$ with the ordinary topology, such that every ordinary open set of $\mathbb{R}_+$ is the union of an increasing sequence of elements of $\mathfrak{G}$.

For every pair (I, F), $I \in \mathfrak{G}$, $F \in \mathfrak{H}$, we choose a countable set $\Delta(I, F)$ dense in I such that the probability

$$\mathbf{P}\{X_t \in F \text{ for all } t \in S \cap I\} \quad (S \text{ countable})$$

is minimal for $S = \Delta(I, F)$.

Next, we set $\Delta(F) = \cup \Delta(I,F)$ for $I \in \mathfrak{G}$. Then, for every ordinary open set I and every countable set S we have

$$\mathbf{P}\{X_t \in F \text{ for all } t \in \Delta(F) \cap I\} \leq \mathbf{P}\{X_t \in F \text{ for all } t \in S \cap I\}. \tag{2.17}$$

Always keeping $F \in \mathfrak{H}$ fixed, we consider for a rational number $r > 0$ the increasing function on $[0, r[$

$$h_r(t) = \inf_S \mathbf{P}\{X_u \in F \text{ for all } u \in S \cap [t, r[\} \quad (S \text{ countable})$$

which we compare to

$$k_r(t) = P\{X_u \in F \text{ for all } u \in \Delta(F) \cap [t, r[\}.$$

We have, by the choice of $\Delta(F)$, $h_r(t+) = k_r(t+)$ for all t and hence h_r and k_r differ only on a countable set N_r. If we enlarge $\Delta(F)$ by replacing it without changing the notation by $\Delta(F) \cup (\cup_r N_r)$, we have, for every rational r and every $t \in [0, r]$, $h_r(t) = k_r(t)$. But then, the same result will hold for all real numbers on passing to the limit. Thus, for every interval $[t, r[$

$$\mathbf{P}\{X_u \in F \text{ for all } u \in \Delta(F) \cap [t, r[, X_t \notin F\} = 0. \tag{2.18}$$

Now let I be an open set under the right topology: I is a countable union of disjoint intervals of the form $]t_i, r_i[$ or $[t_j, r_j[$. The probability

$$\mathbf{P}\{X_u \in F \text{ for all } u \in \Delta(F) \cap I, X_t \notin F\}$$

is zero for all $t \in I$: if t is an inner point of I in the ordinary sense, use (2.16); if t is one of the left-hand end points of intervals $[t_j, r_j[$, use (2.17).

To get the set D of the statement, possessing the above properties for all closed sets, it suffices to take the union of the countable sets $\Delta(F)$, F running through the countable set $\mathfrak{H}$. ∎

Consider the following two examples of processes:

Example 2.3.15 If Ω consist of a single point, then a process is simply a function $f(t)$ on $\mathbb{R}_+$. $f(t)$ may be arbitrarily bad, but (2.15) tells us that there exists some D such that

$$(f(t) \in F \text{ for } t \in D \cap I) \Leftrightarrow (f(t) \in \mathcal{F} \text{ for } t \in I).$$

It follows that f is a right D-separable function and the process f therefore is right-separable. So separability in itself doesn't imply any regularity of the sample functions of a process.

Example 2.3.16 Recall Example (2.14) and consider that for every countable set D we have $\mathbf{P}\{X_u = 0, u \in D\} = 1$; whereas, for almost all ω, the set $\{u : X_u(\omega) = 1\}$ is non-empty. Hence, the process is not separable, and any attempt to make it separable would also make it indistinguishable from 0, and therefore without any interesting paths.

2.3.3.1 Doob's separability theorems

We come to Doob's main theorems, the first one concerning arbitrary processes and the second one is on measurable processes.

Theorem 2.3.17 Every real-valued process $(X_t)_{t \in \mathbb{R}_+}$ has a right separable modification with values in $\bar{\mathbb{R}}$ but may not have one in $\mathbb{R}$.

Proof. Fix $t \in \mathbb{R}_+$ and choose the set D as in Theorem 2.3.14. Denote by $A_t(\omega)$ the non-empty set of cluster values in $\bar{\mathbb{R}}$ of the function $X_.(\omega)$ at the point t from the right and along D,

$$A_t(\omega) = \bigcap_n \overline{\{X_u(\omega), \ u \in D \cap [t, t+1/n[\}}.$$

The set of ω such that $X_t(\omega) \in A_t(\omega)$ is measurable. Let d be a metric defining the topology of $\bar{\mathbb{R}}$. $X_t(\omega) \in A_t(\omega)$ is equivalent to

$$\forall n > 0, \ \forall m > 0, \ \exists u \in [t, t+1/n[\cap D, \ \mathrm{d}(X_t(\omega), X_u(\omega)) < 1/m.$$

So, we claim that $X_t(\omega) \in A_t(\omega)$ for almost all ω.

Suppose otherwise, that $X_t(\omega) \notin A_t(\omega)$, and let's go back to the countable family $\mathfrak{H}$ of closed sets of Theorem 2.3.14; there exists an element F of $\mathfrak{H}$ containing $A_t(\omega)$ such that $X_t(\omega) \notin F$, and, hence a number m such that $\mathrm{d}(X_t(\omega), F) > 1/m$.

If $F_m = \{x : \mathrm{d}(x, F) \leq 1/m\}$, we have for n sufficiently large $X_u(\omega) \in F_m$ for all $u \in D \cap [t, t+1/n[$, because F_m is a neighborhood of the set of cluster values at t along D.

Consequently, for a suitable choice of n, m and $F \in \mathfrak{H}$. We have $\omega \in H(n, m, F)$, where this denotes the set

$$\{\omega : X_u(\omega) \in F_m \text{ for } u \in D \cap [t, t+1/n[, \ X_t(\omega) \notin F_m\}.$$

Since this event has probability zero by the choice of D, so does the union of the $H(n, m, F)$ $(n, m$ integers, $F \in E)$ and we have seen that this union contains the set $\{X_t \notin A_t\}$.

To get the required modification, we finally set

$$X'_t(\omega) = \begin{cases} X_t(\omega) & \text{if } X_t(\omega) \in A_t(\omega), \\ \liminf\limits_{s \downarrow t, s \in D} X_s(\omega) & \text{otherwise.} \end{cases}$$

■

Doob's second theorem concerns the existence of modifications of a process which are both right-separable and progressive. First, let L^0 be the space of classes of real-valued random variables on Ω with the metric of convergence in probability.

Theorem 2.3.18 Let (X_t), $t \in \mathbb{R}_+$ be an $\bar{\mathbb{R}}$ valued process on $(\Omega, \mathcal{F}, \mathbf{P})$ and $\mathring{X}_t$ the class of the r.v. X_t considered as an element of L^0.

Then, (X_t) has a measurable modification, if and only if the mapping $t \mapsto \mathring{X}_t$ is a uniform limit in L^0 of measurable step functions.

Furthermore, if this condition is satisfied, (X_t) has a right-separable and measurable modification. More precisely, if it is satisfied and if (X_t) is adapted to a filtration $(\mathcal{F}_t)$ the modification can be chosen to be right-separable and progressively measurable with respect to the family $(\mathcal{F}_{t+})$.

Proof. It can be shown that the condition of the statement is equivalent to, the function $t \mapsto \mathring{X}_t$ is measurable in the usual sense, that is the inverse image of every Borel set of L^0 is Borel in $\mathbb{R}_+$, and takes its values in a separable subset of L^0. Incidentally, this condition is the correct definition of measurability to be used, for example in the theory of integration with values in Banach spaces.

Since $\bar{\mathbb{R}}$ is homeomorphic to the interval $I = [-1, +1]$, we replace $\bar{\mathbb{R}}$ by I and convergence in probability by convergence in norm in L^1. So, for the rest of the proof, L^1 replaces L^1.

(a) Suppose (X_t) is measurable, we will show that the above condition is satisfied. Let $\mathcal{H}$ be the set of real-valued measurable processes (Y_t) on $(\Omega, \mathcal{F}, \mathbf{P})$ such that (Y_t) is uniformly bounded and the mapping $t \mapsto \mathring{Y}_t$ of $\mathbb{R}_+$ into L^1 is Borel with values in a separable subset of L^1.

Clearly, all processes $Y_t(\omega)$ of the form,

$$\sum_{k \in \mathbb{N}} \mathbf{1}_{[k2^{-n}, (k+1)2^{-n}[}(t) Y^k(\omega)$$

where $n \in \mathbb{N}$ and the Y^k are uniformly bounded random variables, form an algebra contained in $\mathcal{H}$ which generates the σ-field $\mathcal{B}(\mathbb{R}_+) \times \mathcal{F}$.

On the other hand, $\mathcal{H}$ is closed under monotone bounded convergence. The monotone class theorem then implies that every bounded measurable process (X_t) belongs to $\mathcal{H}$, and the condition of the statement follows from Lebesgue approximation (1.2.6).

(b) Conversely, let (X_t) be a process with values in I, satisfying the above condition and adapted to $(\mathcal{F}_t)$; if no family is given, take $\mathcal{F}_t = \mathcal{F}$ for all t. We consider elementary processes (Z_t^n) such that $\|X_t - Z_t^n\| \leq 2^{-n}$ for all t. We can write

$$Z_t^n(\omega) = \sum_k \mathbf{1}_{A_k^n}(t) H_k^n(\omega) \tag{2.19}$$

where the A_k^n form a partition of $\mathbb{R}_+$ and the H_k^n are random variables with values in I. We begin by turning the (Z_t^n) into processes adapted to the family $(\mathcal{F}_{t+})$. Then, let s_k^n be the infimum of A_k^n and let (t_i) be a decreasing sequence of elements of A_k^n converging to s_k^n. The sequence may be constant if $s_k^n \in A_k^n$.

Since the random variables X_{t_i} are uniformly bounded, we can suppose, replacing (t_i) by a subsequence if necessary, that the X_{t_i} converge weakly in L^1 to a $\mathcal{F}_{s_k^n+}$-measurable random variable L_k^n. We have $\|X_{t_i} - H_k^n\|_1 \leq 2^{-n}$ and hence also so $\|L_k^n - H_k^n\|_1 \leq 2^{-n}$, since the norm, being the upper envelope of a family of linear functionals, is a l.s.c. function under the weak topology of L^1. Then, the process

$$Y_t^n(\omega) = \sum_k \mathbf{1}_{A_k^n}(t) L_k^n(\omega) \tag{2.20}$$

is progressive with respect to the family $(\mathcal{F}_{t+})$, and $\|X_t - Y_t^n\|_1 \leq 2 \cdot 2^{-n}$ for all n and t. We set

$$Y_t(\omega) = \liminf_n Y_t^n(\omega). \tag{2.21}$$

This process still is progressive. On the other hand, for each fixed t, Y_t^n converges a.s. to (X_t) and hence (Y_t) is a modification of (X_t).

(c) This modification is not yet right separable. We return to the set D from Theorem 2.3.14 - relative to (X_t) or (Y_t), this amounts to the same, since they are modifications of each other - and set as in Theorem 2.3.17.

$$A_t^n(\omega) = \overline{\{Y_u(\omega), u \in D \cap [t, t+1/n[\}}, \quad A_t(\omega) = \bigcap_n A_t^n(\omega).$$

Let d be the usual metric on I. The process $d(Y_t(\omega), A_t^n(\omega)) = \sup_{s \in D} d(Y_t(\omega), Y_s(\omega)) 1_{]s-1/n,s]}(t)$ is progressive with respect to the family $(\mathcal{F}_{t+1/n})$; hence the process $d(Y_t, A_t)$ is progressive with respect to $(\mathcal{F}_{t+})$. It only remains to define as in 2.3.17

$$X_t' = \begin{cases} Y_t & \text{if } d(Y_t, A_t) = 0, \text{ i.e., if } Y_t \in A_t \\ \liminf_{s \downarrow t, s \in D} Y_t & \text{otherwise.} \end{cases}$$

This is the required modification. ∎

Remark 2.3.19 (1) Let $\mathcal{G}$ be the σ-field generated by the X_t, $t \in \mathbb{R}_+$. If the mapping $t \mapsto \mathring{X}_t$ with values in $L^0(\mathcal{F})$ is Borel and takes its values in a separable subset, it satisfies the same condition relative to $L^0(\mathcal{G})$ and the above proof shows that there exist step processes of type (2.19)

$$Z_t^n = \sum_k 1_{A_k^n}(t) H_k^n(\omega) \tag{2.22}$$

where the H_k^n are $\mathcal{G}$-measurable and converge uniformly to (X_t) in probability. By Theorem 1.2.7, each r.v. H_k^n admits a representation

$$H_k^n = h_k^n \left(\left(X_{t_p^{nk}} \right)_{p \in \mathbb{N}} \right) \tag{2.23}$$

where $(t_p^{nk})_{p \in \mathbb{N}}$ is a sequence in $\mathbb{R}_+$ and h_k^n is a Borel function on $\bar{\mathbb{R}}^{\mathbb{N}}$. But then it is easy to see that the property that the (Z_t^n) given by (2.22) and (2.23) converge uniformly in probability to (X_t) depends only on the time law of the process (X_t). In other words, the existence of a measurable modification is a property of a process which depends only on its time law.

(2) We keep the above notation. The process (X_t), if it is $\mathcal{B}(\mathbb{R}_+) \times \mathcal{F}$-measurable, has $B(\mathbb{R}_+) \times \mathcal{G}$-measurable modification (X_t'). But the σ-field $\mathcal{G}'$ generated by the X_t' may be strictly contained in $\mathcal{G}$ and one cannot be sure that (X_t') is $\mathcal{B}(\mathbb{R}_+) \times \mathcal{G}'$-measurable (that (X_t') is naturally measurable). The same kind of difficulties arise with progressive processes.

2.3.4 Progressive processes of random sets

Up to now we have studied properties of the process $(X_t)_{t \in \mathbb{R}_+}$ that were deduced from knowledge of $(X_t)_{t \in D}$ where D is countable and dense. Here, we will, assuming only that (X_t) is measurable with respect to a family $(\mathcal{F}_t)$, study directly the behavior of paths on $\mathbb{R}_+$ using the theory of capacities (see Theorem 1.3.33).

Definition 2.3.20 Let $(\Omega, \mathcal{F}, \mathbf{P})$ be a probability space with a filtration $(\mathcal{F}_t)_{t \in \mathbb{R}_+}$ and let $A \subseteq \mathbb{R}_+ \times \Omega$. Denote by (a_t) the indicator process of A

$$a_t(\omega) = 1_A(t, \omega). \tag{2.24}$$

A is a measurable random set if $A \in \mathcal{B}(\mathbb{R}_+) \times \mathcal{F}$ or the process (a_t) is measurable, a progressive set if (a_t) is a progressive process, a closed (resp. right, left) set if for all ω the section $A(\omega) = \{t \in \mathbb{R}_+ : (t, \omega) \in A\}$ closed (resp. right, left) in $\mathbb{R}_+$.

It may be useful to recall here that a right-closed interval $]a, b]$ or $[a, b]$ of $\mathbb{R}_+$ is closed under the *left-topology* of $\mathbb{R}_+$ and is hence a *left-closed set*.

Progressive sets form a σ-field on $\mathbb{R}_+ \times \Omega$ known as the progressive σ-field. Progressive processes are precisely those functions on $\mathbb{R}_+ \times \Omega$ which are measurable with respect to the progressive σ-field.

Given a subset A of $\mathbb{R}_+ \times \Omega$, the closure $\bar{A}$ of A is the set whose section $\bar{A}(\omega)$ is, for all $\omega \in \Omega$, the closure of the section $A(\omega)$. We define similarly the right or left closure of A.

A key proposition to the results of this section is as follows:

Theorem 2.3.21 Suppose the space $(\Omega, \mathcal{F}, \mathbf{P})$ is complete ($\mathcal{F}_0$ contains all the negligible sets) and the filtration $(\mathcal{F}_t)$ is right-continuous. Then, the closure (resp. right, left closure) of a progressive set is a progressive set.

Proof. First, let A be a measurable random set. For all $s \geq 0$ define

$$D_s(\omega) = \inf\{t > s : (t, \omega) \in A\} \quad (\inf 0 = +\infty). \tag{2.25}$$

Clearly $D.(\omega)$ is a right-continuous increasing function on $\mathbb{R}_+$. On the other hand, by Theorem 1.3.33, D_s is measurable with respect to the completion of $\mathcal{F}$, that is with respect to $\mathcal{F}$ itself. By Theorem 2.2.6 the process (D_s) is measurable and similarly so is the process $(D_{s-})_{s>0}$.

The set of cluster points of A under the right-topology then is $\{(s, w) : D_s(w) = s\}$ and this set is measurable; so is then the right closure of A. Similarly, the left closure of A is the set $\{(s, w) : s > 0, D_{s-}(\omega) = s\} \cup A$, which is measurable. Taking their union, the closure $\bar{A}$ of A is measurable.

Progressivity follows immediately: what we have done on $[0, \infty[\times \Omega$ carries over to $[0, t[\times \Omega$ using an increasing bijection of $[0, \infty]$ onto $[0, t[$ and it follows that the closure of $A \cap ([0, t[\times \Omega)$ in $[0, t[\times \Omega$ is measurable relative to the σ-field $\mathcal{B}([0, t[) \times \mathcal{F}_t$. But then the remarks of Definition 2.2.4 imply that A is progressive with respect to the family: $\mathcal{F}_{t+} = \mathcal{F}_t$. ∎

The next theorem presents other processes that are progressive:

Theorem 2.3.22 With the same conditions on the probability space and family of fields as in Theorem 2.3.21, let (X_t) be real-valued a progressive process. Then, the following processes are also progressive:
  (a) $X_t^\star = \sup_{s \leq t} X_t$
  (b) $\overline{Y}_t^+ = \limsup_{s \downarrow t} X_s$, $\underline{Y}_t^+ = \liminf_{s \downarrow t} X_s$ and the analogous processes on the left-hand side.
  (c) $\bar{Z}_t^+ = \limsup_{s \downarrow\downarrow t} X_s$, $\underline{Z}_t^+ = \liminf_{s \downarrow\downarrow t} X_s$ and the analogous processes on the left-hand side.

Proof. (a) Write $L_t = \sup_{s<t} X$ for $t > 0$ and $L_0 = -\infty$. As $X_t^{\star} = L_t \vee X_t$, it suffices to show that (L_t) is progressive. But this process is left-continuous: by Theorem 2.2.6, it suffices to show that L_t is $\mathcal{F}_t$-measurable. Since, the set $\{L_t > a\}$ is the projection on Ω of the $(\mathcal{B}(\mathbb{R}_+) \times \mathcal{F}_t)$-measurable set

$$\{(s,\omega) : s < t, \ X_s(\omega) > a\}.$$

then it is $\mathcal{F}_t$-analytic and hence $\mathcal{F}_t$-measurable, $\mathcal{F}_t$ being complete.

For (b) and (c) we are dealing with $\overline{Y}^+$ and $\overline{Z}^+$. Since $\overline{Y}_t^+ = X_t \vee \overline{Z}_t^+$, it suffices to consider only the process $\overline{Z}^+$. But we note that $\overline{Z}_t^+(\omega) \geq a$ if and only if, for all $\epsilon > 0$, t is a right cluster point of $\{s : X_s(\omega) > a - s\}$. Then, we denote by A the set $\{(s,\omega) : X_s(\omega) > a - \epsilon\}$ and return to the discussion of Theorem 2.3.21: we have seen that the set of right cluster points of A can be written as $\{(s.\omega) : D_s(\omega) = s\}$ and that it is measurable and progressive by the end of the proof of Theorem 2.3.21. It follows that $(\overline{Z}_t^+)$ is itself progressive. ∎

Compare the following result to Theorem 2.3.3.

Theorem 2.3.23 Let $(\Omega, \mathcal{F})$ be a measurable space and let (X_t) be a measurable process with values in a separable metrizable space E. Let Ω_{rl}, Ω_r, Ω_c be the subsets of Ω consisting of the ω whose paths are r.c.l.l. resp. right-continuous and continuous.

If E is cosouslin, the complements of these three sets are $\mathcal{F}$-analytic and hence belong to the universal completion σ-field of $\mathcal{F}$.

Proof. We deal with Ω_r and others to the reader. Choose a countable dense subset D of $\mathbb{R}_+$ and return to the proof of Theorem 2.3.3: E is embedded in the cube $I = [0,1]^{\mathbb{N}}$ to which we adjoin an isolated point α. As in Theorem 2.3.3, we define

$$\begin{aligned} X_{t+}(\omega) &= \lim_{s\downarrow t, s\in D} X_s(t) \text{ if this limit exists in } I \\ &= \alpha \text{ otherwise.} \end{aligned}$$

This process is measurable. To say that the path $X.(\omega)$ is right-continuous on $\mathbb{R}_+$ amounts to saying that $X_t(\omega) = X_{t+}(\omega)$ for all t. Thus, Ω_r^c is the projection of the $(\mathcal{B}(\mathbb{R}_+) \times \mathcal{F})$-measurable set $\{(t,\omega) : X_t(\omega) \neq X_{t+}(\omega)\}$, hence, it is $\mathcal{F}$-analytic. ∎

2.3.5 Almost equivalence

Consider the process $(X_t)_{t\in\mathbb{R}_+}$ on the probability space $(\Omega, \mathcal{F}, \mathbf{P})$ taking values in a metrizable separable space E. Assume the mapping $(t,\omega) \mapsto X_t(\omega)$ of $\mathbb{R}_+ \times \Omega$ into E is measurable where $\mathbb{R}_+ \times \Omega$ is endowed with the completed σ-field $\mathcal{B}(\mathbb{R}_+) \times \mathcal{F}$ with respect to the measure $dt \otimes d\mathbf{P}(\omega)$, a property which we express by saying (X_t) is Lebesgue measurable.

Throughout this section we adopt the point of view of Remark 2.1.5 according to which we do not have access to the r.v. X_t themselves, but only to functions on Ω of the form

$$\mathsf{M}_\phi^X(\omega, g) = \int_0^\infty g(X_t(\omega))\phi(t)dt \tag{2.26}$$

where g is Borel on E and positive or bounded and ϕ is Borel, positive and integrable on $\mathbb{R}_+$. Measurability in the Lebesgue sense guarantees both the existence of integrals (2.26) and the fact that the functions $\mathsf{M}_\phi^X(\cdot, g)$ are r.v. on the completed space $(\Omega, \mathcal{F})$. Now, we give a definition that will help us choose the best possible almost-modification.

Definition 2.3.24 Two processes (X_t) and (Y_t), $t \in \mathbb{R}_+$ with values in E defined on possibly different probability spaces and are Lebesgue measurable, are said to be *almost-equivalent* if, for every finite system of pairs (ϕ_i, g_i) $(1 \le i \le n$, where ϕ_i are Borel, positive and integrable on $\mathbb{R}_+$ and g_i are Borel and positive on E), the following random variables with values in $\mathbb{R}^n$

$$(\mathsf{M}_{\phi_1}^X(\cdot, g_1), \ldots, \mathsf{M}_{\phi_n}^X(\cdot, g_n)) \text{ and } (\mathsf{M}_\phi^Y(\cdot, g_1), \ldots, \mathsf{M}_{\phi_n}^Y(\cdot, g_n)). \tag{2.27}$$

have the same law. In particular, if (X_t) and (Y_t) have the same law and are defined on the same probability space and for almost all t,

$$X_t = Y_t \quad \text{a.s.} \tag{2.28}$$

we then say that these processes are *almost-modifications* of each other.

In studying almost-equivalences, it is natural that we try to use topologies on $\mathbb{R}$ that ignore sets of measure zero. This was done in analysis in the theory of differentiation where many results depend on the notion of approximate limit (limit along a set of density 1). This procedure was also introduced to probability theory by Itô [26]. But It was Doob, Chung and Walsh who did a systematic study of topologies on $\mathbb{R}$ and discovered the essential topology, a topology that is better adapted to the needs of probability theory. Here we will present the notion of essential topologies on $\mathbb{R}$ and dependent results.

Recall that if f is a real-valued function on $\mathbb{R}$ and I is an interval, then the essential supremum of f, $ess\sup_{s \in I} f(s)$, is the greatest lower bound of the numbers c such that the set $\{s : s \in I, f(s) > c\}$ is negligible relative to Lebesgue measure. This definition does not require f to be measurable.

We will not change $ess\sup_{s \in I} f(s)$ by modifying f on a set of measure zero, but since $ess\sup_{s \in I} f(s)$ is an increasing function of I, we make the following definitions:

Definition 2.3.25 Let f be a real-valued function on $\mathbb{R}$, measurable or not, and let $f \in \mathbb{R}$. We define for any t

$$ess \lim_{s \downarrow\downarrow t} \sup f(s) = \lim_{\epsilon \downarrow\downarrow 0} \, ess \sup_{t < s < t+\epsilon} f(s) \tag{2.29}$$

$$ess \lim_{s \to t, s \neq t} \sup f(s) = \lim_{s \downarrow\downarrow 0} \, ess \sup_{t-\epsilon < s < t+\epsilon} f(s) \tag{2.30}$$

$$ess \lim_{s \to t} \sup f(s) = f(t) \vee ess \lim_{s \to t, s \neq t} \sup f(s). \tag{2.31}$$

$ess \lim \sup$ for "$s \uparrow\uparrow t$", "$s \downarrow t$", "$s \uparrow t$", and $\lim \inf$ can be defined analogously.

The limit (2.30) can be written as $\limsup_{s \to t, s \neq t} f(s)$ relative to essential topology on $\mathbb{R}$ described as an essential neighborhood of t is a (not necessarily measurable) set containing $\{t\}$ and a set $]t - \epsilon, t + \epsilon[\backslash N$, where ϵ is > 0 and N is a negligible Borel set.

Similarly, (2.29) can be written as $\limsup_{s \to t, s \neq t} f(s)$ under right essential topology on $\mathbb{R}$, whose twin sister is the left essential topology.

Next we present fundamental properties of essential limits. The first theorem in this regard concerns functions on $\mathbb{R}$ and the second processes.

Theorem 2.3.26 For every function f on $\mathbb{R}$ we denote by $\overline{f}$ the function $t \mapsto ess \limsup_{s \downarrow \downarrow t} f(s)$.

(a) If $f = g$ a.e., then $\overline{f} = \overline{g}$ everywhere.
(b) The function $\overline{f}$ is Borel and

$$\overline{f}(t) = ess \limsup_{s \downarrow \downarrow t} f(s) = \limsup_{s \downarrow \downarrow t} \overline{f}(s). \tag{2.32}$$

(c) $f \leq \overline{f}$ almost everywhere on $\mathbb{R}_+$.
(d) If f is continuous under the right essential topology, f is right continuous under the ordinary topology.

Proof. First we need to establish a lemma that the right topology on $\mathbb{R}_+$ has the LL property of [19], III.63. To do so, let $(L_i)_{i \in I}$ be a family of intervals $L_i = [a_i, b_i[$. There exists a countable subfamily $(L_i)_{i \in J}$ such that $\cup_{i \in I} L_i = \cup_{i \in J} L_i$.

We write $L = \cup_{i \in I} L_i$, $K_i =]a_i, b_i[$, $K = \cup_{i \in I} K_i$. Since the usual topology on the line has a countable base, there exists a countable subset J_1 of I such that $\cup_{i \in J_1} K_i = K$. On the other hand, the set $L \backslash K$ is countable, since it is discrete under the right topology: if a belongs to $L \backslash K$, there exists an interval $I = [a, b[\in (L_i)$, then every paint of $]a, b[$ belongs to K and hence $I \cap (L \backslash K) = \{a\}$. Let then J_2 be a countable subset of I such that $L \backslash K \subset \cup_{i \in J_2} L_i$. It only remains to set $J = J_1 \cup J_2$.

We deduce, for the right essential topology, the following consequence: if $I \subset \mathbb{R}_+$ is discrete under the right essential topology, it has outer Lebesgue measure zero. We may indeed associate with every $t \in I$ an interval $L_t = [t, t + \epsilon[$ such that $L_t \cap I$ has outer measure zero. According to the lemma, we can cover I with a countable infinity of such intervals, and this implies that I has outer measure zero.

We now prove Theorem 2.3.26. Property (a) is obvious. Similarly, (c) is obvious since the set $\{f \geq \overline{f} + s\}$ is discrete under the right essential topology for all $\epsilon > 0$. To show that $\overline{f}$ is Borel, we note that the property $(\overline{f}(t) \geq a)$ means that, for all $\epsilon > 0$ and all $r > 0$, the set $[t, t+r[\cap (f > a - \epsilon)$ has a strictly positive outer measure. If follows immediately that the set $\{\overline{f} \geq a\}$ is closed under the right topology and hence Borel, and that $\overline{f}$ is Borel. One also sees that $\overline{f}$ is u.s.c. under the right topology and hence that

$$\overline{f}(t) \geq \limsup_{s \downarrow t} \overline{f}(s) \geq \limsup_{s \downarrow \downarrow t} \overline{f}(s) \geq ess \limsup_{s \downarrow \downarrow t} \overline{f}(s)$$

$$\geq ess \limsup_{s \downarrow \downarrow t} f(s) = \overline{f}(t)$$

where the second row of inequalities arises from the fact that $f \leq \overline{f}$ almost everywhere (a.e.). This establishes (2.32). ∎

If f is continuous under the right essential topology, then $f = \overline{f}$ and hence f is u.s.c. under the right topology. Applying this result to $-f$, we see that f is right continuous.

Remark 2.3.27 There is another way of showing that $\overline{f}$ is Borel, when f is Lebesgue measurable. Consider for a positive and bounded function f,

$$ess \sup_{t<s<t+\epsilon} f(s) = \lim_{n} \left(\frac{1}{\epsilon} \int_{t}^{t+\epsilon} f^n(u)du \right)^{1/n} \tag{2.33}$$

The function of t on the right-hand side is continuous and increases with n. The left-hand side is therefore a l.s.c. function under the ordinary topology and $\overline{f}$ is a Baire function of the second class.

Next we consider the real-valued stochastic processes.

Theorem 2.3.28 Let (X_t) be a real-valued process on $(\Omega, \mathcal{F}, \mathbf{P})$ adapted to the raw family $(\mathcal{F}_t)$ and Lebesgue measurable. We set,

$$\overline{X}_t^+(\omega) = ess\limsup_{s\downarrow\downarrow t} X_s(\omega), \qquad \underline{X}_t^+(\omega) = ess\liminf_{s\downarrow\downarrow t} X_s(\omega). \tag{2.34}$$

(a) The process $(\overline{X}_t)$ is indistinguishable from a measurable process in the ordinary sense and is even progressive with respect to $(\mathcal{F}_{t+})$.

(b) (X_t) has a right-continuous almost-modification if and only if $\overline{X}^+$ and $\underline{X}^+$ are indistinguishable.

(c) If $(\Omega, \mathcal{F}, \mathbf{P})$ is complete, then the set of ω such that $X.(\omega)$ is equal a.e. to a right-continuous function is measurable.

Proof. (a) Since the process (X_t) is Lebesgue measurable, there exist two processes (U_t) and (V_t) measurable in the ordinary sense such that $U_t \le X_t \le V_t$ and $\{(t, \omega) : U_t(\omega) < V_t(\omega)\}$ is negligible relative to the measure $dt \otimes d\mathbf{P}(\omega)$.

By Fubini's Theorem, the set A of ω such that $U.(\omega) = V.(\omega)$ does not hold a.e. is $\mathbf{P}$-negligible and, if $\omega \notin A$, then $\overline{U}_t^+(\omega) = \overline{X}_t^+(\omega) = \overline{V}_t^+(\omega)$ for all t by Theorem 2.3.26(a). Hence it suffices to show that the process $(\overline{U}_t^+)$, for example, is progressive with respect to the family $(\mathcal{F}_{t+})$. To this end, we reduce the problem to the case where (U_t) takes its values in the interval $[0, 1]$ and note that the process

$$t \mapsto \left(\frac{1}{\epsilon} \int_{t}^{t+\epsilon} U_s^n ds \right)^{1/n}$$

has continuous paths and is adapted to the family $(\mathcal{F}_{t+\epsilon})$; hence it is progressive with respect to this family. As $n \to \infty$, we deduce that the process $t \mapsto ess \sup_{t<s<t+\epsilon} U_s$ is progressive with respect to the same family and, as $\epsilon \to 0$, it follows that $(\overline{U}_t^+)$ is progressive with respect to the family $(\mathcal{F}_{t+})$ (Definition 2.2.4).

(b) Suppose that (X_t) has a right-continuous almost-modification (Y_t). By Fubini's Theorem, there exists a measurable set $N \subset \Omega$ such that $\mathbf{P}(N) = 0$ and, for $\omega \notin N$, $X_\cdot(\omega) = Y_\cdot(\omega)$ a.e. Then, for $\omega \notin N$, $\overline{X}_t^+(\omega) = \overline{Y}_t^+(\omega) = Y_t(\omega) = \underline{Y}_t^+(\omega) = \underline{X}_t^+(\omega)$ and the two processes are indistinguishable. Conversely, if $\overline{X}_\cdot^+(\omega) = \underline{X}_\cdot^+(\omega)$, their common value is a function which is both u.s.c. and l.s.c. under the right essential topology, hence essentially right-continuous; therefore, right-continuous (Theorem 2.3.26(d)) and a.e. equal to $X_\cdot(\omega)$ (Theorem 2.3.26(c)). If the processes $(\overline{X}_t^+)$ and $(\underline{X}_t^+)$ are indistinguishable, there exists a measurable set N such that $\mathbf{P}(N) = 0$, and for $\omega \notin N$, $\underline{X}_\cdot^+(\omega) = \overline{X}_\cdot^+(\omega)$; the required almost-modification (Y_t) can be taken to be the common value of these processes on N^c and to 0 on N. It must be noted that it is not progressively measurable; it becomes so if the set N of measure zero is adjoined to $\mathcal{F}_0$.

(c) We leave to the reader to verify; as for (b), that the set in the statement is equal to $\{\omega : \overline{X}_\cdot^+(\omega) = \underline{X}_\cdot^+(\omega)\}$. Its complement is the projection of $\{(t,\omega) : \overline{X}_t^+(\omega) \neq \underline{X}_t^+(\omega)\}$; since this set is equal to an element of $\mathcal{B}(\mathbb{R}_+) \times \mathcal{F}$ up to an evanescent set (a), its projection is $\mathcal{F}$-measurable (Theorem 1.3.13 and Section 1.3.4.1). ∎

Remark 2.3.29 Suppose we don't make the completion hypothesis and that (X_t) is measurable, then, the set of all $\omega \in \Omega$ such that $X_\cdot(\omega)$ is equal a.e. to r.c.l.l. mapping is the complement of an $\mathcal{F}$-analytic set. It is natural to ask by analogy to Section 2.3.3 whether this set is $\mathcal{F}$-measurable? It is indeed so, and we sketch a proof below following the definition of essential debut.

Definition 2.3.30 Let A be a measurable set of $\mathbb{R}_+ \times \Omega$. The essential debut of A is the function on Ω

$$E_A(\omega) = \inf\left\{t : \int_0^t \mathbf{1}_A(s,\omega)ds > 0\right\} \quad (\inf \emptyset = +\infty).$$

The essential debut thus appears as the ordinary debut of the measurable set

$$\left\{(t,\omega) : \int_0^t \mathbf{1}_A(s,\omega)ds > 0\right\}.$$

Lemma 2.3.31 The essential debut of a measurable set A is $\mathcal{F}$-measurable.

Proof. This follows immediately from the equality, true for all $t \geq 0$,

$$\{\omega : E_A(\omega) \geq t\} = \left\{\omega : \int_0^t \mathbf{1}_A(s,\omega)ds = 0\right\}.$$

∎

The proof of the property stated above is now similar to that of Section 2.3.3; we simply replace the times T_n^ϵ by essential debuts.

2.3.5.1   Pseudo-Paths

Throughout this section, let $(\Omega, \mathcal{F}, \mathbf{P})$ be a probability space and E a separable metrizable state space which we shall suppose to be Lusin to avoid uninteresting complications. We can then identify E (a measurable but not necessarily topological space) with a Borel subset of the interval $I = [0, 1]$; this permits us to use the tools of real analysis. Further, we assume that $0 \in E$, but note that the role played by 0 could be played by any point of E that is arbitrarily chosen.

We are going to study classes of Lebesgue measurable processes $(X_t)_{t \in \mathbb{R}_+}$ with values in E under the almost-equivalence relation defined in Definition 2.3.24, in particular we are going to construct a canonical representative of each almost-equivalence class.

Remark 2.3.32 If we consider a Lebesgue measurable process (X_t) on Ω with values in E defined as a process with values in I. We know by completion that there exists a process (Y_t) measurable in the ordinary sense, such that, the set $\{(t, \omega) : X_t(\omega) \neq Y_t(\omega)\}$ is negligible relative to the measure $dt \otimes d\mathbf{P}(\omega)$. Replacing Y_t by 0 if $Y_t \notin E$, we can suppose that (Y_t) takes its values in E. But then (Y_t) is an almost-modification of (X_t) by Fubini's Theorem.

Therefore, there is no loss of generality, from the point of view of almost-equivalence classes, in confining attention to measurable processes in the ordinary sense.

Definition 2.3.33 Let f be a Borel mapping of $\mathbb{R}_+$ into E. The image of the Lebesgue measure under the mapping $t \mapsto (t, f(t))$ from $\mathbb{R}_+$ to $\mathbb{R}_+ \times E$ is a measure called the *pseudo-path* associated with f and denoted by $\psi(f)$.

Let $(X_t)_{t \in \mathbb{R}_+}$ be a measurable process defined on Ω with values in E. The pseudo-path associated with the Borel mapping $X_.(\omega)$ is called the pseudo-path of ω and is denoted by $\psi^X(\omega)$ ($\psi(\omega)$ for short).

Our aim next is to provide the set of all pseudo-paths with a *Lusin measurable structure* such that the mapping ψ is measurable. We start by a definition embedding the set of pseudo-paths in a set of positive measures on $\mathbb{R}_+ \times E$:

Definition 2.3.34 We denote by $\mathcal{L}(E)$ the *set of positive measures* on $\mathbb{R}_+ \times E$ whose projection onto $\mathbb{R}_+$ is the Lebesgue measure and by $\Pi(E) \subset \mathcal{L}(E)$ the *set of pseudo-paths* associated with all Borel mappings of $\mathbb{R}_+$ into E.

These spaces are given the *coarsest topology* such that all mappings $\lambda \mapsto\, <\lambda, f \otimes g>$ are continuous where f is continuous with compact support on $\mathbb{R}_+$ and g bounded continuous on E and the corresponding measurable structure.

Remark 2.3.35 The function $(t, s) \longmapsto f(t)g(s)$ is frequently denoted by $f \otimes g$.

We can describe the measurable structure directly: by the monotone class theorem, it is also generated by the mappings $\lambda \mapsto\; <\lambda, \gamma>$ where γ is positive and Borel on $\mathbb{R}_+ \times E$. Hence it depends only on the measurable structure on E and not on the topology chosen on E. We shall leave aside the topological structure which might be studied by embedding E as a topological subspace of a compact metric space.

Theorem 2.3.36 If E is Lusin $\Pi(E)$ is a Lusin measurable space.

Proof. We embed E in I. $\mathcal{L}(I)$ is a set of random measures on the LCC space $\mathbb{R}_+ \times I$ and the topology of $\mathcal{L}(I)$ is the topology of vague convergence. On the other hand, every measure $\mu \in \mathcal{L}(I)$ gives to $[0, n] \times I$ the measure n and hence $\mathcal{L}(I)$ is bounded under the vague topology. Since it is obviously closed, it is compact and metrizable. The function $\mu \mapsto \mu(E^c)$ is Borel on $\mathcal{L}(I)$ and hence the set $\mathcal{L}(E) = \{\mu \in \mathcal{L}(I) : \mu(E^c) = 0\}$ is Borel in $\mathcal{L}(I)$.

On the other hand, $\Pi(E) = \Pi(I) \cap \mathcal{L}(E)$, for if f denotes a Borel mapping of $\mathbb{R}_+$ into I and if the image of the Lebesgue measure λ on R_+ under $t \mapsto (t, f(t))$ is carried by $\mathbb{R}_+ \times E$, then $f(t) \in E$ for almost all t and there exists a Borel mapping of $\mathbb{R}_+$ into E giving the same image measure. Hence, it suffices to show that $\Pi(I)$ is Borel in $\mathcal{L}(I)$.

To this end we show that $\Pi(I)$ is the set of extreme points of the metrizable compact convex set $\mathcal{L}(I)$. The set of extreme points of such a convex set is $\mathcal{G}_\delta$; this will establish the required result.

Then, let μ be the image of λ under $t \mapsto (t, f(t))$ where f is Borel with values in I and suppose that $\mu = c\mu_1 + (1-c)\mu_2$, with $c \in]0, 1[$ and $\mu_1, \mu_2 \in \mathcal{L}(I)$. We disintegrate μ_1 and μ_2 with respect to their projection λ on $\mathbb{R}_+$

$$\mu_i(dt, dx) = \int \varepsilon_s(dt) \otimes \alpha_s^i(dx)\lambda(ds) \quad (i = 1, 2)$$

where (α_s^i) is a measurable family of probability laws on I. Then, for almost all s, $c\alpha_s^1 + (1-c)\alpha_s^2 = \varepsilon_{f(s)}$, hence $\alpha_s^1 = \alpha_s^2 = \varepsilon_{f(s)}$ and finally $\mu_1 = \mu_2 = \mu$: μ is indeed extremal.

Conversely, let μ be a measure which belongs to $\mathcal{L}(I)$ but not to $\Pi(I)$; we show that it is not extremal in $\mathcal{L}(I)$. We can write $\mu = \int \varepsilon_s \otimes \alpha_s \lambda(ds)$ as above. For all s, we denote by $g(s)$ and $h(s)$ the greatest lower bound and least upper bound of the support of α_s; they are Borel functions of s (for example, $g(s)$ is the supremum of the rationals r such that $\alpha_s([0, r[) > 0)$ and the fact that $\mu \notin \Pi(I)$ implies that $\{s : g(s) < h(s)\}$ is not negligible. We write $j = \frac{g+h}{2}$ and then, for $c \in]0, 1[$,

$$m_s(dx) = \begin{cases} \alpha_s(dx)\mathbf{1}_{[0,j(s)[}(x)\|\alpha_s\left(\mathbf{1}_{[0,j(s)[}\right)\| & \text{if } \alpha_s\left(\mathbf{1}_{[0,j(s)[}\right) > c \\ \alpha_s(dx) & \text{otherwise} \end{cases} .$$

The measure $\mu^1 = \int \varepsilon_s \otimes m_s \lambda(ds)$ belongs to $\mathcal{L}(I)$ and $cm_s \leq \alpha_s$, so that the measure $\mu'' = \mu - c\mu'/(1-c)$ is positive. Since the project on of μ'' on $\mathbb{R}_+$ is equal to the Lebesgue measure, we also have $\mu'' \in \mathcal{L}(I)$ and $\mu = c\mu' + (1-c)\mu''$. It only remains to note that if $g(s) < h(s)$, α_s is non-zero on both intervals $[0, j(s)[$ and $[j(s), 1]$ and hence $m_s \neq \alpha_s$ if c is sufficiently small. For $c = 1/n$ and n sufficiently large, the representation $\alpha = c\mu' + (1-c)\mu''$ is therefore non-trivial and μ is not extremal. ∎

Remark 2.3.37 Vague topology, loosely speaking, is the topology on the set of Borel measures induced by a set of real-valued function on a locally compact Hausdorff space. It is an example of the weak-* topologies. Vague-topology on LCC space is studied in detail in Bourbaki [21].

Remark 2.3.38 (a) We can apply the above argument to E embedded in I, instead of I, and then show that $\Pi(E)$ is always the set of extremal points of the convex (not necessarily be compact) set $\mathcal{L}(E)$. Now, If E is Polish, we embed it as a $\mathcal{G}_\delta$ in some compact metrizable space K. The topology of $\mathcal{L}(E)$ is induced by that of $\mathcal{L}(K)$ and $\mathcal{L}(E)$ is a $\mathcal{G}_\delta$ of $\mathcal{L}(K)$. On the other hand, $\Pi(K)$ is also a $\mathcal{G}_\delta$ of $\mathcal{L}(K)$, by the above remark. Hence $\mathcal{L}(E) \cap \Pi(K) = \Pi(E)$ is also a $\mathcal{G}_\delta$ of $\Pi(K)$ and hence is a Polish space.

(b) There exists a stronger topology on $\Pi(E)$ than that of Definition 2.3.34 inducing the same Borel structure. Let d be a bounded metric defining the topology of E and let μ and μ' be two elements of $\Pi(E)$, which are the images of the Lebesgue measure on $\mathbb{R}_+$ under two mappings $t \mapsto (t, f(t))$, $t \mapsto (t, f'(t))$. Then, define on $\Pi(E)$ a metric $\overline{d}$ by

$$\overline{d}(\mu, \mu') = \int_0^\infty d(f(t), f'(t)) e^{-t} dt. \tag{2.35}$$

This amounts to considering $\Pi(E)$ as a set of classes of measurable mappings of $\mathbb{R}_+$ into E under the topology of convergence in measure. $\Pi(E)$ is then separable because any element of $\Pi(E)$, can be approximated in measure, first by a continuous function, and in turn by a dyadic step function with values in a countable dense set. It is easy to see that the topology of $\Pi(E)$ and the topology of convergence in measure lead to the same Borel σ-field. However, our first topology on $\Pi(E)$ may be better in some respects, it does not seem true that $\Pi(E)$ is Polish under the topology of convergence in measure if E is Polish.

Now we consider on Ω a measurable process (X_t) with values in E. If ϕ is a continuous function with compact support on $\mathbb{R}_+$ and g is a bounded continuous function on E, clearly the function $\omega \mapsto \int g \circ X_s(\omega) \phi(s) ds = < \psi^X(\omega), \phi \otimes g >$ is measurable on Ω. In other words, the mapping ψ of Ω into $\Pi(E)$ is measurable and we can consider the image law of $\mathbf{P}$ under ψ (i.e., $\mathbf{P} \circ \psi^{-1}$). Then,

Definition 2.3.39 Let (X_t) be a measurable process defined on $(\Omega, \mathcal{F}, \mathbf{P})$ with values in a Lusin metrizable space E, the image law of $\mathbf{P}$ under the mapping of ψ into $\Pi(E)$ is called the *pseudo-law* of (X_t).

Remark 2.3.40 Suppose Ω is the set of right continuous (resp. r.c.l.l.) mappings of $\mathbb{R}_+$ into E with the σ-field generated by the coordinate mappings. The mapping, which associates with $\omega \in \Omega$ the corresponding pseudo-path $\psi(\omega)$ is injective and we have seen that ψ is measurable. So, we can identify Ω with a subset of $\Pi(E)$.

On the other hand, if $\omega \in \Omega$ and f is bounded and continuous on E, then

$$f(\omega(t)) = \lim_{h \to 0} \int \frac{1}{h} f(\omega(t + s)) ds,$$

so that the coordinate mappings on Ω are measurable relative to the σ-field induced by $\Pi(E)$. Therefore, ψ is a Borel isomorphism and the results stated in (2.3.4(a)) according to which Ω is cosouslin (resp. Lusin if E is Polish) for its usual measurable structure, show that Ω is the complement of an analytic set (resp. a Borel set) in $\Pi(E)$.

We now show that the notion of a pseudo-law solves conveniently the almost equivalence problem of processes.

Theorem 2.3.41 Let (X_t) and (Y_t) be two measurable processes with values in a Lusin space E defined on possibly different probability spaces. The following properties are equivalent:

(a) (X_t) and (Y_t) have the same pseudo-law $\mathbf{Q}$;

(b) (X_t) and (Y_t) are almost-equivalent according to (Definition 2.3.24);

(c) there exists a Borel subset $\mathbb{T}$ of $\mathbb{R}_+$ of full Lebesgue measure, such that, the processes $(X_t)_{t\in\mathbb{T}}$ and $(Y_t)_{t\in\mathbb{T}}$ have the same time law.

Further, there exists a measurable stochastic process $(Z_t)_{t\in\mathbb{R}_+}$ defined on $\Pi(E)$ such that for every law $\mathbf{J}$ on $\Pi(E)$, the pseudo-law of the process (Z_t) under $\mathbf{J}$ is $\mathbf{J}$ itself. In particular, if $\mathbf{J} = \mathbf{Q}$, (Z_t) belongs to the pseudo-equivalence class of (X_t).

Proof. From before we know that, if g is positive and Borel on E, and ϕ is positive and Borel on $\mathbb{R}_+$, then

$$\int g(X_t(\omega))\phi(t)dt = <\psi^X(\omega), \psi \otimes g>.$$

It follows immediately that (a)⇔(b).

We show that (a)⇒(c); again consider E as embedded in $I = [0,1]$ with $0 \in E$. By the Lebesgue differentiation theorem, we have a.e. γ denoting the identity mapping of E into $[0,1]$,

$$X_t(\omega) = \lim_n n \int_t^{t+1/n} X_s(\omega)dt = \lim_n <\psi^X(\omega), e_n(t+) \otimes \gamma>. \tag{2.36}$$

where $e_n(s) = n\mathbf{1}_{[0,1/n]}(s)$. We set, for every measure $\mu \in \Pi(E)$

$$Z_t(\omega) = \begin{cases} \liminf_n <\mu, e_n(t+) \otimes \gamma> & \text{if this number belongs to } E \\ 0 & \text{otherwise.} \end{cases} \tag{2.37}$$

Clearly, the process (Z_t) on$\Pi(E)$ is measurable. On the other hand, by (2.36), for all ω, $Z.(\psi^X(\omega)) = X.(\omega)$ a.e. consequently, by Fubini's theorem, the set of all t such that $X_t = Z_t \circ \psi^X$ a.s. does not hold is negligible. Then, let T denote the set of all t such that both $X_t = Z_t \circ \psi^X$ a.s. and $Y_t = Z_t \circ \psi^Y$ a.s. By the definition of image laws, the time law of the process $(X_t)_{t\in\mathbb{T}}$ (resp. $(Y_t)_{t\in\mathbb{T}}$) is the law of the process $(Z_t)_{t\in\mathbb{T}}$ under $\mathbf{Q}$. Thus (a) implies (c).

Next, we show that (c)⇒(b); replacing X_t and Y_t by 0 if $t \notin \mathbb{T}$, we may assume that both processes have the same time law on the whole of $\mathbb{R}_+$. Then, it suffices to show the theorem on a finite interval $[0, \Lambda[$, which enables us to reduce it to the case where (X_t) and (Y_t) are periodic on $\mathbb{R}$ with period Λ. In the argument below we take $\Lambda = 1$.

We now follow Doob [5], Stochastic Processes, p. 63, to which we refer the reader for full detail. Let g be a bounded Borel function on E and ϕ be a bounded Borel function on $\mathbb{R}$ of period 1. We want to compute, for all ω belonging to the space Ω on which (X_t) is defined, the integral

$$\mathsf{M}^X(\omega) = \mathsf{M}^X_\phi(\omega, g) = \int_0^1 g(X_t(\omega))\phi(t)dt \tag{2.38}$$

and to this end we compare it to the following Riemann sum, where 0 is a parameter belonging to $[0, 1[$,

$$\mathsf{R}_i^X(\sigma, \omega, \phi, g) = \mathsf{R}_i^X(\sigma, \omega) = 2^{-i} \sum_{k < 2^i} g(X_{\sigma + k 2^{-i}}(\omega))\phi(\sigma + k2^{-i}).$$

It is known from real analysis that for all ω

$$\int |\mathsf{M}^X(\omega) - \mathsf{R}_i^X(\sigma, \omega)d\sigma \xrightarrow[i \to \infty]{} 0.$$

We integrate with respect to ω, thus getting convergence in L^1 relative to the measure $d\mathbf{P}(\omega) \otimes d\sigma$, and choose a sequence (i_n) such that $\mathsf{R}_{i_n}^X(\sigma, \omega) \to \mathsf{M}^X(\omega)$ a.s. in σ, ω. Then, there exists by Fubini's theorem, a set N of measure zero such that for $\sigma \notin N$.

$$\mathsf{R}_{i_n}^X(\sigma, \omega') \to \mathsf{M}^X(\omega') \text{ a.s. in } \omega' \tag{2.39}$$

Since these Riemann sums are uniformly bounded, convergence in fact takes place in every L^p $(p < \infty)$. Extracting a subsequence from (i_n) and enlarging the set N of measure zero if necessary, we can also assert that, on the space Ω' where (Y_t) is defined,

$$\mathsf{R}_{i_n}^Y(\sigma, \omega') \to \mathsf{M}^Y(\omega') \text{ a.s. in } \omega' \tag{2.40}$$

This concerns a given pair (ϕ, g). But from successive extractions and enlargements of the set N we get (2.38) and (2.39) for any finite family (ϕ_i, g_i) of pairs as above and this permits us to compute the laws of the random vectors

$$(\mathsf{M}_{\phi_1}^Y(., g_1), \ldots, \mathsf{M}_{\phi_n}^Y(., g_n)), \quad (\mathsf{M}_{\phi_1}^X(., g_1), \ldots, \mathsf{M}_{\phi_n}^X(., g_n))$$

using quantities which depend only on the time law, and which therefore are the same for both processes. Hence these are almost equivalent.

Finally, we leave to the reader the details concerning the last sentence of the statement. ∎

Remark 2.3.42 We have stated in Remark 2.1.5 that the time law of a process is in general an imprecise notion. It is corrected by Theorem 2.3.41 such that two processes that are measurable and have the same law also have the same pseudo-law. This allows us a lot of analysis on the sample functions.

For example, let us show a little more: suppose $g_1, \ldots, g_n$ are bounded Borel functions on the state space E of (X_t) and $\mu_1, \ldots, \mu_n$ be bounded measures on $\mathbb{R}_+$ not necessarily absolutely continuous with respect to Lebesgue measure as in Theorem 2.3.41. We show that the time law completely determines the law of the random vector

$$U = \left(\int g_1 \circ X_s \mu_1(ds), \ldots, \int g_n \circ X_s \mu_n(ds) \right).$$

Since U takes its values in a bounded set of $\mathbb{R}^n$, it suffices to check that we can compute $\mathbf{E}[f(U)]$ for any polynomial $f : \mathbb{R}^n \to \mathbb{R}$. This reduces to the case of monomials, and finally, to that of the product

$$\mathbf{E}\left[\left(\int h_1 \circ X_s \lambda_1(ds) \right) \cdots \left(\int h_k \circ X_s \lambda_k(ds) \right) \right]$$

where k may be larger than n, due to the presence of powers of the coordinates in monomials. And now, thanks to Fubini's theorem, this is just

$$\int_{\mathbb{R}^k} \lambda_1(ds_1)...\lambda_k(ds_k)\mathbf{E}\left[h_1 \circ X_{s_1}...h_k \circ X_{s_k}\right]$$

which depends only on the time law of (X_t).

2.4   Random Times

Here, we start developing fundamental notions of the general theory of processes at least those which can be studied without martingale theory. This theory is not primarily concerned with processes, but rather with structures determined on the measurable space $(\Omega, \mathcal{F})$ or the probability space $(\Omega, \mathcal{F}, \mathbf{P})$ by a filtration $(\mathcal{F}_t)$. The theory, indeed, has two slightly different approaches: the first is a probabilistic theory relative to a given probability measure $\mathbf{P}$; the other is a non-probabilistic theory, depending only on the space and not on any given law.

To begin, we refer the reader to nos. 2.2-2.2.6 for definitions of filtrations. But, we make additional conventions; whenever we speak of a filtration $(\mathcal{F}_t)$ we assume that two additional σ-fields are given: one denoted by $\mathcal{F}_\infty$ – containing all the $\mathcal{F}_t$'s, the other denoted by $\mathcal{F}_{0-}$ contained in $\mathcal{F}_0$. Also, we take the stand that $\mathcal{F}_t = \mathcal{F}_{0-}$ for $t < 0$ and $\mathcal{F}_{\infty-} = \bigvee_t \mathcal{F}_t$. This is very convenient and in no way restricts the generality of the results presented. Furthermore, if not explicitly given we may take $\mathcal{F}_{0-} = \mathcal{F}_0$, $\mathcal{F}_\infty = \mathcal{F}_{\infty-}$ (or $\mathcal{F}_\infty = \mathcal{F}$).

Definition 2.4.1 Let $(\Omega, \mathcal{F}, \mathbf{P})$ be a probability space with a filtration $(\mathcal{F}_t)$, $t \in \mathbb{R}_+$ we say that the filtration is complete if the probability space is complete and if all the $\mathbf{P}$ negligible sets belong to $\mathcal{F}_{0-}$.

We say that the filtration satisfies the usual conditions if it is complete and right continuous (i.e., $\mathcal{F}_t = \mathcal{F}_{t+}$ for all $t \geq 0$).

An arbitrary filtration $(\mathcal{F}_t)$ can always be completed: one completes the space and then adjoins to each σ-field all the negligible sets. If this operation is performed on the family made right-continuous $\mathcal{G}_t = \mathcal{F}_{t+}$ $(\mathcal{G}_{0-} = \mathcal{F}_{0-})$, one gets a family of the family $(\mathcal{F}_t^\star)$. We shall use the symbol $\star$ to identify right-continuous complete filtration.

Remark 2.4.2 Since all the negligible sets are adjoined to $\mathcal{F}_{0-}$ and not only the negligible sets of $\mathcal{F}_t$ to each $\mathcal{F}_t$, the operation of completion strongly affects the *time structure* of the family. For instance, in the theory of Markov processes, completion alone will turn the family $(\mathcal{F}_t)$ into a right continuous one (provided the semi-group is nice enough).

This brings us to the first structural element on the space $(\Omega, \mathcal{F})$, the notion of a random time. The random variable $T : (\Omega, \mathcal{F}) \longrightarrow ([0, \infty] \times \mathcal{B}([0, \infty]))$ is called simply *random time* (r.t.).

2.4.1 Stopping times

The notion of a stopping time is that of a particular type of random time. This seemingly trivial notion of time is due to Doob and is the cornerstone of the general theory of processes.

Definition 2.4.3 Let $(\mathcal{F}_t)$ be a filtration on $(\Omega, \mathcal{F}, \mathbf{P})$. A r.v. T on Ω with values in $\mathbb{R}_+ \cup \{+\infty\}$ is called a stopping time or optional time of the filtration $(\mathcal{F}_t)$ if

$$\text{for all } t \in \mathbb{R}_+, \text{ the event } \{T \le t\} \text{ belongs to } \mathcal{F}_t, \tag{2.41}$$

belong to trivial filtration for all $t \in \mathbb{R}_+$.

Stopping times are also called *Markov time* (m.t.). The process $\mathbf{1}_{T \le t}$ is the simplest process one can consider on the space $(\Omega, \mathcal{F})$ with filtration $\mathcal{F}_t$. Note that we have not said anything concerning the filtration being right-continuous and complete. In general the filtration $\mathcal{F}_t$ upon which random time is defined is the raw filtration unless stated otherwise. Every positive constant is a stopping time. We use both the older terminology (stopping time) and the more recent one (optional time).

Remark 2.4.4 T is called a *wide sense stopping time* if for all t,

$$\{T < t\} \in \mathcal{F}_t. \tag{2.42}$$

Or, we can also say that T a wide sense stopping time is optional in the wide sense if and only if it is optional with respect to the family $(\mathcal{F}_{t+})$.

If the family $(\mathcal{F}_t)$ is right-continuous, stopping times are characterized by (2.42) which is often much easier to verify than (2.41). Another characterization of wide sense stopping times is, for all t,

$$T \wedge t \text{ is } \mathcal{F}_t \text{ measurable.} \tag{2.43}$$

Remark 2.4.5 In the case of a discrete filtration $(\mathcal{F}_n)_{n \in \mathbb{N}}$, stopping times are characterized by any one of the equivalent properties

$$\text{for all } n, \quad \{T \le n\} \in \mathcal{F}_n, \tag{2.44}$$

$$\text{for all } n, \quad \{T = n\} \in \mathcal{F}_n. \tag{2.45}$$

Nothing analogous to (2.45) exists in general in the continuous case.

Recall that the σ-field $\mathcal{F}_t$ is the set of events which we consider as prior to the instant t (2.2.3), in other words, known at time t in the universe of the observer. Let us imagine that the observer watches for the appearance of some random phenomenon and notes the random date $T(\omega)$ at which this phenomenon occurs for the first time. The event $\{T \leq t\}$ means that the phenomenon has occurred at least once before t, hence it belongs to $(\mathcal{F}_t)$, and T is optional. Here is the mathematical content of the preceding discourse:

Theorem 2.4.6 Assume the family $(\mathcal{F}_t^\star)$ satisfies the usual conditions and let A be a progressive subset of $\mathbb{R}_+ \times \Omega$ (Definition 2.3.20). Then, the debut D_A of A is a stopping time.

Proof. Suppose the family $(\mathcal{F}_t)$ does not satisfy the usual conditions. The set $\{D_A < t\}$ is the projection on Ω of the set $\{(s, \omega) : s < t, \ (s, \omega) \in A\}$ which belongs to $\mathcal{B}(\mathbb{R}_+) \times \mathcal{F}_t$ by the definition of progressive sets (2.2.4). Then, (cf. Theorem 1.3.33)

$$\text{for all } t, \ \{D_A < t\} \text{ is } \mathcal{F}_t \text{ analytic.} \tag{2.46}$$

Under the hypotheses of the statement, the σ-field $\mathcal{F}_t^\star$ is complete and hence equal to a $\mathfrak{a}(\mathcal{F}_t)$ (1.3.4.1) and (2.41) implies that D_A is a stopping time of the family $(\mathcal{F}_{t+})$, which is equal to $(\mathcal{F}_t^\star)$ by right-continuity. ∎

Remark 2.4.7 Under the usual conditions, every r.v. which is equal $\mathbf{P}$ a.s. to a stopping time is a stopping time and the theorem applies to every set A which is indistinguishable from a progressive set.

Remark 2.4.8 In an arbitrary filtration $(\mathcal{F}_{t-})$ every stopping time T can be interpreted as the debut of the set $\{(t, \omega) : t > T(\omega)\}$ whose indicator is a right-continuous adapted, hence progressive, process (Theorem 2.2.6).

Similarly, every wide sense stopping time T is the debut of the progressive set $\{(t, \omega) : t > T(\omega)\}$ (on these sets, see Theorem 2.4.20).

Remark 2.4.9 We return to the proof of Theorem 2.4.6 and suppose that A is right-closed. Then, $\{D_A \leq t\}$ is the projection of $\{(s, \omega) : s \leq t, (s, \omega) \in A\}$ on Ω and, as for (2.46),

$$\text{for all } t, \ \{D_A \leq t\} \text{ is } \mathcal{F}_t \text{ analytic.} \tag{2.47}$$

Remark 2.4.10 To illustrate the need for analytic sets in probability theory, let us give an example on measurability of debuts. Consider the set Ω of all r.c.l.l. mappings from $\mathbb{R}_+$ to $\mathbb{R}_+$. For every $\omega \in \Omega$ set $X_t(\omega) = \omega(t)$ and denote by $\mathcal{F}$ the σ-field generated by $(X_t)_{t \geq 0}$. Then, $(\Omega, \mathcal{F}_t)$ is a nice Lusin measurable space, and the simplest hitting time of a closed set in $\mathbb{R}_+$

$$D(\omega) = \inf\{t : X_t(\omega) = 0\}$$

is such that the sets $\{D \leq T\}$ are analytic and not Borel; of course the hitting times $D_\epsilon(\omega) = \inf\{t : X_t(\omega) < \epsilon\}$ are Borel for $\epsilon > 0$ but their limit as $\epsilon \downarrow 0$ isn't D.

Let's consider, now, σ-fields associated with a stopping time.

Definition 2.4.11 Let T be a stopping time of a filtration $(\mathcal{F}_t)$ on Ω. The σ-field of events prior to T, denoted, $\mathcal{F}_T$, consists of all events $A \in \mathcal{F}_\infty$ such that

$$\text{for all } t, \ A \cap \{T \leq t\} \text{ belongs to } t. \tag{2.48}$$

When T is a positive constant r, $\mathcal{F}_T$ is the σ-field $\mathcal{F}_r$ of the filtration; hence the notation and name are reasonable. Let us set $\mathcal{G}_t = \mathcal{F}_{t+}$ for all t $(\mathcal{G}_{0-} = \mathcal{F}_{0-}, \mathcal{G}_\infty = \mathcal{F}_\infty)$ and let T be a stopping time of $(\mathcal{G}_t)$ that is, a wide sense stopping time of $(\mathcal{F}_t)$ (Definition 2.4.3). Then, an event A belongs to $\mathcal{G}_t$ if and only if it belongs to $\mathcal{G}_\infty = \mathcal{F}_\infty$ and

$$\text{for all } t, \ A \cap \{T < t\} \text{ belongs to } \mathcal{F}_t. \tag{2.49}$$

belongs to Remark 2.1.8. It is quite natural to denote this σ-field $\mathcal{F}_T$ by the notation $\mathcal{F}_{T+}$.

The following theorem is a restatement of the previous Definition 2.4.11 by which we introduce an operation on stopping times and some notation that we will use often.

Theorem 2.4.12 Let T be optional relative to $(\mathcal{F}_t)$. Then, A belongs to $\mathcal{F}_T$, if and only if A belongs to $\mathcal{F}_\infty$ and the r.v. T_A defined by

$$T_A(\omega) = T(\omega) \text{ if } \omega \in A, \quad T_A(\omega) = +\infty \text{ if } \omega \notin A \tag{2.50}$$

is optional

We now wish to define, for every stopping time, a σ-field $\mathcal{F}_{T-}$. It would be tempting to introduce, the family of σ-fields $\mathcal{H}_t = \mathcal{F}_{t-}$ then set $\mathcal{F}_{T-} = \mathcal{H}_T$ for every stopping time T of this family. However, this definition would be useless, as it happens frequently that $(\mathcal{F}_t)$ satisfies the usual conditions and further $\mathcal{F}_t = \mathcal{F}_{t-}$ for all t. This definition would lead us to $\mathcal{F}_T = \mathcal{F}_{T-}$ for every stopping time T while the distinction between past and strict-past turns out to be important, even for stopping times of such families. A better definition is as follows:

Definition 2.4.13 Let T be a stopping time of $(\mathcal{F}_t)$. The σ-field of events strictly prior to T, denoted $(\mathcal{F}_{T-})$, is the σ-field generated by $(\mathcal{F}_{0-})$ and the events of the form

$$A \cap \{t < T\}, \ t \geq 0, \ A \in \mathcal{F}_t. \tag{2.51}$$

Note that $\mathcal{F}_{T-}$ is also generated by the following but less convenient sets

$$A \cap \{t \leq T\}, \ t \geq 0, \ A \in \mathcal{F}_{t-}. \tag{2.52}$$

2.4.2   Basic properties of stopping times

In results that follow, all random variables are defined on the same space $(\Omega, \mathcal{F}, \mathbf{P})$ and stopping times are defined relative to the same raw filtration $(\mathcal{F}_t)$ unless otherwise mentioned.

Theorem 2.4.14 Closure. (a) Let S and T be two stopping times, then $S \wedge T$ and $S \vee T$ are stopping times.
(b) Let (S_n) be an increasing sequence of stopping times, then $S = \lim_n S_n$ is a stopping time.
(c) Let (S_n) be a decreasing sequence of stopping times, then $S = \lim_n S_n$ is a stopping time of the family $(\mathcal{F}_{t+})$. It will be a stopping time of $(\mathcal{F}_t)$ if the sequence is stationary, i.e., if for all ω there exists an integer n such that $S_m(\omega) = S_n(\omega)$ for all $m \geq n$.

Proof. (a) $\{S \wedge T \leq t\} = \{S \leq t\} \cup \{T \leq t\}$ belongs to $\mathcal{F}_t$; similarly for $\vee$.
(b) $\{S \leq t\} = \cap_n \{S_n \leq t\}$ belongs to $\mathcal{F}_t$ for an increasing (S_n).
(c) $\{S < t\} = \cup_n \{S_n < t\}$ belongs to $\mathcal{F}_t$ for a decreasing (S_n) and when the sequence is stationary, then also $\{S \leq t\} = \cup_n \{S_n \leq t\}$ belongs to $\mathcal{F}_t$.

From the above we can deduce immediate consequences: the set of stopping times is closed under $(\vee c)$ and the set of stopping times of $(\mathcal{F}_{t+})$ (wide sense stopping times) is closed under the operations $(\vee c, \wedge c)$ and also under countable $\liminf$ and $\limsup$. ∎

Theorem 2.4.15 Events prior to stopping times.
(a) For every stopping time S, $\mathcal{F}_{S-} \subset \mathcal{F}_S$ and S is $\mathcal{F}_{S-}$-measurable.
(b) Let S and T be two stopping times such that $S \leq T$, then, $\mathcal{F}_S \subset \mathcal{F}_T$, $\mathcal{F}_{S-} \subset \mathcal{F}_{T-}$ and if $S < T$ everywhere, then $\mathcal{F}_S \subset \mathcal{F}_{T-}$.
(c) Let S and T be two stopping times. Then,

$$\text{for all } A \in \mathcal{F}_S, \ A \cap \{S < T\} \text{ belongs to } \mathcal{F}_T \tag{2.53}$$

$$\text{for all } A \in \mathcal{F}_S, \ A \cap \{S < T\} \text{ belongs to } \mathcal{F}_{T-}. \tag{2.54}$$

In particular,

$$\{S \leq T\}, \ \{S = T\} \text{ belong to } \mathcal{F}_S \text{ and } \mathcal{F}_T \text{ and } \{S < T\} \text{ to } \mathcal{F}_S \text{ and } \mathcal{F}_{T-}. \tag{2.55}$$

(d) Let (S_n) be an increasing (resp. decreasing) sequence of stopping times and $S = \lim_n S_n$. Then, $\mathcal{F}_{S-} = \vee_n \mathcal{F}_{S_n-}$ $(\mathcal{F}_{S+} = \vee_n \mathcal{F}_{S_n+})$.
(e) Let S be a stopping time and $A \subset \Omega$. If $A \in \mathcal{F}_{\infty-}$ (resp. $A \in \mathcal{F}_\infty$), the set $B \cap \{S = \infty\}$ belongs to $\mathcal{F}_{S-}$ (resp. $\mathcal{F}_S$).

Proof. (a) We must show that every generator B of $\mathcal{F}_{S-}$ belongs to $\mathcal{F}_S$, in other words, B satisfies $B \cap \{S \leq t\} \in \mathcal{F}_t$ for all t. So, we take either $B \in \mathcal{F}_{0-}$ or $B = A \cap \{S > r\}$, $r \geq 0$ and $A \in \mathcal{F}_r$; the verification is obvious. For the second assertion, we note that for all $\{S > t\}$, the sets appear among the generators (2.51) of $\mathcal{F}_{S-}$.

(b) If $S \leq T$ and $A \in \mathcal{F}_S$, then $A = A \cap \{S \leq T\} \in \mathcal{F}_T$; if $S < T$ everywhere $A = A \cap \{S < T\}$ belongs to $\mathcal{F}_{T-}$. Finally, it suffices to verify that the generators (2.51) of $\mathcal{F}_{S-}$ belong to $\mathcal{F}_{T-}$ if $S \leq T$; but, if $B \in \mathcal{F}_t$, then

$$B \cap \{t < S\} = (B \cap \{t < S\}) \cap \{t < T\} \text{ and } B \cap \{t < S\} \in \mathcal{F}_t.$$

(c) Property (2.53) follows from the following equality true for all t:

$$A \cap \{S \leq T\} \cap \{T \leq t\} = [A \cap \{S \leq t\}] \cap \{T \leq t\} \cap \{S \wedge t \leq T \wedge t\}.$$

If A is $\mathcal{F}_S$-measurable, the three events appearing on the right are in $\mathcal{F}_t$. Property (2.54) follows from the following equality, where r runs through the rationals:

$$A \cap \{S < T\} = \cup (A \cap \{S < r\}) \cap \{r < T\}$$

If A belongs to $\mathcal{F}_S$, the events appearing on the right belong to the generating system (2.51) of $\mathcal{F}_T$.

(d) In the case of an increasing sequence, $\cup_n \mathcal{F}_{S_n} \subset \mathcal{F}_{S-}$ by (b). On the other hand, every generator (2.51) of the σ-field $\mathcal{F}_{S-}$ can be written as $A \cap \{t < S\} = \cup_n A \cap \{t < S_n\}$ with $t \geq 0$, $A \in \mathcal{F}_t$, and we see that it belongs to $\cup_n \mathcal{F}_{S_n-}$. In the case of a decreasing sequence, we recall that $\mathcal{F}_{S+}$ is the set of $A \in \mathcal{F}_\infty$ such that $A \cap \{S < t\} \in \mathcal{F}_t$ for all t. As $A \cap \{S < t\} = \cap_n A \cap \{S_n < t\}$, we see that $\mathcal{F}_{S+} \supset \cap_n \mathcal{F}_{S_n+}$ and the converse inclusion follows from (b).

(e) The set of all $A \in \mathcal{F}_{\infty-}$ such that $A \cap \{S = \infty\}$ belongs to $\mathcal{F}_{S-}$ is a σ-field. It suffices then to verify that it contains $\mathcal{F}_t$ for all t. Now if A belongs to $\mathcal{F}_t$, $A \cap \{n < T\}$ is a generator (2.51) of $\mathcal{F}_{S-}$ for all $n \geq t$ and hence $A \cap \{S = \infty\}$ belongs to $\mathcal{F}_{S-}$. Finally, the case of $\mathcal{F}_S$ is trivial. ∎

Remark 2.4.16 (a) Part (a) of the above theorem illustrates an important principle that pertains to stopping times which is all that is known for constant times t and σ-fields $\mathcal{F}_t$ extends to stopping times T and σ-fields $\mathcal{F}_T$. Thus (a) and (b) are extentions to σ-fields of stopping times the monotonicity property of $(\mathcal{F}_t)$ for constant times t. Part (c) is the extension to arbitrary pairs of stopping times of properties (2.49) and (2.51) relating to pairs consisting of a stopping time and a constant. Part (d) is the extension to stopping times of the continuity properties of the families $(\mathcal{F}_{t-})$ and $(\mathcal{F}_{t+})$.

(b) It is not true in general that for all $A \in \mathcal{F}_{S-}$ that $A \cap \{S \leq T\}$ belongs to $\mathcal{F}_{T-}$.

(c) Let S and T be two stopping times. Then, we have

$$\mathcal{F}_{S \wedge T} = \mathcal{F}_S \cap \mathcal{F}_T, \quad \mathcal{F}_{S \vee T} = \mathcal{F}_S \vee \mathcal{F}_T$$

Indeed, if A belongs to $\mathcal{F}_S$ and to $\mathcal{F}_T$, $A \cap \{S \leq S \wedge T\}$ and $A \cap \{T \leq S \wedge T\}$ belong to $\mathcal{F}_{S \wedge T}$ according to (2.53); taking, their union, A belongs to $\mathcal{F}_{S \wedge T}$ so that $\mathcal{F}_S \cap \mathcal{F}_T$; the reverse inclusion is obvious. Similarly, if A belongs to $\mathcal{F}_{S \vee T}$, $A \cap \{S \vee T \leq S\}$ and $A \cap \{S \vee T \leq T\}$ belong to $\mathcal{F}_S \vee \mathcal{F}_T$; taking their union, the same is true for A. The reader may show that $\mathcal{F}_{(S \vee T)-} = \mathcal{F}_{S-} \vee \mathcal{F}_{T-}$ and that $\mathcal{F}_{(S \wedge T)-} = \mathcal{F}_{S-} \cap \mathcal{F}_{T-}$ if S and T are predictable.

Theorem 2.4.17 (a) Let S be a stopping time of the family $(\mathcal{F}_t)$ and T be an $\mathcal{F}_S$ measurable r.v. such that $S \leq T$. Then, T is a stopping time of $(\mathcal{F}_t)$. The same conclusion holds if S is a stopping time of $(\mathcal{F}_{t+})$, T is $\mathcal{F}_{S+}$ measurable $S < T$ on $\{S < \infty\}$. This applies in particular to the r.v. $T = S + t$ $(t > 0)$ and to

$$T = S^{(n)} = \sum_{k \geq 1} k 2^{-n} 1_{\{(k-1)2^{-n} \leq S \leq k2^{-n}\}}. \tag{2.56}$$

(b) Suppose that the family $(\mathcal{F}_t)$ is right continuous. Let S be a stopping time of $(\mathcal{G}_t)$ and the family $(\mathcal{F}_{S+t})$ and T and $\mathcal{F}_\infty$-measurable positive r.v. Then, $U = S + T$ is a stopping time of $(\mathcal{F}_t)$ if and only if T is a stopping time of $(\mathcal{G}_t)$.

Proof. (a) For all u, $\{T \leq u\} = \{T \leq u\} \cap \{S \leq u\}$. As $\{T \leq u\}$ belongs to $\mathcal{F}_S$, this belongs to $\mathcal{F}_u$ by definition of $\mathcal{F}_S$, and T is a stopping time. If $S < T$ on $\{S < \infty\}$, we can replace $\{S \leq u\}$ by $\{S < u\}$ in the argument.

(b) Suppose that T is a stopping time of $(\mathcal{G}_t)$. We write that

$$\{U < t\} = \bigcup_b \{S + b < t\} \cap \{T < b\}$$

on all rationals $b < t$. But $\{T < b\} \in \mathcal{G}_b = \mathcal{F}_{S+b}$, hence, by definition of $\mathcal{F}_{S+b}$, $\{T < b\} \cap \{S + b < t\} \in \mathcal{F}_t$ and $\{U < t\} \in \mathcal{F}_t$. Since the family $(\mathcal{F}_t)$ is right continuous, U is a stopping time of it. Conversely, we suppose for simplicity that S is finite. Then, if U is a stopping time of $(\mathcal{F}_t)$

$$\{T \leq t\} = \{S + T \leq S + t\} \in \mathcal{F}_{S+t} = \mathcal{G}_t \text{ by (2.53)}$$

and T is indeed a stopping time of $(\mathcal{G}_t)$. ∎

Corollary 2.4.18 Every stopping time S of the family $(\mathcal{F}_{t+})$ is the limit of a decreasing sequence of discrete stopping times of the family $(\mathcal{F}_t)$ and can also be represented as the lower envelope of an, in general, non-decreasing sequence of stopping times T of the following type

$$T = a 1_A + (+\infty) 1_{A^c}, \quad a \in \mathbb{R}_+, \quad A \in \mathcal{F}_a. \tag{2.57}$$

Proof. We have $S = \lim_n S^{(n)} = \inf_n S^{(n)}$ (2.56) and $S^{(n)}$ is the infimum of the stopping times

$$k 2^{-n} 1_{A_k} + (+\infty) 1_{A_k^c} \text{ where } A_k = \{S^{(n)} = k2^{-n}\}.$$

∎

Next we give an important application to the completion of σ-fields.

Theorem 2.4.19 Let $\mathcal{H}_t$ be the σ-field obtained by adjoining to $\mathcal{F}_t$ all the **P**-negligible sets. Let T be a stopping time of the family $(\mathcal{H}_t)$. Then, there exists a stopping time U of the family $(\mathcal{F}_{t+})$ such that $T = U$ **P**-a.s. Furthermore, for all $L \in \mathcal{H}_t$, there exists $M \in \mathcal{F}_{U+}$ such that $L = M$ **P**-a.s.

Proof. For the first assertion, it suffices to treat the case of a stopping time of type (2.57) and then pass to the lower envelope. We then write $U = a\mathbf{1}_B + (+\infty)\mathbf{1}_{B^c}$ where B is an element of $\mathcal{F}_a$ **P**-a.s. equal to $A \in \mathcal{H}_a$ (passing to the limit is responsible for the appearance of the family $(\mathcal{F}_{t+})$ instead of $(\mathcal{F}_{t+})$).

The second assertion is a consequence of the first one. Since $L \in \mathcal{H}_\infty$, we choose some $L' \in \mathcal{F}_\infty$ such that $L' = P$-a.s., and then a stopping time V of the family $(\mathcal{F}_{t+})$ such that $V = T_L$ a.s. The required event M then is $(L' \cap \{U = \infty\}) \cup \{V = U < \infty\}$. ∎

2.4.3 Stochastic intervals

Definition 2.4.20 Let U and V be two positive real-valued functions on Ω such that $U \leq V$. Denote by $[\![U, V[\![$ the subset of $\Omega \times \mathbb{R}_+$

$$[\![U, V[\![= \{(\omega, t) : U(\omega) \leq t < V(\omega)\}. \tag{2.58}$$

We define similarly the "stochastic intervals" of types $]\!]U, V]\!]$, $]\!]U, V[\![$ and $[\![U, V]\!]$. In particular, we set $[\![U, U]\!] = [\![U]\!]$, known as, the graph of U.

The use of double brackets $[\![$ aims at distinguishing, when U and V are constant stopping times the stochastic interval $[\![U, V[\![$ in $\mathbb{R}_+ \times \Omega$ from the ordinary interval $[U, V[$ of $\mathbb{R}_+$.

Next we give fundamental definitions relating to a filtration $(\mathcal{F}_t)$ and their relation to stochastic intervals.

2.4.4 Optional and predictable σ-fields

Definition 2.4.21 The σ-field $\mathcal{O}$ on $\mathbb{R}_+ \times \Omega$ generated by the real-valued processes $(X_t)_{t \geq 0}$ adapted to the family $(\mathcal{F}_t)$ and with r.c.l.l. paths (2.3) is called the optional or well measurable σ-field.

The σ-field $\mathcal{P}$ on $\mathbb{R}_+ \times \Omega$ generated by the processes $(X_t)_{t \geq 0}$ adapted to the family $(\mathcal{F}_{t-})$ and with left-continuous paths on $]0, \infty[$ is called the predictable σ-field.

Remark 2.4.22 (a) If we deal with $\mathbb{R}_+ \times \Omega$ the processes in this definition must include an additional variable X_∞. In the optional case, we require that it be $\mathcal{F}_\infty$-measurable; in the predictable case, we require left-continuity at infinity.

(b) It is also possible to consider the two σ-fields as defined on different spaces: the optional σ-field on $[0, \infty[\times \Omega$ but the predictable σ-field on $]0, \infty] \times \Omega$.

(c) If we replace the filtration $(\mathcal{F}_t)$ by the right continuous filtration $\mathcal{G}_t = \mathcal{F}_{t+}$, $\mathcal{G}_{0-} = \mathcal{G}_{0-}$, the optional σ-field is enlarged but the predictable σ-field does not change.

(d) A process (X_t) defined on Ω is said to be optional (resp. predictable) if the function $(t, \omega) \mapsto X_t(\omega)$ on $\mathbb{R}_+ \times \Omega$ is measurable given the optional (resp. predictable) σ-field.

Example 2.4.23 (1) For the stopping times $S \leq T$, all stochastic intervals determined by the pair are optional sets. Since, the stochastic intervals $[\![0, T[\![$ and $[\![S, \infty[\![$ have adapted indicators with r.c.l.l. paths and the same also holds for their intersection $[\![S, T[\![$. Next, consider $[\![S, S + 1/n[\![$ and let n go to infinity we see that $[\![S]\!]$ is optional, out of which we can deduce the optionality of all other stochastic intervals.

Let Z be an $\mathcal{F}_S$-measurable r.v. and let $Z1_{[\![S,T[\![}$ denote the process (X_t) defined by $X_t(\omega) = Z(\omega)1_{[\![S,T[\![}(t, \omega)$. It can be verified that $Z1_{[\![S,T[\![}$ is optional: approximating Z by $\mathcal{F}_S$-measurable elementary r.v., we are immediately reduced to the case where Z is the indicator of an element A of $\mathcal{F}_S$ and then $Z1_{[\![S,T[\![}$ is the indicator of $[\![S_A, T_A[\![$. There are analogous results for the other stochastic intervals.

(2) Suppose that S and T are two stopping times of $(\mathcal{F}_{t+})$ such that $S \leq T$, then the stochastic interval $]\!]S, T]\!]$ is predictable, for that its indicator is the product of the indicators of $]\!]0, T]\!]$ and $[\![S, \infty[\![$ is adapted to $(\mathcal{F}_{t-})$ and has left-continuous paths.

If Z is an $\mathcal{F}_{S+}$-measurable r.v., it can be verified, as above, that the process $Z1_{]\!]S,T]\!]}$ is predictable.

Next consider the following definition which is often used.

Definition 2.4.24 Let $(X_t)_{t \in \mathbb{R}_+}$ be a real-valued process on Ω and let H be a function on Ω with values in $\bar{\mathbb{R}}_+$. Then, we denote by X_H the function defined on $\{H < \infty\}$ by

$$X_H(\omega) = X_{H(\omega)}(\omega) \tag{2.59}$$

state of the process X at time H and by $X_H 1_{\{H < \infty\}}$ the function equal to X on $\{H < \infty\}$ and to 0 on $\{H = \infty\}$.

Of course, for a process $(X_t)_{t \in \bar{\mathbb{R}}_+}$ we can define X_H on the whole of Ω and not only on $\{H < \infty\}$. Next we will begin to study the optional and predictable σ-fields.

Theorem 2.4.25 (a) The optional σ-field is contained in the progressive σ-field.

(b) Let T be a stopping time of $(\mathcal{F}_t)$ and let (X_t) be a progressive process (in particular, an optional process). Then, the function $X_T 1_{T < \infty}$ is $\mathcal{F}_{T^-}$ measurable. Conversely, if Y is an $\mathcal{F}_T$-measurable r.v., there exists an optional process (X_t), $t \in \bar{\mathbb{R}}_+$ such that $Y = X_T$.

(c) The optional σ-field is generated by the stochastic intervals $[\![S, \infty[\![$, where S is a stopping time.

Proof. Part (a) has already been proved in (2.2.6). For (b), let (X_t) be a progressive process; the mapping $(t, \omega) \mapsto X_t(\omega)$ is $\mathcal{B}([0, t]) \times \mathcal{F}_t$-measurable on $[0, t] \times \Omega$ for all t. Since T is a stopping time, the mapping $\omega \mapsto T(\omega) \wedge t$ is $\mathcal{F}_t$-measurable. By composition, we see that $\omega \mapsto X_{T(\omega) \wedge t}(\omega)$ is also $\mathcal{F}_t$-measurable. Let $Y = X_T 1_{T < \infty}$ then for all t $Y 1_{T \leq t} = X_{T \wedge t} 1_{T \leq t}$ is $\mathcal{F}_t$-measurable which means that Y is $\mathcal{F}_T$-measurable (see Equation 2.48). The converse proposition is obvious; if Y is $\mathcal{F}_T$-measurable, the process (X_t) defined for $0 \leq t \leq \infty$ by $X_t = Y 1_{T \leq t}$ is adapted with r.c.l.l. paths and $Y = X_T$.

For part (c) let ϑ_0 be the paving consisting of the stochastic intervals $[\![S, T[\![$ where S and T are stopping times and $S \leq T$. Then, $\vartheta_0 \subset \mathcal{O}$ and we must examine whether every process (X_t) adapted to $(\mathcal{F}_t)$ with r.c.l.l. paths is $\sigma(\vartheta_0)$-measurable. We choose a number $\epsilon > 0$ and introduce inductively the following functions, given that d is a metric on $\overline{\mathbb{R}}$, $T_0^\epsilon = 0$ and

$$T_{n+1}^\epsilon(\omega) = \inf \left\{ t > T_n^\epsilon(\omega) : d(X_{T_n^\epsilon}(\omega), X_t(\omega)) \geq \epsilon \text{ or } d(X_{T_n^\epsilon}(\omega), X_{t-}(\omega))) \geq \epsilon \right\} \tag{2.60}$$

$$Z_n^\epsilon(\omega) = X_{T_n^\epsilon}(\omega) \mathbf{1}_{\{T_n^\epsilon < \infty\}}(\omega). \tag{2.61}$$

The path deviates from $Z_n^\epsilon(\omega)$ by less than ϵ on the interval $[\![T_n^\epsilon(\omega), T_{n+1}^\epsilon(\omega)[\![$ but its oscillation on the closed interval $[\![T_n^\epsilon(\omega), T_{n+1}^\epsilon(\omega)]\!]$ is $\geq \epsilon$ if $T_{n+1}^\epsilon(\omega) < \infty$. Hence, the existence of left-hand limits prevents the $T_n^\epsilon(\omega)$ from accumulating at a finite distance. Thus, (X_t) is the uniform limit as $\epsilon \to 0$, of the processes

$$X_t^\epsilon(\omega) = \sum_{n=0}^{\infty} Z_n^\epsilon(\omega) \mathbf{1}_{[\![T_n^\epsilon, T_{n+1}^\epsilon]\!]}(t, \omega) \tag{2.62}$$

and it suffices to show that (X_t) is $\sigma(\vartheta_0)$-measurable. We first verify that the T_n are stopping times of the family $(\mathcal{F}_t)$: to simplify the notation, we do this only for the first of them and leave the rest to the reader to do by induction on n. We then write

$$T = \inf\{t > 0 : d(X_0, X_t) \geq \epsilon \text{ or } d(X_0, X_{t-}) \geq \epsilon\} \tag{2.63}$$

and check that $\{T \leq t\} \in \mathcal{F}_t$. To this end, we remark that the set between $\{\ \}$ is closed; hence, the relation $T = t$ implies that $d(X_0, X_t) \geq \epsilon$ or $d(X_0, X_{t-}) \geq \epsilon$ and the relation $\{T \leq t\}$ is equivalent to the following, where $\mathbb{Q}_t$ denotes the set consisting of the rationals of $]0, t[$ and the point t:

$$\forall n \in \mathbb{N} \quad \exists r_n \in \mathbb{Q}_t, \quad d(X_0, X_{r_n}) > \epsilon - 1/n.$$

Since $\mathbb{Q}_t$ is countable, clearly, this set belongs to $\mathcal{F}_t$.

Having established this point concerning the T_n^ϵ it remains to remark (leaving out the useless ϵ and n) that processes of the form $Z\mathbf{1}_{[\![S,T[\![}$ where S and T are two stopping times and Z is $\mathcal{F}_S$-measurable, are $\sigma(\vartheta_0)$-measurable (this is obvious from Example 2.4.23(1)). ∎

Theorem 2.4.26 Under the *usual conditions* the optional σ-field is also generated by the right-continuous processes adapted to the family $(\mathcal{F}_t)$.

Proof. First we show that every right-continuous adapted process (X_t) is indistinguishable from an optional process. We employ a variant of the construction in (Theorem 2.4.25), but this time arguing by transfinite induction, since now the left-hand limits may not exist, and therefore accumulation of stopping times can happen at a finite distance. We set $T_0 = 0$ and for every countable ordinal α

$$T_{\alpha+1}^\epsilon = \inf\{t > T_\alpha^\epsilon : d(X_{T_\alpha^\epsilon}, X_t) > \epsilon\},$$

on the other hand, for every limit ordinal β,

$$T_\beta^\epsilon = \sup_{\alpha < \beta} T_\alpha^\epsilon.$$

Z_α^ϵ and X^ϵ are defined as in (2.61) and (2.62), but with the obvious modifications: α instead of n, with the sum over the ordinals. The T_α^ϵ are - thanks to the strict inequality "$> \epsilon$" - stopping times of the family $(\mathcal{F}_{t+})$ and the process (X_t) is approximated uniformly by the (X_t^ϵ). On the other hand, there exists an ordinal α such that $T_\alpha^\epsilon = +\infty$ **P**-a.s. and the process (X_t^ϵ) is therefore **P**-indistinguishable from an optional process (Y_t). The process (Z_t) defined by $Z_t = X_t - Y_t$ then is adapted, right-continuous and evanescent.

Now, we show that it is optional. Define for n the process (Z_n) by writing $Z_0^n = Z_0$ and

$$Z_t^n = Z_{(k+1)/n} \text{ for } t \in]k/n, (k+1)/n]$$

where k runs through the integers. Since (Z_t) is evanescent $Z_{(k+1)/n}$ is **P**-a.s. zero and hence $\mathcal{F}_{k/n}$-measurable (but we are under the *usual conditions*) then, the process (Z_t^n) is optional and so is its limit (Z_t) as n tends to infinity. ∎

Next, we prove that the difference between optional and predictable processes is a countable set.

Theorem 2.4.27 Let (X_t) be an optional process with respect to the family $(\mathcal{F}_t)$. Then, there exists a predictable process (Y_t) such that the set

$$A = \{(t, \omega) : X_t(\omega) \neq Y_t(\omega)\} \tag{2.64}$$

is the union of a sequence of graphs of stopping times. In particular, the section $A(\omega)$ is countable for all ω.

Proof. We note first that it suffices to prove that A is contained in the union of a sequence of graphs of stopping times (this statement is important in itself): for if T is a stopping time, A being progressive, the event $L = \{\omega : (T(\omega), \omega) \in A\}$ belongs to $\mathcal{F}_T$ (Theorem 2.4.25(b)) and the stopping time T_L has its graph contained in A. If A is contained in a union of graphs $[\![T^n]\!]$, A then is equal to the union of the corresponding graphs $[\![T_{L_n}^n]\!]$ where $L_n = \{\omega : (T^n(\omega), \omega) \in A\}$.

On the other hand, we may reduce to bounded real-valued processes. Then, let $\mathfrak{H}$ be the set of optional bounded real processes (X_t) for which the statement is true. By truncating, we may also assume that the corresponding predictable processes are bounded. We then verify that $\mathfrak{H}$ is an algebra closed under monotone convergence (if processes $(X_t^n) \in \mathfrak{H}$ increase to a bounded process (X_t) and predictable processes (Y_t^n) satisfy the statement relative to (X_t^n) then so does the process $Y_t = \liminf_n Y_t^n$ relative to (X_t)). By the monotone class theorem, it suffices to verify the statement for processes generating the optional σ-field. We choose for (X_t) the indicators of intervals $[\![S, T[\![$ where S and T are stopping times, the corresponding predictable processes (Y_t) being the indicator of the intervals $]\!]S, T]\!]$. ∎

We now give the theorem analogous to Theorem 2.4.25 but concerning predictable σ-field.

Theorem 2.4.28 (a) The predictable σ-field is contained in the optional σ-field, hence, in the progressive σ-field.

(b) Let T be a stopping time of the family $(\mathcal{F}_t)$ and let (X_t) be a predictable process. Then, the function $X_T \mathbf{1}_{T < \infty}$ is $\mathcal{F}_T$-measurable. Conversely, if Y is an $\mathcal{F}_T$-measurable r.v. there exists a predictable process $(X_t)_{t \in \bar{\mathbb{R}}_+}$ such that $Y = X_T$.

(c) The predictable σ-field is generated by the sets of the form

$$\{0\} \times A, \ A \in \mathcal{F}_{0-} \ \text{and} \ ]s,t] \times A \ (0 < s < t, \ s \ \text{and} \ t \ \text{rationals}, \ A \in \bigcup_{r<s} \mathcal{F}_r) \qquad (2.65)$$

or also the sets

$$\{0\} \times A, \ A \in \mathcal{F}_{0-} \ \text{and} \ [s,t[\times A \ (0 < s < t, \ s \ \text{and} \ t \ \text{rationals}, \ A \in \bigcup_{r<s} \mathcal{F}_r) \qquad (2.66)$$

Proof. We will begin with the proof of (c). First, we note that, by Example 2.4.23, every stochastic interval $]\!]S, T]\!]$ where S and T are stopping times (even in the wide sense), is predictable. Conversely, let (X_t) be a process adapted to $(\mathcal{F}_{t-})$ with left-continuous paths; X_t is the limit of the processes

$$X^n = X_0 \mathbf{1}_{[\![0]\!]} + \sum_{k \geq 0} X_{k2^{-n}} \mathbf{1}_{]\!]k2^{-n},(k+1)2^{-n}]\!]}. \qquad (2.67)$$

By representing $X_{k2^{-n}}$ as an increasing limit of $\mathcal{F}_{k2^{-n}}$-measurable elementary r.v., we see that (X_t) is measurable with respect to the σ-field generated by the sets $\{0\} \times A$ $(A \in \mathcal{F}_{0-})$ and the stochastic intervals $]\!]S, T]\!]$ of the following special type:

$$\begin{aligned} S &= s \text{ on } A, \ +\infty \text{ on } A^c, \qquad\qquad\qquad\qquad\qquad (2.68) \\ T &= t \text{ on } A, \ +\infty \text{ on } A^c \text{ with } 0 < s < t, \ A \in \mathcal{F}_{s-}. \end{aligned}$$

We can also replace the condition $A \in \mathcal{F}_{s-}$ by $A \in \cup_{r<s}\mathcal{F}_r$. Hence, we get the generating system (2.65), and deduce from it (2.66).

Next is part (a) which is obvious: the predictable σ-field is generated by stochastic intervals and these belong to the optional σ-field.

Finally we prove (b). To simplify some of the work, we suppose that T is finite. The statement then means that $\mathcal{F}_{T-}$ is the inverse image of the predictable σ-field $\mathcal{P}$ under the mapping $f : \omega \mapsto (T(\omega), \omega)$ of Ω into $\mathbb{R}_+ \times \Omega$.

Let us first verify that f is measurable from $\mathcal{F}_{T-}$ into $\mathcal{P}$: it suffices to show that for sets U generating the predictable σ-field $f^{-1}(U)$ belongs to $\mathcal{F}_{T-}$ and to this end we choose $U = \{0\} \times A$ $(A \in \mathcal{F}_{0-})$, $U =]s, \infty[\times A$ $(A \in \mathcal{F}_s)$ whose inverse images are $A \cap \{T = 0\}$, $A \cap \{s < T\}$. These sets belong to $\mathcal{F}_{T-}$: in the first case because T is $\mathcal{F}_{T-}$-measurable, and in the second case by Definition 2.4.13.

We now verify that every $A \in \mathcal{F}_{T-}$ is the inverse image of a predictable set, i.e., $f^{-1}(\mathcal{P})$ contains the generators of $\mathcal{F}_{T-}$; this is immediate by (c) and we leave the details to the reader. ∎

Remark 2.4.29 (a) Provided the σ-fields $\mathcal{F}_{t-}$ are separable, the predictable σ-field is separable; this is seen immediately from the generating system (2.65) or (2.66).

(b) Given an arbitrary $\mathcal{F}$-measurable positive r.v. L, it is natural to define the σ-fields associated with L as follows:

- a r.v. Y is measurable with respect to $\mathcal{F}_L$ $(\mathcal{F}_{L-})$ if and only if there exists an optional (resp. predictable) process $(X_t)_{t \in \mathbb{R}_+}$ such that $Y = X_L$.

- we define similarly $\mathcal{F}_{L+}$ by considering optional processes of the family $(\mathcal{F}_{t+})$.

(c) The lack of symmetry between the two parts of Definition 2.4.21 (r.c.l.l.) process on the one hand and left-continuous on the other is only apparent: the predictable σ-field is also generated by left-continuous processes with right-hand limits! But in fact the predictable σ-field is even generated by the adapted projection processes with continuous paths: for if A belongs to $\mathcal{F}_{0-}$, the process $X_t = t\mathbf{1}_A + \mathbf{1}_{A^c}$ is adapted (to the family $(\mathcal{F}_{t-})$) and has continuous paths and $\{0\} \times A = \{(t, \omega) : X_t(\omega) = 0\}$; similarly, if S is a stopping time, the process $X_t = (t - S)^+$ is adapted with continuous paths and the stochastic interval $]\!]S, \infty[\![$ is equal to $\{(t, \omega) : X_t(\omega) > 0\}$.

2.4.5 Predictable stopping times

The first definition we are going to give for predictable stopping times is not the usual one, but it is convenient for arguments without a probability law. It is, however, equivalent to the definition under the usual conditions (see Theorem 2.4.41). We have seen above that if T is a stopping time, the interval $[\![T, \infty[\![$ is optional. Conversely, if T is a positive function on Ω and the interval $[\![T, \infty[\![$ is optional, then T is a stopping time for that if X denotes the indicator of $[\![T, \infty[\![$, X_t is $\mathcal{F}_t$-measurable (see Theorem 2.4.25(b)) and it is the indicator of $\{T \le t\}$. Now, replacing the optional σ-field by the predictable σ-field, we may define:

Definition 2.4.30 Let T be a mapping of Ω into $\bar{\mathbb{R}}_+$. We say that T is a predictable r.v. or a predictable stopping time of the family $(\mathcal{F}_t)$ if the stochastic interval $[\![T, \infty[\![$ is predictable.

Since $\mathcal{P} \subseteq \mathcal{O}$ predictable r.v.s are stopping times. Constants are obviously predictable stopping times. Also, if we replace the family $(\mathcal{F}_t)$ by the family $\mathcal{G}_t = \mathcal{F}_{t+}$, $\mathcal{G}_{0-} = \mathcal{F}_{0-}$ we do not change the σ-field $\mathcal{P}$ so the predictable stopping times remain the same. The same is true if we replace $\mathcal{F}_t$ by $\mathcal{F}_{t-}$.

In the discrete case the above definition is formulated as follows: If we are given a family $(\mathcal{F}_n)_{n \ge -1}$, $\mathcal{F}_{-1}$ play the role of $\mathcal{F}_{0-}$ in the continuous case. A predictable process then is a process $(X_n)_{n \ge 0}$ such that, for all $n \ge 0$, X_n is $\mathcal{F}_{n-1}$-measurable and a predictable stopping time is a stopping time T such that, for all $n \ge 0$ the set $\{T\} = n$ belongs to $\mathcal{F}_{n-1}$.

Next, we are going to connect the definitions of stopping times to the existence of a sequence of precursory signs about them:

Definition 2.4.31 Let T be a stopping time of the family $(\mathcal{F}_t)$ and (T_n) be an increasing sequence of stopping times (may be only in the wide sense) such that $T_n \le T$ for all n. We say that the sequence (T_n) foretells (announces) T on $A \subset \Omega$ if on the set A,

$$\lim_n T_n = T, \ T_n < T \text{ for all } n. \tag{2.69}$$

Given a probability law $\mathbf{P}$ on $(\Omega, \mathcal{F})$, if 2.69 hold $\mathbf{P}$ a.s. on A, we of course say that the sequence foretells T a.s. on A. Moreover, since the condition $T_n \le T$ implies that $T_n = T$ on $\{T = 0\}$, we say that the sequence T_n foretells T a.s. without mentioning any set A, to mean that (T_n) foretells T a.s. the set $\{T > 0\}$. We also say that T is foretellable (**P**-foretellable) if there exists a sequence (T_n) of stopping times in the wide sense foretelling T **P**-a.s.

Remark 2.4.32 Under the *usual conditions*, every **P**-foretellable stopping time is foretellable. Just let (T_n) be a sequence foretelling T on A^c where A has measure zero; then A belongs to $\mathcal{F}_0$ and the stopping times $T'_n = T_n 1_{A^c} + (T - \frac{1}{n})^+ 1_A$ foretells T everywhere.

Later, we shall see that the almost-converse is true: every predictable stopping time is **P**-foretellable for every law **P**. Next, we give a fundamental example of predictable stopping times as theorems.

Theorem 2.4.33 (a) Let T be a foretellable stopping time such that $\{T = 0\}$ belongs to $\mathcal{F}_{0-}$, then T is predictable.
(b) Every **P** foretellable stopping time T is a.s. equal to a foretellable stopping time and hence to a predictable stopping time if $\mathcal{F}_0 = \mathcal{F}_{0-}$.

Proof. To prove part (a), let (T_n) be a sequence of stopping times in the wide sense foretelling T. Then, the interval $[\![T, \infty[\![$ is the union of the predictable sets $\{0\} \times \{T = 0\}$ and $[\![T_n, \infty[\![$ and is hence predictable. To prove (b), let (T^n) be a sequence which foretells T a.s. We set $T' = \lim_n T^n$ and, for all n, $B_n = \{T = 0\} \cup \{T^n < T'\}$, which belongs to $\mathcal{F}_{T^n+}$ (2.55). Then, the wide sense stopping times $S^n = T^n_{B_n} \wedge n$ are increasing, their limit S is equal to T a.s. and the sequence (S^n) obviously foretells S. ∎

Example 2.4.34 Let S be a wide sense stopping time and T be a $\mathcal{F}_{S+}$-measurable r.v. such that $S < T$ on $\{S < \infty\}$ then T is a stopping time since the sequence $T_n = n \wedge (\frac{1}{n}S + \frac{n-1}{n}T)$ foretells T, T is a predictable stopping time.

Remark 2.4.35 The equivalence between the notions of predictable stopping times and foretellable stopping times at least to within sets of measure zero gives the notion of predictable its intuitive meaning. We have mentioned earlier that the idea of a stopping time is that of the first time some given random phenomenon occurs. The existence of a foretelling sequence means that this phenomenon cannot take us by surprise: we are forewarned by a succession of precursory signs of the exact time the phenomenon will occur.

In the following theorem we group elementary properties of predictable stopping times.

Theorem 2.4.36 (a) The set of predictable stopping times is closed under the operations $\wedge$ and $\vee$.

(b) Let S be a predictable stopping time and T a stopping time. Then,

$$\text{if } A \text{ belongs to } \mathcal{F}_{S-} \text{ then } A \cap \{S \le T\} \in \mathcal{F}_{T-}. \tag{2.70}$$

In particular, the events $\{S < T\}$, $\{S = T\}$ and $\{S \le T\}$ belong to $\mathcal{F}_{T-}$.

(c) Let S be a predictable stopping time and A be an element of $\mathcal{F}_{\infty-}$. In order that $A \in \mathcal{F}_{S-}$, it is necessary and sufficient that S_A be predictable.

(d) Let (S_n) be an increasing sequence of predictable stopping times and let $S = \lim_n S_n$; then S is predictable. The same result is true for a stationary decreasing sequence (S_n) (i.e., such that for all ω there exists N such that $S_{N+k}(\omega) = S_N(\omega)$ for all k).

Proof. (a) We remark that $[\![S \vee T, \infty[\![= [\![S, \infty[\![\wedge [\![T, \infty[\![$ and $[\![S \wedge T, \infty[\![= [\![S, \infty[\![\cup [\![T, \infty[\![$.

(b) We know that $A \cap \{S < T\} \in \mathcal{F}_{T-}$ (2.54) and $A \cap \{T = \infty\} \in \mathcal{F}_{T-}$ (Theorem 2.4.15(e)). So, it suffices to consider $A \cap \{S = T < \infty\}$. Let (X_t) be a predictable process such that $1_A = X_S 1_{S < \infty}$ (Theorem 2.4.28(b)). Let $1_{[\![S]\!]}$ be the indicator of the predictable set $[\![S]\!] = [\![S, \infty[\![\setminus]\!]S, \infty[\![$. Then, $1_{A \cap \{S = T < \infty\}} = X_T 1_{[\![T]\!]} 1_{T < \infty}$ and we conclude using Theorem (2.4.28(b)) that indeed $A \cap \{S = T < \infty\} \in \mathcal{F}_{T-}$.

Taking $A = \Omega$, we get $\{S \le T\} \in \mathcal{F}_{T-}$; we also know that $\{S < T\} \in \mathcal{F}_{T-}$ (2.54) and hence, taking a difference, $\{S = T\} \in \mathcal{F}_{T-}$.

(c) If the stopping time S_A is predictable, let (X_t) be the (predictable) indicator of $[\![S_A, \infty[\![$; we know that $X_S 1_{S < \infty}$ is $\mathcal{F}_{S-}$-measurable (Theorem 2.4.28(b)) and hence $A \cap \{S < \infty\} \in \mathcal{F}_{S-}$. Similarly for $A \cap \{S = \infty\}$ since $A \in \mathcal{F}_{\infty-}$ (Theorem 2.4.15(e)).

Conversely, let $A \in \mathcal{F}_{S-}$; there exists (Theorem 2.4.28(b)) a predictable process (X_t) such that 1_A and X_S coincide on $\{S < \infty\}$. The graph $[\![S_A]\!]$ then is the intersection of $[\![S]\!]$ and of the predictable set $\{(t, \omega) : X_t(\omega) = 1\}$. The graph $[\![S]\!]$ is equal to $[\![S, \infty[\![\setminus]\!]S, \infty[\![$ and is hence predictable; thus $[\![S_A]\!]$ is predictable and so is $[\![S_A, \infty[\![= [\![S_A, \infty[\![\cup [\![S_A]\!]$.

(d) In the first case, $[\![S, \infty[\![= \cap_n [\![S_n, \infty[\![$ and in the second case $[\![S, \infty[\![= \cup_n [\![S_n, \infty[\![$. ∎

Remark 2.4.37 The predictable σ-field is generated by the intervals $[\![S, T[\![$, where S and T are predictable, or even both predictable and foretellable. For among the generators (2.66) those of the second kind are of the form $[\![S, T[\![$ with $S = s_A$ ($A \in \cap_{r<s} \mathcal{F}_r$), $T = t_A$, foretellable by stopping times of the form $n \wedge (s - \frac{1}{n})_A$, $n \wedge (t - \frac{1}{n})_A$ with n sufficiently large. As for the generators of the first kind, if $A \in \mathcal{F}_{0-}$ then $\{0\} \times A = \cap_n [\![0_A, 0_A + 1/n[\![$ and 0_A is foretellable by the sequence $n \wedge 0_A$.

Our next task is to prove the converse of the proposition in Theorem 2.4.33 and establish properties analogous to Theorem 2.4.36, but now for **P**-foretellable stopping times.

Lemma 2.4.38 Let $\mathcal{U}$ be the set of stopping times of $(\mathcal{F}_t)$ which are foretellable a.s. by a sequence of wide sense stopping times:

(a) $\mathcal{U}$ is closed under the operations $\vee$ and $\wedge$ and $0 \in \mathcal{U}$, $+\infty \in \mathcal{U}$.

(b) The limit of an increasing sequence of elements of $\mathcal{U}$ belongs to $\mathcal{U}$.

(c) The limit of a stationary decreasing sequence of elements of $\mathcal{U}$ belongs to $\mathcal{U}$.

(d) If S and T belong to $\mathcal{U}$, so does S_A where $A = \{S < T\}$.

(e) Let $S \in \mathcal{U}$ and T be a stopping time such that $S = T$ a.s., then $T \in \mathcal{U}$.

Proof. (a) Let S and T be two elements of $\mathcal{F}_t$ which are foretold a.s. by sequences (S_n) and (T_n), then $S \wedge T$ and $S \vee T$ are foretold a.s. by the sequences $(S_n \wedge T_n)$ and $(S_n \vee T_n)$.

(b) Let (T_n) be an increasing sequence of elements of $\mathcal{U}$ and for each n let $(T_n^m)_{m \in \mathbb{N}}$ be a sequence foretelling a.s. (T_n). Then, the stopping time $T = \lim_n T_n = \lim_m \lim_n T_n^m$ is a.s. foretold by the sequence $T^m = T_1^m \vee T_2^m \vee ... \vee T_p^m$.

(c) Suppose now that the T_n, still a.s. foretold by the (T_n^m), form a stationary decreasing sequence, and let $T = \lim_n T_n$. We consider some bounded metric d defining the topology on $\overline{\mathbb{R}}_+$ (for example, $\mathrm{d}(x, y) = |e^{-x} - e^{-y}|$) and choose for each n an integer k such that $\mathbf{P}\{\mathrm{d}(T_n^k, T_n) > 1/n\} \le 2^{-n}$. Then we set $U_m = \inf_{n \ge m} T_n$. The U^m are wide sense stopping times and the sequence (U^m) is increasing; denote its limit by U. Since, $T_n^k \le T_n \le T$, we have $U^m \le T$ hence $U \le T$ everywhere. Since $T_n^k < T_n$ a.s. on $\{T_n > 0\}$ $U^m < T_n$ a.s. on $\{T_n > 0\}$ for all $n \ge m$ and hence $U^m < T$ a.s. on $\{T > 0\}$, the sequence (T_n) being stationary.

Finally, $U = T$ a.s. The relation $U < T$ indeed implies that for sufficiently large q, $\mathrm{d}(U, T) > 1/q$; hence, for all sufficiently large m $\mathrm{d}(U^m, T) > 1/q$ and hence for every sufficiently large m there exists $n \ge m$ such that $d(T_n^k, T) > 1/n$. Since the sequence (T_n) is stationary, this is equivalent for large n to $\mathrm{d}(T_n^k, T_n) > 1/n$ and we see that

$$\{U < T\} \subset \limsup_n \{\mathrm{d}(T_n^k, T_n) > 1/n\} \text{ a.s.}$$

which is a negligible set by the Borel-Cantelli lemma. All this means that the sequence (U_m) foretells T a.s.

(d) Let (S^n) and (T^n) be two sequences foretelling S and T a.s. We set

$$U_n^m = n \wedge S_{\{S^n < T^m\}}^n.$$

For fixed m we thus construct an increasing sequence of wide sense stopping times whose limit U^m is equal to $+\infty$ on $\{T^m = 0\}$ and is a.s. equal to $S1_{\{S \le T^m\}}$ on $\{T^m > 0\}$. On the other hand, the sequence (U_n^m) foretells U^m a.s. and U^m belongs to $\mathcal{U}$. But the sequence (U^m) is decreasing and stationary, so that its limit U belongs to $\mathcal{U}$ by (c). Therefore, we denote by (V_n) a sequence which foretells U a.s.; it is then immediate that $U = S_{\{S < T\}}$ a.s., and it only remains to set $V_n' = V_n \wedge S_{S\{\wedge T\}}$ to get a sequence foretelling $S_{\{S \wedge T\}}$ a.s.

(e) If the sequence (S_n) foretells S a.s., the sequence $(S_n \wedge T)$ foretells T a.s. ∎

The following theorem is an important step leading to Theorem 2.4.41. The arguments we are going to use to prove it are exactly those that we are going to use for the proof of the theorems on cross-sections, optional (Theorem 2.4.48) and predictable (Theorem 2.4.52).

Theorem 2.4.39 Every predictable stopping time T belongs to $\mathcal{U}$ (set of stopping times in $(\mathcal{F}_t)$).

Proof. (a) Let ϑ_0 be the paving on $\mathbb{R}_+ \times \Omega$ consisting of all stochastic intervals $[\![S, T[\![$ where S and T are elements of $\mathcal{U}$ such that $S \le T$ and let ϑ be the closure of ϑ_0 under $(\cup f)$.

So, it is immediate that the complement of any element of ϑ_0 is the union of two elements of ϑ_0

$$([\![S, T[\![^c = [\![0, S[\![\cup [\![T, \infty[\![),$$

and, the intersection of two elements of ϑ_0 belongs to ϑ_0

$$([\![S, T[\![\cap[\![U, V[\![= [\![S \vee U, S \vee U \vee (T \wedge V)[\![),$$

so that ϑ is a Boolean algebra. It follows from Remark 2.4.37 that the σ-field $\sigma(\vartheta)$ which is also the monotone class generated by ϑ-contains the predictable σ-field.

(b) Using only Lemma 2.4.38 (a), (b), (c) and (d), we show that the debut of an element B of ϑ_δ is equal a.s. to an element of $\mathcal{U}$.

We begin by showing that the debut of an element C of ϑ belongs to $\mathcal{U}$. By definition on of ϑ, we can write $C = C_1 \cup ... \cup C_i$ where the C_k belong to ϑ_0; then $D_C = D_{C_1} \wedge ... \wedge D_{C_i}$ and it suffices (Lemma 2.4.38(a)) to show that the debut of an interval $C_k = [\![S, T[\![\in \vartheta_0$ belongs to $\mathcal{U}$ – but this debut is equal to $S_{\{S < T\}}$ and this then follows from (Lemma 2.4.38(d)).

Next, we represent B as the intersection of a decreasing sequence (B_n) of elements of ϑ. Let $\mathcal{D}$ be the set of elements of $\mathcal{U}$ smaller than D_B; $\mathcal{D}$ is non-empty ($0 \in \mathcal{D}$), $\mathcal{D}$ is closed under $(\vee c)$; hence there exists an increasing sequence (H_n) of elements of whose upper envelope $H \in \mathcal{U}$, is a.s. equal to the essential upper envelope of $\mathcal{D}$. Let $C_n = B_n \cap [\![H, \infty[\![$ and let S_n be the debut of C_n. C_n belongs to ϑ and hence S_n belongs to $\mathcal{U}$. Since $H \leq D_B$, we have $B_n \supset C_n \supset B$ and hence $\cap_n C_n = B$.

Then, $S_n \geq H$, $S_n \leq D_B$ and hence the fact that H is the essential upper envelope of $\mathcal{D}$ implies that $S_n = H$ a.s. Since the graph of S_n is contained in C_n, the graph of H is contained a.s. in $\cap_n C_n = B$ so $H \geq D_B$ a.s. and finally $H = D_B$ a.s.

(c) For this part, consider the following measure on $\mathbb{R}_+ \times \Omega$

$$\mu(A) = \int \mathbf{1}_A(T(\omega), \omega) \mathbf{1}_{T < \infty}(\omega) \mathbf{P}(d\omega) \quad A \in \mathcal{B}(\mathbb{R}_+) \times \mathcal{F}$$

This is a bounded measure of total mass $\mathbf{P}\{T < \infty\}$, carried by the graph $[\![T]\!]$. This graph is predictable and hence belongs to $\sigma(\vartheta)$. By a classical theorem of measure theory (Remark 2.4.40), for all $\epsilon > 0$ there exists an element B_ϵ of $\mathbf{1}_\delta$ contained in $[\![T]\!]$ and such that $\mu([\![T]\!] \backslash B_\epsilon) \leq \epsilon$. Let T^ϵ be an element of $\mathcal{U}$ equal a.s. to the debut of B ((b) above); the graph of T^ϵ is a.s. contained in that of T, hence the stopping time $T^\epsilon_{\{T = T^\epsilon\}}$ is a.s. equal to T, it belongs to $\mathcal{U}$ by 2.4.38(e) and its graph is entirely contained in that of T. We henceforth denote this stopping time by T^ϵ; we have

$$T^\epsilon \in \mathcal{U}, \ T^\epsilon = T \text{ on } \{T^\epsilon < \infty\}, \ \mathbf{P}\{T^\epsilon < \infty\} \geq \mathbf{P}\{T^\epsilon < \epsilon\} - \epsilon.$$

Let $S_n = T^{1/2} \wedge ... \wedge T^{1/n}$; then $S_n \in \mathcal{U}$, the sequence (S_n) is decreasing and stationary and its limit S is a.s. equal to T. We then conclude using Lemma 2.4.38(c) and (e). $\blacksquare$

Remark 2.4.40 With Choquet's Theorem at our disposal, this is also a result of capacitablity concerning the paving ϑ and the capacity μ^*, where every element of $\mathcal{P} \subset \sigma(\vartheta)$ is ϑ-analytic.

We have come to a definitive result on sequences announcing predictable stopping time, about approximation by elementary predictable times; this result was due to Chung [27].

Theorem 2.4.41 (a) Let T be a predictable stopping time of the family $(\mathcal{F}_t)$. There exists an increasing sequence (T_n) of predictable stopping times which are bounded above by T on the whole of Ω, such that $\lim_n T_n = T$ **P** a.s. and $T_n < T$ **P** a.s. for all n on $\{T > 0\}$. It can further be assumed that each T_n takes its values a.s. in the set of dyadic numbers.

(b) If the *usual conditions* are satisfied, the "a.s." can be omitted in the above statement.

Proof. We shall suppose that T is everywhere > 0; this is not a restriction since the set $\{T > 0\} = \Omega'$ belongs to $\mathcal{F}_{0-}$ and we may argue on Ω' with the induced law and then set $T_n = 0$ on $\{T = 0\}$.

(a) Let (S_n) be an increasing sequence of wide sense stopping times bounded above by T, which foretells T a.s. (Theorem 2.4.39) and for all p let S_n^p be the dyadic approximation of S_n by higher values:

$$S_n^p - (k+1)2^{-p} \Leftrightarrow S_n \in [k2^{-p}, (k+1)2^{-p}[, \quad S_n^p = \infty \Leftrightarrow S_n = \infty.$$

We have noted in Theorem 2.4.17 that S_n^p is a stopping time and in Theorem 2.4.34 that it is even predictable. As $S_n < T$ a.s., $\lim_n S_n^p = S_n$ and for all n there exists an integer $n' > n$ such that

$$\mathbf{P}\{T \le S_n^{n'}\} < 2^{-n} \tag{2.71}$$

Then, we set $U_n = \inf_{m \ge n} S_m^{m'}$. Since $S_m^{m'} \le T + 2^{-m'}$, U_n is everywhere bounded above by T. The sequence (U_n) is increasing and the inequality $S_m^{m'} \ge S_m$ implies $U_n \ge S_n$, hence $\lim_n U_n \ge T$ a.s. and finally $\lim_n U_n = T$ a.s. Finally the Borel-Cantelli Lemma and (2.71) imply that, except on a negligible set, $S_m^{m'} < T$ for all sufficiently large m and hence $U_n < T$ for all n. Thus the sequence (U_n) is a sequence of stopping times in the wide sense which foretells T a.s.

But we can do better: let N be the negligible set outside which $\lim_n U_n = T$, $U_n < T$ for all n and let $\omega \in N^c$. The subset of $\overline{\mathbb{R}}$ consisting of the points $S_m^{m'}(\omega)$ $(m \ge n)$ and $T(\omega)$ is compact, since the sequence $\left(S_p^{p'}\right)$ lying between U_p and $T + 2^{-p'}$ converges to T. Hence it contains its least upper bound $U_n(\omega)$. Since this is not equal to $T(\omega)$ by hypothesis, it is equal to one of the $S_m^{m'}(\omega)$. It follows immediately that $U_n(\omega)$ is a dyadic number.

We now construct a predictable stopping time $V_n \le T$ which is equal a.s. to U_n. First we note that the decreasing sequence of predictable stopping times

$$U^k = \lim_{n+k \ge m \ge n} S_m^{m'}$$

whose limit is U_n, is a.s. stationary by the above. We write $V^k = U_{A_k}^k \wedge T$, where $A_k = \{U^k = U^{k+i}$ for all $i\}$; the V^k are decreasing, are bounded above by T, form an everywhere stationary sequence and are predictable by Theorem 2.4.36(b). Their limit V_n is bounded above by T, is equal a.s. to J_n and is predictable by Theorem 2.4.36(d).

If we now write $T_n = V_1 \vee V_2 \vee \ldots \vee V_n$, we have constructed an increasing sequence of predictable stopping times which are bounded above by T and are equal a.s. to the U_n and hence a.s. have dyadic values and foretell T a.s.

(b) Under the usual conditions, all the negligible sets belong to $\mathcal{F}_{0-}^\star$ (i.e., $(\mathcal{F}_t^\star)$ satisfy the usual conditions). This should simplify all of the above proof, since a stopping time which is a.s. equal to a predictable stopping time is itself predictable. So, let N be the negligible subset of $\{T > 0\}$ where the T_n do not converge to T or are not $< T$ or do not have dyadic values. We modify T_n on N by giving it the value

$$T_n(\omega) = k2^{-n} \text{ if } k2^{-n} < T(\omega) \le (k+1)2^{-n}, \; T_n(\omega) = 2^n \text{ if } T(\omega) = +\infty.$$

Since $\mathcal{F}_{0-}^{\star}$ contains all the negligible sets, T_n still is a predictable time after this modification and the theorem is established. ∎

The best definition of a predictable stopping time of the family $(\mathcal{F}_t)$ is the one that does not depend on the choice of a law **P**. For that if we had defined a predictable stopping time by a sequence of stopping times that foretells it, then this sequence depends on the law **P**. We could have also taken as a definition of predictable stopping time the existence of a sequence of stopping times of $(\mathcal{F}_t)$ which foretell T everywhere on $\{T > 0\}$, a property independent of **P** but then we would have to be contended with a number of "a.s." statements.

The following statement consider the relationship between $(\mathcal{F}_t^{\star})$ and $(\mathcal{F}_t)$ as regards to predictable stopping time; compare its result to Theorem 2.4.19.

Theorem 2.4.42 Let $(\mathcal{F}_t^{\star})$ be the usual augmentation of $(\mathcal{F}_t)$ and let T be a predictable time of $(\mathcal{F}_t^{\star})$. Then there exists a predictable time T' of $(\mathcal{F}_t)$ which is equal to T a.s. Furthermore, for all $A \in \mathcal{F}_{T-}^{\star}$ there exists $A' \in \mathcal{F}_{T-}$ such that $A = A'$ a.s.

Proof. The set $\{T = 0\}$ belongs to $\mathcal{F}_{0-}^{\star}$, so choose a set $H \in \mathcal{F}_{0-}$ which differs from $\{T = 0\}$ by a negligible set and modify T on H^c without changing the notation by replacing it by $+\infty$ on $H^c \cap \{T = 0\}$. Working henceforth in H^c instead of Ω, we can reduce the problem to the case where T is everywhere > 0. We write $T' = 0$ on H and forget about H.

Let (T^n) be a sequence of stopping times of $(\mathcal{F}_t^{\star})$ which foretells T. For all n, let (R^n) be a stopping time of the family $(\mathcal{F}_{t+})$ such that $T^n = R^n$ a.s. (Theorem 2.4.19). Replacing R^n by $R^1 \vee \ldots \vee R^n$ if necessary, we can assume that the sequence (R^n) is increasing and denote its limit by R. Let $A_n = \{R^n = 0\} \cup \{R^n < R\}$. A_n decreases, hence, the stopping times of $(\mathcal{F}_{t+})$, $S_n = R_{A_n}^n \wedge n$ increases and the sequence S_n foretells everywhere its limit T' which is strictly positive. By Theorem 2.4.33 T' is a predictable stopping time of the family $(\mathcal{F}_t)$ and $T' = T$ a.s.

Then an argument using monotone classes based on Remark 2.4.37 shows the following: for every predictable process $(X_t)_{t \in [0,\infty]}$ of the family $(\mathcal{F}_t^{\star})$, there exists a process $(X_t')_{t \in [0,\infty]}$ which is indistinguishable from it and is predictable with respect to $(\mathcal{F}_t)$.

Finally, let $A \in \mathcal{F}_{T-}^{\star}$ and choose (X_t) to be predictable such that $X_T = \mathbf{1}_A$ (Theorem 2.4.28(b)). Then (X_t') is indistinguishable of (X_t) as above; the required event A' is $\{X_{T'}' = 1\}$. ∎

2.4.6 Classification of stopping times

Let $(\mathcal{F}_t^{\star})$ satisfy the usual conditions. Let us introduce some additional notation; given a stopping time T, $\mathcal{S}(T)$ denotes the set of increasing sequences (S_n) of stopping times bounded by T. For every sequence $(S_n) \in \mathcal{S}$, we write

$$A[(S_n)] = \{\lim_n S_n = T, \ S_n < T \text{ for all } n\} \cup \{T = 0\}. \tag{2.72}$$

If we forget about $\{T = 0\}$ (recall that 0 plays the devil in this theory), $A[(S_n)]$ is the set on which the sequence (S_n) foretells T. We note that $A[(S_n)]$ belongs to $\mathcal{F}_{T-}^{\star}$: let $S = \lim_n S_n$;

$A[(S_n)]$ is the union of $\{T = 0\}$, which is $\mathcal{F}_{T-}^{\star}$-measurable by Theorem 2.4.15(a), and of the set $(\cap_n\{S_n < T\})\backslash\{S < T\}$, which is $\mathcal{F}_{T-}^{\star}$-measurable by Theorem 2.4.15(a). We denote by $A(T)$ a representative of the essential union of the sets $A[(S_n)]$ and by $I(T)$ the complement of $A(T)$. Note that $A(T)$ always contains $\{T = 0\}$ and $\{T = +\infty\}$ a.s.

Definition 2.4.43 T is said to be *accessible* if $A(T) = \Omega$ a.s. and *totally inaccessible* if $A(T) = \{T = +\infty\}$ a.s.

In other words, T is totally inaccessible if $T > 0$ a.s. and if for every sequence (T_n) of stopping times that increases to T the event $\{\lim_n T_n = T, T < \infty\}$ has zero probability. This means that one cannot localize T exactly using a sequence of "precursory signs". Hence, totally inaccessible is just the opposite of "predictable". However, there is an important difference whereas the notion of a predictable stopping time can be defined without a probability law, that of a totally inaccessible stopping time is relative to a given law $\mathbf{P}$.

It is also necessary to understand the difference between accessible and predictable: if T is accessible, Ω is a.s. the union of sets A_k, on each of which, T is foretold by some sequence (S_n^k). But outside A_k the sequence (S_n^k) may behave badly and it is impossible – if T is not predictable – to rearrange all these sequences into a single one which foretells T a.s. on the whole of Ω.

We shall see in Section 2.5 examples of totally inaccessible stopping times and of non-predictable accessible times. The following theorem contains the essential results on the classification of stopping times. For more detail and in particular for the study of accessible times, see Dellacherie [28].

Theorem 2.4.44 (a) T is accessible if and only if $[\![T]\!]$ is a.s. contained in a countable union of graphs of predictable stopping times.

(b) T is totally inaccessible, if and only if $\mathbf{P}\{S = T < \infty\} = 0$ for every predictable stopping time S.

(c) The stopping time $T_{A(T)}$ is accessible and the stopping time $T_{I(T)}$ totally inaccessible. This decomposition is unique in the following sense: if U and V are two stopping times with U accessible, V totally inaccessible, $U \wedge V = T$ and $U \vee V = +\infty$, then $U = T_{A(T)}$ and $V = T_{I(T)}$ a.s.

Proof. (a) Suppose that T is accessible. There then exist sequences $(S_n^k)_{n\in\mathbb{N}}$ belonging to $\mathcal{S}(T)$ such that Ω is a.s. the union of the sets $A[(S_n^k)]$. For every k we set $S^k = \lim_n S_n^k$ and then

$$R_n^k = (S_n^k)_{\{S_n^k < S^k\}} \wedge n.$$

We thus define an increasing sequence of stopping times which foretells its limit R^k everywhere on Ω, and $[\![T]\!]$ is a.s. contained in the union of the graphs of the predictable stopping times R^k and 0.

Conversely, suppose that $[\![T]\!]$ is a.s. contained in the union of the graphs of a sequence of predictable stopping times T^k. For each k let (T_n^k) be a sequence which foretells T^k and let $S_n^k = T_n^k \wedge T$; then the sequence (S_n^k) belongs to $\mathcal{S}(T)$, the set $\{T = T^k < \infty\}$ is contained in $A[(S_n^k)]$ and hence T is accessible.

(b) Let S be a predictable time, (S_n) be a sequence foretelling S and let $S'_n = S_n \wedge T$; the sequence (S'_n) belongs to $(\mathcal{S}(T)$ and $\{S = T\} \subset A[(S'_n)]$; since T is totally inaccessible, $A[(S'_n)]$ is a.s. contained in $\{T = \infty\}$ and $\mathbf{P}\{S = T < \infty\} = 0$.

Conversely, suppose that $\mathbf{P}\{R = T < \infty\} = 0$ for every predictable time R. Then in particular $\mathbf{P}[T = 0] = 0$. Let (S_n) be a sequence belonging to $\mathcal{S}(T)$. As in part (a), we set $S = \lim_n S_n$, $R_n = (S_n)_{\{S_n < S\}} \wedge n$ and $R = \lim_n R_n$; R is predictable hence $\mathbf{P}\{R = T < \infty\} = 0$ and this means that $A[(S_n)] \subset \{T = \infty\}$ a.s.; hence T is totally inaccessible.

(c) The argument in (a) shows that $[\![T_{A(T)}]\!]$ is contained in a union of graphs of predictable times and hence $T_{A(T)}$ is accessible. The first part of the argument in (b) shows that, if S is a predictable time, the set $\{S = T < \infty\}$ is a.s. contained in $A(T)$, so that $\mathbf{P}\{T_{I(T)} = S < \infty\} = 0$ and $T_{I(T)}$ is totally inaccessible by (b). We leave uniqueness to the reader. ∎

2.4.7 Quasi-left-continuous filtrations

The left-continuity condition on filtration, $\mathcal{F}_t = \mathcal{F}_{t-}$ for all t, is not sufficient to lead to interesting results. If the same condition holds not only at constant times, but also, at all predictable times, a situation which happens frequently, the classification of stopping times is simplified.

Definition 2.4.45 Let $(\mathcal{F}_t^\star)$ be a filtration satisfying the usual conditions. $(\mathcal{F}_t^\star)$ is said to be *quasi-left-continuous* if $\mathcal{F}_T^\star = \mathcal{F}_{T-}^\star$ for all predictable times T and, in particular, for $T = 0$ and $T = t$.

The quasi in quasi-left-continuous does *not* imply the existence of a stronger notion of a left-continuous family of σ-fields. In particular, equality for all t of the σ-fields $\mathcal{F}_t^\star$ and $\mathcal{F}_{t-}^\star$ is a weaker property than quasi-left-continuity and must not be called left-continuity. Moreover, there exist quasi-left-continuous families such that the equality $\mathcal{F}_{T-}^\star = \mathcal{F}_T^\star$ does not hold at some (non-predictable) stopping times T.

Theorem 2.4.46 Suppose that $(\mathcal{F}_t^\star)$ is quasi-left-continuous. Then,
(a) every accessible stopping time is predictable,
(b) for every increasing sequence (T_n) of stopping times, with $\lim_n T_n = T$, we have

$$\mathcal{F}_T^\star = \vee_n \mathcal{F}_{T_n}^\star$$

Proof. (a) Let T be an accessible stopping time; there exists a sequence of graphs of predictable stopping times R_n whose union contains T. For all n, the set $\{R_n = T\}$ belongs to $\mathcal{F}_{R_n}^\star$, which is equal to $\mathcal{F}_{R_n-}^\star$; we then write $S_n = (R_n)_{\{R_n = T\}}$: S_n is predictable by Theorem 2.4.36 (c). Then, T is the limit of the stationary decreasing sequence $T_n = S_1 \wedge ... \wedge S_n$ of predictable times; hence it is predictable (Theorem 2.4.36 (d)).

(b) Let $H = \{T_n < T$ for all $n\} \in \mathcal{F}_T^\star$ and let $S = T_H$; S is predictable – it is foretold by the sequence $n \wedge S_n$, where $S_n = (T_n)_{\{T_n < T\}}$. Let $A \in \mathcal{F}_T^\star$; we show separately that $A \cap H$ and $A \cap H^c$ belong to $\vee \mathcal{F}_{T_n}^\star$:

1. $A \cap H^c$ is the union of the sets $A \cap \{T \leq T_n\} \in \mathcal{F}_{T_n}^\star$ (2.53).
2. $A \cap H$ belongs to $\mathcal{F}_T^\star$ and hence to $\mathcal{F}_S^\star$ since $T \leq S$. As $\mathcal{F}_S^\star = \mathcal{F}_{S-}^\star$ by hypothesis, $A \cap H = A \cap H \cap \{S \leq T\}$ belongs to $\mathcal{F}_{T-}^\star$ by Theorem 2.4.36 (c) and hence to $\vee_n \mathcal{F}_{T_n-}^\star$ by Theorem 2.4.15 (d). Then it belongs a fortiori to $\vee_n \mathcal{F}_{T_n}^\star$. ∎

Remark 2.4.47 Conversely, if every accessible time is predictable, the filtration is quasi-left-continuous. For if T is predictable and A belongs to $\mathcal{F}_T^\star$, T_A is accessible (Theorem 2.4.44) hence predictable and $A \in \mathcal{F}_{T-}^\star$ (Theorem 2.4.36).

2.4.8 Optional and predictable cross-sections

We come to the most important theorems in this chapter: the optional cross-section and the predictable cross-section theorems. The proofs of these theorems are derived from the ordinary non-filtered cross-section theorem which is an application of Choquet's Theorem.

Theorem 2.4.48 Let $(\Omega, \mathcal{F}, \mathbf{P})$ be a complete probability space with an arbitrary filtration $(\mathcal{F}_t)_{t \geq 0}$ and let A be an optional set. For every $\epsilon > 0$ there exists a stopping time T of the family $(\mathcal{F}_t)$ with the following properties:
(a) for all ω such that $T(\omega) < \infty$, $(T(\omega), \omega) \in A$;
(b) $\mathbf{P}(T < \infty) \geq \mathbf{P}(\pi(A)) - \epsilon$, where $\pi(A)$ is the projection of A onto Ω.

Proof. (a) Using Theorem 1.3.33, we choose a measurable cross-section of A, i.e., an $\mathcal{F}$-measurable real-valued r.v. R (in general not a stopping time) such that

$$R(\omega) < \infty \Longrightarrow (R(\omega), \omega) \in A$$
$$\mathbf{P}(R < \infty) = \mathbf{P}(\pi(A))$$

and we use it to construct a measure μ on $\mathbb{R}_+ \times \Omega$: if G is an element of $\mathcal{B}(\mathbb{R}_+) \times \mathcal{F}$

$$\mu(G) = \int \mathbf{1}_G(R(\omega), \omega) \mathbf{1}_{(R < \infty)}(\omega) \mathbf{P}(d\omega).$$

This measure is carried by A and its mass is equal to $\mathbf{P}(\pi(A))$.

(b) We now copy the proof of Theorem 2.4.39, with $\mathcal{U}$, this time, denoting the set of all stopping times of $\mathcal{F}_t$, ϑ_0 the paving consisting of the intervals $[\![S, T[\![$, with $S \leq T$ and $S, T \in \mathcal{U}$, and ϑ the closure of ϑ_0 under $(\cup f)$. As in Theorem 2.4.39, ϑ is a Boolean algebra which generates the optional σ-field (Theorem 2.4.25) and the debut of an element B of ϑ_δ is a.s. equal to an element of $\mathcal{U}$. Again as in Theorem 2.4.39, by a classical theorem of measure theory, there exists a set $B \in \vartheta_\delta$ contained in A such that $\mu(B) \geq \mu(A) - \epsilon$. We denote by S an element of which is a.s. equal to the debut of B. Because of the "a.s." the graph of S does not pass through B everywhere, so that we take $T = S_L$ where $L = \{\omega : (S(\omega), \omega) \in B\}$, which belongs to $\mathcal{F}_S$ by Theorem 2.4.25. T is then the required stopping time. ∎

Remark 2.4.49 Completeness enables us to write $\mathbf{P}$ here instead of $\mathbf{P}^*$.

Remark 2.4.50 Some authors often say "section theorems" and in topology it is called "selection theorems" or "uniformization theorems".

The same argument gives us the predictable cross-section theorem.

Theorem 2.4.51 Let $(\Omega, \mathcal{F}, \mathbf{P})$ be a complete probability space with an arbitrary filtration $(\mathcal{F}_t)$ and let A be a predictable set. For every number $\epsilon > 0$ there exists a predictable stopping time T of the family $(\mathcal{F}_t)$ with the following properties:
  (a) for all ω such that $T(\omega) < \infty$, $(T(\omega), \omega) \in A$;
  (b) $\mathbf{P}(T < \infty) \geq \mathbf{P}(\pi(A)) - \epsilon$, where $\pi(A)$ is the projection of A onto Ω.

Proof. We construct a measure μ carried by A of mass $\mathbf{P}(\pi(A))$, as in the first part of the preceding proof. Then, we, again, copy the proof of Theorem 2.4.39: $\mathcal{U}$ now denotes the set of all predictable stopping times, ϑ_0 is the set of intervals $[\![S, T[\![$, $(S, T \in \mathcal{U})$ and ϑ is the closure of ϑ_0 under $(\cup f)$, a Boolean algebra which generates the predictable σ-field (Theorem 2.4.36). As in Theorem 2.4.39, it can be verified that the debut of an element of ϑ_δ is a.s. equal to an element of $\mathcal{U}$: the proof of Theorem 2.4.39 is unchanged, as it depends only on the properties (a), (b), (c) and (d) of Lemma 2.4.38, which are satisfied by predictable stopping times (Theorem 2.4.36). Again we choose $B \in \vartheta_\delta$ contained in A and such that $\mu(B) \geq \mu(A) - \epsilon$, we denote by S a predictable time a.s. equal to D_B and we get the required predictable time by taking $T = S_L$ where $L = \{\mu : (S(\omega), \omega) \in B\}$ belongs to $\mathcal{F}_{S-}$ (Theorem 2.4.36). ∎

The following statement is the most frequent application of the cross-section theorems.

Theorem 2.4.52 Let (X_t) and (Y_t) be two optional (resp. predictable) processes. Suppose that for every stopping time (resp. predictable stopping time) T, $X_T = Y_T$ a.s. on $\{T < \infty\}$. Then, the two processes are *indistinguishable*.

Proof. We deal only with the optional case. It is sufficient to show that for all $\epsilon > 0$ the optional set

$$A = \{(t, \omega) : X_t(\omega) > Y_t(\omega) + \epsilon\}$$

is evanescent. But if it were not, it would admit a cross-section by a stopping time T such that $\mathbf{P}\{T < \infty\} > 0$, contradicting the hypothesis. ∎

Remark 2.4.53 (a) This statement extends immediately to processes with values in a separable metrizable space E (consider the real processes $(f \circ X_t)$ and $(f \circ Y_t)$, where f runs through a countable set of Borel functions generating the σ-field $\mathcal{B}(E)$).
  (b) Suppose that for every stopping time (resp. predictable stopping time) T the random variables $X_T \mathbf{1}_{T<\infty}$ and $Y_T \mathbf{1}_{T<\infty}$ are integrable and have the same expectation. Then, the same proof will show that the two processes are indistinguishable.

(c) Let S be a positive r.v. Then, S is a stopping time (resp. predictable stopping time) if (and only if) $[\![S]\!]$ is an optional (resp. predictable) set. Indeed, there then exists an optional (resp. predictable) time T_n which is a cross-section of $[\![S]\!]$, such that $\mathbf{P}\{S \neq T_n\} < 1/n$ and then $[\![S, \infty[\![$ is indistinguishable from $\cup_n [\![T_n, \infty[\![$ which is optional (resp. predictable).

(d) Let H be a predictable set. If the graph of the debut D_H is contained in H (in particular if H is right-closed), D_H is predictable. For then $[\![D_H]\!] = H \backslash [\![D_H, \infty[\![$ which is a predictable set, and we apply the preceding remark.

Another excellent exercise of cross-section theorems is to make precise the structure of some measurable sets with countable sections that often occur as exceptional sets when modifying processes (cf. Theorem 2.4.27, Theorem 2.4.59 below, and Remark 2.5.3(a)).

Theorem 2.4.54 Let $(\Omega, \mathcal{F}, \mathbf{P})$ be a complete probability space with an arbitrary filtration $(\mathcal{F}_t)$.

(a) A progressive set contained in a countable union of graphs of optional times is the union of a sequence of disjoint graphs of optional times, and is therefore optional.

(b) A predictable set contained in a countable union of graphs of predictable (resp. optional) times is equal to (resp. indistinguishable from) the union of a sequence of disjoint graphs of predictable times.

(c) Let H be a subset of $\overline{\mathbb{R}}_+ \times \Omega$ contained in a countable union of graphs of positive random variables. Then, H has an essentially unique (i.e., unique up to indistinguishability) decomposition

$$H = H' \cup H''$$

where H' and H'' are disjoint measurable sets, H' is contained in a countable union of graphs of optional times and H'' intersects every graph of an optional time along an evanescent set. Further, if H is optional (resp. predictable), then H is indistinguishable from the union of a sequence of (disjoint) graphs of optional (resp. predictable) times.

Proof. Let L be a subset of $\mathbb{R}_+ \times \Omega$ contained in the union of the graphs of a sequence (S_n) of positive r.v. We set $T_1 = S_1$ and, for $n \geq 2$, $T_n = S_n$ on $\{S_1 \neq S_n, ..., S_{n-1} \neq S_n\}$ and $T_n = +\infty$ otherwise: the T_n are positive r.v. with disjoint graphs and L is contained in the union of these graphs. If the S_n are optional (resp. predictable) times the T_n are also optional (resp. predictable) times by Theorems 2.4.12 and 2.4.15(c) (resp. Theorem 2.4.36(b) and (c)); if, further, L is progressive (resp. predictable) then, for each n, $[\![T_n]\!] \cap L$ is the graph of an optional (resp. predictable) time. The progressive case has been shown earlier (Theorem 2.4.27); the predictable case can be treated similarly: the indicator (X_t) of L then is indeed a predictable process and $A_n = \{X_{T_n} = 1, T_n < \infty\}$ belongs to $\mathcal{F}_{T-}$ by Theorem 2.4.28(b), so $[\![T_n]\!] \cap L$ is the graph of the predictable time equal to T_n on A_n and to $+\infty$ on A_n^c. We have thus established (a) and part of (b). The other part of (b) follows from (c) which we now prove.

Let $\mathcal{U}$ be the set of optional (resp. predictable) times and for every positive r.v. Z let $V(Z)$ denote a representative of the essential union of the sets $\{Z = T\}$, where T runs through $\mathcal{U}$. We define two positive r.v. Z' and Z'' by $Z' = Z$ on $V(Z)$, $Z' = +\infty$ on $V(Z)^c$ and $Z'' = Z$ on $V(Z)^c$, $Z'' = +\infty$ on $V(Z)$. Paraphrasing Theorem 2.4.44, we could call Z': the $\mathcal{U}$-accessible part of Z and Z'' the $\mathcal{U}$-totally inaccessible part of Z: the graph of Z' is contained in a countable union of graphs of elements of $\mathcal{U}$ and $\mathbf{P}\{Z'' = T < \infty\} = 0$ for all $T \in \mathcal{U}$.

Now, let H be a subset of $\mathbb{R}_+ \times \Omega$ contained in the union of the graphs of a sequence (Z_n) of positive r.v. By the above, with $\mathcal{U}$ the set of optional times there exists a sequence (T_n) of optional times such that $\cup_n [\![Z'_n]\!]$ is contained in $\cup_n [\![T_n]\!]$, and we have seen earlier that the graphs $[\![T_n]\!]$ can be assumed to be disjoint. We then write

$$
\begin{aligned}
H' &= H \cap_n (\cup [\![Z'_n]\!]) = H \cap (\cup_n [\![T_n]\!]), \\
H'' &= H \cap_n (\cup [\![Z''_n]\!]) = H \backslash (\cup_n [\![T_n]\!]).
\end{aligned}
$$

The sets H' and H'' constitute a decomposition of H with the properties required by (c) and clearly this decomposition is unique up to indistinguishability. If H is optional, so are H' and H''; then H'' is evanescent by the cross-section Theorem 2.4.48 and H is indistinguishable from H', which is a countable union of disjoint graphs of optional times by (a). If H is predictable, we return to the decomposition of H with $\mathcal{U}$ this time the set of predictable times. We get another decomposition of H where H' is predictable and is contained in a countable union of graphs of predictable times, and H'' is predictable and intersects every graph of a predictable time along an evanescent set: we then conclude the proof using the cross-section Theorem 2.4.51 and the first part of (b). ∎

Remark 2.4.55 If the filtration $(\mathcal{F}_t)$ is complete, then every function which is positive on Ω and zero $\mathbf{P}$ a.s. is a predictable time. Thus, it is easily deduced in (c) that if the set H is optional (resp. predictable), then it is the union of a sequence of disjoint graphs of optional (resp. predictable) times.

Remark 2.4.56 The above proof says more than the statement of the theorem in assertion (c) that H also has a decomposition relative to predictable times.

Remark 2.4.57 We shall see in Section 2.5 (Remark 2.5.3(b)) that there exist a progressive set contained in a countable union of graphs of (non-optional) random variables, which is not optional and even contains no graph of a stopping time.

Remark 2.4.58 A corollary result to the above theorem that we shall leave to the reader is that if H is a measurable subset of $\mathbb{R}_+ \times \Omega$, such that the section $H(\omega)$ is at most countable for all $\omega \in \Omega$, then H is contained in a countable union of graphs of positive random variables.

The following results in Theorems 2.4.59, 2.4.61 and 2.4.62 are complements to Theorem 2.4.54. The first and last ones are concerned with the explicit representations of some sets with countable sections as unions of graphs of stopping times. The intermediate is a consequence of Theorem 2.4.59, a very convenient criterion for predictability of an optional process.

Theorem 2.4.59 Let (X_t) be a real-valued, adapted r.c.l.l. process. We make the convention $X_{0-} = X_0$ and set

$$
U = \{(t, \omega) : X_t(\omega) \neq X_{t-}(\omega)\}
$$

Then, U is a countable union of disjoint graphs of stopping times, which may be chosen predictable if (X_t) is predictable.

Proof. Let

$$U_n = \{(t,\omega) : |X_t(\omega) - X_{t-}(\omega)| > 2^{-n}\}(n \geq 0),$$

then $V_0 = U_0$, $V_n = U_n \backslash U_{n-1}$ $(n > 0)$; the sets V_n are optional (predictable if so is (X_t)) and disjoint. Next we set

$$
\begin{aligned}
D_n^1(\omega) &= \inf\{t : (t,\omega) \in V_n\}, \\
D_n^{k+1}(\omega) &= \inf\{t > D_n^k(\omega) : (t,\omega) \in V_n\}
\end{aligned}
$$

so that D_n^i is the i-th jump of (X_t) whose size lies between 2^{-n} and 2^{-n+1}. Since (X_t) has r.c.l.l. paths, V_n has no finite cluster point, and the stopping times D_n^i enumerate all points of V_n. If (X_t) is predictable, then the D_n^i are predictable according to Remark 2.4.53(d). Finally, it only remains to reorder the double sequence D_n^i into an ordinary sequence (T_n). ∎

Remark 2.4.60 The conclusion of the theorem still holds for a process taking values in a metrizable separable space E; one just embeds E into $[0,1]^{\mathbb{N}}$ and applies the statement to each coordinate process, then the procedure at the beginning of the proof of Theorem 2.4.54 to turn the stopping times into disjoint ones.

The same remark with the same argument applies to the following application of Theorem 2.4.59:

Theorem 2.4.61 Let $X = (X_t)$ be a real-valued, adapted r.c.l.l. process. Then, X is predictable if and only if the following two conditions are satisfied:

(1) for every totally inaccessible stopping time T, X_T and X_{T-} are a.s. equal on $\{T < \infty\}$,

(2) for every predictable time T, X_T is $\mathcal{F}_{T-}$ measurable on $\{T < \infty\}$.

Proof. Assume X is predictable. Then (2) is satisfied for every (optional) T according to Theorem 2.4.28. On the other hand, from Theorem 2.4.59 the set $U = \{(t,\omega) : X_t(\omega) \neq X_{t-}(\omega)\}$ is the union of countably many graphs of predictable times, whence (1) follows at once.

Conversely, assume (1) and (2) are satisfied. We represent U as a countable union of graphs of optional times S_n, then decompose S_n into its totally inaccessible part S_n^i and its accessible part S_n^a (Theorem 2.4.44(c)). According to (1) S_n^i is a.s. equal to $+\infty$, so that $S_n = S_n^a$ is accessible. Then the graph of S_n is contained in a countable union of graphs of predictable times S_{nk} $(k \in \mathbb{N})$ by virtue of Theorem 2.4.44(a). Let V be the union of the graphs $[\![S_{nk}]\!]$. From Theorem 2.4.54 we may represent V as a union $[\![T_m]\!]$ of disjoint graphs of predictable times.

For every m, X_{T_m} and X_{T_m-} both are $\mathcal{F}_{T_m-}$ measurable: the first one according to (2), the second one from Theorem 2.4.28. So the same is true of $\triangle X_{T_m} = X_{T_m} - X_{T_m-}$. According to Theorem 2.4.28 there exists a predictable process (Y_t^m) such that $Y_{T_m}^m = \triangle X_{T_m}$ on $\{T_m < \infty\}$. On the other hand, the graph $[\![T_m]\!]$ is predictable. Denoting by X_- the process $(X_{t-})_{t \geq 0}$ $(X_{0-} = X_0)$, which is predictable by left-continuity, we have

$$X = X_- + \sum_m Y^m \mathbf{1}_{[\![T_m]\!]}$$

and X is therefore predictable. ∎

The following result looks like Theorem 2.4.59 but for r.c. instead of r.c.l.l. processes; however, it is less useful.

Theorem 2.4.62 Let (X_t) be real-valued, adapted and right-continuous. We make the convention that $X_{0-} = X_0$ and set

$$U = \{(t, \omega) : X_{t-} \text{ doesn't exist or } X_{t-}(\omega) \neq X_t(\omega)\}.$$

Then, U is a countable union of disjoint graphs of stopping times, which may be chosen predictable if (X_t) is predictable.

Proof. We prove the predictable case, the optional case can be done in a similar way. First, we show that U is predictable. To this end, we define

$$Y_t^+ = \limsup_{s \uparrow\uparrow t} X_s, \quad Y_t^- = \liminf_{s \uparrow\uparrow t} X_s$$

which are predictable. Then, U is the union of the predictable sets $\{Y^+ \notin X\}$, $\{Y^- \neq X\}$.

To conclude, we need to prove (according to Theorem 2.4.54) that U is contained (up to evanescent sets) in a countable union of graphs of positive random variables. Such r.v.s (which turn out to be optional, but not necessarily predictable) are constructed as follows: we choose $\epsilon > 0$ and set by transfinite induction

$$
\begin{aligned}
T_0^\epsilon &= 0, \quad T_{\alpha+1}^\epsilon = \inf\left\{t > T_\alpha^\epsilon : \left|X_t - X_{T_\alpha^\epsilon}\right| > \epsilon\right\}, \\
T_\beta^\epsilon &= \sup_{\alpha < \beta} T_\alpha^\epsilon \text{ if } \beta \text{ is a limit ordinal.}
\end{aligned}
$$

Since X is right-continuous, we have $T_\alpha^\epsilon < T_{\alpha+1}^\epsilon$ on the set $\{T_\alpha^\epsilon < \infty\}$. There exists a countable ordinal γ_ϵ such that $T_{\gamma_\epsilon}^\epsilon = +\infty$ a.s. Then, U is contained in the union of all graphs $[\![T_\alpha^\epsilon]\!]$ for $\epsilon = 1/n$ ($n \in \mathbb{N}$) and $\alpha < \gamma_\epsilon$ times t where the left-hand limit does't exist appear among all T_β^ϵ corresponding to limit ordinals, if ϵ is small enough, while jump times occur among all $T_{\alpha+1}^\epsilon$ for ϵ small enough. ∎

Remark 2.4.63 The result extends to processes with values in a metrizable separable space E, but the argument is slightly more delicate. We embed E into $F = [0,1]^{\mathbb{N}}$, and remark that the result is trivial when X is considered as a F-valued process, but that the left limit may exist in F without existing in E. Then we decompose U into

$$
\begin{aligned}
U &= \{X_{t-} \text{ doesn't exist in } E\} \cup \{X_{t-} \text{ exists in } E \text{ and } X_{t-} \neq X_t\} \\
&= \{X_{t-} \text{ doesn't exist in } F\} \cup \{X_{t-} \text{ exists in } F \text{ and } X \notin E\} \cup \{X_{t-} \text{ exists in } E \text{ and } X_{t-} \neq X_t\} \\
&= \{X_{t-} \text{ doesn't exist in } F\} \cup \{X_{t-} \text{ exists in } F \text{ and } X_{t-} \neq X_t\}
\end{aligned}
$$

since (X_t) is an E-valued process. So we are really reduced to the same problem about F, which was shown above to be trivial.

2.5 Optional and Predictable Processes

We return to results in Section 2.2 on measurability of stochastic processes, intending now to establish that processes are optional or predictable instead of just progressive. The essential result is the following one where we establish two statements one for sets and one for processes, but they are really the same theorem.

Theorem 2.5.1 Let $(\Omega, \mathcal{F}, \mathbf{P})$ be a complete probability space with a filtration $(\mathcal{F}_t^\star)$ which satisfies the usual conditions (with the convention $\mathcal{F}_{0-}^\star = \mathcal{F}_0^\star$). Let L be a progressive random set. Then,
   (a) the set L_1 of left accumulation points of L is predictable on $]0, \infty]$;
   (b) the closure $\bar{L}$ of L is optional;
   (c) the set L_2 of right accumulation points of L is progressive on $[0, \infty[$.

Theorem 2.5.2 Let $(X_t)_{t \in \mathbb{R}_+}$ be a real-valued, progressive process. Then, with the same hypotheses stated above,
   (a) the process $U_t = \limsup\limits_{s \uparrow\uparrow t} X_s$ is predictable on $]0, \infty]$;
   (b) the process $V_t = \limsup\limits_{s \to t} X_s$ is optional on $[0, \infty[$;
   (c) the process $W_t = \limsup\limits_{s \downarrow\downarrow t} X_s$ is progressive on $[0, \infty[$.

Proof. Assertions (c) of both statements are repetitions of Theorems 2.3.21-2.3.22. We first prove assertions (a) and (b) about sets. We set $A_t(\omega) = \sup\{s < t : (s, \omega) \in L\}$ ($t > 0$; with the convention $\sup \emptyset = 0$). Since the σ-fields $(\mathcal{F}_t^\star)$ are complete, and analyticity argument which we have often used, in Theorem 2.4.6 for example, shows that A_t is $\mathcal{F}_t^\star$-measurable. On the other hand, the paths $A.(\omega)$ are left-continuous and increasing. It first follows that the process $(A_t)_{t>0}$ is predictable. Then, the right-hand limits (A_{t+}) exist everywhere and constitute a r.c.l.l. process adapted to the family $(\mathcal{F}_{t+}^\star) = (\mathcal{F}_t^\star)$ – and hence an optional process. We conclude the proof by noting that

$$L_1 = \{(t, \omega) : 0 < t < \infty, \ A_t(\omega) = t\} \tag{2.73}$$
$$\bar{L} = \{(t, \omega) : 0 < t < \omega, \ A_{t+}(\omega) = t\} \cup [\![D_L]\!]; \tag{2.74}$$

where D_L takes care of what happens for $t = 0$. We now pass easily from sets to processes: $U_t(\omega) \geq a$ if and only if for all $\epsilon > 0$, t is all left accumulation point of the set $L_\epsilon = \{(s, \omega) : X_s(\omega) > a - \epsilon\}$. The argument for (V_t) is analogous. $\blacksquare$

Remark 2.5.3 (a) Here are some supplements to the above statement: (1) If (X_t) is predictable, the process

$$U_t' = \limsup\limits_{s \uparrow t} X_s$$

is predictable ($U'_t = U_t \vee X_t$ for $t > 0$ and $U'_0 = X_0$). By way of symmetry, we note that if (X_t) is progressive, the process

$$W'_t = \limsup_{s \downarrow t} X_s = W_t \vee X_t$$

is progressive. This is of no interest, unlike the assertion about predictable processes; (2) if (X_t) is optional the process

$$V'_t = \lim_{s \to t, s \neq t} \sup X_s$$

is optional. For $V'_t = U_t \vee W_t$ and

$$V'_t = \limsup_{s \to t} V'_s.$$

From the first equality, it follows that (V'_t) is progressive and from the second that it is then optional. Note incidentally that the set $\{V \neq V'\}$, which is optional, is contained in the union of the graphs of stopping times

$$D^{a,b}_r = \inf\{t \geq r : V'_t \leq a < b \leq V_t\}$$

where a, b run through all rationals and r through the positive rationals. This follows easily from the fact that V' is bounded above by V and that for fixed ω, a and b the set $\{t : V'_t(\omega) \leq a < b \leq V_t(\omega)\}$ is discrete. It then follows from Theorem 2.4.54 that $\{V \neq V'\}$ is the union of a sequence of disjoint graphs of stopping times.

(b) It is quite natural to ask whether statements (c) in the above theorems can be improved. we show that this is impossible, even when (X_t) is a predictable process. Let (B_t) be a Brownian motion starting from 0, with continuous paths, and $(\mathcal{F}^*_t)$ the filtration $(\sigma(B_s, s \leq t))$ suitably augmented. Let (X_t) be the indicator of the random set $M = \{(t, \omega) : B_t(\omega) = 0\}$ since (B_t) is continuous hence predictable, the set M is closed and predictable – it can be shown that it is a.s. a perfect set without interior points. The process (W_t) then is the indicator of the set L of points of M which are not isolated from the right and the set $K = V \backslash L$ is the indicator of the set of points of M which are isolated from the right. K is therefore a progressive set which is discrete under the right topology and has countable sections (it is very easy to represent K explicitly as a union of a sequence of graphs).

Now, let T be a stopping time whose graph passes through K; its graph also passes through $M' = \{t : B_t = 0\}$ and it follows from the strong Markov property of (B_t) that T is a.s. a right accumulation point of M' on the set $\{T < \infty\}$. In other words, by the definition of K, every stopping time whose graph passes through K is a.s. infinite. Since K is not evanescent, it follows from Theorem 2.4.48 that K is not optional and hence $L = M \backslash K$ is progressive and not optional and (W_t) is not optional although (X_t) is predictable.

There exists a progressive set with a.s. uncountable sections, and still containing no graph of a stopping time: for example, the set

$$H = \left\{ (t, \omega) : \limsup_{h \downarrow \downarrow 0} \frac{|B_{t+h}(\omega) - B_t(\omega)|}{\sqrt{2 \log \log \frac{1}{h}}} < 1 \right\}$$

see Knight [29], and for another example see Dellacherie [30].

Now, under the usual conditions, we make Theorems 2.3.1 and 2.3.28 more precise but without giving proofs.

Theorem 2.5.4 Let D be a countable dense subset of $\mathbb{R}_+$ and let $(X_t)_{t\in D}$ be a real-valued process defined on D such that X_t is $\mathcal{F}_t$-measurable for all $t \in D$. Then
    (a) the process
$$U_t = \limsup_{s\uparrow\uparrow t, s\in D} X_s$$

is predictable on $]0, \infty]$;
    (b) the process
$$V_t = \limsup_{s\to t, s\in D} X_s$$

is optional on $[0, \infty[$.

Theorem 2.5.5 Let $(X_t)_{t\in\mathbb{R}_+}$ to a real progressive process. Then,
    (a) the process
$$U_t = ess\limsup_{s\uparrow\uparrow t} X_s$$

is predictable on $]0, \infty]$;
    (b) the process
$$V_t = ess\limsup_{s\to t} X_s$$

is optional on $[0, \infty[$.

These two theorems above reduce to Theorem 2.5.2; to prove Theorem 2.5.4, Theorem 2.5.2 is applied to the progressive process (Z_t) where $Z_t = X_t$ for $t \in D$ and $Z_t = -\infty$ for $t \notin D$; for Theorem 2.5.5, we known by Theorem 2.3.28 that the processes concerned are progressive, and we apply Theorem 2.5.2 to them. But in both cases assertions (a), and this is important, can also be shown true, directly, even without the usual conditions on the family of σ-fields.

Chapter 3

Martingales

In this chapter we learn about martingales – the most pervasive stochastic processes. We start with martingale theory in discrete time, then work our way to continuous time martingales. The chapter is classical and most of its results are familiar to probabilists and can be found in many stochastic processes and probability books. So, we are going to be economical in our presentation, simply stating results without proofs except for those propositions that we think their proofs are informative and would bring about a needed insight for other chapters of this book.

3.1 Discrete Parameter Martingales

In this section we cover theorems of martingale theory in discrete time: inequalities, stopping theorems and some convergence theorems.

We denote by Π a set with an order relation denoted by $\leq$ and we will initially work with this set. Then, we will restrict ourselves to the case where Π is an interval of the set of integers; whereas, in the continuous case Π will be set to $\mathbb{R}_+$. Note that the notions of an increasing family of σ-fields and an adapted process, in later sections, extend to arbitrary ordered sets in the obvious way.

Definition 3.1.1 Let $(\Omega, \mathcal{F}, \mathbf{P})$ be a probability space, $(\mathcal{F}_t)_{t \in \Pi}$ an increasing family of sub-σ-fields of $\mathcal{F}$ and $X = (X_t)_{t \in \not\geq}$ a real-valued process adapted to the family $(\mathcal{F}_t)$. X is called a martingale (resp. supermartingale, submartingale) with respect to the family $(\mathcal{F}_t)$ if
   (1) each random variable X_t is integrable;
   (2) for every ordered pair (s, t) of elements of Π such that $s \leq t$,

$$\mathbf{E}[X_t | \mathcal{F}_s] = X_s \text{ a.s. (resp. } \mathbf{E}[X_t | \mathcal{F}_s] \leq X_s, \geq X_s). \tag{3.1}$$

Remark 3.1.2 (a) Ville [31] introduced the notion of a martingale, and Snell [32] introduced that of a submartingale. However, it was Doob who proved almost all the fundamental results in martingale theory and used them in all of probability theory, in such a way that no probabilist can afford to ignore martingale theory.

(b) A process X is a submartingale if and only if $-X$ is a supermartingale. Therefore, we shall restrict ourselves to studying one of the two classes of processes – usually that of supermartingales, which is more frequently used in potential theory.

(c) A stochastic process X given without reference to a family of σ-fields is called a martingale (resp. supermartingale) if it satisfies Definition 3.1.1 with respect to its natural family of σ-fields $\mathcal{F}_t = \sigma(X_s, s \leq t)$.

(d) Definition 3.1.1 has a number of more or less interesting generalizations. The heart of martingale theory consists of results about real-valued processes indexed by the integers or the reals, defined on a probability space whose random variables are integrable and satisfy (3.1).

It is possible to relax one or an other of these hypotheses and get generalized theories, for example, the theory of vector-valued martingales; the theory of martingales whose time sets are not linearly ordered, such as, $\mathbb{R}_+^n$ and $\mathbb{N}_+^n$ and martingale theory over a σ-finite measured space.

Remark 3.1.3 An adapted positive process is called a generalized martingale (supermartingale, submartingale) if it satisfies the corresponding relation (3.1) without necessarily being integrable.

3.1.1   Basic properties

Property 3.1.4 Let X be a supermartingale of the family $(\mathcal{F}_t)$. If $s \leq t$, then $\mathbf{E}[X_t|\mathcal{F}_t] \leq X_s$ a.s.; for these two r.v.s to be equal, it is necessary and sufficient that they have the same expectation. Thus, X is a martingale if and only if the function $t \to \mathbf{E}[X_t]$ is constant.

Property 3.1.5 Let X and Y be two martingales of the same family of σ-fields. Then, obviously $aX + bY$ is a martingale. There is an analogous result for supermartingales if the coefficients a, b are positive. If X and Y are supermartingales, so is $X \wedge Y$.

Property 3.1.6 Let X be a martingale of the family $(\mathcal{F}_t)$ and let f be a concave function $\mathbb{R}$. Then, the process $Y_t = f \circ X_t$ a is a supermartingale, provided that the Y_t are integrable. The conclusion is the same if X is a supermartingale and f is concave and increasing. We consider for example the latter case. Let s and t be such that $s \leq t$. Then $\mathbf{E}[X_t|\mathcal{F}_s] \leq X_s$ and hence, using first the fact that f is increasing and then Jense inequality,

$$Y_s = f \circ X_s \geq f \circ \mathbf{E}[X_t|\mathcal{F}_s] \geq \mathbf{E}[f \circ X_t|\mathcal{F}_s] = \mathbf{E}[Y_t|\mathcal{F}_s].$$

A much used consequence: if (X_t) is a martingale and p is an exponent ≥ 1, the process $(|X_t|^p)$ is a submartingale, provided integrability holds as above.

There are many examples of martingales, but here we confine ourselves to a few quite elementary examples and one that is not.

Theorem 3.1.7 (a) Let $(\Omega, \mathcal{F}, \mathbf{P})$ be a probability space with an increasing family of σ-fields $(\mathcal{F}_t)_{t \in \Pi}$. Let Y be an integrable real r.v. Then, the process

$$Y_t = \mathbf{E}[Y|\mathcal{F}_t] \quad (t \in \Pi)$$

is a martingale.

(b) Let $(\Omega, \mathcal{F}, \mathbf{P})$ be a probability space. Let Π be the set of finite sub-σ-fields of $\mathcal{F}$ ordered by inclusion. Denote by $\mathcal{F}_t$, for $t \in \Pi$, the σ-field "t" itself (for convenience and to avoid confusion). Thus, for each $t \in \Pi$ (the σ-field $\mathcal{F}_t = t$) is generated by a finite partition P_t of Ω. Let $\mathbf{Q}$ be a positive additive set function defined on $\mathcal{F}$. We denote by X_t the following, obviously, $\mathcal{F}_t$-measurable function,

$$X_t(\omega) = \sum_{A \in P_t} \frac{\mathbf{Q}(A)}{\mathbf{P}(A)} \mathbf{1}_A(\omega) \quad \text{(with the convention that } 0/0 = 0\text{).} \tag{3.2}$$

Then, X_t is integrable if $\mathbf{Q}(\Omega) < \infty$. If $\mathbf{Q}(A) = 0$ for all A such that $\mathbf{P}(A) = 0$ and the process $(X_t)_{t \in \Pi}$ is a martingale; otherwise, it would only be a supermartingale.

(c) Let $X_1,..., X_n,...$ be integrable and independent random variables of zero expectation. Then, the sum process

$$S_n = X_1 + ... + X_n \tag{3.3}$$

is a martingale.

(d) Let $(B_t)_{t \in \Pi}$ be the process of Brownian motion in $\mathbb{R}^n$ starting from an arbitrary point x and let h be a real-valued function on $\mathbb{R}^n$. The real-valued process $X_t = h \circ B_t$ is, under suitable integrability conditions, a martingale if h is harmonic, a supermartingale if h is superharmonic, and a submartingale if h is a subharmonic.

3.1.2 Right and left closed supermartingales

Definition 3.1.8 Let $(X_t)_{t \in \Pi}$ be a martingale (resp. supermartingale) and suppose that Π has no largest element – if Π has one, what follows is of no interest. We say that (X_t) is *right closed* by a r.v. Y if Y is integrable and for all $t \in \Pi$

$$X_t = \mathbf{E}[Y|\mathcal{F}_t] \quad (\text{resp. } X_t \geq \mathbf{E}[Y|\mathcal{F}_t]). \tag{3.4}$$

Remark 3.1.9 (a) The above statement can be interpreted as follows: we adjoin a largest element denoted ∞ to Π and write $\bar{\Pi} = \Pi \cup \{\infty\}$. Then, we choose a σ-field $\mathcal{F}_\infty$ containing all the σ-fields $\mathcal{F}_t$, $t \in \Pi$, and with respect to which Y is measurable – the whole σ-field $\mathcal{F}$, for example. Thus, the relation (3.4) means that the process $(X_t)_{t \in \Pi}$ obtained by setting $X_\infty = Y$ is also a martingale (resp. supermartingale) with respect to $(\mathcal{F}_t)_{t \in \Pi}$.

(b) Since a martingale X is at the same time a supermartingale, it is necessary to say precisely whether Y closes X as a martingale or as a supermartingale. For example, if X is a positive martingale, the r.v. $Y = 0$ always closes X as a supermartingale, but the only case it closes X as a martingale is $X = 0$.

(c) We draw the reader's attention to the possibility of choosing different r.v. Y to close a martingale (or supermartingale) X on the right. We shall see later that in many cases a martingale indexed by $\mathbb{N}$ can be right closed using its a.s. limit $\xi = \lim_n X_n$. Given that in this chapter we denote by X_∞ every r.v. closing X on the right, it would be dangerous to denote this limit systematically by X_∞, as do many authors; we usually denote it by ξ as above, or by $X_{\infty-}$. Similarly, since there is a choice about the σ-field $\mathcal{F}_\infty$, we shall avoid using this to denote the σ-field $\vee_n \mathcal{F}_n$ which we shall generally denote by $\mathcal{F}_{\infty-}$.

3.1.3 Doob's stopping theorem

For the rest of this section we shall concern ourselves with the discrete case where Π is the set of integers: in principle, the set $\mathbb{Z} = \{0, 1, 2, ...\}$ – not $\mathbb{N}$ since $\mathbb{N}$ denoted $\{1, 2, 3, ...\}$. Unless otherwise stated, the processes are all defined on the same probability space $(\Omega, \mathcal{F}, \mathbf{P})$ with an increasing family of σ-fields $(\mathcal{F}_n)_{n \geq 0}$. A process $(V_n)_{n \geq 0}$ is called predictable if V_0 is $\mathcal{F}_0$-measurable and V_n is $\mathcal{F}_{n-1}$-measurable for $n \geq 0$ – this is the notion in the discrete case. If X is an adapted process and V a predictable process, we shall denote by $V \cdot X$ the process defined by

$$(V \cdot X_n) = V_0 X_0 + V_1(X_1 - X_0) + \cdots + V_n(X_n - X_{n-1}). \tag{3.5}$$

The process $V \cdot X$ is sometimes called the transform of X by V (cf. Burkholder's martingale transforms [33]). This very elementary notion is just the discrete form of the stochastic integral $\int V dX$. In particular, if we let T be a stopping time we can define, a predictable process V, by

$$V_n = \begin{cases} 1 & \text{if } n \leq T, \\ 0 & \text{if } n > T \end{cases}. \tag{3.6}$$

By analogy with the continuous case, this process is denoted by $\mathbf{1}_{[\![0,T]\!]}$. Then, the transform $V \cdot X$ is X stopped at time T denoted by X^T as in continuous time $(X_n^T = X_{T \wedge n})$.

Following these definitions we now give three results which express the fundamental property of martingales and supermartingales and deduce from it all, *Doob's inequalities*.

Theorem 3.1.10 Let $X = (X_n)_{n \geq 0}$ be a martingale and $V = (V_n)_{n \geq 0}$ a predictable process. If the r.v. $(V \cdot X)_n$ are integrable, $V \cdot X$ is a martingale. Similarly, if X is a supermartingale, V is positive and predictable and the r.v. $(V \cdot X)_n$ are integrable, $V \cdot X$ is a supermartingale.

Proof. E$[(V \cdot X)_{n+1} - (V \cdot X)_n | \mathcal{F}_n] = E[V_{n+1}(X_{n+1} - X_n)|\mathcal{F}_n] = V_{n+1} E[X_{n+1} - X_n | \mathcal{F}_n] = 0$ (≤ 0 in the second case). ∎

In particular, if V is given by 3.6, $(V \cdot X)_n \leq X_{T \wedge n}$ is integrable, we obtain:

Theorem 3.1.11 Let X be a martingale (supermartingale) and T a stopping time. Then, the stopped process X^T is a martingale (supermartingale).

The following stopping theorem is proved in the case of bounded stopping times. But in nos. 3.1.15-3.1.18 we shall give easy extensions.

Theorem 3.1.12 Let X be a martingale (supermartingale) and S and T two bounded stopping times such that $S \leq T$. Then, X_S and X_T are integrable and we have a.s.

$$X_S = \mathbf{E}[X_T|\mathcal{F}_S] \quad (\text{resp. } X_S \geq \mathbf{E}[X_T|\mathcal{F}_S]). \tag{3.7}$$

Proof. Since T is bounded, we may choose an integer $k \geq T$. Then, $|X_S| \leq |X_0| + ... + |X_k|$, hence, X_S is integrable. Let $A \in \mathcal{F}_S$; for all $j \leq k$, $A \cap \{S = j\} \in \mathcal{F}_j$ and

$$\int_{A\cap\{S=j\}} (X_k - X_S)\mathbf{P} = \int_{A\cap\{S=j\}} (X_k - X_j)\mathbf{P} = 0 \quad (\text{resp. } \leq 0). \tag{3.8}$$

Summing over j we obtain $\mathbf{E}[X_k - X_S|\mathcal{F}_S] = 0$ (resp. ≤ 0). To get (3.7), it is sufficient to apply this result to the stopped process X^T. ∎

Corollary 3.1.13 Let X be a supermartingale and T a stopping time bounded by k. Then,

$$\mathbf{E}[X_0] \geq \mathbf{E}[X_T] \geq \mathbf{E}[X_k] \tag{3.9}$$

$$\mathbf{E}[|X_T|] \leq \mathbf{E}[X_0] + 2\mathbf{E}[X_k^-] \leq 3\sup_{n\leq k} \mathbf{E}[|X_n|] \tag{3.10}$$

Proof. (3.9) is obvious: apply Theorem 3.1.12 to the stopping times 0, T, k. To establish (3.10), we write $\mathbf{E}[|X_T|] = \mathbf{E}[X_T] + 2\mathbf{E}[X_T^-]$; $\mathbf{E}[X_T]$ is less than $\mathbf{E}[X_0]$ by Theorem 3.1.12. On the other hand, the process $(X_n^+ = X_n \wedge 0)$ is a supermartingale, so that (X_n^-) is a submartingale and we have $\mathbf{E}[X_T^-] \leq \mathbf{E}[X_k^-]$ by Theorem 3.1.12. ∎

These theorems deserve some comments.

Remark 3.1.14 The r.v. X_n can be interpreted as a gambler's fortune at time n, and so, successive gains are given by the r.v. $x_n = X_n - X_{n-1}$ for $n \geq 1$. For example, X_1 is the gamble's wealth at the first step of game, whilst X_0 is his initial fortune, then his gain is $x_1 = X_1 - X_0$ – positive or negative.

The gambler is in a casino, where he may choose between all sorts of games, move from one table to another, and bet the way he wishes. Though, it is understood that he has no prophetic gifts, his decisions can only be taken as functions of the past and not the future, with the convention that the present, the game which is in the process of being played, forms part of the future.

The supermartingale inequality $\mathbf{E}[X_n|\mathcal{F}_{n-1}] \leq 0$ means that, whatever decisions are taken by the gambler just before the n-th game, the average profit from that will be negative. In other words, the game favors the casino. The martingale equality corresponds to the case of an equitable casino.

Imagine now that the gambler, fearing that he is under the influence of an evil spirit, confides his fortune to a luckier colleague, who is also unprophetic, and goes out of the casino's floor to breathe some air between random times $S_0 \leq S_1....$ The stopping Theorem 3.1.12 tells us that what he observes at these random instants is also a game favorable to the casino, or merely equitable in the martingale case. In other words, things are no better!

The existence of a restriction on the length of the stopping times S_i has the following meaning: the gambler may tell his colleague "call me at the first moment S_1 when my total gain $X_{S_1} - X_0$ is positive". At such a moment the mean gain is also positive and hence Theorem 3.1.12 seems to be contradicted! But in fact, what Theorem 3.1.12 affirms is that S_1 is not bounded.

Theorem 3.1.10 corresponds to the situation where the gambler makes an agreement with another gambler, or with the casino, by which his gain x_n at the n-th game will be multiplied by a coefficient V_n, to be determined just before playing (i.e., the predictability of the process V, where certainly, if it could be determined after the game, it would be easy to win!). The simplest case, for the choice of V, corresponds to $V = 0$ or 1 (0 for when one chooses to skip some games). Here again, Theorem 3.1.10 expresses the "sad" integrability restriction as the restriction on the length of the stopping times in Theorem 3.1.12.

The same is true in a fair game of heads or tails, doubling the stake each time and then stopping after the first positive gain is a well-known method of beginners, which only appears to contradict the statement of what Theorem 3.1.10 affirms is that the stopping time S (i.e., the first n such that $x_n > 0$) is unbounded – and here again it can be shown that $\mathbf{E}[S] = +\infty$.

3.1.3.1　Extension to unbounded stopping times

We will consider first a right-closed martingale (Definition 3.1.8) of the form $X_n = \mathbf{E}[Y|\mathcal{F}_n]$. As in Definition 3.1.8, we adjoin a σ-field $\mathcal{F}_\infty$ containing the $\mathcal{F}_n$ and making Y measurable and adopt the convention $X_\infty = Y$, so that, $(X_n)_{n \leq \infty}$ is a martingale. If S is a stopping time, finite or otherwise, then X_S is well defined ($X_S = X_\infty$ on $\{S = \infty\}$), as is $\mathcal{F}_S$ (Definition 2.4.11).

Theorem 3.1.15 Let X be a right-closed martingale, with the above notation, then,
(a) the set of r.v. X_S, where S is a stopping time finite, or otherwise, is uniformly integrable;
(b) if S and T are two stopping times such that $S \leq T$, we have a.s.

$$X_S = \mathbf{E}[X_T|\mathcal{F}_S] \quad (= \mathbf{E}[X_\infty|\mathcal{F}_S]). \tag{3.11}$$

Proof. It is sufficient to show that the set of X_S, where S runs through the set of bounded stopping times, is uniformly integrable. For suppose this holds, then for arbitrary S,

$$X_S = \lim_k \left(X_{s \wedge k} \mathbf{1}_{\{S \leq k\}} + X_\infty \mathbf{1}_{\{S > k\}} \right).$$

Let $\mathcal{H}$ be the set of r.v. appearing in the bracket for all possible choices of S and k; $\mathcal{H}$ is uniformly integrable by the result about bounded stopping times: X_S is the a.s. limit of a sequence of elements of $\mathcal{H}$ and hence belongs to the closure of $\mathcal{H}$ in L^1 which is a uniformly integrable set.

Then, let S be a stopping time bounded by k; we have by Theorem 3.1.12

$$X_S = \mathbf{E}[X_k|\mathcal{F}_S] = \mathbf{E}[X_\infty|\mathcal{F}_k|\mathcal{F}_S] = \mathbf{E}[X_\infty|\mathcal{F}_S],$$

then

$$\int_{\{|X_S|>c\}} |X_S|\mathbf{P} \leq \int_{\{|X_S|>c\}} |X_\infty|\mathbf{P},$$

$$\mathbf{P}\{|X_S| > c\} \leq \frac{1}{c}\mathbf{E}[|X_S|] \leq \frac{1}{c}\mathbf{E}[|X_\infty|].$$

It follows that $\mathbf{P}\{|X_n| > c\}$ and then $\int_{\{X_S>c\}} |X_S|\mathbf{P}$ tend to 0 uniformly in S as $c \to +\infty$, which proves (a).

To establish (b) it is sufficient to prove that $X_S = \mathbf{E}[X_\infty|\mathcal{F}_S]$, for we shall then also have $X_S = \mathbf{E}[X_\infty|\mathcal{F}_T|\mathcal{F}_S] = \mathbf{E}[X_T|\mathcal{F}_S]$. Let $A \in \mathcal{F}_S$; obviously, we have

$$\int_{A\cap\{S=\infty\}} X_S\mathbf{P} = \int_{A\cap\{S=\infty\}} X_\infty\mathbf{P}$$

and on the other hand $A \cap \{S \leq k\} = A \cap \{S \leq S \wedge k\} \in \mathcal{F}_{S\wedge k}$ (2.53) and hence by Theorem 3.1.12

$$\int_{A\cap n\{S\leq k\}} X_S\mathbf{P} = \int_{A\cap\{S\leq k\}} X_\infty\mathbf{P}.$$

It only remains to let k tend to $+\infty$ since X_S is integrable by (a). ∎

Remark 3.1.16 (a) The proof of uniform integrability establishes some additional result: if Y is an integrable r.v., the set of all r.v. of the form $\mathbf{E}[Y|\mathcal{G}]$, where $\mathcal{G}$ is an arbitrary sub-σ-field of $\mathcal{F}$, is uniformly integrable.

(b) A process (X_n) is said to belong to class D if the set of all r.v. X_S, where S is an arbitrary finite stopping time, is uniformly integrable – this is an important notion, especially in continuous time. Part (a) of Theorem 3.1.15 therefore amounts to saying that every right-closed martingale belongs to class D.

We now consider the extension of the stopping theorem to arbitrary stopping times for the case of positive supermartingales. Since every positive supermartingale X is closed on the right by the r.v. 0, and we take that $X_\infty = 0$, then we have:

Theorem 3.1.17 If X is a positive supermartingale, with the above convention, then (a) for every stopping time S finite or otherwise, X_S $(= X_S\mathbf{1}_{\{S<\infty\}})$ is integrable, (b) if S and T are two stopping times finite or otherwise, such that, $S \leq T$, then

$$X_S \geq \mathbf{E}[X_T|\mathcal{F}_S] \quad \text{a.s.} \tag{3.12}$$

Proof. By Corollary 3.1.13, $\mathbf{E}[X_{S\wedge k}] \leq \mathbf{E}[X_0]$ for all k; (a) then follows by letting k tend to $+\infty$ and using Fatou's lemma.

Similarly, by Theorem 3.1.12, if A belongs to $\mathcal{F}_S$ (and hence $A \cap \{S \leq k\} \in \mathcal{F}_{S\wedge k}$),

$$\int_{A\cap\{S\leq k\}} X_{S\wedge k}\mathbf{P} \geq \int_{A\cap\{S\leq k\}} X_{T\wedge k}\mathbf{P} \geq \int_{A\cap\{T\leq k\}} X_{T\wedge k}\mathbf{P}$$

In the left-hand integral, we can replace $X_{S\wedge k}$ by X_S and, as X_S is integrable and zero on $\{S = \infty\}$, the left-hand side converges to $\int_A X_S\mathbf{P}$ as $k \to \infty$. In the right-hand side, we have, similarly, convergence to $\int_A X_T\mathbf{P}$ and the theorem is established. ∎

The two Theorems 3.1.15 and 3.1.17 imply a more general result concerning supermartingales:

Theorem 3.1.18 Let X be a supermartingale closed on the right by a r.v. X_∞. Then,
 (a) for every stopping time S finite or otherwise, $X_S(= X_{S\mathbf{1}_{\{S<\infty\}}} + X_\infty\mathbf{1}_{\{S=\infty\}})$ is integrable;
 (b) if S and T are two stopping times such that $S \leq T$, then

$$X_S \geq \mathbf{E}[X_T|\mathcal{F}_S] \quad \text{a.s.} \tag{3.13}$$

Proof. For (3.13) it is implicitly assumed that some σ-field $\mathcal{F}_\infty$ has been chosen, as earlier. Theorem 3.1.18 reduces immediately to Theorems 3.1.15 and 3.1.17. For writing $X'_n = \mathbf{E}[X_\infty|\mathcal{F}_n]$, a martingale closed by $X'_\infty = X_\infty$, $X''_n = X_n - \mathbf{E}[X_\infty|\mathcal{F}_n]$, a positive supermartingale closed on the right by $X''_\infty = 0$, we apply Theorem 3.1.15 to X' and Theorem 3.1.17 to X'' and combine the two. ∎

Remark 3.1.19 We now return to martingales; let S and T be two arbitrary stopping times – here we do not assume that $S \leq T$ – and let Y be an integrable r.v. We introduce the martingale $X_n = \mathbf{E}[Y|\mathcal{F}_n]$, which we close on the right by the σ-field $\mathcal{F}_\infty = \mathcal{F}$ and the r.v. $X_\infty = Y$. The stopped process X^T has the value at time n

$$\begin{aligned}
X_{T\wedge n} &= \mathbf{E}[X_\infty|\mathcal{F}_{T\wedge n}] = \mathbf{E}[X_\infty|\mathcal{F}_T|\mathcal{F}_{T\wedge n}] = \mathbf{E}[X_T|\mathcal{F}_{T\wedge n}] \\
&= X_T\mathbf{1}_{\{T<n\}} + \mathbf{E}[X_T|\mathcal{F}_n]\mathbf{1}_{\{T\geq n\}} = \mathbf{E}[X_T|\mathcal{F}_n].
\end{aligned}$$

Thus X^T is the martingale $\mathbf{E}[X_T|\mathcal{F}_n]$. Stopping at time S, we obtain

$$X_{S\wedge T} = X^T_S = \mathbf{E}[X_T|\mathcal{F}_S] = \mathbf{E}[X_\infty|\mathcal{F}_T|\mathcal{F}_S].$$

The left-hand side is symmetric in S and T and moreover takes the value $\mathbf{E}[X_\infty|\mathcal{F}_{S\wedge T}]$. As $X_\infty = Y$ is arbitrary, we have proved: the operators $\mathbf{E}[\cdot|\mathcal{F}_S]$ and $E[\cdot|\mathcal{F}_T]$ commute and their product is $\mathbf{E}[\cdot|\mathcal{F}_{S\wedge T}]$. It follows that if Y is measurable with respect to both $\mathcal{F}_S$ and $\mathcal{F}_T$, it is a.s. measurable with respect to $\mathcal{F}_{S\wedge T}$. But this result is obvious without the precaution of adding a.s. Let $A \in \mathcal{F}_S \cap \mathcal{F}_T$, then for all t

$$A \cap \{S \wedge T \leq t\} = (A \cap \{S \leq t\}) \cup (A \cap \{T \leq t\}) \in \mathcal{F}_t$$

and this means that $A \in \mathcal{F}_{S\wedge T}$. The converse inclusion is obvious and we have $\mathcal{F}_S \cap \mathcal{F}_T = \mathcal{F}_{S\wedge T}$.

3.1.4 Fundamental inequalities

We begin by giving the following notations for the maximum and minimum of a process X:

$$\overline{X}_k = \sup_{n \le k} X_n, \quad \overline{X} = \overline{X}_\infty = \sup_n X_n \tag{3.14}$$

$$\underline{X}_k = \inf_{n \le k} X_n, \quad \underline{X} = \underline{X}_\infty = \inf_n X_n \tag{3.15}$$

$$X_k^* = \sup_{n \le k} |X_n|, \quad X^* = X_\infty^* = \sup_n |X_n|. \tag{3.16}$$

Furthermore, we now introduce the norm

$$||X||_p = \sup_n ||X_n||_p \quad (1 \le p \le \infty) \tag{3.17}$$

which is finite if and only if the sequence (X_n) is bounded in L^p.

For a martingale (positive or otherwise) or a positive submartingale, we have already seen that $\mathbf{E}[|X_S|^p] \le \mathbf{E}[|X_k|^p]$ for every stopping time S bounded by k ((3.9); $|X_n|^p$ is a submartingale by Jensen. Hence, by Fatou's lemma

$$||X_S||_p \le ||X||_p \tag{3.18}$$

for every finite stopping time S.

Similarly, for a supermartingale $||X_S||_1 \le 3||X||_1$ for every finite stopping time S (3.10). In these special cases the norm $||\cdot||_p$ thus controls the behavior of the process at random times well enough. On the other hand, it seems to be of no benefit for more general processes.

3.1.4.1 Maximal lemma

Estimations of the type in Theorem 3.1.20 play a fundamental role in analysis, see for example the Hardy-Littlewood maximal lemma in differentiation theory and Stein "maximal ergodic lemma" [34] and its applications by Neveu [35]. These theorems then allow, using interpolation, the determination of more precise upper bounds in p ($p > 1$) and results on almost everywhere convergence.

Theorem 3.1.20 Let X be a supermartingale and λ a positive constant. Then,

$$\lambda \mathbf{P}\{\overline{X}_k \ge \lambda\} \le \mathbf{E}[X_0] - \int_{\{\overline{X}_k < \lambda\}} \overline{X}_k \mathbf{P} \le \mathbf{E}[X_0] + \mathbf{E}[X_k^-] \tag{3.19}$$

$$\lambda \mathbf{P}\{\underline{X}_k \le -\lambda\} \le -\int_{\{\underline{X}_k \le -\lambda\}} X_k \mathbf{P} \le \mathbf{E}[X_k^-]. \tag{3.20}$$

Proof. To establish (3.19), we write

$$T(\omega) = \inf\{n \le k : X_n(\omega) \ge \lambda\}$$

where $T(\omega) = k$ if this set is empty. Clearly T is a stopping time and by (3.9) we have

$$\mathbf{E}[X_0] \ge \mathbf{E}[X_T] \ge \lambda \mathbf{P}\{\overline{X}_k > \lambda\} + \int_{\{\overline{X}_k < \lambda\}} X_k \mathbf{P}$$

which is equivalent to (3.19). For the second formula we write similarly

$$T(\omega) = \inf\{n \le k : X_n(\omega) \le -\lambda\}$$

where $T(\omega) = k$ if this set is empty. Again by (3.9)

$$\mathbf{E}[X_k] \le \mathbf{E}[X_T] \le -\lambda \mathbf{P}\{\underline{X}_k \le -\lambda\} + \int_{\{\underline{X}_k \ge -\lambda\}} X_k \mathbf{P}$$

which is equivalent to (3.20). ∎

Corollary 3.1.21 If X is a supermartingale, then

$$\lambda \mathbf{P}\{X^* \ge \lambda\} \le 3\|X\|_1. \tag{3.21}$$

(If X is positive, the right-hand side can be replaced by $\mathbf{E}[X_0]$.)

Proof. Add (3.19) and (3.20) and let $k \to +\infty$. ∎

Corollary 3.1.22 If X is a martingale or a positive submartingale, then

$$\lambda^p \mathbf{P}\{X^* \ge \lambda\} \le \|X\|_p^p \quad (1 \le p < \infty). \tag{3.22}$$

Proof. Apply 3.20 to the supermartingale $-|X_n|^p$ – taking the limit as $k \to \infty$ must be performed over $\lambda \mathbf{P}\{X_k^* > \lambda\}$, after which we obtain (3.21) or (3.22) by taking the sup over $\lambda' < \lambda$. We shall see later a better result (Theorem 3.1.24) when $p > 1$, so that only the case $p = 1$ is of real interest here. ∎

Remark 3.1.23 (a) Let (S_n) be a martingale of the form (3.3)

$$S_n = X_0 + \dots + X_n$$

where the X_i's are independent, square integrable and of zero expectation. Then by (3.22)

$$\lambda^2 \mathbf{P}\{S_k^* \ge \lambda\} \le \sum_{i=0}^{k} \mathbf{E}[X_i^2].$$

This is Kolmogorov's famous inequality in the theory of addition of independent r.v.

(b) Let $A \in \mathcal{F}_0$ be such that $\mathbf{P}(A) > 0$. If we replace the law $\mathbf{P}$ by the conditional law $\mathbf{P}_A(B) = \frac{\mathbf{P}(AB)}{\mathbf{P}(A)}$, the process X remains a supermartingale, to which the inequalities of Theorem 3.1.20 apply. This gives us the following improvements to (3.19) and (3.20):

$$\mathbf{P}\{\overline{X}_k \geq x | \mathcal{F}_0\} \leq X_0 + \mathbf{E}[X_k^- | \mathcal{F}_0] \quad \text{a.s.} \tag{3.23}$$

$$\lambda \mathbf{P}\{\underline{X}_k \leq -\lambda | \mathcal{F}_0\} \leq \mathbf{E}[X_k^- | \mathcal{F}_0] \quad \text{a.s.} \tag{3.24}$$

Here, the inequalities are conditioned by the initial situation, an essential fact in martingale theory.

3.1.4.2   Domination in L^p

Theorem 3.1.24 Let X be a positive submartingale. Then, for all $p > 1$, with q denoting the exponent conjugate to p

$$\|X^*\|_p \leq q\|X\|_p. \tag{3.25}$$

The result applies in particular to $|X|$ for every martingale X.

Proof. We set $x = X_k$ and $y = \sup_{n \leq k} X_n$; these are positive r.v. and (3.20) can then be written – for a submartingale:

$$\lambda \mathbf{P}\{y \geq \lambda\} \leq \int_{\{y \geq \lambda\}} x\mathbf{P} \quad (\lambda \geq 0). \tag{3.26}$$

We shall show that this inequality is sufficient in itself to imply $\|y\|_p \leq q\|x\|_p$. After that it only remains to let k tend to infinity.

Let Φ be a function on $\mathbb{R}_+$ which is increasing, right-continuous and such that $\Phi(0) = 0$. By Fubini's Theorem we have

$$
\begin{aligned}
\mathbf{E}[\Phi \circ y] &= \mathbf{E}\left[\int_0^y d\Phi(\lambda)\right] = \int_0^\infty \mathbf{P}\{y \geq \lambda\}d\Phi(\lambda) \\
&\leq \int_0^\infty d\Phi(\lambda)\frac{1}{\lambda}\int_{\{y \geq \lambda\}} x\mathbf{P} = \int x \left(\int_0^y \frac{d\Phi(\lambda)}{\lambda}\right)\mathbf{P}.
\end{aligned}
$$

We take $\Phi(\lambda) = \lambda^p$. The inner integral has the value $\int_0^y p\lambda^{p-2}d\lambda = \frac{p}{p-1}y^{p-1}$. We recall that $\frac{p}{p-1} = q$ and apply Holder's inequality

$$\mathbf{E}[y^p] \leq q\int_x xy^{p-1}\mathbf{P} \leq q\|x\|_q\|y^{p-1}\|_q = q\|x\|_p\|y\|_p^{p/q}.$$

The left-hand side has the value $\|y\|_p^p$, whence dividing by $\|y\|_p^{p/q}$, as $p - \frac{p}{q} = 1$,

$$\|y\|_p \leq q\|x\|_p. \tag{3.27}$$

But this argument is not entirely legitimate, since we cannot be sure that the quantity by which we have divided is finite. Should that not hold, we set $y_n = y \wedge n$ and apply the above result to $y_n \in \mathrm{L}^p$ and to x which obviously satisfies (3.26); we then let n tend to infinity. ∎

Remark 3.1.25 (a) The analytic theorem above is often used outside of martingale theory. The argument falls down in the last line on a space of infinite measure and (3.26) does not then imply (3.27) in general, as the following example shows:

take $\Omega = \mathbb{R}$ with the Lebesgue measure dt, $x(t) = \frac{1}{1+|t|} \in L^2$ and $y(t) = |t| \notin L^2$. However, if $\{y \geq \lambda\}$ is of finite measure for all $\lambda > 0$ we can apply the result of Theorem 3.1.24 to the space of finite measure $\{y \geq 1/n\}$ and let n tend to infinity.

(b) Concerning the above theorem, again, a variation of the proof allows us to obtain the result more quickly but with a poorer constant. We write (3.26) with 2λ in place of λ

$$
\begin{aligned}
2\lambda \mathbf{P}\{y \geq 2\lambda\} &\leq \int_{\{y \geq 2\lambda\}} X\mathbf{P} \\
&\leq \int_{\{X \geq \lambda\}} X\mathbf{P} + \int_{\{X < \lambda, y \geq 2\lambda\}} X\mathbf{P} \\
&\leq \int_{\{X \geq \lambda\}} X\mathbf{P} + \lambda\mathbf{P}\{y \geq 2\lambda\}
\end{aligned}
$$

whence a simplification in comparison with the left-hand side. Integrating with respect to $d\Phi(\lambda)$ we obtain

$$
\mathbf{E}\left[\Phi \circ \frac{y}{2}\right] \leq \mathbf{E}\left[x \int_0^\cdot \frac{d\Phi(\lambda)}{\lambda}\right]. \tag{3.28}
$$

The calculation when $\Phi(t) = t^p$ is immediate.

(c) Let $\Phi(t) = (t-1)^+$. The calculation in Theorem 3.1.24 gives us

$$
\mathbf{E}[y-1] \leq \mathbf{E}[\Phi \circ y] \leq \mathbf{E}[x \log^+ y]
$$

but if a and b are positive, then $a \log b \leq a \log^+ a + \frac{b}{e}$, whence

$$
\mathbf{E}[y-1] \leq \mathbf{E}[x \log^+ x] + \frac{1}{e}\mathbf{E}[y]
$$

whence we deduce the useful inequality

$$
\mathbf{E}[X^\star] \leq A(1 + \sup_n \mathbf{E}\left[X_n \log^+ X_n\right]) \quad (A = \frac{e}{e-1} < 2) \tag{3.29}
$$

This inequality has been studied many times (Blackwell and Dubins [36], Gundy [37], Chou [38], for example). One conclusion of these works is that it is impossible to improve on (3.29) as far as the integrability of $X^\star$ is concerned.

3.1.4.3 Martingales upcrossings and downcrossings

The numbers of upcrossings and downcrossings were defined in Section 2.3.2. Here we restrict ourselves to a direct proof of Doob's main inequality regarding upcrossing and downcrossing of martingales.

Theorem 3.1.26 Let X be a supermartingale and let $M_a^b(\omega)$ be the number of upcrossings (2.3.2) of the path $X.(\omega)$ on the interval $[a, b]$. Then,

$$\mathbf{E}[M_a^b] \leq \frac{1}{b-a} \sup k\mathbf{E}[(X_k - a)^-] \leq \frac{1}{b-a}(|a| + \sup k\mathbf{E}[|X_k|]). \tag{3.30}$$

Proof. We need only consider the case where X is stopped at an integer k (we can then let k tend to $+\infty$).

We make the following observations: if S and T are two stopping times such that $S \leq T$ and A is an element of $\mathcal{F}_S$ and B a measurable subset of A, then

$$\int_B (X_T - X_S)\mathbf{P} \leq \int_{A\backslash B} (X_S - X_T)\mathbf{P}.$$

For

$$\int_A (X_T - X_S)\mathbf{P} = \int_B (X_T - X_S)\mathbf{P} + \int_{A\backslash B} (X_T - X_S)\mathbf{P} \leq 0$$

We now introduce the stopping times

$$\begin{aligned} S_1 &= (\inf\{n : X_n \leq a\}) \wedge (k+1), \\ T_1 &= (\inf\{n > S_1 : X_n \geq b\}) \wedge (k+1) \end{aligned}$$

$$\begin{aligned} S_2 &= (\inf\{n > T_1 : X_n \leq a\}) \wedge (k+1), \\ T_2 &= \ldots \end{aligned}$$

with the usual convention $\inf \emptyset = +\infty$. Let $A_i = \{S_i < k+1\} \in \mathcal{F}_{S_i}$ and $B_i = \{T_i < k+1\} \subset A_i$.

By the above observation, we have

$$(b-a)\mathbf{P}(B_i) \leq \int_{B_i} (X_{T_i} - X_{S_i})\mathbf{P} \leq \int_{A_i / B_i} (X_{S_i} - X_{T_i})\mathbf{P} \leq \int_{A_i \backslash B_i} (a - X_k)\mathbf{P}$$

for on A_i we have $X_{S_i} \leq a$ and on B_i^c we have $X_{T_i} = X_{k+1} = X_k$. We now sum over i from 1 to k. On the left-hand side, the number of B_i containing ω is equal to the number of upcrossings $M_a^b(\omega)$ and thus $\sum_i \mathbf{P}(B_i) = \mathbf{E}[M_a^b]$. On the right-hand side, the $A_i \backslash B_i$ are disjoint and we have

$$(b-a)\mathbf{E}[M_a^b] \leq \mathbf{E}[(a - X_k)^+] = \mathbf{E}[(X_k - a)^-].$$

∎

Dubins proved in [39] a stronger inequality, which gives (3.30) on summing over n. If X is stopped at k, then

$$\mathbf{P}\{M_a^b > n\} \leq \frac{1}{b-a} \int_{\{M_a^n = n\}} (X_k - a)^- \mathbf{P} \tag{3.31}$$

in particular, for a positive supermartingale with a and b positive, we have $(X_k - a)^- \leq a$:

$$\mathbf{P}\{M_a^b > n\} \leq \frac{a}{b-a} \mathbf{P}\{M_a^b = n\} \tag{3.32}$$

which can be written

$$\mathbf{P}\{M_a^b > n\} \leq \frac{a}{b}(\mathbf{P}\{M_a^b > n\} + \mathbf{P}\{M_a^b = n\}) = \frac{a}{b}\mathbf{P}\{M_a^b > n-1\}. \tag{3.33}$$

This has an amusing interpretation: $\mathbf{P}(M_a^b > n | M_a^b > n-1) \leq \frac{a}{b}$. Descending through the integers as far as 0, we have

$$\mathbf{P}\{M_a^b > n\} \leq \left(\frac{a}{b}\right)^n \mathbf{P}\{M_a^b > 0\} \leq \left(\frac{a}{b}\right)^n \frac{\mathbf{E}[X_0 \wedge a]}{b}. \tag{3.34}$$

We shall give no more details of this inequality, since we are proving nothing. Dubins and Freedman [40] have shown that (3.34) cannot be improved.

On the other hand, we reproduce, in terms of the number of downcrossings of a supermartingale, the Doob inequality for the number of upcrossings of a submartingale stopped at k. The proof is almost identical with that of (3.30).

$$\mathbf{E}[D_d^b] \leq \frac{1}{b-a}\mathbf{E}[(X_k - b)^- - (X_0 - b)^-] = \frac{1}{b-a}\mathbf{E}[X_0 \wedge b - X_k \wedge b]. \tag{3.35}$$

Similarly, there exists a Dubins [39] inequality for the number of downcrossings of a positive supermartingale

$$\mathbf{P}\{D_a^b > n\} \leq \left(\frac{a}{b}\right)^n \frac{\mathbf{E}[X_0 \wedge a]}{b}. \tag{3.36a}$$

Finally, we recall the principle of conditioning by $\mathcal{F}_0$ mentioned in Remark 3.1.23(b). It is possible to replace all the expectations and probabilities by conditional expectations or probabilities, for example

$$\mathbf{E}[M_a^b | \mathcal{F}_0] \leq \frac{1}{b-a} \sup_k \mathbf{E}[(X_k - a)^- | \mathcal{F}_0].$$

3.1.5   Convergence and decomposition theorems

In this section we study the most frequently used results in martingale theory. First, we show the two almost sure convergence results for supermartingales for the positive integers and then for the negative integers. Then, we study – mainly for martingales – the problem of convergence in L^1 and in L^p $(p > 1)$. Following this, we study Riesz and Krickeberg decompositions but leave the Doob's decomposition to the section where we study martingales in continuous time.

3.1.5.1 Almost sure convergence of supermartingales

Theorem 3.1.27 Let $(X_n)_{n \geq 0}$ be a supermartingale relative to the family $(\mathcal{F}_n)_{n \geq 0}$. Suppose that the family (X_n) is bounded in L^1 that is

$$\sup_n \mathbf{E}[|X_n|] < +\infty \tag{3.37}$$

a condition equivalent, here, to

$$\sup_n \mathbf{E}[X_n^-] < +\infty. \tag{3.38}$$

Then, the sequence (X_n) converges a.s. to an integrable r.v.

Proof. We first prove the equivalence of (3.37) and (3.38). Clearly (3.37)$\Rightarrow$(3.38). Conversely, the relation $|X_n| = X_n + 2X_n^-$ and the inequality $\mathbf{E}[X_n] \leq \mathbf{E}[X_0]$ imply $\sup_n \mathbf{E}[|X_n|] \leq \mathbf{E}[X_0] + 2\sup_n \mathbf{E}[X_n^-]$. This inequality will also imply the integrability of ξ when the a.s. convergence has been established. For by Fatou's lemma, we shall have $\mathbf{E}[|\xi|] \leq \liminf_n \mathbf{E}[|X_n|] \leq \sup_n \mathbf{E}[|X_n|]$.

We now prove the a.s. convergence. If the path $X.(\omega)$ has no limit at infinity, we can find two rational numbers a, b such that $\liminf_n X_n(\omega) < a < b < \limsup_n X_n(\omega)$ and then the number of upcrossings $M_a^b(\omega)$ of the path $X.(\omega)$ on the interval $[a, b]$ is equal to $+\infty$. Hence, it is sufficient to prove that for every ordered pair of rational numbers $a < b$ the r.v. M_a^b is a.s. finite. This follows immediately from (3.37) and Doob's inequality (3.30)

$$\mathbf{E}[M_a^b] \leq \sup_k \frac{\mathbf{E}[(X_k - a)^-]}{b - a} \leq \frac{\sup_k \mathbf{E}[|X_k|] + a}{b - a}.$$

∎

Next, we examine whether the inequality for supermartingales holds when passing to the limit. We use the notation of Theorem 3.1.27.

Theorem 3.1.28 (a) Suppose that the sequence (X_n^-) is uniformly integrable which is the case in particular if the sequence (X_n) is uniformly integrable or positive. Then, condition (3.38) is satisfied and $X_n \geq \mathbf{E}[\xi | \mathcal{F}_n]$ a.s., in other words, ξ closes X on the right.

(b) Conversely, if there exists an integrable r.v. X_∞ closing X on the right, the sequence (X_n^-) is uniformly integrable, hence ξ also closes X on the right and $\xi \geq \mathbf{E}[X_\infty | \mathcal{F}_{\infty-}]$ a.s.

Proof. If the sequence (X_n^-) is uniformly integrable, then (3.38) is satisfied by 1.2.20 and X_n^- converges in L^1 to ξ^- by 1.8. Let $A \in \mathcal{F}_n$. In the supermartingale inequality,

$$\int_A X_n^+ \mathbf{P} - \int_A X_n^- \mathbf{P} = \int_A X_n \mathbf{P} \geq \int_A X_{n+m} \mathbf{P} = \int_A X_{n+m}^+ \mathbf{P} - \int_A X_{n+m}^- \mathbf{P}$$

we apply this convergence in L^1 to the X_{n+m}^- and Fatou's lemma to the X_{n+m}^+ to obtain

$$\int_A X_n \mathbf{P} \geq \int \xi \mathbf{P}$$

i.e., ξ closes X.

Conversely, if X_∞ closes X we have $X_n \geq \mathbf{E}[X_\infty|\mathcal{F}_n] \geq \mathbf{E}[-X_\infty^-|\mathcal{F}_n]$ and hence $X_n^- \leq \mathbf{E}[X_\infty^-|\mathcal{F}_n]$. We saw in Theorem 3.1.15 that the r.v. $\mathbf{E}[X_\infty^-|\mathcal{F}_n]$ are uniformly integrable and so therefore also are the X_n^-.

We shall see in a moment (the martingale case, Theorem 3.1.30) that $\mathbf{E}[X_\infty|\mathcal{F}_n]$ converges a.s. to $\mathbf{E}[X_\infty|\mathcal{F}_{\infty-}]$. Hence the inequality $X_n \geq \mathbf{E}[X_\infty|\mathcal{F}_n]$ gives us, on taking a.s. limits, the inequality $\xi \geq \mathbf{E}[X_\infty|\mathcal{F}_{\infty-}]$ a.s. ∎

We now state a theorem without proof for the case where the family $(\mathcal{F}_n)$ and the processes (X_n) are indexed by the negative integers where we set $\mathcal{F}_{-\infty} = \cap \mathcal{F}_n$. The theorem is just as important as Theorem 3.1.27 with the most remarkable part being assertion (a) on uniform integrability.

Theorem 3.1.29 Let $(X_n)_{n\leq 0}$ be a supermartingale relative to the family $(\mathcal{F}_n)_{n\leq 0}$. Suppose at the family (X_n) is bounded in L^1

$$\sup_n \mathbf{E}[|X_n|] < +\infty \tag{3.39}$$

a condition equivalent here to

$$\lim_{n\to-\infty} \mathbf{E}[X_n] < +\infty. \tag{3.40}$$

(a) The family (X_n) is then uniformly integrable. (b) The r.v. X_n converge a.s. and in L^1 to an integrable r.v. ξ and for all n

$$\xi \geq \mathbf{E}[X_n|\mathcal{F}_{-\infty}] \quad \text{a.s.} \tag{3.41}$$

3.1.5.2 Uniform integrability and martingale convergence

We return to Theorem 3.1.27 but with indexing set infinite on the right. In this case almost sure convergence result is no better for martingales than for supermartingales: every martingale (X_n) bounded in L^1 converges a.s. to an integrable r.v. ξ. The problem consists of knowing when ξ closes X as a martingale, i.e., when $X_n = \mathbf{E}[\xi|\mathcal{F}_n]$ for all n. The answer is very simple and corresponds to Theorem 3.1.28: this property holds if and only if the martingale is uniformly integrable. The new result compared with Theorem 3.1.28 is concerned rather with convergence in L^p ($p > 1$).

Theorem 3.1.30 Let $(X_n)_{n\geq 0}$ be a martingale relative to the family $(\mathcal{F}_n)_{n\geq 0}$. (a) If (X_n) is uniformly integrable, it is bounded in L^1 and the r.v. $\xi = \lim_n X_n$ (which exists and is integrable by 3.1.27) closes X on the right, i.e., $X_n = \mathbf{E}[\xi|\mathcal{F}_n]$ for all n. Further, X_n converges to ξ in L^1. (b) Conversely, if there exists an integrable r.v. X_∞ which closes X on the right, the martingale (X_n) is uniformly integrable and $\xi = \mathbf{E}[X_\infty|\mathcal{F}_{\infty-}]$ a.s. (c) If (X_n) is bounded in L^p ($1 < p < \infty$), X_n converges to ξ in L^p.

Proof. (a) The condition $\sup_n \mathbf{E}[|X_n|] < \infty$ follows from uniform integrability (1.2.20). As convergence to ξ and uniform integrability imply convergence in L^1 (Theorem 1.2.22). If A belongs to $\mathcal{F}_n$, the relation $\int_A X_n \mathbf{P} = \int_A X_{n+m} \mathbf{P}$ then holds in the limit as $m \to \infty$ to give $X_n = \mathbf{E}[\xi|\mathcal{F}_n]$ a.s. If the sequence (X_n) is bounded in L^p, $(p > 1)$, the r.v. $X^* = \sup_n |X_n|$ belongs to L^p (Theorem 3.1.24) and hence X_n converges to ξ in L^p by dominated convergence, if $1 < p < \infty$.

Having proved (a) and (c), we now pass to (b). Uniform integrability of the r.v. $X_n = \mathbf{E}[X_\infty|\mathcal{F}_n]$ was shown in Theorem 3.1.15. By (a), for $A \in \mathcal{F}_n$, $\int_A X_n \mathbf{P} = \int_A \xi \mathbf{P} = \int_A X_\infty \mathbf{P}$. Hence, the bounded measures $X_\infty \mathbf{P}$ and $\xi \mathbf{P}$ coincide on the Boolean algebra $\cup_n \mathcal{F}_n$ which generates $\mathcal{F}_{\infty-}$. By the monotone class theorem, (1.2.8), they coincide on $\mathcal{F}_{\infty-}$ and hence $\xi = \mathbf{E}[X_\infty|\mathcal{F}_{\infty-}]$ a.s. ∎

Remark 3.1.31 (a) There are many proofs of convergence theorems for martingales and super-martingales that do not use the numbers of upcrossings. For us this is of no interest, for that the number of upcrossings is necessary in continuous time.

(b) A simple example of a martingale which is bounded in L^1 but does not converge in L^1 is: let $\Omega = [0, 1[$ with the Lebesgue measure, $\mathcal{F}_n$ be the σ-field generated by the dyadic intervals $I_k = [k2^{-n}, (k+1)2^{-n}[$ and let $X_n = 2^n$ on I_0, $X_n = 0$ on I_0^c. Then, (X_n) is a positive martingale of expectation 1 (it is a martingale of the type of in 3.1.7, (b), with $\mathbf{Q}$ the unit mass ϵ_0) and X_n obviously converges to 0 away from the origin.

The result analogous to Theorem 3.1.30 for martingales indexed by the negative integers is almost obvious.

Theorem 3.1.32 Let $(X_n)_{n \leq 0}$ be a martingale relative to the family $(\mathcal{F}_n)_{n \leq 0}$. Then (X_n) is uniformly integrable and converges a.s. and in L^1 to the r.v. $\xi = \mathbf{E}[X_0| \cap_n \mathcal{F}_n]$. If X_0 belongs to L^p $(p < \infty)$, convergence holds in L^p.

Proof. Let $\mathcal{F}_{-\infty} = \cap_n \mathcal{F}_n$. Uniform integrability follows from Theorem 3.1.15 (or Theorem 3.1.29) and a.s. convergence and convergence in L^1 from Theorem 3.1.29. If A belongs to $\mathcal{F}_{-\infty}$, then $\int_A X_n \mathbf{P} = \int_A X_0 \mathbf{P}$, whence, on taking limits in L^1, $\int_A \xi \mathbf{P} = \int_A X_0 \mathbf{P}$. As ξ is $\mathcal{F}_{-\infty}$-measurable, this implies that $\xi = \mathbf{E}[X_0|\mathcal{F}_{-\infty}]$. Finally, convergence in L^p $(1 < p < \infty)$ follows from Theorem 3.1.24 and the dominated convergence theorem. ∎

3.1.5.3 Riesz decompositions of supermartingales

Here we state two closely related decomposition results. The first is obvious that we only give its formal statement and present the proof of the second one.

Theorem 3.1.33 Let $X = (X_n)_{n \geq 0}$ be a supermartingale such that $\sup_n \mathbf{E}[X_n^-] < \infty$. Then we know by (3.1.27) that the limit $\lim_n X = \xi$ exists a.s. and is integrable and we can write $X = U + V$ where

$$U_n = \mathbf{E}[\xi|\mathcal{F}_n], \quad V_n = X_n - \mathbf{E}[\xi|\mathcal{F}_n].$$

The first process is a uniformly integrable martingale and the second a supermartingale such that $V_{\infty-} = 0$. This decomposition is unique, for a uniformly integrable martingale which is zero at infinity.

Provided the X_n^- are uniformly integrable and in particular if X is positive, we know that ξ closes the supermartingale X on the right and hence that V is a positive supermartingale.

Theorem 3.1.34 Riesz Decomposition. Every supermartingale $X = (X_n)_{n \geq 0}$ which is positive or more generally of $\lim_n \mathbf{E}[X_n] > -\infty$ has a decomposition of the form $X = Y + Z$, where Y is a martingale and Z a potential. This decomposition is unique and Y is the greatest submartingale bounded above by X.

Proof. We have

$$X_n \geq \mathbf{E}[X_{n+m}|\mathcal{F}_n] \geq \mathbf{E}[X_{n+m+1}|\mathcal{F}_{n+m}|\mathcal{F}_n] = \mathbf{E}[X_{n+m+1}|\mathcal{F}_n] \qquad (3.42)$$

hence the r.v. $\mathbf{E}[X_{n+m}|\mathcal{F}_n]$ decrease as m increases. Let Y_n be their limit: as all the r.v. $\mathbf{E}[X_{n+m}|\mathcal{F}_n]$ are bounded above by X_n which is integrable, Y_n is integrable if and only if $\lim_m \mathbf{E}[\mathbf{E}[X_{n+m}|\mathcal{F}_n] > -\infty$, a condition which is independent of n and means simply that $\lim_m \mathbf{E}[X_m] > -\infty$. If this condition holds, then taking limits in L^1

$$Y_n = \lim_m \mathbf{E}[X_{n+m+1}|\mathcal{F}_n] = \lim_m \mathbf{E}[X_{n+m+1}|\mathcal{F}_{n+1}|\mathcal{F}_n] = \mathbf{E}[Y_{n+1}|\mathcal{F}_n].$$

In other words, the process Y is a *martingale* and the process $Z = X - Y$ a *positive supermartingale*. On the other hand,

$$\mathbf{E}[Z_n] = \mathbf{E}[X_n] - \mathbf{E}[Y_n] = \mathbf{E}[X_n] - \lim_m \mathbf{E}[X_m]$$

whence it follows that $\lim_n \mathbf{E}[Z_n] = 0$. Consequently, Z is a *potential*.

If H is a submartingale bounded above by X, then,

$$H_n \leq \mathbf{E}[H_{n+m}|\mathcal{F}_n] \leq \mathbf{E}[X_{n+m}|\mathcal{F}_n]$$

whence $H_n \leq Y_n$ as $m \to \infty$. Thus, the martingale Y is the greatest submartingale bounding X from below.

To establish the uniqueness, it is sufficient to establish that a martingale M which is also the difference of two potentials is zero. But the process $(|M_n|)$ is a submartingale and $\mathbf{E}[|M_n|] \leq \lim_k \mathbf{E}[|M_{n+k}|] = 0$, which implies $M_n = 0$ a.s. for all n. ∎

A potential, for reasons that will be explained in Remark 3.1.36(b), is any positive supermartingale, V, such that $\lim_n \mathbf{E}[V_n] = 0$, which by Fatou's lemma, $V_{\infty-} = 0$ a.s. So, if in the decomposition 3.1.33, X is uniformly integrable, so are U and $V = X - U$ and V is a potential; decomposition 3.1.33 then coincides with decomposition 3.1.34.

Corollary 3.1.35 Let X be a supermartingale. The following conditions are equivalent: (a) $\lim_n \mathbf{E}[X_n] > -\infty$, (b) X is bounded below by a submartingale, and (c) X is bounded below by a martingale.

   Proof. (b)$\Rightarrow$(a)$\Rightarrow$(c)$\Rightarrow$(b). ∎

Remark 3.1.36 (a) Let X be a supermartingale such that $\sup_n \mathbf{E}[X_n^-] < \infty$. We apply the decomposition $X = U + V$ of Theorem 3.1.33 to X and then the Riesz decomposition of Theorem 3.1.34 to V, $V = W + Z$. Then the Riesz decomposition of $X = (U + W) + Z$ where $U + W = Y$ is a martingale. But this decomposition into three parts is more interesting: we have a uniformly integrable martingale U, a potential Z and a positive martingale W which is zero at infinity. In continuous time we can further decompose Z as a local martingale and a potential of class D.

(b) We now recall some results about the classical Riesz decomposition in the unit ball in $\mathbb{R}^n$, for example. The symbol $\approx$ is to be read as "translates as" or "corresponds to": Let f be a superharmonic function ($\approx$ supermartingale X) with a harmonic minorant ($\approx$ bounded below by a (sub)martingale). Then, f can be written as $g + h$, where g is the greatest harmonic minorant of f and h is a potential ($\approx X = Y + Z$, where Y is the greatest martingale dominated by X and Z is a potential).

3.1.5.4 Krickeberg decomposition of martingales

We recall that a martingale $X = (X_n)_{n \geq 0}$ is bounded in L^1 if and only if the quantity

$$\|X\|_1 = \lim_n \mathbf{E}[|X_n|] = \sup_n \mathbf{E}[|X_n|] \tag{3.43}$$

is finite.

Theorem 3.1.37 A martingale $X = (X_n)_{n \geq 0}$ is bounded in L^1 if and only if it is the difference of two positive martingales, and it then has a unique decomposition of the form

$$X = Y - Z \tag{3.44}$$

where Y and Z are two positive martingales such that $\|X\|_1 = \|Y\|_1 + \|Z\|_1$. Moreover, Y is the smallest positive martingale bounding X above and Z the smallest positive martingale bounding $-X$ above.

Proof. Every positive martingale is obviously bounded in L^1 and so is every difference of positive martingales. Conversely, let X be a martingale which is bounded in L^1. The process $(-X_n^+)$ is a supermartingale such that $\lim_n \mathbf{E}[-X_n^+] > -\infty$. Hence it has a Riesz decomposition which we write as $X_n^+ = Y_n - A_n$ (Y a martingale, A a potential) where the martingale Y is given by $Y_n = \lim_m \mathbf{E}[X_{n+m}^+|\mathcal{F}_n]$ (increasing limit). Similarly, considering the supermartingale $(-X_n)$, we can write

$$X_{\overline{n}} = Z_n - B_n$$

where the martingale Z is given by $Z_n = \lim_m \mathbf{E}[X_{n+m}^-|\mathcal{F}_n]$ (increasing limit). Clearly, from their explicit expression Y and Z are positive martingales and we have

$$Y_n - Z_n = \lim_m \mathbf{E}[X_{n+m}^+ - X_{n+m}^-|\mathcal{F}_n] = \lim_m \mathbf{E}[X_{n+m}|\mathcal{F}_n] = X_n$$

so that $X = Y - Z$, the difference of two positive martingales. We have $\|Y\|_1 + \|Z\|_1 = \mathbf{E}[Y_0 + Z_0] = \lim \mathbf{E}[X_m^+ + Y_m^+] = \|X\|_1$. It follows from the Riesz decomposition theory that Y is the smallest martingale bounding X^+ above, that's the smallest positive martingale bounding X above - the analogous assertion for Z follows by replacing X by $-X$. We now consider the uniqueness: if we write X as the difference of two positive martingales, $X = U - V$, then $U \geq X^+$, $V \geq X^-$ and hence by Theorem 3.1.34 we get $U \geq Y$, $V \geq Z$. If $\|X\|_1 = \|\cup\|_1 + \|V\|_1 = \|Y\|_1 + \|Z\|_1$, then $\|U\|_1 = \|Y\|_1$ and $\|V\|_1 = \|Z\|_1$. But on the other hand, as U, Y and $U - Y$ are positive martingales, $\|U\|_1 = \|Y\|_1 + \|U - Y\|_1$, hence $U - Y = 0$ and finally $U = Y$ and $V = Z$. ∎

Remark 3.1.38 Martingales bounded in L^1 correspond to harmonic functions which can be represented as Poisson integrals of bounded measures. Then, Krickeberg decomposition corresponds to the decomposition of a bounded measure as a difference of two positive measures.

3.1.6   Some applications of convergence theorems

The following lemma enables us to reduce certain problems on a.s. convergence of conditional expectations, where functions and σ-fields both vary, to convergence theorem for martingales.

Theorem 3.1.39 Let $(\Omega, \mathcal{F}, \mathbf{P})$ be a probability space with an increasing family of σ-fields $(\mathcal{F}_n)$ and let $\mathcal{F}_{\infty-} = \vee \mathcal{F}$. Let (X_n) be a sequence of n random variables bounded above in absolute value by an integrable r.v. Y which converges a.s. to a r.v. X. Then

$$\lim_n \mathbf{E}[X_n | \mathcal{F}_n] = \mathbf{E}[X | \mathcal{F}_{\infty-}] \tag{3.45}$$

a.s. and in L^1.

Proof. Convergence in L^1 follows from the fact that $X_n \to X$ in L^1 and the inequality

$$\|\mathbf{E}[X_n | \mathcal{F}_n] - \mathbf{E}[X | \mathcal{F}_{\infty-}]\|_1 \leq \|\mathbf{E}[X_n - X | \mathcal{F}_n]\|_1 + \|\mathbf{E}[X | \mathcal{F}_n] - \mathbf{E}[X | \mathcal{F}_{\infty-}]\|_1$$
$$\leq \|X_n - X\|_1 + \|\mathbf{E}[X | \mathcal{F}_n] - \mathbf{E}[X | \mathcal{F}_{\infty-}]\|_1.$$

Then, we apply Theorem 3.1.30 to the last term. We now prove a.s. convergence.

Let $u_m = \inf_{n \geq m} X_n$ and $v_m = \sup_{n \geq m} X_n$. The difference $v_m - u_m$ tends to 0 a.s. and in L^1 as $m \to \infty$. On the other hand, we have $\mathbf{E}[u_m | \mathcal{F}_n] \leq \mathbf{E}[X_n | \mathcal{F}_n] \leq \mathbf{E}[v_m | \mathcal{F}_n]$ if $n \geq m$ and hence, applying the convergence theorem for martingales to the left- and right-hand sides,

$$\mathbf{E}[u_m | \mathcal{F}_{\infty-}] \leq \liminf \mathbf{E}[X_n | \mathcal{F}_n] \leq \limsup \mathbf{E}[X_n | \mathcal{F}_n] \leq \mathbf{E}[v_m | \mathcal{F}_{\infty-}]. \tag{3.46}$$

For sufficiently large m, $\mathbf{E}[\mathbf{E}[v_m | \mathcal{F}_{\infty-}] - \mathbf{E}[u_m | \mathcal{F}_{\infty-}]] = \mathbf{E}[v_m - u_m]$ is arbitrarily small, so that $\liminf$ and $\limsup$ in (3.46) are a.s. equal. Finally $u_m \leq X \leq v_m$ a.s. and the limit can be identified as $\mathbf{E}[X | \mathcal{F}_{\infty-}]$. ∎

The next result is an extension of the classical Borel-Cantelli lemma which one can re-cover when X_n are indicators of independent events. It is, also, an example of the so-called local convergence theorem – namely, a theorem asserting a.s. convergence, not on the whole of Ω, but on the subset of Ω where some explicitly given "controlling process" remains bounded.

Theorem 3.1.40 Let $(\Omega, \mathcal{F}, \mathbf{P})$ be a probability space with an increasing family of σ-fields $(\mathcal{F}_n)_{n \geq 0}$ and let (x_n) be a sequence of positive random variables. Suppose that x_n is $\mathcal{F}_n$-measurable for all n. We set

$$X_n = x_0 + ... + x_n, \quad X_\infty = \lim_n X_n$$

$$\tilde{X}_n = x_0 + \mathbf{E}[x_1|\mathcal{F}_0] + ... + \mathbf{E}[x_n|\mathcal{F}_{n-1}], \quad \tilde{X}_\infty = \lim_n X_n.$$

Then, the set $\tilde{A} = \{\tilde{X}_\infty < \infty\}$ is a.s. contained in the set $A = \{X_\infty < \infty\}$ and if the x_n are bounded by a constant M these two sets are a.s. equal.

Proof. The process $Z = X - \tilde{X}$ is a martingale and the r.v. $T = T_N = \inf\{n : \tilde{X}_{n+1} > N\}$ is a stopping time. Hence the process $Y_n = Z_{T \wedge n}$ is a martingale. As Y_n^- is bounded by N, Y converges a.s. But for $\omega \in \tilde{A}$, $T_N(\omega) = +\infty$ for sufficiently large N, hence $Z.(\omega) = Y.(\omega)$ for sufficiently large N and the a.s. convergence of Y implies that of Z. It follows that $X_n(\omega)$ converges a.s. If the x_i are uniformly bounded by M, we can use an analogous argument with $T'_N = \inf\{n : X_n > N\}$; the corresponding martingale Y satisfies $Y_n^+ \leq N + N$ and the conclusion is the same. ∎

3.2 Continuous Parameter Martingales

In this section we cover some parts of martingale theory in continuous time: supermartingales on countable sets, existence of right and left limits of martingales, right-continuous supermartingales, projection and dual projection theorems and decomposition of supermartingales.

3.2.1 Supermartingales on countable sets

On the probability space $(\Omega, \mathcal{F}, \mathbf{P})$ let $(\mathcal{F}_t)_{t \geq 0}$ be an arbitrary (raw) filtration – neither right-continuous nor complete. On it, consider the supermartingale $X = (X_t)_{t \in \mathbb{R}_+}$ and a countable set D dense in $\mathbb{R}_+$. We will show the existence of limits of X along D. However, we begin by reproducing several inequalities that were established for martingales in the discrete case and whose extension to the continuous case is obvious, simply by taking sups over finite subsets of D.

3.2.1.1   Fundamental inequalities

First, the *maximal lemma* for the supermartingale X on D, we have

$$\lambda \mathbf{P}\left(\sup_{t \in D} |X_t| \geq \lambda\right) \leq 3 \sup_{t \geq 0} \mathbf{E}[|X_t|]. \tag{3.47}$$

Consequently, if X is right-continuous on $\mathbb{R}_+$, then

$$\lambda \mathbf{P}\left(\sup_{t \geq 0} |X_t| \geq \lambda\right) \leq 3 \sup_{t \geq 0} \mathbf{E}[|X_t|]. \tag{3.48}$$

This follows immediately from Corollary 3.1.21. And, If the supermartingale X is positive, the right-hand side can be replaced by $\mathbf{E}[X_0]$.

Similarly, by Theorem 3.1.24, if p and q are conjugate exponents with $0 < p < \infty$ and if X is a martingale, then

$$\|\sup_{t \in D} |X_t|\|_p \leq q \sup_{t \in D} \||X_t|\|_p. \tag{3.49}$$

Again, if X is right-continuous, then

$$\|\sup_{t \geq 0} |X_t|\|_p \leq q \sup_{t \geq 0} \||X_t|\|_p. \tag{3.50}$$

Next we state the inequality on the number of up-crossings. Let H be a subset of $\mathbb{R}_+$ and $M_a^b(H, \omega)$ be the number of up-crossings of the path $X.(\omega)$, considered on H, above the interval $[a, b]$. Then, by Theorem 3.1.26, for all $T < +\infty$,

$$\mathbf{E}[M_a^b(D \cap [0, N])] \leq \frac{1}{b-a}(|a| + \sup_{t \leq T} \mathbf{E}[|X_t|]) \tag{3.51}$$

and if X is right-continuous

$$\mathbf{E}[M_a^b([0, N])] \leq \frac{1}{b-a}(|a| + \sup_{t \leq T} \mathbf{E}[|X_t|]). \tag{3.52}$$

As in Remark 3.1.23, all these inequalities can be given in a conditional form with $\mathbf{E}$ replaced by $\mathbf{E}[.|\mathcal{F}_s]$, $s \geq 0$.

Next, we state a fundamental theorem on limits of martingales.

3.2.1.2 Existence of right and left limits

Theorem 3.2.1 (1) For almost all $\omega \in \Omega$ the mapping $t \to X_t(\omega)$ of D into $\mathbb{R}$ has at every point t of $\mathbb{R}_+$ a finite right-hand limit,

$$X_{t+}(\omega) = \lim_{s \in D,\ s \downarrow\downarrow t} X_t(\omega). \tag{3.53}$$

and at every point t of $\mathbb{R}_+ \backslash \{0\}$ a finite left-hand limit,

$$X_{t-}(\omega) = \lim_{s \in D,\ s \uparrow\uparrow t} X_s(\omega). \tag{3.54}$$

(2) For all $t \in \mathbb{R}_+$, the r.v. X_{t+} is integrable and

$$X_t \geq \mathbf{E}[X_{t+}|\mathcal{F}_t] \quad \text{a.s.} \tag{3.55}$$

with equality if the function $t \to \mathbf{E}[X_t]$ is right-continuous, and in particular if X is a martingale. The process $X_+ = (X_{t+})_{t \geq 0}$ is a supermartingale with respect to the family $(\mathcal{F}_{t+})$ and is a martingale if X is.

(3) For all $t > 0$, X_{t-} is integrable and

$$X_{t-} \geq \mathbf{E}[X_t|\mathcal{F}_{t-}] \quad \text{a.s.} \tag{3.56}$$

with equality if the function $t \mapsto \mathbf{E}[X_t]$ is left-continuous, and in particular if X is a martingale. The process $X_- = (X_{t-})_{t \geq 0}$ is a supermartingale with respect to the family $(\mathcal{F}_{t-})$ and is a martingale if X is.

Proof. Note that the processes (X_{t+}) and (X_{t-}) are only defined outside an evanescent set; if it is required that they be defined everywhere, it is sufficient to replace lim by lim sup on the right-hand side of (3.53) and (3.54).

For every ordered pair (a, b) of rationals such that $a < b$, let $\tau_a^b(\omega)$ be the greatest lower bound of the rationals r such that $M_a^b(D \cap [0, r], \omega) = \infty$; τ_a^b is a stopping time of the family $(\mathcal{F}_{t+})$. Also let σ be the greatest lower bound of the rationals r such that $\sup_{s \leq r,\ s \in D} |X_s| = \infty$; σ is a stopping time of $(\mathcal{F}_{t+})$ and so finally is

$$p = \sigma \wedge \inf_{a,b} \tau_a^b. \tag{3.57}$$

The set H of ω such that $X.(\omega)$ has a finite right limit along D at every point of $\mathbb{R}_+$ and a finite left limit at every point of $]0, \infty[$, is then equal to $\{p = \infty\}$. It is more or less obvious that $H \subset \{p = \infty\}$ and this is sufficient to establish (3.53) if the σ-fields are complete. It is, in particular, $\mathcal{F}$-measurable. On the other hand, inequality (3.51) tells us that $\tau_a^b = \infty$ a.s., inequality (3.47) that $\sigma = \infty$ a.s. and property (3.53) is established.

Let $t \in \mathbb{R}_+$ and t_n be a sequence of elements of D such that $t_n \downarrow\downarrow t$. We set $\mathcal{G}_{-n} = \mathcal{F}_{t_n}$, $Y_{-n} = X_{t_n}$; since the process $(Y_k, \mathcal{G}_k)_{k \leq 0}$ is a supermartingale and $\mathbf{E}[Y_k]$ remains bounded by $\mathbf{E}[X_0]$ as $k \to -\infty$, the r.v. Y_k are uniformly integrable by Theorem 3.1.29 and the sequence (X_{t_n}) converges to X_{t+}, not just a.s., but in L^1. In particular, X_{t+} is integrable. We take the limit of the inequality $X_t \geq \mathbf{E}[X_{t_n}|\mathcal{F}_t]$ in L^1 and obtain $X_t \geq \mathbf{E}[X_{t+}|\mathcal{F}_t]$ a.s.; on the other hand $\mathbf{E}[X_{t+}] = \lim_n \mathbf{E}[X_{t_n}]$. Thus, if $\lim_n \mathbf{E}[X_{t_n}] = \mathbf{E}[X_t]$, the positive r.v. $X_t - \mathbf{E}[X_{t+}|\mathcal{F}_t]$ has zero expectation and we have $X_t = \mathbf{E}[X_{t+}|\mathcal{F}_t]$ a.s.

Preserving the notation t and t_n, we consider $s < t$ and $s_n \in D$ with $s_n \downdownarrows s$. We can also assume $s_n < t$ and then, by what has just been proved,

$$X_{s_n} \geq \mathbf{E}[X_t | \mathcal{F}_{s_n}] \geq \mathbf{E}[X_{t+} | \mathcal{F}_t | \mathcal{F}_{s_n}] = \mathbf{E}[X_{t+} | \mathcal{F}_{s_n}]$$

and consequently $X_{s+} \geq \mathbf{E}[X_{t+} | \mathcal{F}_{s+}]$ by Theorem 3.1.29, and the process (X_{t+}) is a supermartingale with respect to $(\mathcal{F}_{t+})$. If X is a martingale, we have seen that $\mathbf{E}[X_{t+}] = \mathbf{E}[X_t]$, a constant function of t, and X_+ is also a martingale.

We now come to part (3) concerning left limits, forewarning the reader that it is much less important than (2) (its extension to predictable stopping times, in Theorem 3.1.15, is however very important). Let $t > 0$ and t_n be a sequence of elements of D such that $t_n \uparrow\uparrow t$ and set $\mathcal{G}_n = \mathcal{F}_{t_n}$, $\mathcal{G}_\infty = \mathcal{F}_t$, $Y_n = X_{t_n}$ and $Y_\infty = X_t$. The martingale $(Y_n, \mathcal{G}_n)$ is right closed by Y_∞ and converges a.s. to $y = X_{t-}$ and Theorems 3.1.27-3.1.28 then tell us that y is integrable and closes the supermartingale (Y_n), and that $y \geq \mathbf{E}[Y_\infty | \mathcal{G}_\infty -]$, that is (3.56).

Now let $s \in]0, t[$ we can assume that $t_0 = s$ and the fact that y closes (Y_n) can be written

$$X_s \geq \mathbf{E}[X_{t-} | \mathcal{F}_s] \tag{3.58}$$

in particular $\mathbf{E}[X_s] \geq \mathbf{E}[X_{t-}]$ and hence $\lim_{s \uparrow\uparrow t} \mathbf{E}[X_s] \geq \mathbf{E}[X_{t-}]$. We deduce that $\mathbf{E}[X_t] = \mathbf{E}[X_{t-}]$ if the function $t \to \mathbf{E}[X_t]$ is left-continuous and equality holds in (3.56). Returning to the general case and combining (3.57) and (3.56), we have

$$X_{s-} \geq \mathbf{E}[X_s | \mathcal{F}_{s-}] \geq \mathbf{E}[X_{t-} | \mathcal{F}_s | \mathcal{F}_{s-}] = \mathbf{E}[X_{t-} | \mathcal{F}_{s-}]$$

so that (X_{t-}) is a supermartingale with respect to $(\mathcal{F}_{t-})$. If X is a martingale, the function $t \to \mathbf{E}[X]$ is constant and X_- is a martingale. ∎

We now give two fundamental corollaries of Theorem 3.2.1.

Theorem 3.2.2 Let $X = (X_t)_{t \geq 0}$ be a supermartingale with respect to $(\mathcal{F}_t)$ such that for almost all ω the path $X.(\omega)$ is right-continuous on $\mathbb{R}_+$. Then

(1) X is a supermartingale with respect to the family $(\mathcal{F}_{t+})$ and with respect to the usual augmentation $(\mathcal{F}_t^\star)$ of $(\mathcal{F}_t)$.

(2) For almost all ω the path $X.(\omega)$ is cadlag.

Proof. We apply Theorem 3.2.1 with D any countable dense set. The processes X and X_+ are indistinguishable and (1) follows from Theorem 3.2.1 part (2), since X is adapted to $(\mathcal{F}_{t+})$; the adjunction to $\mathcal{F}_0$ of the $\mathbf{P}$-negligible sets of the completion clearly does not alter the supermartingale inequality. Property (2) then follows from Theorem 3.2.1, part 1. ∎

Theorem 3.2.3 If the supermartingale X is such that $t \to \mathbf{E}[X_t]$ is right-continuous and if $(\mathcal{F}_t)$ satisfies the usual conditions, then X has a modification all of whose paths are r.c.l.l.

Proof. We take D to be an arbitrary countable dense set and return to the proof of Theorem 3.2.1: the required modification is

$$Y_t(\omega) = X_{t+}(\omega) \text{ if } \omega \in H, \ Y_t(\omega) = 0 \quad \text{if} \quad \omega \notin H. \tag{3.59}$$

For H^c is negligible and hence belongs to $\mathcal{F}_0$ and Y is therefore an adapted cadlag process. On the other hand (3.55) here can be written as $X_t \to \mathbf{E}[X_{t+}|\mathcal{F}_t]$ since $\mathbf{E}[X_s]$ is a right-continuous function of s; since the family $(\mathcal{F}_s)$ is right-continuous, this can be written simply as $X_t = X_{t+}$ a.s. and Y is indeed a modification of X. ∎

Remark 3.2.4 The proof of Theorem 3.2.1 gives us a slightly more precise result than Theorem 3.2.3. We suppose that the family $(\mathcal{F}_t)$ is right-continuous but make no assumption with regard to completion. Then, the process (3.59) is no longer adapted and we must construct another modification. We take (in the notation of the proof of Theorem 3.2.1)

$$Y_t = X_{t+} \quad \text{if} \quad 0 \leq t < \rho, \quad Y_t = 0 \quad \text{if} \quad t \geq \rho. \tag{3.60}$$

This time Y is adapted (since ρ is a stopping time of the family $(\mathcal{F}_{t+}) = (\mathcal{F}_t)$) and has right-continuous paths, but the left limit $Y_{t-}(\omega)$ does not necessarily exist for all $t > 0$: it exists for $t < \rho(\omega)$ and also for $t > \rho(\omega)$, but if $\rho(\omega) < \infty$ there is an "explosion" of the path at the instant $t = \rho(\omega)$, either it diverges to infinity or it oscillates without a limit.

We now verify that the process (Y_t) thus constructed is optional with respect to the family $(\mathcal{F}_t)$ (without completion, right-continuous adapted processes are no longer necessarily optional!). We return to the proof of Theorem 2.4.26, writing for all $\epsilon > 0$, $T_0^\epsilon \to 0$ and then

$$T_{\alpha+1}^\epsilon = \inf\{t > T_\alpha^\epsilon : |Y_t - Y_{T_\alpha^\epsilon}| > \epsilon\}$$

for every countable ordinal α,

$$T_\beta^\epsilon = \sup_{\alpha < \beta} T_\alpha^\epsilon$$

for every limit ordinal β. The T_α^ϵ are stopping times of the family $(\mathcal{F}_{t+})$ and the argument of Theorem 2.4.26 shows us that a right-continuous, adapted process Y is optional so long as there exists, for all $\epsilon > 0$, a countable ordinal γ such that $T_\gamma^\epsilon = \infty$. Here the stopping times T_n^ϵ can only accumulate at finite distance at the instant ρ and hence we have $T_\gamma^\epsilon \geq \rho$ from the first infinite ordinal γ onwards. Since the process Y is constant after ρ we have $T_{\gamma+1}^\epsilon = +\infty$.

It is convenient to work with a well-defined process with left limits. First we set

$$\bar{Y}_{t-} = \limsup_{s\uparrow\uparrow t} Y_s = \limsup_{s\uparrow\uparrow t, s\in D} X_s$$

and similarly for $\underline{Y}_{t-}$ with $\liminf$, and then

$$Y_{t-} = \bar{Y}_{t-} \quad \text{if} \quad Y_t = \bar{Y}_{t-}, \; Y_{t-} = 0 \quad \text{otherwise.} \tag{3.61}$$

Returning to the arguments of Theorems 2.5.1-2.5.2, it can be checked without difficulty that this process is predictable.

Remark 3.2.5 Let X be a supermartingale which is progressively measurable with respect to a filtration $(\mathcal{F}_t)$ for simplicity, we shall assume that it satisfies the *usual conditions*. By Theorem 2.3.18, X has a modification Y which is progressively measurable and right D-separable, where D is a suitable countable dense set. By Theorem 3.2.1, for almost all ω the path $Y.(\omega)$ has right limits along D; as Y is D-separable, these are also limits along $\mathbb{R}_+$. By Fubini's theorem, the set $\{(t, \omega) : X_t(\omega) \neq Y_t(\omega)\}$ is negligible for the measure $\lambda \otimes \mathbf{P}$, where λ is Lebesgue measure and hence the set $\{t : X_t(\omega) \neq Y_t(\omega)\}$ is λ-negligible for almost all ω. Hence, the paths $X.(\omega)$ and $Y.(\omega)$ have the same cluster values in the right essential topology; but $Y.(\omega)$ has right limits in the ordinary topology and a fortiori in the right essential topology. Hence for almost all ω,

$$Y_{t+}(\omega) = \lim ess_{s \downarrow \downarrow t} X_s(\omega). \tag{3.62}$$

Consequently, the process $\overline{X}_t = \liminf ess_{s \downarrow \downarrow t} X_s$ (whose definition involves neither Y nor D) is indistinguishable from a cadlag supermartingale. There are analogous considerations for left limits.

3.2.2   Right-continuous supermartingale

Here we consider right-continuous supermartingales as *extensions* of the discrete case. We assume henceforth that the family $(\mathcal{F}_t^\star)$ satisfies the usual conditions. By Theorem 3.2.2, this implies no loss of generality in the study of right-continuous supermartingales. We begin by repeating the convergence theorems. The notion of a right closed supermartingale (martingale) is defined as in the discrete case (Definition 3.1.8). Since cadlag processes on usual augmentation are the most studied process in stochastic calculus, and, our focus is ladlag processes on *un*usual probability spaces, in this section, we are going to state results, many without proofs and refer the reader back to the source [20].

Theorem 3.2.6 Let X be a right-continuous supermartingale (martingale). Suppose that $\sup_t \mathbf{E}[|X_t|] < \infty$, a condition equivalent to $\lim_{t \to \infty} \mathbf{E}[X_t^-]$. Then the r.v. $\lim_{t \to \infty} X_t = \xi$ exists a.s. and is integrable.

The above condition is satisfied if the supermartingale (martingale) X is closed on the right by a r.v. X_∞; the r.v. then closes the supermartingale (martingale) X and $A \geq \mathbf{E}[X_\infty | \mathcal{F}_{\infty-}^\star]$ ($\xi = \mathbf{E}[X_\infty | \mathcal{F}_{\infty-}^\star]$). If the supermartingale (martingale) X is uniformly integrable, ξ exists and closes X on the right.

Proof. Similar to the discrete case of Theorems 3.1.27, 3.1.28 and 3.1.30. ■

The continuous parameter analogue of supermartingales indexed by the negative integers is provided by supermartingales over the interval $]0, \infty[$, open at 0.

Theorem 3.2.7 Let $X = (X_t)_{t>0}$ be a right-continuous supermartingale such that $\sup_{t \leq 1} \mathbf{E}[|X_t|] < \infty$, a condition equivalent to $\lim_{t \to 0} \mathbf{E}[|X_t|] < \infty$. Then, the limit $\lim_{t \to 0} \mathbf{E}[|X_t|] = \xi$, exists a.s.; moreover X_t converges to ξ in L^1 as $t \to \infty$.

Proof. The proof is identical to that of the discrete case of Theorem 3.1.29. ■

Next we sill state Riesz decompositions 3.1.33-3.1.35.

Theorem 3.2.8 Riesz Decompositions. Let X be a right-continuous supermartingale such that $\sup_t \mathbf{E}[X_t^-] = \lim_{t \to \infty} \mathbf{E}[X_t^-] < \infty$. Then, we know that the limit $\xi = \lim_{t \to \infty} X_t$ exists a.s. and is integrable and X has the decomposition $X = U + V$, where $U = (U_t)$ is a right-continuous version of the uniformly integrable martingale $\mathbf{E}[\xi | \mathcal{F}_t]$ and $V_t = X_t - U_t$ is a right-continuous supermartingale which is zero a.s. at infinity (and is positive if the r.v. X_t^- are uniformly integrable).

Definition 3.2.9 Potentials. A positive supermartingale Z which is right-continuous and such that $\lim_{t \to \infty} \mathbf{E}[Z_t] = 0$ is called a potential.

Theorem 3.2.10 Every right-continuous supermartingale X which is positive (or more generally such that $\lim_{t \to \infty} \mathbf{E}[X_t] > -\infty$) can be decomposed as a sum $X = Y + Z$, where Y is a right-continuous martingale and Z is a potential. This decomposition is unique except on an evanescent set and Y is the greatest right-continuous martingale (or submartingale) bounded above by X.

The proof given for Corollary 3.1.35 extends immediately, from the discrete case to the continuous case for the above theorem. Further, there is no difficulty concerning right-continuity; the martingales $Y_t^m = \mathbf{E}[X_{t+m} | \mathcal{F}_t^\star]$ decrease to a martingale (Y_t) and the process $Z_t = X_t - Y_t$ is a potential. Taking right limits along the rationals, we have $X = Y_+ - Z_+$ and we have indeed obtained a decomposition of X as a right-continuous positive martingale and a right-continuous potential.

Now, we have come to the stopping theorem; it should be noted that the theorem applies to bounded stopping times without the condition that X is closed. In order to give the most general definition of the σ-fields $\mathcal{F}_T^\star$, we provide an additional σ-field $\mathcal{F}_\infty^\star$.

Theorem 3.2.11 Let X be a right-continuous supermartingale (martingale) which is closed on the right by a r.v. X_∞ and is $\mathcal{F}_\infty^\star$-measurable and integrable. Let S and T be two stopping times such that $S \leq T$. Then X and X_T are integrable and

$$X_S \geq \mathbf{E}[X_T | \mathcal{F}_S^\star] \quad \text{a.s.}$$

with equality if X_∞ closes X as a martingale.

Theorem 3.2.12 If X is a right-continuous supermartingale (martingale) and U a stopping time, the stopped process $X^U = (X_{t \wedge U})$ is a supermartingale (martingale).

Now we come to an interesting and perhaps a useful consequence of the stopping theorem, which, is only true under the usual conditions and that there is no comparable result for supermartingales.

Theorem 3.2.13 Let (Z_t), $t \in [0, \infty]$ be an optional process. Suppose that for every stopping time T the r.v. Z_T is integrable and that $\mathbf{E}[Z_T]$ does not depend on T. Then, (Z_t) is a uniformly integrable, right-continuous martingale (except on an evanescent set).

Theorems 3.2.14-3.2.16 have no analogues in the discrete case either because they concern left limits of processes, or because they involve local behavior of paths. We begin with the predictable form of the stopping theorem. We recall the convention $\mathcal{F}_{0-}^{\star} = \mathcal{F}_{0}^{\star}$.

Theorem 3.2.14 Let X be a right-continuous supermartingale (martingale) which is closed on the right by a r.v. X_{∞}, We adopt the convention $X_{0-} = X_0$. Let S and T be two predictable stopping times such that $S \leq T$. Then X_S and X_T are integrable and

$$X_{S-} \geq \mathbf{E}\left[X_{T-} | \mathcal{F}_{S-}^{\star}\right] \geq \mathbf{E}\left[X_T | \mathcal{F}_{S-}^{\star}\right] \quad \text{a.s.} \tag{3.63}$$

with equality if X_{∞} closes X as a martingale.

From the stopping theorem we deduce a result on the zeros of positive supermartingales, which is a probabilistic version of the minimum principle of potential theory (i.e., a positive superharmonic function which is zero at one point is zero everywhere).

Theorem 3.2.15 Let X be a positive right-continuous supermartingale. We set

$$T(\omega) = \inf\{t : X_t(\omega) = 0 \text{ or } t > 0, \ X_{t-}(\omega) = 0\}. \tag{3.64}$$

Then, for almost all ω, the function $X.(\omega)$ is zero on $[T(\omega), \infty[$.

In potential theory it is shown that the limit of an increasing sequence of superharmonic functions is also superharmonic. We now give the probabilistic analogue of this result.

Theorem 3.2.16 Let (X^n) be an increasing sequence of right-continuous supermartingales. We set

$$X_t = \sup_n X_t^n. \tag{3.65}$$

Then, the process X is indistinguishable from a cadlag process. We say process and not supermartingale, as X_t is not necessarily integrable.

We go back to the raw filtration $(\mathcal{F}_t)$, without the usual conditions, and state the following definition and theorem:

Definition 3.2.17 Let X be a measurable process. X is said to be bounded in L^1 (with respect to $(\mathcal{F}_t)$) if the number

$$||X||_1 = \sup_T \mathbf{E}[|X_T|\mathbf{1}_{\{T<\infty\}}] \tag{3.66}$$

is finite, where T runs through the set of all stopping times of $(\mathcal{F}_t)$. Moreover, if all the r.v. $X_T\mathbf{1}_{\{T<\infty\}}$ are uniformly integrable, X is said to belong to class D.

Theorem 3.2.18 The space of optional processes (resp. optional processes with cadlag, r.c., continuous, paths) X such that $||X||_1 < \infty$ is complete with respect to the norm $||\cdot||_1$.

3.2.3 Projections theorems

This section contains complements to the general theory of processes. In substance, it does not involve delicate results, at least under the usual conditions, but simple variations on the theme of the existence of cadlag versions of martingales. We begin by defining the optional and predictable projection of measurable processes and say a few words about the regularity of their paths. Then, a digression on increasing processes which is preparatory not only for the notion of a dual projection, but for the whole decomposition theory of supermartingales. Finally, we define the notion of a dual projection of a process.

We adopt the usual conditions for $(\mathcal{F}_t^\star)$, with the convention $\mathcal{F}_{0-}^\star = \mathcal{F}_0^\star$. For the case under the raw filteration $(\mathcal{F}_t)$, see Chapter 4. The optional and predictable projections are given by the following theorem.

Theorem 3.2.19 Let X be a positive or bounded measurable process. There exist an optional process Y and a predictable process Z such that

$$\mathbf{E}[X_T\mathbf{1}_{T<\infty}|\mathcal{F}_T^\star] = Y_T\mathbf{1}_{T<\infty} \quad \text{a.s.} \tag{3.67}$$

for every stopping time T,

$$\mathbf{E}[X_T\mathbf{1}_{T<\infty}|\mathcal{F}_{T-}^\star] = Z_T\mathbf{1}_{T<\infty} \quad \text{a.s.} \tag{3.68}$$

for every predictable time T. Y and Z are unique to within evanescent processes. They are called the optional projection and the predictable projection of X and they are denoted by ${}^o X$ and ${}^p X$.

Proof. Note first that since we didn't require integrability condition on X, Z and Y may take the value $+\infty$. Uniqueness follows from optional and predictable cross-section Theorems 2.4.48 and 2.4.51; we also get: (a) If X and $\overline{X}$ have projections $Y, \overline{Y}$ and $Z, \overline{Z}$, and $X \leq \overline{X}$, then also, $Y \leq \overline{Y}$ and $Z \leq \overline{Z}$ outside an evanescent set. In particular, if K is a constant and $|X| \leq K$, then $|Y| \leq K$ and $|Z| \leq K$ outside an evanescent set. (b) If positive or uniformly bounded processes X^n have projections Y^n and Z^n and increase to a process X, then X has the projections $Y = \liminf_n Y^n$ and $Z = \liminf_n Z^n$.

There is an analogous statement when bounded X^n (admitting projections) converge uniformly to X. Finally, clearly if X and $\overline{X}$ have projections $Y, Z, \overline{Y}, \overline{Z}$, then $\lambda X + \mu \overline{X}$ has projections $\lambda Y + \mu \overline{Y}$ and $\lambda Z + \mu \overline{Z}$ for $\lambda, \mu \in \mathbb{R}$.

These properties will enable us to argue using monotone classes (Theorem 1.2.10) to establish the existence of projections for all bounded measurable processes, it is sufficient to produce explicitly a set of processes admitting predictable projections, which is stable under multiplication and generates the σ-field $\mathcal{B}(\mathbb{R}_+) \times \mathcal{F}$. We shall choose processes of the form

$$X_t(\omega) = \mathbf{1}_{[0,u[}(t)H(\omega), \quad 0 < u \le \infty, \ H \in L^\infty(\mathcal{F}).$$

Let (H_t) be a cadlag version of the martingale $\mathbf{E}[H|\mathcal{F}_t^\star]$ (with the convention $H_{0-} = H_0$). Then, Theorems 3.2.11 and 3.2.14 imply that the processes

$$Y_t(\omega) = \mathbf{1}_{[0,u[}(t)H_t(\omega), \quad Z_t(\omega) = \mathbf{1}_{[0,u[}(t)H_{t-}(\omega) \tag{3.69}$$

satisfy (3.67) and (3.68). To deal with the case of positive measurable processes X, we apply the above result to $X \wedge n$ and let n tend to infinity. ∎

The two projections do not differ by much:

Theorem 3.2.20 Let X be a positive or bounded measurable process. The set of (t, ω) such that

$$^oX_t(\omega) \neq {}^pX_t(\omega)$$

is a countable union of graphs of stopping times of $(\mathcal{F}_t^\star)$.

How regular are the paths of projection processes? The following theorem tells us something about that.

Theorem 3.2.21 Let X be a bounded measurable process and Y and Z be, respectively, the optional projection and the predictable projection of X.
   (1) If X is right-continuous (resp. cadlag), Y is right-continuous (resp. cadlag).
   (2) If X is left-continuous (resp. caglad), Z is left-continuous (resp. caglad).
   Note also that if X is continuous, Y and Z are not necessarily so.

The theorem below is an immediate consequence of the definition of projections:

Theorem 3.2.22 Let Y be a bounded optional process. Then,
   (a) Suppose that for every decreasing sequence (T_n) of bounded stopping times, $\lim_n \mathbf{E}[Y_{T_n}]$ exists. Then, the paths of Y have right limits. If further $\lim_n \mathbf{E}[Y_{T_n}] = \mathbf{E}[Y_{\lim_n T_n}]$ for every sequence (T_n) as above, Y is right-continuous.
   (b) Suppose that for every uniformly bounded increasing sequence (S_n) of stopping times $\lim_n \mathbf{E}[Y_{S_n}]$ exists. Then, the paths of Y have left limits.

Here is the corresponding result for the predictable case.

Theorem 3.2.23 Let Z be a bounded predictable process. Suppose that for every uniformly bounded increasing sequence (S_n) of predictable times $\lim \mathbf{E}[Z_{S_n}]$ exists. Then the paths of Z have left limits. If further $\lim_n \mathbf{E}[Z_{S_n}] = \mathbf{E}\left[Z_{\lim_n S_n}\right]$ for every sequence (S_n) as above, Z is left-continuous.

Before proceeding to the notion of a dual projection, we need some results on increasing processes.

Definition 3.2.24 An increasing process is any process $(A_t)_{t \geq 0}$ adapted to the family $(\mathcal{F}_t^\star)$, whose paths are positive, increasing, finite, and right-continuous functions on $[0, \infty[$.

Every increasing process is therefore optional. An essential idea behind an increasing process A is that it is a mean of representing a certain random measure on $\mathbb{R}_+$, the measure $dA_t(\omega)$ for which $A_.(\omega)$ is its distribution function. For this reason we always make the convention $A_{0-} = 0$, so that A_0 represents the mass of the measure at the point 0. We denote by $A_{\infty-}$ the limit $\lim_{t \to \infty} A_t$ which is the total mass of the measure.

An increasing process A is called integrable if $\mathbf{E}[A_{\infty-}] < \infty$. The differences of increasing processes (resp. integrable increasing processes) are called processes of *finite variation* (resp. integrable variation). We sometimes consider increasing processes $(A_t)_{0 \leq t \leq \infty}$ representing measures on $[0, \infty]$. These measures may have a jump at infinity, equal to $A_\infty - A_{\infty-}$. Of course, such a process is then called integrable if $\mathbf{E}[A_\infty] < \infty$. An increasing process A is called continuous if its paths are continuous, including at 0 (i.e., if $A_0 = 0$). The measure $dA_.(\omega)$ then has no atom on $[0, \infty[$.

Now, we elucidate the structures of increasing processes and predictable increasing processes by the following theorem:

Theorem 3.2.25 Let A be an increasing process. There exist a continuous increasing process A^c, a sequence (T_n) of stopping times with graphs in general not disjoint, and a sequence (λ_n) of constants > 0, such that

$$A_t = A_t^c + \sum_n \lambda_n \mathbf{1}_{\{T \leq t\}} \tag{3.70}$$

If A is predictable, then T_n can be chosen predictable. The process A^c is unique and is called the continuous part of A. The process $A - A^c$ is denoted by A^d and is called the (purely) discontinuous part or jump part of A.

Remark 3.2.26 (a) For the proof of the above theorem we must employ the decomposition,

$$A_t^d = \sum_{0 \le s \le t} \triangle A_s = \sum_n H_n 1\{S_n \le t\}$$

which is quite useful in itself, where the S_n are optional (predictable) stopping times with disjoint graphs and the H_n are r.v. which are measurable with respect to $\mathcal{F}_{S_n}^\star$ ($\mathcal{F}_{S_n-}^\star$).

(b) The proof applies just as well to a process, A of finite variation, giving decomposition

$$A_t = A_t^c + A_t^d = A_t^c + \sum_n H_n 1\{t \ge T_n\} \tag{3.71}$$

where the graphs of the T_n are disjoint. Then, we also have

$$\int_0^t |dA_s^d| = \sum_n |H_n| 1\{T_n \le t\}$$

$$\int_0^t |dA_s^c| = \lim_n \sum_{k2^n} \left| A_{(k+1)2^{-n}t}^c - A_{k2^{-n}t}^c \right|$$

so that the two increasing processes on the left are adapted (predictable). Thus if A is a predictable adapted process of finite variation, the process $\int_0^t |dA_s|$ is adapted (predictable).

(c) Let S be a predictable time and H a positive $\mathcal{F}_{S_-}^\star$-measurable r.v. Then, the increasing process $A_t = H1_{S \le t}$ is predictable. It is sufficient to return to the dyadic decomposition of H given above, $H = \sum_k c_k 2^{-k}$, and write $A_t = \sum_k 2^{-k} 1_{\{S_{C_k} \le t\}}$ where $C_k = \{c_k \ne 0\}$.

(d) Here is an interesting consequence of (b) and (c): let (A_t) be an optional (predictable) increasing process decomposed in the above form

$$A_t = A_{t-}^c + \sum_n H_n 1\{S_n \le t\}$$

and let (K_t) be an optional (predictable) process, which we at first assume to be positive and bounded. Then, the increasing process

$$B_t = \int_{[0,t]} K_s dA_s$$

has the decomposition

$$B_t = \int_{[0,t]} K_t dA_s^c + \sum_n H_n K_{S_n} 1\{S_n \le t\}.$$

The first process on the right-hand side is adapted and continuous, and on the other hand K_{S_n} is measurable with respect to $\mathcal{F}_{S_n}^\star$ ($\mathcal{F}_{S_n-}^\star$). Hence B is optional (predictable). We pass without difficulty to the case where K is not necessarily bounded, or indeed, by taking a difference, to the case where A is of finite variation. Note that all this enables us the possibility of developing a Radon-Nikodym theory for increasing processes.

(e) The same result applies to increasing processes, or processes of finite variation, indexed by $[0, \infty]$. Let us give a universal method of reducing problems relative to $[0, \infty]$ to the corresponding problems on $[0, \infty[$:

For $t \in [0,1[$ we set $\alpha(t) = \frac{t}{1-t}$; for $u \in [0,\infty[$ we set $\beta(u) = \frac{u}{1+u}$; α and β are two inverse isomorphisms between the ordered sets $[0,1[$ and $[0,\infty[$ which can be extended to $[0,1]$ and $[0,\infty]$. For all $t \in \mathbb{R}$ we set

$$\mathcal{G}_t = \mathcal{F}^\star_{\alpha(t)} \text{ for } 0 < t < 1, \mathcal{G}_t = \mathcal{F}^\star_\infty \text{ for } t \geq 1;$$

when considering increasing processes $(A_t)_{0 \leq t \leq \infty}$, we must specify the σ-field $\mathcal{F}^\star_\infty$ to which A_∞ is adapted. Then,

$$B_t = A_{\alpha(t)} \text{ for } 0 \leq t < 1, B_t = A_\infty \text{ for } t \geq 1.$$

We then apply Theorem 3.2.25 to (B_t). As B is constant after 1, we can if necessary replace T_n by $T_n \wedge 1$. Finally we return to the original problem by means of the isomorphism β of $[0,\infty]$ onto $[0,1]$.

Next, we present Lebesgue's lemma on the transformation of Stieltjes integrals to the ordinary Lebesgue integral. This lemma and the next are essential for establishing optional (predictable) dual projections.

Theorem 3.2.27 Suppose that a is finite on $[0,\infty[$. Let f be a positive Borel function on $[0,\infty[$. Then,

$$\int_{[0,\infty[} f(s)da(s) = \int_0^\infty f(c_-(s))\mathbf{1}_{c_-(s)<\infty}ds \tag{3.72}$$

$$= \int_0^\infty f(c(s))\mathbf{1}_{c(s)<\infty}ds$$

More generally, if G is a finite, positive, right-continuous function on $[0,\infty[$ (with the convention $G(0-) = 0$) then

$$\int_{[0,\infty[} f(s)dG(a(s)) = \int_{[0,\infty[} f(c_-(s))\mathbf{1}_{c_-(s)<\infty}dG(s). \tag{3.73}$$

Where a is a function on $[0,\infty[$, which is positive, increasing and right-continuous. We require neither that $a(0) = 0$ nor that a be finite; we adopt the conventions $a(0-) = 0$ and $a(\infty) = \infty$, but $a(\infty-) = \lim_t a(t)$ may be finite. We define the second increasing function c by setting $c(s) = \inf\{t : a(t) > s\}$ $(0 \leq s < \infty)$ with the conventions $c(0-) = 0$, $c(\infty) = \infty$; c is increasing, right-continuous and with left limits.

Now, we introduce the random version of the preceding result by way of the following definition:

Definition 3.2.28 Let (A_t) be a (not necessarily finite) increasing process. The change of time associated with (A_t) is the process

$$C_t = \inf\{s : A_s > t\}. \tag{3.74}$$

In general, a change of time in the theory of processes is any right-continuous increasing family $(C_t)_{t \geq 0}$ of stopping times. Furthermore, $\{A_t \geq s\} = \{C_{s-} \leq t\}$ belongs to $\mathcal{F}_t^\star$ for all s, hence (A_t) is not necessarily finite increasing process.

The notion of a change of time is a technical tool with which one can prove the following theorems.

Theorem 3.2.29 Let X be a positive measurable process and Y and Z its optional and predictable projections. Let A be an increasing process. Then,

$$\mathbf{E}\left[\int_{[0,\infty[} X_s dA_s\right] = \mathbf{E}\left[\int_{[0,\infty[} Y_s dA_s\right] \tag{3.75}$$

and if A is predictable

$$\mathbf{E}\left[\int_{[0,\infty[} X_s dA_s\right] = \mathbf{E}\left[\int_{[0,\infty[} Z_s dA_s\right]. \tag{3.76}$$

Theorem 3.2.30 Let (A_t) be a raw integrable increasing process (or one of integrable variation).
    (a) Suppose that for every bounded measurable process X, we have

$$\mathbf{E}\left[\int_{[0,\infty[} X_s dA_s\right] = \mathbf{E}\left[\int_{[0,\infty[} {}^o X_s dA_s\right] \tag{3.77}$$

Then, A is optional.
    (b) If under the same conditions, we have

$$\mathbf{E}\left[\int_{[0,\infty[} X_s dA_s\right] = \mathbf{E}\left[\int_{[0,\infty[} {}^p X_s dA_s\right], \tag{3.78}$$

A is predictable.

Remark 3.2.31 A process which is not necessarily adapted but whose paths have this property will be called a not necessarily adapted increasing process or a raw increasing process. Raw increasing processes are not more general objects than increasing processes: they are the increasing processes for the family of σ-fields $\mathcal{F}_t = \mathcal{F}$ ($\mathcal{F}_t^\star = \mathcal{F}$ under usual conditions).

Next, we shall show how to construct increasing processes by disintegration of measures on $\mathbb{R}_+ \times \Omega$, which enables us to construct the dual projections of a raw increasing process.

Theorem 3.2.32 For every positive **P**-measure (resp. every bounded **P**-measure) μ on $\mathfrak{M}^\star$ there exists a raw integrable increasing process (resp. a raw process of integrable variation) A, which is unique to within an evanescent process, such that for every bounded measurable process X,

$$\mu(X) = \mathbf{E}\left[\int_{[0,\infty[} X_s dA_s\right] \tag{3.79}$$

A is called the raw integrable increasing process (resp. process of integrable variation) associated with.

For A to be optional (resp. predictable), it is necessary and sufficient that for every bounded measurable process X

$$\mu(X) = \mu(\,^o X) \tag{3.80}$$
$$\mu(X) = \mu(\,^p X) \tag{3.81}$$

The theorem above affirms that every **P**-measure can be obtained by a pair (A, X). $\mathfrak{M}$ to denote the σ-field of all measurable sets $(\mathcal{B}(\mathbb{R}_+) \times \mathcal{F})$ not necessarily completed by **P**, $\mathcal{O}$ the optional σ-field generated by the cadlag processes adapted to $(\mathcal{F}_t)$) and $\mathcal{P}$ the predictable σ-field generated by the adapted left-continuous processes. Augmenting these σ-fields with all the **P**-evanescent sets, we obtain σ-fields denoted by $\mathfrak{M}^\star$ (resp. $\mathcal{O}^\star$ and $\mathcal{P}^\star$).

We now come to the applications of Theorem 3.2.32. We first obtain very easily a Radon-Nikodym Theorem for processes of integrable variation.

Theorem 3.2.33 Let A and B be two optional (resp. predictable) processes of integrable variation. Suppose that for almost all ω the measure $dB.(\omega)$ on $\mathbb{R}_+$ is absolutely continuous with respect to $dA.(\omega)$. Then, there exists an optional (resp. predictable) process $(H_t)_{t \geq 0}$ such that

$$B_t = \int_{[0,t]} H_s dA_s \text{ and } \mathbf{E}\left[\int_{[0,\infty[} |H_s||dA_s|\right] < \infty \tag{3.82}$$

Another consequence of Theorem 3.2.32 can be interpreted as the uniqueness theorem for the Doob decomposition supermartingales, which we cover in the next section.

Theorem 3.2.34 (a) Let A and B be two (optional) processes of integrable variation. If for every stopping time T,

$$\mathbf{E}\left[A_\infty - A_{T-}|\mathcal{F}_T^\star\right] = \mathbf{E}\left[B_\infty - B_{T-}|\mathcal{F}_T^\star\right] \quad \text{a.s.} \quad (A_{0-} = B_{0-} = 0)$$

then A and B are indistinguishable.

(b) Similarly, if A and B are predictable and $A_0 = B_0$ and for all t

$$\mathbf{E}\left[A_\infty - A_t|\mathcal{F}_t^\star\right] = \mathbf{E}\left[B_\infty - B_t|\mathcal{F}_t^\star\right] \quad \text{a.s.}$$

then A and B are indistinguishable.

Next we give a definition of dual projections and some trivial consequences.

Definition 3.2.35 Let A be a raw integrable increasing process. The optional (resp. predictable) *dual projection* of A is the optional (resp. predictable) increasing process B defined by

$$\mathbf{E}\left[\int_{[0,\infty[} X_s dB_s\right] = \mathbf{E}\left[\int_{[0,\infty[} {}^oX_s dA_s\right] \quad \text{resp. } \mathbf{E}\left[\int_{[0,\infty[} {}^pX_s dA_s\right]. \tag{3.83}$$

Remark 3.2.36 Ordinary projections are denoted by ${}^oX$, ${}^pX$, the notation for the dual projections is A^o and A^p. Definition 3.2.35 can be extended to raw processes of integrable variation or again to processes indexed by $[0,\infty]$. Moreover, we emphasize the difference between a projection and a dual projection: formally, the projection operation consists of suitably defining the process $\mathbf{E}[A_t|\mathcal{F}_t^\star]$, whereas the dual projection operation consists of defining the symbolic integral $\int_0^t \mathbf{E}[dA_s|\mathcal{F}_s^\star]$. Note also that the dual projection of a bounded process of integrable variation is not in general a bounded process, unlike the ordinary projection.

Now, we present some trivial but useful results:

Theorem 3.2.37 Let A and B be two raw increasing processes of integrable variation.

(a) For A and B to have the same optional dual projection, it is necessary and sufficient that for every stopping time T

$$\mathbf{E}[A_\infty - A_{T-}|\mathcal{F}_T^\star] = \mathbf{E}[B_\infty - B_{T-}|\mathcal{F}_T^\star] \text{ a.s. } (A_{0-} = B_{0-} = 0). \tag{3.84}$$

(b) For A and B to have the same predictable dual projection it is necessary and sufficient that

$$\mathbf{E}[A_0|\mathcal{F}_{0-}^\star] = \mathbf{E}[B_0|\mathcal{F}_{0-}^\star] \text{ a.s.;} \tag{3.85}$$
$$\mathbf{E}[A_\infty - A_t|\mathcal{F}_t^\star] = \mathbf{E}[B_\infty - B_t|\mathcal{F}_t^\star] \text{ a.s. } (t \geq 0).$$

Remark 3.2.38 We denote by B an integrable raw increasing process, which is zero at 0 and may have a jump at infinity; A is its *predictable dual projection* (the case where B is predictable is not excluded; then $B = A$). We denote by N the cadlag version of the martingale $\mathbf{E}[A_\infty|\mathcal{F}_t^\star]$, ($t \in [0,\infty]$) and by Z the positive cadlag supermartingale $Z_t = N_t - A_t$. By Theorem 3.2.37 we have for every stopping time T

$$Z_T = \mathbf{E}[A_\infty|\mathcal{F}_T^\star] - A_T = \mathbf{E}[A_\infty - A_T|\mathcal{F}_T^\star] = \mathbf{E}[B_\infty - B_T|\mathcal{F}_T^\star] \tag{3.86}$$

so that Z is the optional projection of the processes $(B_\infty - B_t)$ and $(A_\infty - A_t)$. Note that $Z_\infty = 0$ and that Z belongs to class D; Z is a potential of class D if and only if $Z_{\infty-} = \mathbf{E}[B_\infty - B_{\infty-}|\mathcal{F}_{\infty-}^\star] = 0$ a.s., in other words if and only if B is continuous at infinity. However, at the risk of some confusion, we always call Z the potential generated by B (and by A).

Let T be a predictable stopping time. Then $Z_{T-} = N_{T-} - A_{T-}$ whence applying Theorems 3.2.14 and 3.2.37

$$Z_{T-} = \mathbf{E}[A_\infty|\mathcal{F}^\star_{T-}] - A_{T-} = \mathbf{E}[A_\infty - A_{T-}|\mathcal{F}^\star_{T-}] = \mathbf{E}[B_\infty - B_{T-}|\mathcal{F}^\star_{T-}] \tag{3.87}$$

so that Z_- is the predictable projection of the processes $(B_\infty - B_{t-})$ and $(A_\infty - A_{t-})$

For the optional case; let B be an integrable raw increasing process which may have a jump at 0 and at infinity always with the convention $B_{0-} = 0$; A is its *optional dual projection*, N the cadlag version of the martingale $\mathbf{E}[A_\infty|\mathcal{F}^\star_t]$ and Z denotes the positive supermartingale $Z_t = N_t - A_{t-}$. The process Z is generally neither right- nor left-continuous, but simply *laglad*; it belongs to the class of strong supermartingales, which we study in Chapter 4.

By Theorem 3.2.37, we have for every stopping time T

$$Z_T = \mathbf{E}[A_\infty|\mathcal{F}^\star_T] - A_{T-} = \mathbf{E}[A_\infty - A_{T-}|\mathcal{F}^\star_T] = \mathbf{E}[B_\infty - B_{T-}|\mathcal{F}^\star_T] \tag{3.88}$$

so that Z is the optional projection of the processes $(B_\infty - B_{t-})$ and $(A_\infty - A_{t-})$. Z is called ([41]) the left potential generated by B (and by A), a dangerous terminology since we no longer have even $Z_\infty = 0$!

It follows from Theorem 3.2.34 that in both cases Z determines A. Suppose that $B_0 = 0$. Then, it is possible to associate both a potential and a left potential with B; we denote them by Z and Z^g. Then (3.87) means that Z_- is the predictable projection of Z^g.

Having introduced this notation, we can now say precisely what our aim is to find an upper bound for A_∞ from estimates concerning either the r.v. B_∞ or the r.v. $Z^\star = \sup_s Z_s$. Besides formula (3.83) of the definition of dual projections, which occurs on numerous occasions, we shall use some Stieltjes integral technique, which we now present.

Theorem 3.2.39 Let Z be the potential (left potential) generated by the predictable (optional) integrable increasing process A. Then in the predictable case

$$\mathbf{E}[A^2_\infty] = \mathbf{E}\left[\int_{[0,\infty]} (Z_s + Z_{s-})dA_s\right], \tag{3.89}$$

and in the optional case

$$\mathbf{E}[A^2_\infty] = \mathbf{E}\left[\int_{[0,\infty]} (Z_{s+} + Z_s)dA_s\right] \tag{3.90}$$

Remark 3.2.40 The quantity $\frac{1}{2}\mathbf{E}[A^2_\infty]$ (finite or otherwise) is denoted by $\mathfrak{e}(Z)$ or $\mathfrak{e}(A)$ and is called the energy of Z or A and the formulas (3.89) or (3.90) are the energy formulas.

Some interesting inequalities are: consider A and B be two predictable integrable increasing processes and X and Y their respective potentials and suppose that $X \leq Y$. Then

$$\mathbf{E}[A_\infty^2] \leq 4\mathbf{E}[B_\infty^2],$$

so that X is of finite energy if Y is of finite energy. The optional case is the same. Moreover, the same argument shows that, if A is the predictable (optional) dual projection of a raw increasing process B, then we have the same formula as above $\mathbf{E}[A_\infty^2] \leq 4\mathbf{E}[B_\infty^2]$. We shall see in (Theorem 3.2.42) a more general inequalities of this type. Similarly, if Z denotes the potential (left potential) generated by a predictable (optional) A, we have the following result:

$$\frac{1}{2}||Z^\star||_2 \leq ||A_\infty||_2 \leq 2||Z^\star||_2.$$

The next theorem is useful for studying convergence.

Theorem 3.2.41 Let A be a predictable (optional) process of integrable variation whose potential (left-potential) Z is bounded above in absolute value by a cadlag martingale $M_t = \mathbf{E}[M_\infty|\mathcal{F}_t]$. Then,

$$\mathbf{E}[A_\infty^2] \leq 2\mathbf{E}\left[M_\infty \int_{[0,\infty]} |dA_s|\right]$$

and if the right-hand side is finite (3.89) or (3.90) also holds.

We associate the number

$$||X||_\phi = \inf\left\{\lambda > 0 : \mathbf{E}\left[\phi\left(|X|/\lambda\right)\right] \leq 1\right\}. \tag{3.91}$$

with the r.v. X, finite or otherwise. When $\phi(x) = x^p$ we recover the usual seminorm $||X||_p$ defining L^p.

Theorem 3.2.42 Let A be a predictable (optional) increasing process whose potential (left potential) Z is bounded above by a cadlag martingale $M_t = \mathbf{E}[M_\infty|\mathcal{F}_t]$. Then

$$\mathbf{E}[\phi(A_\infty)] \leq \mathbf{E}[M_\infty\phi(A)] \tag{3.92}$$

$$\mathbf{E}[A_\infty\phi(A_\infty)] \leq \mathbf{E}[pM_\infty\phi(A_\infty)] \tag{3.93}$$

where p is the constant associated with ϕ in Equation 3.91,

$$\mathbf{E}[\phi(A_\infty)] \leq \mathbf{E}[\phi(pM_\infty)] \tag{3.94}$$

$$||A_\infty||_\phi \leq p||M_\infty||_\phi. \tag{3.95}$$

3.2.4 Decomposition of supermartingales

Here we present existence and uniqueness of Doob's decomposition of a supermartingale of class D. The method we employ for the proof is a variant of the old proof by C. Doléans-Dade given by Dellacherie and Meyer in [20]. It depends on a functional analytic lemma which is of interest in itself used to determine the dual of the space H^1.

3.2.4.1 Functional analytic decomposition theorem

We are given a probability space $(\Omega, \mathcal{F}, \mathbf{P})$ with a filtration $(\mathcal{F}_t^\star)$ satisfying the usual conditions. Use $\mathcal{D}$ to denote a vector space of processes over Ω, which is $\wedge$-stable, contains the constants, and has the following properties:

Property 3.2.43 Every $X \in \mathcal{D}$ is bounded, adapted and cadlag including a limit at infinity denoted by $X_{\infty-}$.

Property 3.2.44 For every stopping time T, the process $\mathbf{1}_{[T,\infty[}$ belong to $\mathcal{D}$.

The elements of $\mathcal{D}$ are processes indexed by $[0,\infty[$. If $X \in \mathcal{D}$ we denote by X_- the process $(X_{t-})_{0<t\leq\infty}$ indexed by $]0,\infty]$ and $\mathfrak{D}$ will denote the space of processes X_-, where X runs through $\mathcal{D}$; this is all so a $\wedge$-stable vector space and $\mathcal{D}$ and $\mathfrak{D}$ will later play symmetric roles. It is sometimes convenient in both cases to consider processes indexed by $[0,\infty]$ with the convention that $X_{0-} = 0$, $X_\infty = 0$. If $X \in \mathcal{D}$, we set as usual $X^* = \sup_t |X_t|$.

Properties (3.2.43) and (3.2.44) are chosen so as to apply to the following examples: (a) $\mathcal{D}$ is the space of all adapted cadlag bounded processes and (b) $\mathcal{D}$ is the space of processes of the form

$$X = X_0 \mathbf{1}_{[0,T_1[} + X_1 \mathbf{1}_{[T_1,T_2[} + ... + X_n \mathbf{1}_{[T_n,\infty[} \tag{3.96}$$

where n is an integer, the T_i are stopping times such that $0 \leq T_1 \leq ... \leq T_n \leq \infty$ and, for all $i \leq n$, X_i is bounded and $\mathcal{F}_{T_i}$-measurable.

Next is a functional analytic theorem on which all later proofs in this section are based.

Theorem 3.2.45 Let J be a positive linear form on $\mathcal{D}$ with the following property:

$$\text{For every decreasing sequence } (X^n) \text{ of elements of } \mathcal{D} \text{ such that} \tag{3.97}$$
$$\lim_n (X_t^n)^* = 0 \text{ a.s.,} \quad \lim_n J(X_t^n) = 0.$$

Recall $X_t^* = \sup_t |X_t|$. Then, there exist two integrable increasing processes: A^+, which is optional and may have a jump at 0 but none at infinity, and A^-, which is predictable and may have a jump at infinity but none at 0, such that for $X \in \mathcal{D}$

$$J(X) = \mathbf{E}\left[\int_{]0,\infty]} X_{s-} dA_s^- + \int_{[0,\infty[} X_s dA_s^+\right]. \tag{3.98}$$

We can impose on A^+ the property of being purely discontinuous and the representation is then unique.

Proof. Consider two disjoint copies of Ω denoted by Ω_+ and Ω_- and their union by $\hat{\Omega}$, so that, every $\omega \in \Omega$ is "split" into two elements ω_+, ω_- of Ω. We set

$$\mathrm{W}_+ = [0, \infty[\times\Omega_+, \ \mathrm{W}_- =]0, \infty] \times \Omega_-, \ \hat{\mathrm{W}} = \mathrm{W}_+ \cup \mathrm{W}_-.$$

We associate with every process $X \in \mathcal{D}$ a function $\hat{X}$ on $\hat{\mathrm{W}}$ defined by

$$\hat{X}(t, \omega_+) = X_t(\omega) \text{ and } \hat{X}(t, \omega_-) = X_{t-}(\omega)$$

and we denote by $\hat{\mathcal{D}}$ the set of all $\hat{X}$, where X runs through $\mathcal{D}$; it is a $\wedge$-stable vector space of functions on $\hat{\mathrm{W}}$, containing the constants, and clearly $X \mapsto \hat{X}$ is a bijection of $\mathcal{D}$ onto $\hat{\mathcal{D}}$. Thus we can define a positive linear form $\hat{\mathrm{J}}$ on $\hat{\mathcal{D}}$: $\hat{\mathrm{J}}(\hat{X}) = \mathrm{J}(X)$ for $X \in \mathcal{D}$. Next we shall show that $\hat{\mathrm{J}}$ is a Daniell's integral using the following lemma:

On $[0, \infty]$, let (f_n) be a decreasing sequence of positive cadlag functions which tend pointwise to 0 as do their left limits. Then the sequence (f_n) tends uniformly to 0.

Let $\epsilon > 0$ and associate with each $t \in [0, \infty]$ a neighborhood $I_t \subset [0, \infty]$ and an integer $n(t)$ such that $f_{n(t)}(s) \le \epsilon$ for $s \in I_t$. Then, covering $[0, \infty]$ with a finite number of interval's I_{t_k} and denoting the greatest of the integers $n(t_k)$ by N, we shall have $f_n \le s$ on $[0, \infty]$ for all $n \ge N$.

To construct I_t and $n(t)$, suppose that $0 < t < \infty$. Choose m such that $f_m(t) < \epsilon$, $f_m(t-) < \epsilon$ and then $a < t$ such that $f_m(s) < \epsilon$ for $s \in]a, t[$ and $b > t$ such that $f_m(s) < \epsilon$ for $s \in [t, b[$. Then, take $n(t) = m$, $I_t =]a, b[$. The case $t = 0$ or $+\infty$ is similar.

Now let $(\hat{X}^n)$ be a decreasing sequence of elements of $\hat{\mathcal{D}}$, converging pointwise to 0 in $\hat{\mathrm{W}}$. By the lemma, we also have $\lim_n (X^n)^\star = 0$ on Ω. By (3.97) we have $\lim_n \hat{\mathrm{J}}(\hat{X}^n) = \lim_n d(X^n) = 0$ and $\hat{\mathrm{J}}$ is a Daniell's integral on $\hat{\mathcal{D}}$.

By Daniell's theorem [19] III.35, there exists a bounded positive measure λ on $\hat{\mathrm{W}}$ with the (σ-field $\hat{\mathcal{U}}$ generated by $\hat{\mathcal{D}}$ and such that

$$\mathrm{J}(X) = \hat{\mathrm{J}}(\hat{X}) = \int_{\hat{\mathrm{W}}} \hat{X}\lambda \text{ for all } X \in \mathcal{D}. \tag{3.99}$$

Let $\hat{Z}$ be a bounded positive $\hat{\mathcal{U}}$-measurable function and let $\epsilon > 0$. There exists a function $\hat{X}$ on $\hat{\mathrm{W}}$, the limit of a decreasing sequence $(\hat{X}^n)$ of elements of $\hat{\mathcal{D}}$, such that

$$\hat{Y} \le \hat{Z}, \quad \lambda(\hat{Y}) \ge \lambda(\hat{Z}) - \epsilon.$$

Suppose in particular that $\hat{Z}$ is evanescent, i.e., that the set $\{\omega : \exists t, \ Z(t, \omega_+) \ne 0 \text{ or } Z(t, \omega_-) \ne 0\}$ is **P**-negligible. Then $\hat{Y}$ is evanescent and the lemma implies that $\lim_n (X^n)^\star = 0$ a.s. Hence $\lambda(\hat{Y}) = \lim_n \lambda(\hat{X}^n) = \lim_n \mathrm{J}(X^n) = 0$, finally $\lambda(\hat{Z}) \le \epsilon$ and hence $\lambda(\hat{Z}) = 0$.

Now let H be a subset of $\hat{\mathrm{W}}$ not necessarily belonging to $\hat{\mathcal{U}}$, but evanescent, i.e., $\{\omega; \exists t, (t, \omega_-) \in$ H or $(t, \omega_-) \in$ H$\}$ is **P**-negligible.

It follows from the preceding paragraph that H is then internally negligible on the measurable space $(\hat{\mathrm{W}}, \hat{\mathcal{U}}, \lambda)$. Since a countable union of evanescent sets is evanescent, by Definition 1.2.23 we may adjoin to $\hat{\mathcal{U}}$ all the evanescent sets and extend the measure λ to the σ-field thus generated, so that they all become negligible. We shall continue to use $\hat{\mathcal{U}}$ and λ to denote the σ-field and measure thus constructed.

One last extension remains to be accomplished, in order to render the two pieces $\hat{\mathrm{W}}_+, \hat{\mathrm{W}}_-$ of $\hat{\mathrm{W}}$ measurable. Let K be an element of $\hat{\mathcal{U}}$ containing $\hat{\mathrm{W}}_-$ and of measure equal to the outer measure of $\hat{\mathrm{W}}_-$; then the set $K \backslash \hat{\mathrm{W}}_-$ is internally λ-negligible and we extend the measure by making it

λ-negligible without changing the notation $\widehat{\mathcal{U}}$ and λ. We can then represent λ as the sum of two measures: λ_+ on $\hat{\mathrm{W}}_+ = [0, \infty[\times \Omega_+$, with the trace σ-field of $\widehat{\mathcal{U}}$, which by (Property 3.2.44) is the optional σ-field (by identifying Ω_+ with Ω), and λ_- on $\hat{\mathrm{W}}_- =]0, \infty] \times \Omega_-$, with the trace σ-field of $\widehat{\mathcal{U}}$ on $\hat{\mathrm{W}}_-$, which by (Property 3.2.44) is the predictable σ-field. Hence we have proved:

There exist two bounded positive measures, λ_- on $(]0, \infty] \times \Omega, \mathcal{P})$ and λ_+ on $([0, \infty[\times \Omega, \mathcal{O})$, such that for all $X \in \mathcal{D}$

$$J(X) = \lambda_-(X_-) + \lambda_+(X). \tag{3.100}$$

As λ_- and λ_+ have no mass on evanescent sets, we take respectively their predictable extensions and their optional extensions ([20] VI.72(b)) and then apply Theorem 3.2.32 to reduce the measures to the case of increasing processes. Thus, we obtain representation (3.98).

Now we go on to show uniqueness. In representation (3.98), we may remove from the left-hand side the continuous part of the increasing process A^+, since the processes X and X_- differ only on a set with countable sections. So, we may assume that A^+ to be purely discontinuous.

Now consider two representations of type (3.98)

$$J(X) = \mathbf{E}\left[\int X_{s-} dA_s^- + X_s dA_s^+\right] = \mathbf{E}\left[\int X_{s-} dB_s^- + X_s dB_s^+\right] \tag{3.101}$$

with A^+ and B^+ purely discontinuous. Taking $X = \mathbf{1}_{[T, \infty[}$, we obtain

$$\mathbf{E}[A_\infty^+ - A_{T-} + A_\infty^- - A_T^-] = \mathbf{E}[B_\infty^+ - B_{T-} + B_\infty^- - B_T^-]. \tag{3.102}$$

We replace T by $T + 1/n$, take the difference and let n tend to ∞. Then, as A^+ and B^+ have no jumps at infinity,

$$\mathbf{E}[\triangle A_T^+ \mathbf{1}_{T<\infty}] = \mathbf{E}[\triangle B_T^+ \mathbf{1}_{T<\infty}].$$

Replacing T by T_U where U runs through $\mathcal{F}_T$, we obtain

$$\mathbf{E}[\triangle A_T^+ | \mathcal{F}_T] = \mathbf{E}[\triangle B_T^+ | \mathcal{F}_T]$$

a.s. and hence $\triangle A_T^+ = \triangle B_T^+$ a.s. As A^+ and B^+ are purely discontinuous and T is an arbitrary stopping time, A^+ and B^+ are indistinguishable. Then (3.102) gives $\mathbf{E}[A_\infty^- - A_T^-] = \mathbf{E}[B_\infty^- - B_T^-]$ for every stopping time T. Replacing T by T_U as earlier, the expectations can be replaced by conditional expectations, and we then conclude (Theorem 3.2.34(b)) that the predictable increasing processes A^- and B^- are equal. Theorem 3.2.45 has been proved. ∎

Remark 3.2.46 (a) If we impose on A^- (instead of A^+) the condition of being purely discontinuous there is likewise a uniqueness result. To see this, start with (3.101) assuming that A^- and B^- are purely discontinuous, write (3.102) with T predictable and > 0 and a sequence of T_n foretelling T, and deduce that $\mathbf{E}[\triangle A_T^- | \mathcal{F}_{T-}] = \mathbf{E}[\triangle B_T^- | \mathcal{F}_{T-}]$ for T predictable and > 0; as A^- and B^- are purely discontinuous, predictable and zero at 0, it follows that $A^- = B^-$ and so on.

(b) Mertens [3] showed (Mertens Decomposition) that we could have taken, instead of an ordered pair (A^-, A^+) of increasing processes, a unique predictable increasing process, which is neither right- nor left-continuous,

$$A_t = A_t^- + A_{t-}^+ \quad (A_{0-} = A_0 = 0, \; A_{0+} = A_0^+). \tag{3.103}$$

For $\triangle_+ A_t = A_{t+} - A_t = A_t^+ - A_{t-}^+ = \triangle A_t^+$, which, since A^+ is purely discontinuous, enables us to recover A^+

$$A_t^+ = \sum_{0 \le s \le t} \triangle_+ A_s \tag{3.104}$$

from which we recover $A_t^- = A_t - A_{t-}^+$, which is right-continuous, predictable and zero at 0. Note that

$$\triangle_- A_t = A_t - A_{t-} = \triangle A_t^-. \tag{3.105}$$

The potential associated with A (or the ordered pair (A^-, A^+)) is the unique optional process Z such that, for every stopping time T, (3.105) $Z_T = \mathbf{E}[A_\infty - A_T | \mathcal{F}_T]$ and it is easy to see that $\mathbf{E}[Z_T \mathbf{1}_{\{T<\infty\}}] = J(\mathbf{1}_{[\![T,\infty[\![})$ for every stopping time T, whence it follows without difficulty that Z determines uniquely the functional J and hence the ordered pair (A^-, A^+). We return to this representation in Chapter 4.

However, note that this representation conflicts with the custom of representing measures by right-continuous increasing processes: if J is of the form $J(X) = \mathbf{E}\left[\int_{[0,\infty[} X_s dB_s\right]$, with B optional, the Mertens increasing process associated with J is the left-continuous process (B_{t-}).

Remark 3.2.47 Variants of Theorem 3.2.45. The same method applies to the representation of positive linear forms J satisfying (3.97) on other spaces of processes:

(a) The space of bounded adapted cadlag processes indexed by $[0, \infty]$. The same representation also allows a non-zero jump $\triangle A_\infty^+$.

(b) The space of bounded optional processes X indexed by $[0, \infty]$ with laglad paths. We need three integrable increasing processes

$$J(X) = \mathbf{E}\left[\int X_{s-} dA_s^- + X_s dA_s + X_{s+} dA_s^+\right] \tag{3.106}$$

where A^- and A^+ are purely discontinuous. We shall not give any more details.

(c) The space of bounded adapted processes indexed by $[0, \infty]$ with continuous paths. We then have a representation of the form

$$J(X) = \mathbf{E}\left[\int_{[0\infty]} X_s dC_s\right] \tag{3.107}$$

using a single predictable integrable increasing process C, which may have jumps at 0 and $+\infty$.

The last representation can be put in a more abstract form. Let $\mathbb{T}$ be a metrizable compact space and C a $\wedge$-stable space (or an algebra) of bounded $\mathcal{B}(\mathbb{T}) \times \mathcal{F}$-measurable functions $X(t, \omega)$ on $\mathbb{T} \times \Omega$ such that, for all $\omega \in \Omega$, $X(., \omega)$ is continuous on $\mathbb{T}$. Suppose that C contains the functions of the form $f \otimes \mathbf{1}_A$ (f continuous on T, $A \in F$). For $X \in$ C we set $X^\star(\omega) = \sup_t |X(t, \omega))|$. Then let J be a positive linear form on C satisfying (3.97); J has a representation

$$J(X) = \mathbf{E}\left[\int_{\mathbb{T}} \mu(\omega, dt) X(t, \omega)\right] \tag{3.108}$$

where $(\mu, (t, .))_t$ is an $\mathcal{F}$-measurable family of bounded positive measures on $\mathbb{T}$ (a kernel). The proof can begin as in Theorem 3.2.45, but it can also be deduced from the bimeasures theorem [19] III.74, or rather from a somewhat more general form of that theorem (Morando [42], p. 221, Horowitz [43]). For the mapping $(A, f) \to \mathrm{J}(f \otimes \mathbf{1}_A)$ can be interpreted as a bimeasure, with which we associate a measure θ on $\mathbb{T} \times \Omega$ and (3.108) gives the disintegration of θ with respect to $\mathbf{P}$.

3.2.4.2   Extension to non-positive functionals

Consider functionals on the space $\mathcal{D}$ of Theorem 3.2.45, which are not necessarily positive; however, they satisfy the following property that is a little stronger than (3.97):

Property 3.2.48 For every, not necessarily decreasing, uniformly bounded sequence of positive elements have X^n of $\mathcal{D}$ such that $\lim_n (X^n)^* = 0$ a.s., we have $\lim_n \mathrm{J}(X^n) = 0$.

We will show that J is the difference of two positive linear forms, for which Theorem 3.2.45 holds.

(a) Let $\mathcal{D}^+$ be the positive cone of $\mathcal{D}$. For $X \in \mathcal{D}^+$ we set

$$\mathrm{J}^+(X) = \sup_{Y \in \mathcal{D}+, Y \leq X} \mathrm{J}(Y). \tag{3.109}$$

Then $\mathrm{J}^+(X) \geq 0$ (take $Y = 0$) and $\mathrm{J}^+(X) < \infty$ (otherwise, there would exist $Y^n \leq X$ such that $\mathrm{J}(Y^n) \geq n$ and the processes Y^n/n would contradict (3.2.48). Then, obviously $\mathrm{J}^+(tX) = t\mathrm{J}^+(X)$ for all $t \in \mathbb{R}_+$ and $\mathrm{J}^+(X + X') \geq \mathrm{J}^+(X) + \mathrm{J}^+(X')$. On the other hand, every $Z \in \mathcal{D}^+$ bounded above by $X + X'$ can be written as $Z = Y + Y'$ with $Y \in \mathcal{D}^+$, $Y' \in \mathcal{D}^+$, $Y \leq X$, $Y' \leq X'$ (take $Y = Z \wedge X$, $Y' = Z' - Y'$); hence $\mathrm{J}^+(X + X') \leq \mathrm{J}^+(X) + \mathrm{J}^+(X')$ and finally $\mathrm{J}^+(X + X') = \mathrm{J}^+(X) + \mathrm{J}^+(X')$. Hence J^+ can be extended to a positive linear form on $\mathcal{D} = \mathcal{D}^+ - \mathcal{D}^+$; we also denote this extension by J^+.

We now show that J^+ satisfies (3.97). Let (X^n) be a decreasing sequence of elements of $\mathcal{D}^+$ such that $\lim_n (X^n)^* = 0$ a.s.; for all n, let $Y^n \in \mathcal{D}^+$ be such that $Y^n \leq X^n$ and $\mathrm{J}(Y^n) \geq \frac{1}{2}\mathrm{J}^+(X^n)$. We also have $\lim_n (Y^n)^* = 0$ a.s., hence $\lim_n \mathrm{J}(Y^n) = 0$ by (3.2.48) and finally $\lim_n \mathrm{J}^+(X^n) = 0$, the required result. Thus we can apply Theorem 3.2.45.

(b) Recall that the set of linear forms on $\mathcal{D}$ satisfying (3.2.48) is obviously a vector space, ordered by the cone of positive linear forms. Clearly the linear form J^+ which we constructed is the smallest positive linear form bounding J above, in other words $\mathrm{J}^+ = \mathrm{J} \vee 0$. Then, the space is a lattice and $\mathrm{J} = \mathrm{J}^+ - \mathrm{J}^-$ where $\mathrm{J}^- = (-\mathrm{J}) \vee 0$ is given by

$$\mathrm{J}^-(X) = \sup_{Y \in \mathcal{D}^+, Y \leq X} (-\mathrm{J}(Y)) \quad (X \in \mathcal{D}^+) \tag{3.110}$$

and we have $|\mathrm{J}| = \mathrm{J} \vee (-\mathrm{J}) = \mathrm{J}^+ + \mathrm{J}^-$, a linear form given also by

$$|\mathrm{J}|(X) = \sup_{Y \in \mathcal{D}, |Y| \leq X} \mathrm{J}(Y) \quad (X \in \mathcal{D}^+). \tag{3.111}$$

(c) Let's return to the proof of Theorem 3.2.45; associate with $J, J^+, J^-, |J|$ linear forms on $\hat{\mathcal{D}}$ which we denote by $\hat{J}, \hat{J}^+, \hat{J}^-, |\hat{J}|$; since this is done simply by transport of structure, using an isomorphism of $\mathcal{D}$ onto $\hat{\mathcal{D}}$, we still have, for example, $\hat{J}^+ = \hat{J} \vee 0$ in the space of linear forms on $\hat{\mathcal{D}}$ which satisfy Daniell's condition. If we now pass from linear forms satisfying Daniell's condition to measures on $(\hat{W}, \hat{\mathcal{U}})$, we construct four measures $\lambda, \lambda^+, \lambda^-, |\lambda|$, again related by the relations $\lambda^+ = \lambda \vee 0$, etc., since Daniell's theorem provides an isomorphism of ordered vector spaces between spaces of linear forms and measures. In particular, we have $\lambda = \lambda^+ - \lambda^-$, $|\lambda| = \lambda^+ + \lambda^-$ and λ^+ and λ^- are mutually singular (i.e., carried by two disjoint $\hat{\mathcal{U}}$-measurable sets). These properties are preserved in later extensions, and in particular when adjoining the inner $|\lambda|$-negligible set $K \backslash \hat{W}_-$ to the σ-field $\hat{\mathcal{U}}$.

To finish, we obtain a representation of J of the form

$$J(X) = \mathbf{E}\left[\int X_{s-} dA_{\overline{s}} + \int X_s dA_s^+\right] \tag{3.112}$$

where A^+ and A^- here are not necessarily increasing processes of integrable variation, with A^- predictable and A^+ optional and purely discontinuous. On the other hand, if X belongs to $\mathcal{D}^+$ we have

$$|J|(X) = \mathbf{E}\left[\int X_{s-} |dA_s^-| + \int X_s |dA_s^+|\right] \tag{3.113}$$

and this expresses the fact that the measure associated with $|J|$ is $|\lambda|$.

3.2.4.3 Decomposition of positive supermartingale of class **D**

All supermartingales, martingales, and other processes are henceforth assumed to be cadlag unless otherwise mentioned. Let Z be a positive supermartingale; we adopt the convention $Z_\infty = 0$ but do not assume that $Z_{\infty-} = 0$. We use G to denote the space of adapted, bounded, cadlag processes indexed by $]0, \infty]$ of the form

$$X = X_0 \mathbf{1}_{]0,T_1]} + X_1 \mathbf{1}_{]T_1,T_2]} + ... + X_n \mathbf{1}_{]T_n,\infty]} \tag{3.114}$$

where the T_i are a finite number of stopping times such that $0 \le T_1 \le ... \le T_n < \infty$ and for each i, X_i is a bounded $\mathcal{F}_{T_i}$-measurable r.v. We set

$$J(X) = \mathbf{E}\left[X_0(Z_0 - Z_{T_1}) + X_1(Z_{T_1} - Z_{T_2}) + ... + X_n Z_{T_n}\right] \tag{3.115}$$

where the r.v. under the $\mathbf{E}$ sign is obviously integrable. This functional can also be written

$$J(X) = \mathbf{E}\left[-\int_{]0,\infty]} X_s dZ_s\right]$$

for each ω,

$$\int_{]0,\infty]} X_s(\omega) dZ_s(\omega)$$

is the trivial integral of all left-continuous step functions with respect to a right continuous function, so that J depends only on X and not on the representation (3.114) chosen. Thus, J is linear and the fact that Z is a supermartingale implies that J is positive. In particular, if X is a positive element of $\mathfrak{D}$, bounded above by a constant c, then

$$J(X) \leq J(c) = c\mathbf{E}[Z_0].$$

The mapping $X \mapsto X_+$ is an isomorphism of $\mathfrak{D}$ onto the space $\mathcal{D}$ of Section 3.2.4.1, Example (b) (representation (3.96)). Hence we can transport J to a linear form on $\mathcal{D}$ and apply Theorem 3.2.45 to it.

The following result is crucial; it argues for the direct use of $\mathfrak{D}$ without the need to make the "transports" explicit.

Lemma 3.2.49 If Z belongs to class D, J satisfies (3.97).

Proof. Let X be an element of $\mathfrak{D}^+$ bounded above by a constant c and let $\epsilon > 0$. We set

$$T = \inf\{t : X_t \geq \epsilon\} \tag{3.116}$$

and let $Y = \epsilon \mathbf{1}_{\rrbracket 0,T \rrbracket} + c\mathbf{1}_{\rrbracket T,\infty \rrbracket}$. Then $X \leq Y$ and hence

$$J(X) \leq \epsilon \mathbf{E}[Z_0] + c\mathbf{E}[Z_T \mathbf{1}_{T<\infty}].$$

Let (X^n) be a decreasing sequence of elements of $\mathfrak{D}$ such that $\lim_n (X^n)^* = 0$ a.s.; let c be a constant bounding X^0 above and let $\epsilon > 0$. We associate with each X^n a stopping time T_n by (3.116). The sequence T_n is increasing and

$$\lim_n J(X^n) \leq \epsilon \mathbf{E}[Z_0] + c \lim_n \mathbf{E}[Z_{T_n} \mathbf{1}_{T_n<\infty}].$$

The condition $\lim_n J(X^n)^* = 0$ implies that $\mathbf{P}\{T_n < \infty\} \to 0$, and the fact that Z belongs to class D (i.e., that the Z_{T_n} are uniformly integrable) implies that the last limit is zero. As ϵ is arbitrary, $\lim_n J(X^n) = 0$. $\blacksquare$

Next we give the fundamental decomposition theorem for cadlag positive supermartingales under the usual conditions. The same proof will apply in Chapter 4 to the most general situation of ladlag supermartingales without the usual conditions.

Theorem 3.2.50 Let Z be a positive supermartingale of class D. There exists a predictable integrable increasing process A indexed by $[0, \infty]$, which is zero at 0 but may also have a jump at infinity, such that

$$Z_t = \mathbf{E}[A_\infty - A_t | \mathcal{F}_t^*] \text{ a.s. for } t \in [0, \infty[. \tag{3.117}$$

This process is unique to within an evanescent process. Conversely, if Z has a representation 3.117, Z belongs to class D.

Proof. We begin with the latter point. Let (M_t) be a cadlag version of the martingale $\mathbf{E}[A_\infty|\mathcal{F}_t^\star]$. We know that M belongs to class D ([20] VI.23). On the other hand, $Z_t \leq M_t$ a.s. and hence $Z \leq M$ except on evanescent sets by right-continuity. Hence Z belongs to class D.

Uniqueness has already been established in Theorem 3.2.34(b). We will prove existence. By Lemma 3.2.49 and Theorem 3.2.45, the linear form J has a representation

$$J(X) = \mathbf{E}\left[\int_{]0\infty]} X_s dA_s + \int_{[0,\infty[} X_{s+} dA_s^+\right] \tag{3.118}$$

where A^- is predictable and zero at 0 and A^+ is optional and purely discontinuous. We show $A^+ = 0$, which amounts to saying (since A^+ is the sum of its jumps) that $\mathbf{E}[\triangle A_T^+] = 0$ for every stopping time T. We take $X^n = 1_{]\!]T,T+1/n]\!]} \in \mathfrak{D}$, so that $X^n \downarrow 0$ and $X_+^n \downarrow 1_{[\![T]\!]}$ and then $J(X^n) \downarrow \mathbf{E}[\triangle A_T^+]$. But we have $J(X^n) = \mathbf{E}[Z_T - Z_{T+1/n}]$ which tends to 0 by the right continuity of Z and the uniform integrability of the r.v. $Z_{T+1/n}$. Thus (3.118) reduces to

$$J(X) = \mathbf{E}\left[\int_{]0,\infty]} X_s dA_s\right] \tag{3.119}$$

where we have written A instead of A^-. Then, taking $X = 1_{]\!]T,\infty]\!]}$, we obtain

$$\mathbf{E}[Z_T 1_{T<\infty}] = \mathbf{E}[A_\infty - A_T];$$

and for any $U \in \mathcal{F}_T^\star$, replacing T by T_U, we obtain $Z_T = \mathbf{E}[A_\infty - A_T|\mathcal{F}_T^\star]$ and (3.117). ∎

Remark 3.2.51 (a) As we have said in no. 3.2.38, Z is called the potential generated by A, even if Z is not a potential in the martingale theoretic sense (nos. 3.2.8-3.2.10). Also, A is called the increasing process associated with Z.

Since Z belongs to class D, to investigate under what condition Z is indeed a potential it is sufficient to express the fact that $Z_{\infty-} = 0$ a.s. But by (3.117) and (3.2.6) we have that, as $t \to \infty$,

$$Z_{\infty-} = \mathbf{E}[A_\infty|\mathcal{F}_{\infty-}^\star] - A_{\infty-}.$$

Since A is predictable, this is simply $A_\infty - A_{\infty-} = \triangle A_\infty$. Thus, Z is a potential if and only if A has no jump at infinity.

(b) The explicit formula (3.117) implies that the process $(Z_t + A_t)$ is equal to the uniformly integrable martingale $M_t = \mathbf{E}[A_\infty|\mathcal{F}_t^\star]$. Thus every positive supermartingale Z of class D can be written on $[0,\infty[$ as

$$Z_t = M_t - A_t \tag{3.120}$$

where M is a martingale and A a predictable (integrable) increasing process which is zero at 0. In this form (called the *Doob decomposition* of Z) Theorem 3.2.50 will be considerably generalized in Theorem 3.2.55.

(c) It is interesting to give the decomposition (3.120) explicitly in the discrete case. In that case we write

$$Z_n = \sum_{k=0}^n z_k \text{ with } z_0 = Z_0, z_k = Z_k - Z_{k-1} \text{ for } k \geq 1,$$

$$A_n = \sum_{k=0}^{n} a_k \text{ with } a_0 = 0, \ a_k = Z_{k-1} - \mathbf{E}[Z_k|\mathcal{F}_{k-1}] \text{ for } k \geq 1,$$

$$M_n = \sum_{k=0}^{n} m_k \text{ with } m_0 = 0, \ m_k = Z_k - \mathbf{E}[Z_k|\mathcal{F}_{k-1}] \text{ for } k \geq 1.$$

Then

$$z_k = m_k - a_k, \ \mathbf{E}[m_k|\mathcal{F}_{k-1}] = 0, \ a_k \geq 0 \text{ and } a_k \text{ is } \mathcal{F}_{k-1}\text{-measurable for } k \geq 1.$$

This gives the intuitive interpretation of the increasing process A in the continuous case: it is the symbolic integral

$$A_t = \int_0^t -\mathbf{E}[dZ_s|\mathcal{F}_s^\star].$$

The following theorem is as important as Theorem 3.2.50; it addresses the issue of jumps of increasing process. When we extend Theorem 3.2.50 to arbitrary supermartingales we shall also extend the results in Theorem 3.2.52 (cf. Theorem 3.2.57).

Theorem 3.2.52 We use the notation of Equation 3.117.

(1) The process $\triangle A = A - A_-$ is equal to $Z_- - {}^p Z$, where ${}^p Z$ is the predictable projection of Z.

(2) In particular, A is continuous if and only if $Z_- = {}^p Z$ (Z is regular, cf. [20] VI.50). This condition is equivalent to the following: for every increasing sequence of stopping times of (T_n) of limit T, we have $\mathbf{E}[Z_T] = \lim_n \mathbf{E}[Z_{T_n}]$.

Proof. (1) We have $Z = M - A$ and hence $Z_- = M_- - A_-$ and ${}^p Z = {}^p M - {}^p A$. As A is predictable, we have ${}^p A = A$; on the other hand, since M is a uniformly integrable martingale, we have ${}^p M = M_-$ ([20] VI.45(a)). Thus, we have $Z_- - {}^p Z = A - A_- = \triangle A$.

If A is continuous and $T_n \uparrow T$, we have $A_{T_n} \uparrow A_T$ and hence

$$\mathbf{E}[Z_{T_n}] = \mathbf{E}[A_\infty - A_{T_n}] \downarrow \mathbf{E}[A_\infty - A_T] = \mathbf{E}[Z_T].$$

Conversely, if this property holds, let T be a predictable stopping time > 0 and (T_n) a sequence foretelling T. The relation $\mathbf{E}[Z_T] = \lim_n \mathbf{E}[Z_{T_n}]$ implies $\mathbf{E}[A_T] = \lim_n \mathbf{E}[A_{T_n}] = \mathbf{E}[A_{T-}]$ and hence $\triangle A_T = 0$ a.s.; by the predictable cross-section theorem, the process $\triangle A$ is evanescent and A is continuous. ■

Remark 3.2.53 (a) We have worked with the index set $[0, \infty]$; thus the relation $\triangle A = Z_- - {}^p Z$ gives in particular $\triangle A_\infty = Z_{\infty-} - \mathbf{E}[Z_\infty|\mathcal{F}_{\infty-}^\star] = Z_{\infty-}$ already seen in Remark 3.2.51(a). The continuity of A on $[0, \infty]$ can easily be expressed by property (2), written for uniformly bounded increasing sequences of stopping times.

(b) The following, more explicit, form of (1) is often used:

$$\triangle A_T = Z_{T-} - \mathbf{E}[Z_T|\mathcal{F}_{T-}] \tag{3.121}$$

a.s. for every predictable stopping time T.

In Chapter 4 we shall see how the proof of Theorem 3.2.50 can be extended to optional strong supermartingales of class D with no usual hypothesis on the family of σ-fields $(\mathcal{F}_t)$. But here we will continue with the usual conditions and make some remarks on an important special case.

Remark 3.2.54 Let Z be a positive optional process of class D indexed by $[0, \infty]$ with paths having right and left limits. Suppose that we have the following properties:

(1) If S and T are stopping times with $S \leq T$, then $Z_S \geq \mathbf{E}[Z_T | \mathcal{F}_S^\star]$ a.s. (with the convention $Z_\infty = 0$).

(2) If (T_n) is a uniformly bounded increasing sequence of stopping times and $T = \lim_n T_n$, then $\mathbf{E}[Z_T] = \lim_n \mathbf{E}[Z_{T_n}]$.

Then, Z is called a regular (optional) strong supermartingale of class D.

We now return to the definitions of Section 3.2.4.3, Lemma 3.2.49 and the beginning of the proof of Theorem 3.2.50; not a word needs to be changed but in the proof of Theorem 3.2.50, $\triangle A_T^-$ is zero for every bounded predictable stopping time T, by property (2) above, and hence

$$J(X) = \mathbf{E}\left[\int_{[0,\infty]} X_{s+} dA_s\right] \tag{3.122}$$

(with the convention $X_{\infty+} = X_\infty$), where we have regrouped into the measure dA: the purely discontinuous optional measure dA^+ and the measure dA^-, which consists of a continuous part (for which it does not matter whether we write X_{s+} or X_s and a mass at infinity. This is zero if and only if $Z_{\infty-} = 0$.

But then (3.122) gives us the following representation of Z,

$$Z_T = \mathbf{E}[A_\infty - A_{T-} | \mathcal{F}_T^\star] \tag{3.123}$$

for every stopping time T, a representation which we mentioned in Definition 3.2.38: Z is the left potential of A. Finally, the computation of jumps of Remark 3.2.53 becomes

$$\triangle A_T = Z_T - Z_{T+} \tag{3.124}$$

for every bounded stopping time T, $\triangle A_\infty = Z_{\infty-}$. For, if T is bounded (Property 3.2.44) applied to $T + 1/n$ gives us $Z_{T+} = E[A_\infty - A_T | \mathcal{F}_T^\star]$ whence (3.96) follows immediately by taking differences. Note that the optional cross-section theorem enables us to write simply $\triangle A = Z - Z_+$ on $[0, \infty[$, instead of the first equation of (3.124).

3.2.4.4   The general case of Doob decomposition

Theorem 3.2.55 Doob decomposition. Let Z be a supermartingale. Then Z has an unique decomposition on $[0, \infty[$ of the form

$$Z = M - A \tag{3.125}$$

where M is a local martingale and A is a predictable increasing process which is zero at 0. Then for all finite t,

$$\mathbf{E}[A_t] \leq \mathbf{E}[Z_0 - Z_t]$$

and in particular the condition $\lim_{t \to \infty} \mathbf{E}[Z_t] > -\infty$ implies that A is integrable.

This decomposition is called the Doob decomposition of Z, and A is called the increasing process associated with Z.

Proof. First we prove *uniqueness* by considering two decompositions $M - A$ and $M' - A'$ for Z. Then, $A - A'$ is a predictable process of finite variation, which is zero at 0 and is a local martingale; then it is zero by [20] VI.80, and we get $M = M'$. Now we deduce the following property: let S and T be two stopping times, and suppose that we have found two predictable increasing processes A and B, which are zero at 0 such that $Z^S - A = M$ and $Z^T - B = N$ are local martingales. Then, $Z^S - A^S = M^S$ is a local martingale and A^S is predictable, hence $A = A^S$ and A is stopped at S. Then, $Z^{S \wedge T} - A^T = M^T$ and $Z^{S \wedge T} - B^S = N^S$ are local martingales, hence $A^T = B^S$; in other words, $A = B$ up to the instant $S \wedge T$.

We now prove *existence*. It is sufficient to construct, for every integer n, a predictable increasing process A^n which is zero at 0 such that the process $Z_{t \wedge n} + A_t^n$ is a local martingale and $\mathbf{E}[A_t^n] \leq \mathbf{E}[Z_0 - Z_{t \wedge n}]$. By the preceding remarks, for all t we shall have $A_t^n = \lim_k A_{t \wedge k}^n$ and there exists a process A such that $A_t^n = A_{t \wedge n}$ for all n. As $A_t = \lim_n A_t^n$, A is obviously increasing, zero at 0 and predictable and satisfies the statement. Thus, we are reduced to the case where the supermartingale Z is stopped at instant n. But then the process $Y_t = Z_t - \mathbf{E}[Z_n | \mathcal{F}_t^\star]$ is a positive supermartingale (zero on $[n, \infty[$) with $\mathbf{E}[Y_0 - Y_t] = \mathbf{E}[Z_0 - Z_t]$ for any n, and we are done. ∎

Theorem 3.2.56 Let Y be a positive supermartingale. There exists a predictable integrable increasing process B which is zero at 0, such that $Y + B$ is a local martingale. Then for every ordered pair (r, u) such that $0 \leq r \leq u < \infty$

$$\mathbf{E}[B_u - B_r | \mathcal{F}_r^\star] \leq \mathbf{E}[Y_r - Y_u | \mathcal{F}_r^\star] \leq Y_r \qquad (3.126)$$

(valid also for $u = \infty-$). In particular, if we write

$$U_t = \mathbf{E}[B_{\infty-} | \mathcal{F}_r^\star] - B_t, \quad N_t = Y_t + B_t - \mathbf{E}[B_{\infty-} | \mathcal{F}_r^\star]$$

U is a potential of class D, N is a positive local martingale and we have a Riesz type decomposition

$$Y = U + N \qquad (3.127)$$

which is unique if we impose the above properties on U and N.

Proof. Set $T_k = \inf\{t : Y_t \geq k\} \wedge k$ for any integer k. The stopping times T_k increase to $+\infty$ as $k \to \infty$. The supermartingale Y^{T_k} is bounded above by k on $[0, T_k[$ and by Y_{T_k} on $[T_k, \infty[$ hence by $k \vee Y_{T_k}$ everywhere. But this r.v. is integrable (3.2.11), T_k hence Y^{T_k} belongs to class D and by Theorem 3.2.50 there exists a predictable increasing process B^k which is zero at 0 and such that $Y^{T_k} + B^k$ is a uniformly integrable martingale. By the remarks in the proof of Theorem 3.2.55, we can construct by pasting together a process B such that $B^k = B^{T_k}$ for all k; B is obviously increasing and zero at 0. As $B = \lim_k B^k$, B is predictable. Finally $Y + B$ is a local martingale, reduced by the stopping times T_k and it remains to establish (3.126).

Recall that we are working on $[0, \infty[$ so we must write a representation of Y^{T_k} without involving a jump at infinity, in the form

$$Y_t^{T_k} = \mathbf{E}[B_{\infty-}^k - B_t^k | \mathcal{F}_t^\star] + \mathbf{E}[Y_\infty^{T_k} | \mathcal{F}_t^\star]$$

and in particular for $t = 0$

$$Y_0 \geq \mathbf{E}[B^k_{\infty-}|\mathcal{F}^\star_0] + \mathbf{E}[Y_{T_k}|\mathcal{F}^\star_0]$$

and as $k \to \infty$, by Fatou's lemma

$$Y_0 - \mathbf{E}[Y_{\infty-}|\mathcal{F}^\star_0] \geq \mathbf{E}[B_{\infty-}|\mathcal{F}^\star_0].$$

Applying this to $Y_{t \wedge s}$, whose associated increasing process is $B_{t \wedge s}$, we obtain

$$Y_0 \geq \mathbf{E}[Y_0 - Y_s|\mathcal{F}^\star_0] \geq \mathbf{E}[B_s|\mathcal{F}^\star_0].$$

Applying this to $Y'_u = Y_{r+u}$ relative to the σ-fields $\mathcal{F}^{\star\prime}_u = \mathcal{F}^\star_{r+u}$, whose associated increasing process is $B'_u = B_{r+u} - B_r$, we obtain

$$Y_n \geq \mathbf{E}[Y_r - Y_{r+s}|\mathcal{F}^\star_r] \geq \mathbf{E}[B_{r+s} - B_r|\mathcal{F}^\star_r]$$

that is (3.126) in another notation. This implies as $s \to \infty$ that, in the decomposition (3.127), the local martingale N is positive.

As for uniqueness of decomposition (3.127), it follows from the fact that any local martingale which belongs to class D is a uniformly integrable martingale ([20] VI.30(f)). Then, if we have two decompositions $Y = U + N = U' + N'$, then $U - U'$ and $N - N'$ are uniformly integrable martingales which are zero at infinity, hence zero. ∎

Next, we state an extended version of Theorem 3.2.52 on jumps of the increasing process associated with a positive supermartingale to decompositions of arbitrary supermartingales.

Theorem 3.2.57 Let Z be a supermartingale. For every finite predictable stopping time T the conditional expectation $\mathbf{E}[Z_T|\mathcal{F}^\star_{T-}]$ exists, hence the predictable projection ^pZ is well defined ([20] VI.44(f)). If A is the increasing process associated with Z, then on $[0, \infty[$ we have

$$\triangle A = Z_- - {}^pZ. \tag{3.128}$$

Proof. We have $Z = M - A$, where M is a local martingale. We know ([20] VI.32) that $\mathbf{E}[|M_T||\mathcal{F}_{T-}]$ is a.s. finite; as A_T is $\mathcal{F}^\star_T$-measurable,

$$\mathbf{E}[|Z_T||\mathcal{F}^\star_{T-}] \leq \mathbf{E}[|M_T||\mathcal{F}^\star_{T-}] + |A_T|$$

is a.s. finite, whence our first assertion. Then $^pM = M_-$ ([20] VI.32) and the proof is completed as in Theorem 3.2.52. ∎

Chapter 4

Strong Supermartingales

The theory of strong supermartingales under the *un*usual conditions is studied here. The notion of strong supermartingale was first introduced by Mertens [44] under the usual conditions. Strong supermartingales are optional processes that satisfy the supermartingale inequality for all finite stopping times, rather than just for regular time parameter, hence the term "strong."

Fundamental results on the theory of strong supermartingales without the usual conditions on the family of σ-fields, and with no right-continuity of paths of processes, are presented here. In particular, we deduce optional and predictable projections theorems, Mertens decomposition and the theory of optimal stopping.

4.1 Introduction

Throughout this chapter, let $(\Omega, \mathcal{F}, \mathbf{F}, \mathbf{P})$ denote probability space with a filtration $\mathbf{F} = (\mathcal{F}_t)$ not necessarily complete or right-continuous with the convention $\mathcal{F}_{0-} = \mathcal{F}_0$. Then, the family $\mathbf{F}^+ = (\mathcal{F}_{t+})$ is right-continuous but not complete. Let $\mathbf{F}^\star = (\mathcal{F}_t^\star)$ denote the usual augmentation of $\mathbf{F}$, obtained by completing $\mathbf{F}^+$ by adjoining all the negligible sets of $\mathcal{F}$ to $\mathcal{F}_{t+}$; to emphasis the measure $\mathbf{P}$ in the completion of $\mathbf{F}^+$ we may write, sometimes, $\mathbf{F}^{\mathbf{P}}$ for $\mathbf{F}^\star$.

Recall Definition 2.4.21, that the optional field $\mathcal{O}(\mathbf{F})$, or $\mathcal{O}$ for short, on $\mathbb{R}_+ \times \Omega$ is generated by the processes that are adapted to $\mathbf{F}$ with *cadlag* paths (we emphasize that the processes are adapted to $(\mathcal{F}_t)$ not necessarily to $(\mathcal{F}_t^\star)$). Moreover, note that the optional cross-section theorem gives the same result in the *un*usual case, as in (Theorem 2.4.48) under the usual conditions. For clarity, we keep the word stopping time without qualifications for stopping time of $\mathbf{F}$. The predictable cross-section Theorem 2.4.51 requires no adjustment for the *un*ususal case, as predictable times are the same for the filtrations $\mathbf{F}$ and $\mathbf{F}^+$.

Let's start by giving the definition of strong supermartingale.

Definition 4.1.1 An optional process $X = (X_t)$ with $t \in \mathbb{R}_+$ is an optional strong supermartingale (resp. martingale) if

(1) for every bounded stopping time T, X_T is integrable, i.e., $\mathbf{E}[X_T] < \infty$,
(2) for every pair of bounded stopping times S, T such that $S \leq T$,

$$X_S \geq \mathbf{E}[X_T | \mathcal{F}_S] \quad (\text{resp.} \quad X_S = \mathbf{E}[X_T | \mathcal{F}_S]) \quad \text{a.s.} \tag{4.1}$$

In general, optional strong supermartingales are not necessarily cadlag, even under the usual conditions; for example, the optional projection of a decreasing process that is not necessarily right-continuous is an optional strong supermartigale that is not cadlag – consider the case of left potentials in Definition 3.2.38. Similarly, the limit of a decreasing sequence of cadlag positive supermartingales is an optional strong supermartingale that is, in general, not cadlag.

Furthermore, a cadlag supermartigale is an optional strong supermartingale where the usual conditions are not needed. However, under the usual conditions, every optional strong martingale is the optional projection of a constant process on every bounded intervals, hence cadlag. Besides optional strong supermartingales, we also have predictable strong supermartingales:

Definition 4.1.2 A predictable process $X = (X_t)$ with $t \in \mathbb{R}_+$ is a predictable strong supermartigale (resp. martingale) if
(1) for every bounded predictable time T, X_T is integrable,
(2) for every pair of predictable times S, T such that $S \leq T$,

$$X_S \geq \mathbf{E}[X_T | \mathcal{F}_{S-}] \quad (\text{resp.} =) \text{ a.s..} \tag{4.2}$$

Some inequalities of the usual supermartingale theory follow by the stopping theorem to the *un*usual case, we give an example below. Doob's inequalities on up-crossings and down-crossings are still valid for strong supermartingales but are harder to prove.

Corollary 4.1.3 Maximal Inequality. Let X be an optional or predictable strong supermartingale. For any $\tau \in \mathbb{R}_+$ and $\lambda \geq 0$ we have

$$\lambda \mathbf{P}\left(\sup_{t \leq \tau} |X_t| \geq \lambda\right) \leq 3 \sup_{t \leq \tau} \mathbf{E}\left[|X_t|\right]. \tag{4.3}$$

Proof. Let X_τ^* denote the $\mathcal{F}$-measurable $\sup_{t \leq \tau} |X_t|$ and h the sup of the expected value on the right-hand side of Equation 4.3. For every stopping time $S \leq \tau$; $\mathbf{E}[|X_S|] \leq 3h$ (3.10). Let $\epsilon > 0$ and choose a stopping time T such that $T(\omega) < \infty$, in such a way that, whenever $T(\omega) \leq \tau$ and $|X_T(\omega)| \geq \lambda$, then

$$\mathbf{P}(T < \infty) \geq \mathbf{P}(\exists t \leq \tau : |X_t(\omega)| \geq \lambda) - \epsilon$$

by the optional cross-section theorem.
Set $S = T \wedge \tau$, then

$$\begin{aligned}
\lambda \mathbf{P}(X_T^* > \lambda) &\leq \lambda \mathbf{P}(\exists t \leq \tau : |X_t| \geq \lambda) \leq \lambda(\mathbf{P}(|X_T| \geq \lambda) + \epsilon) \\
&\leq \mathbf{E}[|X_T|] + \lambda\epsilon \leq 3h + \lambda\epsilon.
\end{aligned}$$

∎

The above Corollary is also true for predictable strong supermartingale. Other inequalities of the type (Equations 3.19, 3.20 and 3.25) can be shown analogously.

Next is a fundamental theorem due to Mertens about the path properties of strong supermartingales under the usual conditions. The proof is essentially the same as that of Mertens but modified for the *unusual* case.

Theorem 4.1.4 Let X be an optional or predictable strong supermartingale. For almost all $\omega \in \Omega$, the path $X.(\omega)$ is *bounded* on every compact interval of $\mathbb{R}_+$ and has *right and left limits*.

Proof. First note that the σ-fields $\mathbf{F} = (\mathcal{F}_t)$ on which X is adapted are not complete and the set of all $\omega \in \Omega$ such that $X.(\omega)$ does not possess the above property of right and left limits can still be enclosed in a negligible set, not necessarily $\mathbf{F}$-measurable.

We will treat the optional case and leave the predictable case to the reader. We will prove the existence of right and left limits in $\overline{\mathbb{R}}_+$ then by using inequality (4.3) we will be able to conclude that these limits are in $\mathbb{R}_+$ too. We begin by setting

$$Y = \frac{X}{(1 + |X|)}.$$

For every sequence (S_n) of uniformly bounded stopping times that are either increasing or decreasing, the $\lim_n X_{S_n}$ exists a.s. by supermartingales convergence theorem, then, so does $\lim_n Y_{S_n}$, and $\lim_n \mathbf{E} Y_{S_n}$ exists by dominated convergence. Hence, the conditions for applying Theorem 3.2.22 (Theorem 3.2.23 for the predictable case) hold except that the usual conditions do not hold for the family $\mathbf{F}$.

So, let us take up the proof of Theorem 3.2.22 and adjust it for the σ-fields without the usual conditions. In the proof, the processes U and V are progressive with respect to the augmented family $\mathbf{F}^\star$, hence the debut S is a stopping time of this family, but we can use Theorem 2.4.19 (without change of notation) to make it into a stopping time of the family $\mathbf{F}^+$. Then, the stochastic interval $]\!]S, \infty[\![$ is optional with respect to the family $\mathbf{F}$ and the corresponding stopping times $S_0, S_1, \dots$ are constructed using the optional cross-section theorem applied to the family $\mathbf{F}$. Therefore, these are true stopping times and it is no longer difficult to prove that $S = \infty$ a.s. However, the second part of (a) (criterion for right-continuity) does not generalize to arbitrary filtrations.

We now move to part (b). Here again, the processes U and V are predictable only with respect to the family $\mathbf{F}^\star$, similarly for K, but then we modify the predictable stopping time T (without changing the notation) using Theorem 2.4.42 to make it into a stopping time of the family $\mathbf{F}$ and the proof then proceeds without difficulty. ∎

Remark 4.1.5 (a) Let D be a countable dense subset of $\mathbb{R}_+$. By Theorem 4.1.4 every result on convergence at infinity proved for the process $(X_t)_{t \in D}$ extends to the whole of the process $(X_t)_{t \in \mathbb{R}_+}$. Thus, it is unnecessary here to reproduce the statements of the convergence theorems.

(b) There exists a process X_+, which is optional with respect to the family $\mathbf{F}^+$ and indistinguishable from the process of right limits of X, and a predictable process X_-, which is indistinguishable from the process of left limits of X. The construction of such a process X_- is very simple: it is sufficient to take

$$X_{t-} = \liminf_{s \in D, s \uparrow\uparrow t} X_s \quad \text{for} \quad t > 0, \quad X_{0-} = X_0 \tag{4.4}$$

but the construction of X_+ is much more delicate. It was first explained in Remark 3.2.4. The processes X_+ and X_- constructed here have much more precise properties: X_+ is optional with respect to $\mathbf{F}^+$ and right-continuous. There exists a stopping time ρ of the family $\mathbf{F}^+$ such that $\mathbf{P}(\rho < \infty) = 0$ and X_+ has finite left limits on $]0, \rho[$ and is zero on $[\rho, \infty[$, but at instant ρ, on $\{\rho < \infty\}$, either the left limit of X_+ does not exist or it is infinite. With this description, it is moreover easily verified that ρ is predictable. As for X_-, in Equation 3.60 the following definition was proposed

$$X_- = \text{the left limit of } X_+ \text{ if it exists, } X_- = 0 \text{ otherwise.}$$

(c) Let X be an optional or predictable strong supermartingale. Then, X_- is a predictable strong supermartingale and X_+ an optional strong supermartingale with respect to the family $\mathbf{F}^+$. If T is a predictable time (resp. stopping time), then

$$X_{T-} \geq \mathbf{E}[X_T | \mathcal{F}_{T-}] \quad (\text{resp.} \quad X_T \geq \mathbf{E}[X_{T+}^\circ | \mathcal{F}_T]). \tag{4.5}$$

(d) Theorem 4.1.4 is interesting even under the usual conditions. For example, let (X^n) be a decreasing sequence of right-continuous positive supermartingales and let X be its limit. The process X is a strong supermartingale and hence it has a process of right limits X_+ and $X \geq X_+$ by right upper semi-continuity (this also follows from the second inequality (4.5), which can be written as $X_T \geq X_{T+}$ under the usual conditions). For all $E > 0$, the set $\{X \geq X_+ + \epsilon\}$ has no accumulation point at finite distance. This little remark is the abstract form of a convergence theorem of potential theory which can be stated as follows: let (f_n) be a decreasing sequence of excessive functions and let f be its limit; f is a strongly supermedian function (the analogue of strong supermartingale in potential theory) which has an excessive regularization $\hat{f} \leq f$. For all $\epsilon > 0$, the set $\{f \geq \hat{f} + \epsilon\}$ has no regular point and the set $\{f > \hat{f}\}$ is semi-polar.

(e) Let X be a positive time optional strong supermartingale which is completed with $X_\infty = 0$. The strong supermartingale inequality and the optional cross-section theorem give easily: If S is a stopping time such that $X_S = 0$ a.s., X is indistinguishable from 0 on $[S, \infty[$. Then, let T be the essential lower envelope of the set of stopping times S such that $X_S = 0$ a.s.; T is equal a.s. to a stopping time of the family $(\mathcal{F}_{t+})$, X is indistinguishable from 0 on $]T, \infty[$ and the optional cross-section theorem shows that X is > 0 on $[0, T[$ except on an evanescent set. Let S be a predictable stopping time such that $X_{S-} = 0$ a.s. on $(S < \infty)$; the inequality $X_{S-} \geq \mathbf{E}[X_S | \mathcal{F}_{S-}]$ a.s. (4.5) implies $X_S = 0$ on $(S < \infty)$ and hence everywhere, and hence $S \geq T$ a.s. The set $(X_- = 0) \cap [0, T]$ is predictable with countable cross-sections, hence it is indistinguishable from a sequence of predictable graphs S_n and, applying the above to each one, it follows that $X_- > 0$ on $[0, T[$ except on an evanescent set. Similarly, let U be a stopping time of the family $\mathbf{F}^+$ such that $X_{U+} = 0$ a.s. on $\{U < \infty\}$ (and hence everywhere), the equality $X_{U+} \geq \mathbf{E}[X_{U+\frac{1}{n}} | \mathcal{F}_{U+}]$ a.s. implies $X_{U+\frac{1}{n}} = 0$ a.s., hence $U + \frac{1}{n} \geq T$ and finally $U \geq T$. Arguing as above on the set $(X_+ = 0) \cap [0, T]$, we see that $X_+ > 0$ on $[0, T[$ except on an evanescent set. Thus, we obtain a version of Theorem 3.2.15, which is valid for positive optional strong supermartingales.

4.2 Projection Theorems

Our aim here is to understand optional and predictable projections on the raw filtration $(\mathcal{F}_t)$.

Theorem 4.2.1 (Projection) Let X be a positive or bounded measurable process. Then,
(a) There exists a predictable process Y such that

$$Y_T = \mathbf{E}[X_T|\mathcal{F}_{T-}] \quad \text{a.s.} \tag{4.6}$$

for every bounded predictable time T.
(b) There exists an optional process Z such that

$$Z_T = \mathbf{E}[X_T|\mathcal{F}_T] \quad \text{a.s.} \tag{4.7}$$

for every bounded stopping time T. Y and Z are unique within n evanescent processes. They are called the predictable and optional projections of X and are denoted by $Y = {}^pX$, $Z = {}^oX$.

Proof. The uniqueness follows from the cross-section theorems (2.4.48-2.4.51). To prove the existence, we shall reduce it to the case of the usual conditions using Lemma 4.2.3, which is of interest in itself.

We choose versions of projections of X relative to $\mathbf{F}^\star = (\mathcal{F}_t^\star)$: by Lemma 4.2.3 we can choose a version of the predictable projection which is predictable relative to $(\mathcal{F}_t)$ and denote it by Y. Since every predictable time T of $(\mathcal{F}_t)$ is a predictable time of $(\mathcal{F}_t^\star)$, Y satisfies (4.6) and that proves (a). In the optional case we choose a version which is optional relative to $(\mathcal{F}_{t+})$, which we denote by ζ; it satisfies (4.7) but in general it is not optional relative to $(\mathcal{F}_t)$ and we must modify it.

By Theorem 3.2.22, the set $H = \{(t, u)) : Y_t(u)) \neq \zeta_t(\omega)\}$ is contained in a countable union of graphs of random variables. We apply Theorem 2.4.20 to it in order to write it in the form $H = K \cup L$ where K is such that for every stopping time T of $(\mathcal{F}_t)$, $\mathbf{P}\{\omega : (T(\omega), \omega) \in K\} = 0$ and L is contained in a countable union of graphs of stopping times T_n. These graphs can be assumed to be disjoint in $\mathbb{R}_+ \times \Omega$. We use x_n to denote a version of $\mathbf{E}[X_{T_n}\mathbf{1}_{(T_n<\infty)}|\mathcal{F}_{T_n}]$ and for all (t, ω) we set

$$\begin{aligned}
Z_t(\omega) &= Y_t(\omega) \text{ if } (t, \omega) \notin \cup_n[\![T_n]\!], \tag{4.8}\\
Z_t(\omega) &= x_n(\omega) \text{ if } (t, \omega) \in [\![T_n]\!].
\end{aligned}$$

Clearly the process Z is optional. Let T be a bounded stopping time of the family $(\mathcal{F}_t)$ and let $A \in \mathcal{F}_T$. We set $A_n = A \cap \{T = T_n\}$, which belongs to $\mathcal{F}_{T_n}$ and $\mathcal{F}_T$, and $B = A \backslash \cup_n A_n$, which belongs to $\mathcal{F}_T$.

On A_n we have

$$Z_T = x_n = \mathbf{E}[X_T|\mathcal{F}_{T_n}] = \mathbf{E}[X_{T_n}\mathbf{1}_{A_n}|\mathcal{F}_{T_n}] = \mathbf{E}[X_T\mathbf{1}_{A_n}|\mathcal{F}_T]$$

and hence (4.9)

$$\int_{A_n} Z_T d\mathbf{P} = \int_{A_n} \mathbf{E}[X_T\mathbf{1}_{A_n}|\mathcal{F}_T]d\mathbf{P} = \int_{A_n} X_T d\mathbf{P} \tag{4.9}$$

On B, $T \neq T_n$ for all n and hence $(T(\omega), \omega) \notin L$. As $\mathbf{P}\{(T(\omega), \omega) \in K\} = 0$, we have a.s. $(T(\omega), \omega) \notin H$, in other words, $Y_T(\omega) = \zeta_T(\omega)$ a.s. On the other hand, $Z_T = Y_T$ on B. Thus

$$\int_B Z_T d\mathbf{P} = \int_B \zeta_T d\mathbf{P} = \int_B \mathbf{E}[X_T|\mathcal{F}_{T+}] d\mathbf{P} = \int_B X_T d\mathbf{P}. \tag{4.10}$$

To obtain (4.7) it only remains to add the relations (4.9) and (4.10). ∎

Remark 4.2.2 The restriction to bounded stopping times is intended only to avoid unnecessary details. The extension to finite stopping times or arbitrary stopping times when the processes are indexed by $[0, \infty]$ is immediate.

Lemma 4.2.3 (a) Every process which is predictable relative to the family relative $\mathbf{F}^\star = (\mathcal{F}_t^\star)$ is indistinguishable from a process which is predictable relative to $\mathbf{F} = (\mathcal{F}_t)$.

(b) Every process which is optional relative to the family $\mathbf{F}^\star = (\mathcal{F}_t^\star)$ is indistinguishable from a process which is optional relative to $\mathbf{F}^+ = (\mathcal{F}_{t+})$.

Proof. Using monotone classes, we are reduced to proving the lemma for stochastic intervals and we then apply Theorem 2.4.42 in the predictable case and Theorem 3.2.30 in the optional case. ∎

Remark 4.2.4 (a) If $X_t(\omega) = X(\omega)$, $X \in L^1(\mathbf{F})$, we obtain the existence of versions of the conditional expectations $\mathbf{E}[X|\mathcal{F}_T]$ or $\mathbf{E}[X|\mathcal{F}_{T-}]$ which are optional or predictable strong martingales. Denoting these versions by Z and Y, it is immediately verified that $Y = Z_-$ as under the usual conditions, and that $Z = {}^o(Z_+) = {}^o(Y_+)$.

(b) As under the usual conditions, the set $\{{}^oX \neq {}^pX\}$ is a countable union of graphs of stopping times. No proof is necessary as it follows by the very construction of ${}^oX$ above.

We give two examples to illustrate Theorem 4.2.1. The first example is a consequence due to Kunita [45]. The second example will come later.

We begin by defining the notion of precise density:

Definition 4.2.5 Let $(\Omega, \mathcal{F}, \mathbf{P})$ be a probability space and let $\mathbf{Q}$ be another law on $(\Omega, \mathcal{F})$. A positive r.v. U is called a *precise density* of $\mathbf{Q}$ relative to $\mathbf{P}$ if (1) the absolutely continuous part of $\mathbf{Q}$ relative to $\mathbf{P}$ is $U \cdot \mathbf{P}$ (thus U is $\mathbf{P}$-integrable and hence $\mathbf{P}$ a.s. finite) and (2) the singular part of $\mathbf{Q}$ is carried by $\{U = +\infty\}$.

It is easy to see that U is determined to within a $(\mathbf{P} + \mathbf{Q})$-negligible function. Also, one can compute a precise density, for example, as follows: choose a density V of $\mathbf{P}$ with respect to the law $(\mathbf{P} + \mathbf{Q})/2$, such that $0 \leq V \leq 2$, then set $U = \frac{2}{V} - 1$.

Theorem 4.2.6 Granted the definition of precise density, let $(\Omega, \mathcal{F}, \mathbf{P})$ be endowed with the raw filtration $(\mathcal{F}_t)$. Then, there exists a unique optional strong supermartingale U_t indexed by $[0, \infty]$ such that for every stopping time T, U_T is a precise density of $\mathbf{Q}$ relative to $\mathbf{P}$ on the σ-field $\mathcal{F}_T$.

Proof. By Theorem 4.2.1 applied to the law $(\mathbf{P} + \mathbf{Q})/2$ there exists an optionally strong martingale (V_t) under this law such that V_∞ is a density of $\mathbf{P}$ with respect to $(\mathbf{P} + \mathbf{Q})/2$ on $\mathcal{F}_\infty$. Then V_T is a density of $\mathbf{P}$ with respect to $(\mathbf{P} + \mathbf{Q})/2$ on $\mathcal{F}_T$ for every stopping time T. If we set $U_t = 2/V_t - 1$, the process U is optional and U_T is a precise density of $\mathbf{Q}$ with respect to $\mathbf{P}$ on $\mathcal{F}_T$ for every stopping time T. It follows immediately that U is a strong supermartingale and the uniqueness follows from the cross-section theorem. ∎

Remark 4.2.7 It can be easily shown, similarly, that the process U^{-1} provides precise densities of $\mathbf{P}$ with respect to $\mathbf{Q}$.

Here is the second example (Dellacherie [46]) that connects supermartingales to optional strong supermartingales and is of profound significance.

Theorem 4.2.8 Every supermartingale X has a modification Z which is an optional strong supermartingale.

Proof. Let D_n denote the set of dyadic numbers of the form $k/2^n$ where k and n are integers. Let D be the set of all dyadic numbers and H the countable set of points t where the decreasing function $\mathbf{E}[X_t]$ is not right-continuous.

Let $U = X_+$ be the process of right limits of X. From Theorem 4.1.4 and Remark 4.1.5, we know that U is a strong supermartingale of the family $(\mathcal{F}_{t+})$ and that $X_T \geq \mathbf{E}[U_T | \mathcal{F}_T]$ a.s. for every bounded stopping time T. Let $V = {}^oU$ be the optional projection of U; if reassurance is needed that this projection exists, note that V is bounded from below on every finite interval $[0, n]$ by the strong martingale $(\mathbf{E}[U_n | \mathcal{F}_t])$. Also bounded by X from above $X \geq V$. We set

$$Z_t = X_t \mathbf{1}_{\{t \in H\}} + V_t \mathbf{1}_{\{t \notin H\}}$$

and show that Z is a modification of X and that it is an optional strong supermartingale.

(1) It is sufficient to check that $Z_t = V_t = X_t$ a.s. at a point $t \notin H$. But at such a point

$$\mathbf{E}[X_t] = (\mathbf{E}[X_.])_+ = \mathbf{E}[X_{t+}] = \mathbf{E}[V_t],$$

which with the inequality $X_t \geq V_t$ a.s. implies $X_t = V_t$ a.s.; in the preceding equalities we have used the uniform integrability property in Theorem 3.1.29.

(2) Z is an optional process. To show that it is a strong supermartingale, we consider two bounded stopping times S and T such that $S \leq T$ and let $A \in \mathcal{F}_S$. We set

$$T_n = \begin{cases} T & \text{if } T \in H \\ \inf\{t \in D_n : t > T\} & \text{if } T \notin H \end{cases},$$

T_n is a stopping time and it is not difficult to check that T_n decreases to T: through strictly greater values when $T \notin H$ and stationarily when $T \in H$. We define S_n similarly from S; $S_n \leq T_n$ does not necessarily hold on the set $\{S \notin H, T \in H\}$, but to ensure that this inequality holds without altering the manner in which S_n converges to S, described above, it suffices to replace S_n by $S_n \wedge T_n$ which we do without changing the notation.

We then have, taking the limit under the $\int$ sign being justified by the uniform integrability property (Theorem 3.1.29),

$$\lim_n X_{S_n} = U_S 1_{(S \notin H)} + X_S 1_{(S \in H)}$$

and similarly for T

$$\int_A Z_S \mathbf{P} = \lim_n \int_A X_{S_n} \mathbf{P} \geq \lim_n \int_A X_{T_n} \mathbf{P} = \int_A Z_T \mathbf{P}$$

which is the strong supermartingale inequality. ∎

Remark 4.2.9 Let Y be another optional strong supermartingale which is a modification of X. We have to within an evanescent process:

$$Y_+ = U, Y \geq {}^o(Y_+), \text{ hence } Y \geq 0 = V \text{ and finally } Y \geq Z \text{ on } H^c \times \Omega.$$

On the other hand, Y and Z are both modifications of Y and hence Y and Z are indistinguishable on the set $H \times \Omega$, because H is countable. Thus $Y \geq Z$ to within an evanescent process. Thus the modification which we have constructed is the smallest possible.

Next, we define projections of $\mathbf{P}$ measures and random measures and dual projections of integrable increasing processes. Moreover, the relation between optional (predictable) $\mathbf{P}$ measures and optional (predictable) increasing processes will be examined in some detail. Since the usual definition of increasing processes requires that their paths all be increasing, right-continuous and finite, which is incompatible with adaptation to the family $(\mathcal{F}_t)$ when the usual conditions do not hold, thus something must be lost on one side or the other. We begin with the analogue of proposition 3.2.29.

Theorem 4.2.10 Dual Projection. Let A be an integrable increasing process which is indistinguishable from a process B which is predictable (optional) relative to $(\mathcal{F}_t)$. Then the measure μ

$$\mu(\cdot) = \mathbf{E}\left[\int_0^\infty (\cdot) A(\omega, dt)\right]$$

associated with A is predictable (optional).

Proof. We shall use the results from Theorem 4.2.1, where we have worked on $\mathbb{R}_+ \times \Omega$ not on $\bar{\mathbb{R}}_+ \times \Omega$. For simplicity, as a result of being on $\mathbb{R}_+ \times \Omega$, we assume that A does not jump at infinity; this is not a real restriction to generality (see Remark 3.2.26(e) for an explanation).

The predictable case is straightforward. For that if A is predictable relative to $(\mathcal{F}_t)$, then $\mu(X) = \mu({}^pX)$ for every bounded measurable process X where ${}^pX$ denotes the predictable projection relative to $(\mathcal{F}_t)$. But by Lemma 4.2.3 ${}^pX$ is indistinguishable from the predictable projection relative to $(\mathcal{F}_t)$ and on the other hand μ has no mass on evanescent sets.

We now treat the optional case. We decompose A into A^c and A^d, its continuous and discontinuous parts; A^c is predictable relative to $(\mathcal{F}_t^\star)$ and hence indistinguishable from a predictable process relative to $(\mathcal{F}_t)$ by Lemma 4.2.3. We denote this process by B^c and set $B^d = B - B^c$, which is optional relative to $(\mathcal{F}_t)$. Since the measure associated with A^c is predictable by the preceding paragraph, it is sufficient to study the measure associated with A^d, in other words, we can reduce it to the case where A is purely discontinuous and suppress all the ^{d}.

We set $C_t = \sup_{r \in \mathbb{Q}, r < t} B_t$ if $t > 0$, $C_0 = 0$. This process is predictable relative to $(\mathcal{F}_t)$ and indistinguishable from the process (A_{t-}). Hence the set $\{C \neq B\}$ is indistinguishable from the set $\{A \neq A_-\}$. But the first is optional, whereas the second is a union of graphs of r.v. Then by the end of Theorem 2.4.54 (c) we see that $\{B \neq C\}$ and hence $\{A \neq A_-\}$ is indistinguishable from the union of a sequence of graphs $[\![U_n]\!]$ of stopping times of $(\mathcal{F}_t)$. On the other hand, the r.v. $\delta_n = \triangle A_{U_n}$ is a.s. equal to $(B_{U_n} - C_{U_n}) \mathbf{1}_{(U_n < \infty)}$ and hence it is a.s. $\mathcal{F}_{U_n}$-measurable. We conclude by noting that, for every bounded measurable process X with optional projection Z relative to $(\mathcal{F}_t)$,

$$
\begin{aligned}
\mu(X) &= \mathbf{E}\left[\sum_n X_{U_n}\delta_n\right] = \mathbf{E}\left[\sum_n \mathbf{E}\left[X_{U_n}|\mathcal{F}_{U_n}\right]\delta_n\right] \\
&= \mathbf{E}\left[\sum_n Z_{U_n}\delta_n\right] \\
&= \mu(Z)
\end{aligned}
$$

where all the terms are meaningful since $\delta_n = 0$ on $\{U_n = \infty\}$. ∎

Remark 4.2.11 In practice, the results of the theorem are used in a slightly different way. It is B that is defined first as an optional process such that, for almost all $\omega \in \Omega$, the path $B.(\omega)$ is finite, increasing and right-continuous. Then, the measure

$$
\mu(X) = \mathbf{E}\left[\int_0^\infty X_s dB_s\right]
$$

is defined, where X is positive and measurable and the function under the $\mathbf{E}$ sign is only defined a.s. and μ is assumed to be bounded. Then, let N be a negligible set such that, for $\omega \notin N$, $B.(\omega)$ is increasing, finite and right-continuous. We set

$$
A_t(\omega) = B_t(\omega) \quad \text{if} \quad \omega \in N^c, \quad A_t(\omega) = 0 \quad \text{if} \quad \omega \in N.
$$

As μ is also the measure associated with the integrable increasing process A, Theorem 4.2.10 tells us that μ is indeed optional.

Conversely, suppose that μ is a predictable or optional bounded $\mathbf{P}$ measure on $\mathbb{R}_+ \times \Omega$; we wish to represent it using a process with properties as precise as possible.

Theorem 4.2.12 The predictable (optional) $\mathbf{P}$ measure μ has a representation (4.11)

$$
\mu(X) = \mathbf{E}\left[\int_{[0,\infty[} X_s dB_s\right] \tag{4.11}
$$

where B is a predictable (optional) process all of whose paths are increasing and right-continuous, with the possible exception of a single value of μ for which $B_t < \infty$ and $B_{t+} = \infty$.

Moreover, B can be represented as a sum $\sum_n B^n$ of true predictable (optional) integrable increasing processes.

Proof. We begin with the measure μ predictable relative to $(\mathcal{F}_t)$. Hence it has a representation (Theorem 3.2.25)

$$\mu(X) \; = \; \mathbf{E}\left[\int_{[0,\infty[} X_s dA_s\right]$$

$$A_t \; = \; \alpha_t + \sum_n \lambda_n \mathbf{1}_{(t \geq T_n)}$$

where α is a continuous, integrable, increasing process that is 0 at 0 and adapted to the family $(\mathcal{F}_t)$, the λ_n are positive constants and the T_n are predictable times of the family $(\mathcal{F}_t)$. We also introduce the stopping times of the family $(\mathcal{F}_t)$, $S_n = \inf\{t : \alpha_t \geq n\}$ which are predictable (S_n is the debut of a right closed predictable set), and the increasing processes bounded by 1

$$\alpha_t = \alpha_{t \wedge S_{n+1}} - \alpha_{t \wedge S_n}, \quad \beta^n_t = \mathbf{1}_{(t \geq T_n)}.$$

We now regularize these processes. For β^n we choose a predictable T_n of the family $(\mathcal{F}_t)$ such that $\overline{T}_n = T_n$ a.s. (2.4.42) and we set $\overline{\beta}^n_t = \mathbf{1}_{(t \geq \overline{T}_n)}$. For α^n we choose for r rational r.v. ξ^n_r which is $\mathcal{F}_{r+}$-measurable and equal a.s. to α^n_r and then for t real > 0 we set $\eta^n_t = \sup_{r<t} \xi^n_r$ and then for t real ≥ 0, $\zeta^n_t = \eta^n_{t+}$ an increasing process which is adapted to $(\mathcal{F}_{t+})$, bounded by 1 and indistinguishable from α^n. We then denote the continuous part of ζ^n by $\bar{a}^n$ (at the same time removing the jump at 0); this increasing process is still indistinguishable from α^n, still adapted to $(\mathcal{F}_{t+})$ and, as it is continuous, it is predictable relative to $(\mathcal{F}_t)$. It only remains to set

$$B_t = \sum_n \overline{\alpha}^n_t + \sum_n \lambda_n \overline{\beta}^n_t. \tag{4.12}$$

This process is indistinguishable from A, it is predictable with respect to $(\mathcal{F}_t)$ and its paths are all increasing. If $B_t(\omega) < \infty$, the series (4.12) is uniformly convergent on $[0, t]$ and hence the path $B_{\cdot}(\omega)$ is right-continuous on $[0, t[$, hence B can have only the type of right-discontinuity indicated in the statement. As $\mathbf{E}[B_\infty] < \infty$, note all so that this lack of right-continuity can occur only on a set of measure zero.

We now come to the optional case, which is a little more delicate. Let μ be a bounded $\mathbf{P}$ measure on $\mathbb{R}_+ \times \Omega$ such that $\mu(X) = \mu(\,^{\circ}X)$ for every bounded measurable process X. We decompose μ as a sum $\mu = v + \pi$ of two positive measures with the following properties:

- π is carried by a countable union of graphs of stopping times of $(\mathcal{F}_T)$ and

- v has no mass on any graph of a stopping time (it is sufficient to define π as the least upper bound of the measures bounded above by μ carried by a countable union of graphs of stopping times and to set $v = \mu - \pi$).

It is immediately verified that there exists a set H which is a countable union of graphs of stopping times such that $\pi = \mathbf{1}_H \cdot \mu$; hence, π is optional and the same is true of v by taking the difference.

As the set where ^oX and ^pX difference is a countable union of graphs of stopping times, we have $v(X) = v(\,^oX) = v(\,^pX)$; hence is predictable and we can apply the first part of the proof to it. Hence, it remains to consider π.

We represent H as a union of disjoint graphs of stopping times U_n. Then $\pi = \mathbf{1}_H \cdot \mu$ is the sum of the measures $\mathbf{1}_{[\![U_n]\!]} \cdot \mu = \pi^n$ and it is sufficient to represent each measure π^n. As π^n is carried by the graph of U^n, we can write $\pi^n(X) = \mathbf{E}[X_{U_n}\delta_n]$ where δ_n is an integrable positive r.v. On the other hand, π^n is optional; taking $X = 0$ outside the graph of U_n, we see that δ_n is a.s. equal to a positive $\mathcal{F}_{U_n}$-measurable r.v. It only remains to write $\delta_n = \sum_m \lambda_{nm} \mathbf{1}_{H_{mn}}$ a.s. where the λ_{nm} are constants and the H_{nm} elements of $\mathcal{F}_{U_n}$, and then to set $T_{nm} = U_n$ on H_{nm} and $= +\infty$ on H_{nm}^C, to obtain the representation

$$\pi(X) = \sum_n \pi^n(X) = \sum_{nm} \lambda_{nm} \mathbf{E}\left[X_{T_{nm}} \mathbf{1}_{(T_{nm} < \infty)}\right]$$

and hence the "increasing process" associated with π which is $\sum_{nm} \lambda_{nm} \mathbf{1}_{(t \geq T_{nm})}$. ∎

Remark 4.2.13 From the results of Theorems 4.2.10 and 4.2.12, we can obtain an interesting result in itself: if A is an integrable increasing process which is indistinguishable from a predictable process, then there exists a process B which is indistinguishable from A and is a sum of predictable integrable increasing processes (all the properties hold simultaneously for each term in the sum). Similarly for the optional case.

4.3 Special Inequalities

Let X be a strong supermartingale indexed by $[0, \infty]$. We associate with it a measurable process X_+ indistinguishable from the process of right limits of X (for example, the version constructed in Remark 4.1.5(b)) and a predictable process X_- indistinguishable from the process of left limits of X. The relations (4.5) can be written

$$X \geq {}^pX, \quad X \geq {}^o(X_+). \tag{4.13}$$

We call X *regular* if $^pX = X_-$ including at ∞. If X belongs to class D this means, as in (Theorem 3.2.52), that X is left-continuous in expectation: if (T_n) is an increasing sequence of stopping times and $\lim_n T_n = T$, then $\lim_n \mathbf{E}[X_{T_n}] = \mathbf{E}[X_T]$. But the second inequality (4.5) introduces a new notion:

Lemma 4.3.1 With the above notation, the following properties are equivalent:
(a) $X = {}^oX_+$.
(b) for every stopping time T,

$$\mathbf{E}[X_T] = \lim \mathbf{E}\left[X_{T + \frac{1}{n}}\right].$$

(c) for every stopping time T and every decreasing sequence of stopping times $T_n \downarrow T$,

$$\mathbf{E}[X_{T_n}] = \lim_n \mathbf{E}[X_{T_n}].$$

If these properties hold, X is called continuous in expectation.

Proof. Clearly (c)$\Rightarrow$(b). Since the r.v. $X_{T+1/n}$ are uniformly integrable (Theorem 3.1.29), (b) can be written as $\mathbf{E}[X_T] = \mathbf{E}[X_{T+}]$. On the other hand $X_T \geq \mathbf{E}[X_{T+}|\mathcal{F}_{T-}]$, whence the equality $X_T = \mathbf{E}[X_{T+}|\mathcal{F}_T]$. Since this holds for every stopping time T, we see that (b)$\Rightarrow$(a). Finally, in the notation of (c),

$$\lim_n \mathbf{E}[X_{T_n}] = \mathbf{E}[\lim_n X_{T_n}] = \int_A X_T d\mathbf{P} + \int_{A^c} X_{T+} d\mathbf{P},$$

where A is the set $\{\exists n : T_n = T\}$, which belongs to $\mathcal{F}_T$. Given (a), this sum is equal to $\mathbf{E}[X_T]$ and the last implication (a)$\Rightarrow$(c) is proved. $\blacksquare$

Remark 4.3.2 Under the usual conditions, properties (a), (b), and (c) mean simply that X is right-continuous, but this is not true in general.

The notion of splitting stopping time (s.s.t.) is due to Bismut [47] who called it quasi-stopping time (q.s.t.). It is especially useful for studying strong supermartingales without the usual conditions. A generalization of this notion can be found in [48].

Definition 4.3.3 A *splitting stopping time* (abbreviated s.s.t.) is an ordered pair $\tau = (H, T)$ where T is a stopping time and H an element of $\mathcal{F}_T$ such that T_H is predictable, where $T_H(\omega) = T(\omega)$ if $\omega \in H$, $T_H(\omega) = +\infty$ if $\omega \in H$.

Since T_H is predictable, then H belongs to $\mathcal{F}_{T-}$. Furthermore, τ is called bounded, finite, etc. if T is bounded, finite, etc. The intuitive idea behind such an ordered pair is the following: divide each point x in $\overline{\mathbb{R}}_+$ into two points, one, again, denoted by x and the other by $x-$ and extend the order relation by the conventions $x- < x$, $(y- \leq x-$ if and only if $y \leq x)$ and $(y \leq x-$ if and only if $y < x)$. Then, τ can be considered as a function with values in $\overline{\mathbb{R}}_+$ doubled in this way which takes the value $T-$ on H and the value T on H^c.

If X is a strong supermartingale or more generally a process which a.s. admits left limits, we set

$$X_\tau = X_{T-} \text{ on } H, \quad X_\tau = X_T \text{ on } H^c \tag{4.14}$$

with the convention that $X_{0-} = X_0$ if X_{0-} has not been specified otherwise (if τ is not a.s. finite, it is necessary for X to be indexed by $[0, \infty]$).

If X is optional, X_τ is measurable with respect to the σ-field $\mathcal{F}_\tau$ consisting of the $A \in \mathcal{F}_T$ such that $A \cap H$ belongs to $\mathcal{F}_{T-}$.

Let $\sigma = (G, S)$ and $\tau = (H, T)$ be two s.s.t.; we write $\sigma < \tau$ if $S \leq T$ and $H \cap \{S = T\} \subset G \cap \{S = T\}$, conforming with the order relation defined above on the split half-line. It is easy to check that $\mathcal{F}_\sigma \subset \mathcal{F}_\tau$.

With every stopping time T can be associated with the s.s.t. $(\emptyset, T)$ which we identify with T. If T is predictable, we can associate with it the s.s.t. (Ω, T), which we denote by $T-$. Then the following theorem generalizes the two stopping Theorems 3.2.11 and 3.2.14.

Theorem 4.3.4 Let X be an optional strong supermartingale indexed by $[0, \infty]$. We adopt the convention $X_{0-} = X_0$. Let $\sigma = (G, S)$ and $\tau = (H, T)$ be two s.s.t. such that $\sigma \leq \tau$. Then,

$$X_\sigma \geq E[X_\tau | \mathcal{F}_\sigma] \quad \text{a.s.} \tag{4.15}$$

with equality if X is an optional strong martingale on $[0, \infty]$.

Proof. First, suppose that X is an optional strong martingale, then it is immediate that $X_\tau = \mathbf{E}[X_\infty | \mathcal{F}_\tau]$, $X_\sigma = \mathbf{E}[X_\infty | \mathcal{F}_\sigma]$ and (4.15) with equality. For X an optional strong supermartingale, consider $X - \mathbf{E}[X_\infty | \mathcal{F}_t]$, and we are reduced to the case where X is positive for which we wish to establish the inequality

$$\int_A X_\sigma \mathbf{P} \geq \int_A X_\tau \mathbf{P} \quad \text{for} \quad A \in \mathcal{F}_\sigma.$$

We begin by treating the case where $A = \Omega$. We are free to truncate X at n and then take the limit, and hence we may assume that X is bounded. Let (R_n) be a sequence which a.s. foretells S_G (2.4.41) and let $S_n = R_n \wedge S$; then $X_\sigma = \lim_n S_{S_n}$ a.s. and hence it is sufficient to check that

$$\int X_{S_n} \mathbf{P} \geq \int X_\tau \mathbf{P}.$$

Let (U_m) be a sequence which a.s. foretells T_H; their relation $\sigma \leq \tau$ implies that the sequence $T_m = S_m \vee (U_m \wedge T)$ is then such that $X_{T_m} \to X_\tau$ a.s.; on the other hand, for all n and all m we have $\int X_{S_n} d\mathbf{P} \geq \int X_{T_m}^m d\mathbf{P}$ by the definition of strong supermartingales, whence the required inequality.

To deal with the general case, we proceed as for ordinary stopping times: we apply the above to the s.s.t. $\sigma_A = (G \cap A, S_A)$ and $\tau_A = (H \cap A, T_A)$. ∎

Now, we aim to extend the inequalities of Section (3.2) to general situations and at the same time to unify them. To accomplish this, consider the following:

Remark 4.3.5 We consider two integrable increasing processes A^- and A^+; the first is predictable relative to $(\mathcal{F}_t)$ and zero at 0 but may jump at infinity, the second is optional relative to $(\mathcal{F}_t)$ may jump at 0 but not at infinity and is purely discontinuous. We denote by A the predictable integrable increasing process $A = A^+ + A^-$ which is neither right- nor left-continuous. The potential Z generated by A is by definition the optional projection of the process $(A_\infty - A_t)$, that is the process

$$Z_t = \mathbf{E}[A_\infty | \mathcal{F}_t] - A_t \tag{4.16}$$

where the process $\mathbf{E}[A_\infty|\mathcal{F}_t]$ must be interpreted as an optional strong martingale. Our aim is to estimate A_∞ as a function of Z.

Under the usual conditions, the situation includes the two cases of Definition 3.2.38: the predictable case corresponds to a right-continuous process A and the optional case to a left-continuous process A. Under the *un*usual conditions, our hypotheses are rather too restrictive, but see Remark 4.3.8.

We saw in Remark 3.2.46 that A determines A^+ and A^-.

Remark 4.3.6 We apply formula [20], VI.91.4, to the right-continuous process $A_+ = A^- + A^+$

$$A_\infty^2 = \int ((A_\infty - A_{t-}) + (A_\infty - A_{t+}))dA_{t+}$$
$$= 2\int (A_\infty - A_t)dA_{t+} + \int ((A_t - A_{t-}) + (A_t - A_{t+}))dA_t^+.$$

The latter integral is equal to $\sum \left((\triangle A_t^-)^2 - (\triangle A_t^+)^2 \right)$. We thus obtain

$$A_\infty^2 = \int (A_\infty - A_t)dA_{t+} + \int (A_\infty - A_{t-})dA_t^- + \int (A_\infty - A_{t+})dA_t^+.$$

We now integrate; we can replace $(A_\infty - A_t)$ by its optional projection (Z_t), $(A_\infty - A_{t-})$ by its predictable projection (Z_{t-}) and $(A_\infty - A_{t+})$ by its optional projection (Z_{t+}) relative to $(\mathcal{F}_{t+})$. We obtain,

$$\mathbf{E}[A_\infty^2] = \mathbf{E}\left[\int Z_s dA_{s+} + \int Z_{s-} dA_s^- + \int Z_{s+} dA_s^+ \right]. \tag{4.17}$$

We thus generalize the two energy formula of Theorem 3.2.39.

Remark 4.3.7 We shall now generalize the fundamental Theorem 3.2.42. As noted in [20] VI.100, it all amounts to generalizing the "Maximal Lemma", VI.100.1, and to this end we argue directly, using the s.s.t introduced in Definition 4.3.3. It will be convenient to assume that A does not jump at infinity, but it is not restricted to its generality (see Remark 3.2.26(c) and [20] VII.15).

This time we set $\alpha_t = A_t^+ + A_t^-$; α is a right-continuous optional increasing process and the optional set $J = \{(t,\omega) : \alpha_t(\omega) \ge \lambda\}$ is a stochastic interval $[\![T, \infty[\![$, whose debut $T(\omega) = \inf (t : t \in \mathbb{Q}, \, \alpha_t(\omega) \ge \lambda)$ is a stopping time of $(\mathcal{F}_{t+})$. Let $H = \{\omega : A_T(\omega) \ge \lambda\}$; as $\alpha_- \le A \le \alpha$, the sets $\{A \ge \lambda\}$ and $\{\alpha \ge \lambda\}$ differ only by a subset of the graph of T and hence

$$[\![T_H]\!] = \{(t,\omega) : A_t(\omega) \ge \lambda\} \backslash]\!]T, \infty[\![$$

is a predictable set. On the other hand, T_H is a stopping time of $(\mathcal{F}_{t+})$, and hence $]\!]T_H, \infty[\![$ is predictable and finally $[\![T_H, \infty[\![= [\![T_H]\!] \cup]\!]T_H, \infty[\![$ is predictable, which means that T_H is a predictable time (Definition 2.4.30) and hence that $\tau = (H, T)$ is a s.s.t.

The set $\{A_\infty \ge \lambda\} = \{T < \infty\} \cup \{T = \infty, A_\infty \ge \lambda\}$ belongs to $\mathcal{F}_T$ and its intersection with H is equal to H and hence is $\mathcal{F}_{T-}$ measurable; hence it belongs to $\mathcal{F}_T$. Finally, $A_T \le \lambda$, whence the inequality

$$\int_{(A_\infty \ge \lambda)} (A_\infty - \lambda)\mathbf{P} \le \int_{(A_\infty \ge \lambda)} (A_\infty - A_\tau)\mathbf{P} = \int_{(A_\infty \ge \lambda)} Z_\tau \mathbf{P}$$

by Theorem 4.3.4 applied to the strong martingale $(Z_t + A_t)$. If Z is bounded above by a strong martingale $(\mathbf{E}[M_\infty|\mathcal{F}_t])$, we deduce that

$$\int_{(A_\infty \geq \lambda)} (A_\infty - \lambda)\mathbf{P} \leq \int_{(A_\infty \geq \lambda)} M_\infty \mathbf{P} \tag{4.18}$$

that is VI.100.1 and all the consequences of this inequality proved in Chapter VI of [20] remain true in the general case.

Remark 4.3.8 We return to the hypotheses of Remark 4.3.5; in fact, if we are working under the *un*usual conditions, it was pointed out earlier that A^+ and A^- are only very rarely true optional or predictable increasing processes: they are constructed from $\mathbf{P}$ measures by the procedure of Theorem 4.2.12; in other words A^- and A^+ are both processes with increasing paths, the first is predictable and the second optional, but right-continuity does not hold identically there may exist ω-forming a set of measure zero for which $A^+_\cdot(\omega)$ or $A^-_\cdot(\omega)$ is not right-continuous. How do we overcome this difficulty?

By Theorem 4.2.12, A^+ and A^- are sums of analogous processes but which are identically right-continuous and hence A itself is a sum of the corresponding processes A^n and Z the sum of the corresponding processes Z^n. We set

$$\alpha^k = \sum_1^k A^n, \quad \zeta^k = \sum_1^k Z^n.$$

If Z is bounded above by the strong martingale $(\mathbf{E}[M_\infty|\mathcal{F}_t])$, a fortiori so is ζ^k and, slightly modifying the integration sets, we obtain

$$\int_{(\alpha^k_\infty > \lambda)} (\alpha^k_\infty - \lambda)\,\mathbf{P} \leq \int_{(\alpha^k_\infty > \lambda)} M_\infty \mathbf{P}$$

then, letting k tend to infinity,

$$\int_{(A_\infty > \lambda)} (A_\infty - \lambda)\mathbf{P} \leq \int_{(A_\infty > \lambda)} M_\infty \mathbf{P}$$

and finally, taking limits gives inequality ([20] VI.100.1) in complete generality

$$\int_{(A_\infty > \lambda)} (A_\infty - \lambda)\mathbf{P} \leq \int_{(A_\infty > \lambda)} M_\infty \mathbf{P}. \tag{4.19}$$

Moreover, the same results are valid if the hypotheses on A^+ and A^- are slightly modified: it can be assumed that these two processes are true integrable increasing processes which are indistinguishable, the first from an optional process and the second from a predictable process. To see this, we must use Remark 4.2.13, which reduces these hypotheses to the preceding ones.

4.4 Mertens Decomposition

Theorem 4.4.1 General Mertens Decomposition. Let Z be a positive optional strong super-martingale of class D with the convention $Z_{0-} = Z_0$, $Z_\infty = 0$. Then, there exists a predictable process A with increasing but not necessarily right- or left-continuous paths such that

$$\mathbf{E}[A_\infty] < \infty, \quad Z_T = \mathbf{E}[A_\infty | \mathcal{F}_T^\star] - A_T \tag{4.20}$$

for every stopping time T. A is unique and the following inequality between processes holds:

$$\triangle A = A - A_- = Z_- - {}^p Z; \quad \triangle^+ A = A_+ - A = Z - {}^o(Z_+). \tag{4.21}$$

In particular, A is right-continuous if and only if Z is right-continuous and left-continuous if and only if $Z_- = {}^p Z$ (i.e., if Z is regular). Under the usual conditions ${}^o(Z_+) = Z_+$.

Proof. Here we ask the reader to refer to Section 3.2.4.3 and follow our modifications of 3.2.49 and Theorem 3.2.50. It is actually sufficient to suppress cadlag throughout. In Theorem 3.2.50 the proof shows that $A^+ = 0$ if and only if Z is right-continuous; in the general case, we have, simply,

$$J(X) = \mathbf{E} \left[\int_{]0,\infty]} X_s dA_s^- + \int_{[0,\infty[} X_{s+} dA_s^+ \right]$$

which, when X is the indicator of $[\![T, \infty[\![$, gives

$$\begin{aligned} J(X) &= \mathbf{E}[A_\infty^+ - A_{T-}^+ + A_\infty^- - A_T^-] = \mathbf{E}[A_\infty - A_T] \\ &= \mathbf{E}[(A_\infty - A_T)\mathbf{1}_{T<\infty}] = \mathbf{E}[Z_T \mathbf{1}_{T<\infty}] \end{aligned}$$

that is (4.20).

Theorem 3.2.52 gives without modification the first equality of (4.21), and the second follows similarly.

Finally the uniqueness of A follows from the fact that being given A is equivalent to being given the ordered pair (A^-, A^+) by 3.2.46, which is itself equivalent to being given the linear form J (3.2.45) and finally to being given Z.

We now consider the general *unusual* conditions, where $\mathcal{F}_T^\star$ is replaced by $\mathcal{F}_T$ in 4.20. First we must carefully re-read the proof of Theorem 3.2.45: all is well with the steps of the proof, where constructing measures is concerned, but, when it comes to increasing processes, nos. 4.2.10-4.2.12 of this chapter leave us the choice between two possibilities:

(1) to take A^+ and A^- to be two true integrable increasing processes, but which are only indistinguishable from optional or predictable processes, or

(2) to take A^+ and A^- to be two processes which are respectively optional and predictable, with increasing paths, but not identically right-continuous and finite.

The first choice leads to the simpler result. For it enables us to set $A = A^- + A_-^+$, a process with finite values which is indistinguishable from a predictable process, and then $A_+ = A^- + A^+$ which is right-continuous. Then, the measurability conditions on A can merely be expressed as follows: A is indistinguishable from a predictable process and A_+ from an optional process.

That aside, there is no important modification. Note, however, that right-continuity of A is no longer equivalent to right-continuity of Z in expectation. ∎

4.5   Snell Envelope

Here we present elements of the theory of the Snell envelope and its connections to strong supermartingales. The raison dêtre of this theory is the problem of optimal stopping: given an arbitrary process X, find a stopping time T for which $\mathbf{E}[X_T]$ is as large as possible. Snell [32] introduced the notion of a submartingale and the corresponding stopping theorem along with the envelope in order to study optimal stopping in the discrete time. Mertens [3, 49] extended Snell's works to the continuous time case and introduced the notion of a strong supermartingale, Mertens decomposition and the notion of a strongly supermedian function in potential theory. The theory of optimal stopping and Snell envelope is a subject of enormous literature to mention a few important figures such as Bismut [47], Bismut and Skalli [50] and Maingueneau [48]. The reader can consult Lipster and Shiryaev [51] for review.

We now give the existence theorem of Snell's envelope. For a proof under the usual conditions see Maingueneau [48], p. 458. As we are without the usual conditions, we are going to use a "sledge hammer" theorem in our proof.

Let Y denote an optional process, indexed by $[0, \infty[$ belonging to class D. The r.v. Y_T, where T uniformly runs through the set of all bounded stopping times, forms a uniformly integrable set. We adopt the convention $Y_\infty = 0$. Then Y_T is defined for every stopping T finite or otherwise, and the set of Y_T, where T runs through the set I of all stopping times, is also uniformly integrable. Then,

Theorem 4.5.1 Snell Envelope. (a) There exists a positive optional strong supermartingale Z with the following property: $Z \geq Y$ and for every positive optional strong supermartingale Z' which bounds Y above, $Z' \geq Z$ (the inequalities hold as usual except on evanescent sets). Z is unique (to within evanescent sets). It is called the Snell envelope of Y. Moreover, Z belongs to class D. (b) For every stopping time T we have a.s.

$$Z_T = \operatorname*{ess\,sup}_{S \in I, S \geq T} \mathbf{E}[Y_S | \mathcal{F}_T] \tag{4.22}$$

Proof. We fix T and consider the set of r.v. $\mathbf{E}[Y_S | \mathcal{F}_T]$, where S runs through the set of stopping times $S \geq T$. We make the fundamental remark that this set is directed and increasing. For if S_1 and S_2 are stopping time $S \geq T$, let $U_i = \mathbf{E}[Y_{S_i} | \mathcal{F}_T]$, $i = 1, 2$. We have $U_1 \vee U_2 = \mathbf{E}[Y_R | \mathcal{F}_T]$, where $R = S_1$ if $U_1 \geq U_2$, $R = S_2$ if $U_1 < U_2$.

Then let $Z(T)$ denote the supess on the right-hand side of (4.22). We carefully write $Z(T)$ and not Z_T because a class of r.v. is involved and we do not yet know whether we can paste together the $Z(T)$ into a process Z. Note some properties of these r.v.:

(a) $Z(T) \geq 0$ (take $S = +\infty$ in (4.22)); $Z(T)$ is $\mathcal{F}_T$-measurable.

(b) For all $A \in \mathcal{F}_T$,

$$\int_A Z(T)\mathbf{P} = \sup_{S \geq T} \int_A Y_S \mathbf{P} \tag{4.23}$$

(an immediate consequence of the fact that the r. v. $\mathbf{E}[Y_S | \mathcal{F}_T]$ form an increasing directed set).

(c) If T and U are two stopping times with $T \leq U$, then $Z(T) \geq \mathbf{E}[Z(U)|\mathcal{F}_T]$ a.s. and $Z(T) = Z(U)$ a.s. on $\{T = U\}$ (an easy consequence of (4.23)).

(d) The r.v. $Z(T)$ are uniformly integrable. For let $M = \sup_S \mathbf{E}[Y_S] < \infty$; by (4.23) with $A = \Omega$, we have $\mathbf{E}[Z_T] \leq M$ and hence $\mathbf{P}(Z(T) \geq c) \leq \frac{1}{c}M$. Then by (4.23) again

$$\int_{\{Z(T) \geq c\}} Z_T \mathbf{P} \leq \sup_{S \geq T} \int_{\{Z(T) \geq c\}} Y_S \mathbf{P}$$

As $\mathbf{P}(Z(T) \geq c) \leq \frac{1}{c}M$ independently of T and the r.v. Y_S are uniform integrable, the right-hand side tends to 0 as $c \to \infty$, uniformly in T and S. Hence the stated result.

And now we return to Section 3.2.4.3. If X is a cadlag process which is bounded on $]0, \infty]$ and of the form (6.1)

$$X = X_0 1_{]0,T_1]} + \cdots + X_n 1_{]T_n,\infty]}$$

we set

$$J(X) = \mathbf{E}[X_0(Z(0) - Z(T_1)) + \cdots + X_n Z(T_n)]$$

and we take the whole of the proof of the Mertens decomposition Theorem 4.4.1 above. We obtain a representation of J in the form

$$J(X) = \mathbf{E}\left[\int_{]0,\infty]} X_s dA_s^- + \int_{[0,\infty[} X_{s+} dA_s^+\right]$$

as in Theorem 4.4.1. Then, if we denote by Z the process which is the optional projection of the process $(A_\infty^- + A_\infty^+ - A_t - A_{t-})$, which is a positive optional strong supermartingale, we find that the measure associated with Z is J, and hence $Z(T) = Z_T$ a.s. for every stopping time T. Thus, we have established the existence of a positive optional strong supermartingale of class D, which satisfies (4.22).

The rest is obvious. Taking $S = T$ in (4.22), we see that $Z_T \geq Y_T$ and $Z \geq Y$ by the optional cross-section theorem. If Z' is another positive optional strong supermartingale bounding Y above, we have for every stopping time T and every stopping time $S \geq T$

$$Z_T' \geq \mathbf{E}[Z_S'|\mathcal{F}_T] \geq \mathbf{E}[Y_S|\mathcal{F}_T] \quad \text{a.s.}$$

and taking supess in S we obtain $Z_T' \geq Z_T$ a.s and hence $Z' \geq Z$ by the optional cross-section theorem. Finally, the Snell envelope Z is characterized uniquely as the smallest positive optional strong supermartingale bounding Y above (except on evanescent sets). ∎

Remark 4.5.2 (a) Instead of describing Z in the statement as a positive optional strong supermartingale, Z could be described as an optional strong supermartingale on $[0, \infty]$, bounding Y above on $[0, \infty]$. In this form the result extends to processes Y indexed by $[0, \infty]$. Instead of going through the whole argument again for this case, the simplest thing to do is to introduce the optional strong martingale $M_t = \mathbf{E}[Y_\infty|\mathcal{F}_t]$, the optional process $\hat{Y}_t = Y_t - M_t$ (zero at infinity) and its Snell envelope $(\hat{Z}_t)$; then the smallest strong supermartingale bounding Y above on $[0, \infty]$ is $Z = \hat{Z} + N$.

(b) The following is a consequence of the proof of Theorem 4.5.1. For every stopping time $T \le +\infty$ let $Z(T)$ be a class under a.s. equality, belonging to $L^1(\mathcal{F}_T)$, such that, for $S \le T$, $Z(S) \ge \mathbf{E}[Z(T)|\mathcal{F}_S]$ a.s. and $Z(S) = Z(T)$ a.s. on $\{S = T\}$. Then there exists a unique optional strong supermartingale Z such that $Z_T = Z(T)$ a.s. for every stopping time T (reduce it to the positive case by subtracting $\mathbf{E}[Z(\infty)|\mathcal{F}_T]$, truncate at n to reduce it to the bounded case, construct the functional J and proceed as at the end of Theorem 4.5.1, then let n tend to $+\infty$). There is an analogous result when limiting to bounded T.

(c) We adopt the usual conditions. It is not difficult to check that, for every finite stopping time T,

$$Z_{T+} = ess \sup_{S \in I, S > T} \mathbf{E}[Y_S|\mathcal{F}_T]$$

and to deduce that $Z = Y \vee Z_+$. In particular if Y is right-continuous Z is cadlag.

An interesting consequence of the existence of Snell envelopes, which is due to Mertens is as follows:

Theorem 4.5.3 Let Y be an optional process. For Y to belong to class D, it is necessary and sufficient that there exists an optional strong martingale $M_t = \mathbf{E}[M_\infty|\mathcal{F}_t]$ such that $|Y| \le M$ except on evanescent sets.

Proof. The condition is obviously sufficient. To show that it is necessary, we note that the positive process $|Y|$ belongs to class D, and hence has a Snell envelope which itself belongs to class D. Thus Z has a Mertens decomposition $Z_T = \mathbf{E}[A_\infty|\mathcal{F}_T] - A_T$ where A is a predictable increasing process with no right- or left-continuity. It is then sufficient to take $M_\infty = A_\infty$. ∎

Remark 4.5.4 Note that $\mathbf{E}[A_\infty] = \mathbf{E}[Z_0] = \sup_T \mathbf{E}[|Y_T|]$ and that the martingale M is the optional projection of the non-adapted process $U_t = A_\infty$ such that $\mathbf{E}[U^\star] = \mathbf{E}[A_\infty]$.

In [52] Emery gave another proof of Theorem 4.5.3 (under the usual conditions) and made it more precise as follows: Let Y be an optional process of class D. Then there exists a measurable process U such that Y is the optional projection of U and $\mathbf{E}[U^\star] = \sup_T \mathbf{E}[|Y_T|]$. If Y is cadlag, U be can be taken to be cadlag.

This result may be compared to a more difficult theorem of Bismut [53]: under the usual conditions, Let Y be a regular cadlag process of class D. Then for all $\epsilon > 0$ there exists a measurable process U with continuous paths such that Y is the optional projection of U and $\mathbf{E}[U^\star] \le \sup_T \mathbf{E}[|Y_T|] + \epsilon$.

Chapter 5

Optional Martingales

In this chapter we study optional martingales on probability space without the usual conditions. We begin by proving existence and uniqueness of optional martingales and give many additional results on the structures of stopping times on *un*usual probability space. Furthermore, we present many results on optional increasing and finite variation processes and projections theorems.

Under the usual conditions, Doléans-Dade and Meyer showed that every right-continuous local martingale X can be represented in the form $X = X^c + X^d$, where X^c is continuous local martingale and X^d is a right-continuous local martingale that is the sum of compensated jumps of the form $\triangle X_T = X_T - X_{T-}$. Moreover, they obtained a formula for a change of variables, i.e., a representation of the process $F(Y)$ for the twice continuously differentiable function F of the semimartingale Y.

In the same way, we prove that every optional local martingale X is representable in the form $X = X^c + X^d + X^g$ where X^c is a continuous local martingale, X^d is right-continuous local martingale and X^g is left-continuous local martingales which are the sum of compensated jumps of the form $\triangle^+ X_T = X_{T+} - X_T$.

Moreover, this chapter contains many results concerning square-integrable optional martingales. Also, we define the increasing processes $\langle X, X \rangle$ and $[X, X]$ corresponding to an optional square-integrable martingales and prove the Kunita-Watanabe inequalities for optional processes. Most importantly, we introduce a stochastic integral with respect to optional martingales and study its properties.

This chapter and the next two are based mostly on the work of Galchuk [9, 10, 54] and Abdelghani and Melnikov [18].

5.1 Introduction

Let $\left(\Omega, \mathcal{F}, \mathbf{F} = (\mathcal{F}_t)_{t \geq 0}, \mathbf{P} \right)$, $t \in \mathbb{R}_+ = [0, \infty)$ be a complete but *un*usual (raw or nonstandard) probability space, where $\mathcal{F}_t \subseteq \mathcal{F}$, $\mathcal{F}_s \subseteq \mathcal{F}_t$, $s \leq t$. The space is complete because $\mathcal{F}$ contains all its **P**-null sets and **P** a completed probability measure. The space is *unusual* in the sense that the family **F** is not assumed to be complete, right- or left-continuous.

We shall see that probability spaces of this nature are interesting, as they bring about new mathematical phenomena of stochastic processes and offer a versatile set of tools for modeling financial markets and some physical processes. In this short introduction, we present basic concepts that are built upon later sections of this chapter, but we also give a summary of definitions and theorems that will appear later in the chapter. Also, the reader will notice that some of the concepts we present here were studied in the previous chapters. However, revisited definitions are given here; we bring together those concepts that are necessary for the development of the stochastic calculus of optional processes.

Let us introduce the σ-algebras: $\mathbf{F}_+ = (\mathcal{F}_{t+})$ where $\mathcal{F}_{t+} = \cap_{t \le s} \mathcal{F}_s$, for any t, $\mathbf{F}^{\mathbf{P}}_+ = (\mathcal{F}^{\mathbf{P}}_{t+})$ and $\mathbf{F}^{\mathbf{P}} = (\mathcal{F}^{\mathbf{P}}_t)$. $\mathbf{F}^{\mathbf{P}}_+$ ($\mathbf{F}^{\mathbf{P}}$ respectively) is obtained from $\mathbf{F}_+$ by adding **P**-null sets to $\mathcal{F}_{0+}$ (respectively from **F** by adjoining to $\mathcal{F}_0$). We shall also use the notation $\mathbf{F}^\star := \mathbf{F}^{\mathbf{P}}_+$ for the usual filtration that is right-continuous and complete.

A r.v. $T : (\Omega, \mathcal{F}) \longrightarrow ([0, \infty] \times \mathcal{B}([0, \infty]))$ is called a *random time* (r.t.). The r.v. T with values in $[0, \infty]$ is a *Markov time* (m.t.) if the set $(T \le t) \in \mathcal{F}_t$ for all t.

Definition 5.1.1 (a) We say the σ-algebra $\mathcal{O}(\mathbf{F})$ on $[0, \infty) \times \Omega$ is *optional* if it is generated by all right-continuous $\mathcal{F}_t$-adapted processes having limits on the left. Optional σ-algebra can also be generated by the sets $\{(t, \omega) : S(\omega) \le t < T(\omega)\}$, where S, T run through the set of all Markov times.

(b) We say the σ-algebra $\mathcal{P}(\mathbf{F})$ on $\Omega \times [0, \infty)$ is predictable if it is generated by all left-continuous $\mathcal{F}_t$-adapted processes having limits on the right or by sets $\{(\omega, t) : S(\omega) < t \le T(\omega)\}$, where S, T run through the set of all Markov times.

Definition 5.1.2 (a) A random process $X = (X_t)$, $t \in [0, \infty)$, is said to be *optional* if it is $\mathcal{O}(\mathbf{F})$-measurable. Optional processes are progressively measurable, and thus clearly measurable. In general, optional processes have right and left limits but are not necessarily continuous on the right, left or otherwise.

(b) A random process $Y = (Y_t)$, $t \in [0, \infty)$, is said to be *predictable* if it is $\mathcal{P}(\mathbf{F})$-measurable. Predictable processes are also optional, i.e., $\mathcal{P}(\mathbf{F}) \subseteq \mathcal{O}(\mathbf{F})$.

For optional stochastic processes, we can define the following:

(a) The left process $X_- = (X_{t-})_{t \ge 0}$ with $X_{0-} = X_0$ and $X_{t-} = \lim_{s \uparrow t} X_s$, and the right processes $X_+ = (X_{t+})_{t \ge 0}$.

(b) The regular differential process $\triangle X = (\triangle X_t)_{t \ge 0}$ where $\triangle X_t = X_t - X_{t-}$ and the forward differential process $\triangle^+ X = (\triangle^+ X_t)_{t \ge 0}$ where $\triangle^+ X_t = X_{t+} - X_t$.

A measurable set D in the product space $(\Omega \times \mathbb{R}_+, \mathbf{F} \times \mathcal{B}(\mathbb{R}_+))$ is called *negligible* if $\mathbf{P}(\pi(D)) = 0$, where $\pi(D)$ is the projection of D on Ω. According to this definition $\pi(D) \in \mathcal{F}$ since $\mathcal{F}$ is a complete

σ-algebra. We say that the measurable processes $X = (X_t)$ and $Y = (Y_t)$ are *indistinguishable* if the set $\{(\omega, t) : X_t(\omega) \neq Y_t(\omega)\}$ is *negligible*. An example of indistinguishable processes is, for every $\mathcal{P}(\mathbf{F}_+^{\mathbf{P}})$-predictable process X there exists a $\mathcal{P}(\mathbf{F})$-predictable process X' indistinguishable from it. Let us agree to the convention that, when we talk of uniqueness or assert equality or inclusion, we will mean that they are satisfied to within indistinguishability unless otherwise stated.

Definition 5.1.3 (a) The r.v. T with values in $[0, \infty]$ is an **F**-stopping time (s.t., **F**-s.t.) if it is a Markov time $((T \leq t) \in \mathcal{F}_t$ for any $t \in \mathbb{R}_+)$ and the $\mathbf{P}(T < \infty) = 1$.

(b) T is a stopping time in the wide (broad) sense if $\mathbf{F}_+$-stopping time $((T \leq t) \in \mathcal{F}_{t+}$ for any $t \in \mathbb{R}_+)$ and we write T is a wide sense stopping time (w.s.s.t.) or a stopping time in the broad sense (s.t.b.).

Let $\mathcal{T}$ be the class of all s.t.'s with respect to the family **F** and $\mathcal{T}_+$ the class of all s.t.b. or the set of stopping times with respect to the family $\mathbf{F}_+$. A s.t.b. or a broad sense stopping time is also referred to as optional time as we have described in Section 2.4. We will use both terms interchangeably. We shall denote by $\mathcal{T}^p$ ($\mathcal{T}^i$) the set all of predictable (totally inaccessible, respectively) s.t.'s with respect to **F**.

We say the **F**-s.t. is predictable if there exists an $\mathbf{F}_+$-s.t. sequence (S_n), $n \in \mathbb{N}$, such that $\lim S_n = T$ a.s. and $S_n < T$ a.s. on the set $(T > 0)$ for all $n \in \mathbb{N}$. The **F**-s.t. T is said to be totally inaccessible if $\mathbf{P}(S = T < \infty) = 0$ for every predictable **F**-s.t. S.

Definition 5.1.4 A random process $X = (X_t)$, $t \in \mathbb{R}_+$, whose trajectories have limits from the right is said to be predictable if $\mathcal{P}(\mathbf{F})$-measurable and *strongly predictable* if (a) $(X_t) \in \mathcal{P}(\mathbf{F})$, and (b) $(X_{t+}) \in \mathcal{O}(\mathbf{F})$. In particular this means that, for every stopping time T the random variables $X_{T+}\mathbf{1}_{T<\infty}$ is $\mathcal{F}_T$-measurable, and $X_{T-}\mathbf{1}_{T<\infty}$ is $\mathcal{F}_{T-}$-measurable. We denote by $\mathcal{P}_s(\mathbf{F})$ or $\mathcal{P}_s$ the set of strongly predictable processes.

Let $\mathcal{J}$ be a space of processes, a process $X = (X_t)$, $t \in \mathbb{R}_+$, belongs to the space $\mathcal{J}_{loc}$ if there exists a sequence of s.t.b. (R_n), $R_n \in \mathcal{T}_+$, $n \in \mathbb{N}$, $R_n \uparrow \infty$ a.s., such that $X\mathbf{1}_{[0,R_n]} \in \mathcal{J}$ for all n.

We begin with some important results about the nature of stopping times on the raw space $(\Omega, \mathcal{F}, \mathbf{F}, \mathbf{P})$.

Theorem 5.1.5 For every $\mathbf{F}^{\mathbf{P}}$-s.t. T there exists an $\mathbf{F}_+$-s.t. U such that $T = U$ a.s. Moreover, for every event $L \in \mathcal{F}_T^{\mathbf{P}}$ there exists an $M \in \mathcal{F}_{U+}$ such that $L = M$ a.s.

Proof. The proof is given in Theorem 2.4.19 or in [19] (Chapter IV, Theorem 59). ∎

Theorem 5.1.6 For every predictable $\mathbf{F}_+^{\mathbf{P}}$ s.t. T there exists a predictable **F** s.t. T' such that $T = T'$ a.s. Moreover, for every $\mathcal{P}(\mathbf{F}_+^{\mathbf{P}})$-predictable process X there exists a $\mathcal{P}(\mathbf{F})$-predictable process X' indistinguishable from it.

Proof. The proof is given in Theorem 2.4.42 or in [19] (Chapter IV, Theorem 78). ∎

Theorem 5.1.7 For every $\mathbf{F}_+$ s.t. T there exist a sequence of $\mathbf{F}$ s.t. (S_n), $n \in \mathbb{N}$, with non-intersecting graphs and an $\mathbf{F}_+$ s.t. T' such that $[\![T]\!] \subseteq (\cup_n[\![S_n]\!]) \cup [\![T']\!]$ and $\mathbf{P}(T' = U < \infty) = 0$ for every $\mathbf{F}$ s.t. U. From the preceding theorem, it follows that T' is a totally inaccessible s.t.b.

Proof. The proof is given in [4] (Lemma 2). ∎

Theorem 5.1.8 Suppose T is a s.t. There exist a totally inaccessible s.t. T' and a sequence of predictable s.t. (T_n) with pairwise non-intersecting graphs such that $[\![T]\!] \subseteq [\![T']\!] \cup (\cup_n[\![T_n]\!])$.

Proof. T is also a s.t. with respect to $\mathbf{F}^{\mathbf{P}}_+$. According to [28] (Chapter III, Theorem 41), there exist a sequence of $\mathbf{F}^{\mathbf{P}}_+$ predictable s.t. $(\tilde{T}_n)$ and a totally inaccessible $\mathbf{F}^{\mathbf{P}}_+$-s.t. $\tilde{T}$ such that $[\![T]\!] \subseteq [\![\tilde{T}]\!] \cup (\cup_n[\![\tilde{T}_n]\!])$.

Let (T_n) denote a sequence of $\mathbf{F}$-predictable s.t., $\tilde{T}_n = T_n$ a.s., $n \in \mathbb{N}$, which exists according to Theorem 5.1.6. We immediately assume that the graphs of (T_n) do not intersect.

Let $B = [\![T]\!] \setminus \cup_n [\![T_n]\!]$. The set $B \in \mathcal{O}(\mathbf{F})$, as well as each of its ω-sections, contains not more than one point. Suppose the sequence of positive numbers (ϵ_n) decreases to zero. According to the Theorem 2.4.48 (see [19] Chapter IV, Theorem 84) for every $\epsilon_n > 0$ there is an $\mathbf{F}$-s.t. T'_n such that $[\![T'_n]\!] \subseteq B$ and $\mathbf{P}(T'_n < \infty) > \mathbf{P}(\pi(B)) - \epsilon_n$.

Let $T' = \min T'_n$. We have $[\![T']\!] \subseteq B$, and T' is an $\mathbf{F}$-s.t. $(T' \leq t) = \cup_n (T'_n \leq t) \in \mathcal{F}_t \, \forall t \in \mathbb{R}_+$. It is easy to see that, to within a negligible set, $B = [\![T']\!]$, and that T' is a totally inaccessible s.t. ∎

Definition 5.1.9 The sequence of s.t.b. T_n, $n \in \mathbb{N}$, absorbs the jumps of the process $X = (X_t)$, $t \in \mathbb{R}_+$, if for every s.t.b. T for which the set $[\![T]\!] \cap (\cup_n[\![T_n]\!]) = \emptyset$ we have $\triangle X_T = \triangle^+ X_T = 0$ a.s. on $(T < \infty)$, where $\triangle X_T = X_T - X_{T-}$ and $\triangle^+ X_T = X_{T+} - X_T$. Let us agree below to assume that $X_{0-} = 0$ for every process (X_t), $t \in \mathbb{R}_+$. Then $\triangle X_0 = X_0$.

Theorem 5.1.10 Suppose $X = (X_t)$, $t \in \mathbb{R}_+$, is an optional process whose trajectories have limits from the left and right a.s. Then there exist sequences (S_n), (T_n) and (U_n), $n \in \mathbb{N}$, of predictable s.t., totally inaccessible s.t., and totally inaccessible in the broad sense s.t.b., respectively, absorbing all jumps of the process X and having the following properties: the graphs of these s.t.'s are mutually non-intersecting within each sequence and $\mathbf{P}(U_n = T < \infty) = 0$ for all n and all s.t. T.

Proof. Consider the sets

$$D = \{(\omega, t) : X_{t-}(\omega)) \neq X_t(\omega)\}, \quad D^+ = \{(\omega, t) : X_t(\omega) \neq X_{t+}(\omega)\}.$$

They are $\mathcal{O}(\mathbf{F})$- and $\mathcal{O}(\mathbf{F}_+)$-optional respectively. For both sets every ω-section is not more than countable. According to [19] (Chapter IV, Theorem 117) there exist sequences (τ_n) and (σ_n), $n \in \mathbb{N}$, of s.t. and s.t.b., respectively, such that $D = \cup_n[\![\tau_n]\!]$ and $D^+ = \cup_n[\![\sigma_n]\!]$. By Theorem 5.1.8 there are sequences (T'_n) and (S'_n) of totally inaccessible and predictable s.t., respectively, such that

$$D = \cup_n[\![\tau_n]\!] \subseteq (\cup_n[\![T'_n]\!]) \cup (\cup_n[\![S'_n]\!]).$$

From Theorems 5.1.7 and 5.1.8, it follows that there are sequences (S_n''), (T_n'') and (U_n), $n \in \mathbb{N}$, of predictable s.t., totally inaccessible s.t., and totally inaccessible s.t.b., respectively, such that

$$D^+ = \cup_n [\![\sigma_n]\!] \subseteq (\cup_n [\![S_n'']\!]) \cup (\cup_n [\![T_n'']\!]) \cup (\cup_n [\![U_n]\!]) .$$

Here, on the strength of Theorem 5.1.7, $\mathbf{P}(U_n = T < \infty) = 0$ for every s.t. T and $n \in \mathbb{N}$. Now, re-indexing the sequences (S_n') and (S_n'') into one and making their graphs non-intersecting, we obtain the desired sequence of predictable s.t. (S_n). Proceeding analogously with (T_n') and (T_n''), we obtain a sequence of totally inaccessible s.t. (T_n). ∎

Definition 5.1.11 We shall say that the sequence $(T_n) \subset \mathcal{T}_+$ absorbs the jumps of $X = (X_t)$ if, for any $T \in \mathcal{T}_+$ for which the set $[\![T]\!] \cap (\cup_n [\![T_n]\!])$ is negligible, $\triangle X_T = \triangle^+ X_T = 0$ a.s. on $(T < \infty)$. Let us agree to assume that $X_{0-} = 0$ for any process (X_t), $t \in \mathbb{R}_+$. Then $\triangle X_0 = X_0$.

Theorem 5.1.12 Let D be an $\mathcal{O}(\mathbf{F}_+^{\mathbf{P}})$-measurable rare set (i.e., a set each ω-section of which is finite or countable). Then, there exist sequences $(S_n) \subset \mathcal{T}^p$, $(T_n) \subset \mathcal{T}^i$, $(U_n) \subset \mathcal{T}_+^i$, such that, to within a negligible set, $D \subseteq \cup_n([\![S_n]\!] \cup [\![T_n]\!] \cup [\![U_n]\!])$, and $\mathbf{P}(U_n = T < \infty) = 0$ for any $T \in \mathcal{T}$, $n \in \mathbb{N}$.

Proof. According to [28], Chapter VI, Theorem 10, $D = \cup_n [\![\tau_n]\!]$, where (τ_n) is the sequence of s.t.'s with respect to $\mathbf{F}_+^{\mathbf{P}}$. For each n there exists $\tau_n' \in \mathcal{T}_+$ such that $\tau_n = \tau_n'$ a.s. (see [19], Chapter IV, Theorem 59). Further, for each $\tau_n' \in \mathcal{T}_+$ there exist a sequence $(\tau_{nm}) \subset \mathcal{T}$ and $U_n \in \mathcal{T}_+$ such that $\mathbf{P}(U_n = T < \infty) = 0$ for any $T \in \mathcal{T}$ and $[\![\tau_n']\!] \subseteq [\![U_n]\!] \cup (\cup_m [\![\tau_{nm}]\!])$ (see [4], Lemma 2). Finally, for each $\tau_{nm} \in \mathcal{T}_+$ there exist a sequence $(\tau_{nmk}) \subset \mathcal{T}^p$ and s.t.'s $T_{nm} \in \mathcal{T}^i$ such that

$$[\![\tau_{nm}]\!] \subseteq [\![T_{nm}]\!] \cup (\cup_k [\![\tau_{nmk}]\!])$$

(see Theorem 5.1.8 or [54] Theorem 1.7). If we renumber the $(\tau_{nmk}) \subset \mathcal{T}^p$ and $(\tau_{nm}) \subset \mathcal{T}^i$, we get the required sequences $(S_n) \subset \mathcal{T}^p$ and $(T_n) \subset \mathcal{T}^i$. ∎

A martingale is a stochastic process where knowledge of the past history of the process does not in any way help us predict its future average. In other words, the expectations of the future are not different from the present given knowledge of all prior observations. Martingales and their variants super- and submartingale and local versions are well understood in the general theory of stochastic processes on the usual basis. Here we provide an extension of those ideas to *un*usual probability spaces. For this end, we prove next the existence of optional martingales on the *un*usual stochastic basis.

5.2 Existence and Uniqueness

Here we prove a theorem, on existence and uniqueness of optional martingales which became the foundation of stochastic calculus on probability spaces without the usual conditions. Before proving the existence and uniqueness theorem, we need to give a generalization of the well-known Doob inequality.

Lemma 5.2.1 Optional Doob Inequality. Let (Y_t) be an optional submartingale for which there exists an integrable r.v. Y such that for any Markov time T,

$$Y_T \mathbf{1}_{(T<\infty)} \leq \mathbf{E}\left[Y \mathbf{1}_{(T<\infty)} | \mathcal{F}_T\right] \quad \text{a.s.}$$

Then, for any $c > 0$,

$$\mathbf{P}\left(\sup_t Y_t > c\right) \leq c^{-1}\mathbf{E}Y^+ \text{ where } Y^+ = Y \vee 0.$$

Proof. First, we need to show that the set $B = \{\omega : \sup_t Y_t(\omega) > c\}$ is $\mathcal{F}$-measurable. The set $A = \{(\omega, t) : Y_t(\omega) > c\}$ is $\mathcal{O}(\mathbf{F})$-measurable. Thus, it is measurable with respect to $\mathcal{F} \times \mathcal{B}([0, \infty))$. The projection $\pi(A) \in \mathcal{F}$ and since $\pi(A) = B$ then $B \in \mathcal{F}$.

By the theorem on cross-section (see [4], Theorem 1), there exists, for any $\epsilon > 0$, a Markov time S such that

$$S(\omega) < \infty \Rightarrow (\omega, S(\omega)) \in A, \tag{5.1}$$

$$\mathbf{P}(S < \infty) \geq \mathbf{P}(\pi(A)) - \epsilon. \tag{5.2}$$

From Equation (5.2) and the fact that the sets $\pi(A)$ and B coincide, it follows that

$$\mathbf{P}(B) \leq \mathbf{P}(S < \infty) + \epsilon. \tag{5.3}$$

On the set $(S < \infty)$ we have $c^{-1}Y_S > 1$; therefore,

$$\mathbf{P}(S < \infty) \leq c^{-1}\mathbf{E}Y_S \mathbf{1}_{(S<\infty)}.$$

Hence from Equation (5.3) and the condition of the lemma we find that

$$\mathbf{P}\left(\sup_t Y_t > c\right) \leq c^{-1}\mathbf{E}Y_S \mathbf{1}_{(S<\infty)} + \epsilon \leq c^{-1}\mathbf{E}Y^+ + \epsilon,$$

from which, due to arbitrariness of $\epsilon > 0$, the required inequality follows. ∎

Theorem 5.2.2 Existence and Uniqueness of Optional Martingales. Let X be an integrable r.v. Then, there exists a modification X_t of the martingale $(\mathbf{E}[X|\mathcal{F}_t])$ such that X_t is an optional process and for any Markov time T

$$X_T \mathbf{1}_{(T<\infty)} = \mathbf{E}[X \mathbf{1}_{(T<\infty)} | \mathcal{F}_T] \quad \text{a.s.,} \tag{5.4}$$

If another optional modification, $(\tilde{X}_t)$ exists satisfying (5.4), then (X_t) and $(\tilde{X}_t)$ are indistinguishable.

Proof. Existence. Note that this theorem was proved for the case when X is a bounded r.v. (see [4], Theorem 2).

Let $X^n = X\mathbf{1}_{|X| \leq n}$, $n > 0$. Then, as we have noted, the martingale $(\mathbf{E}[X^n|\mathcal{F}_t])$ has a unique (to within indistinguishability) optional modification denoted by (X_t^n). This modification has the following property: for each Markov time T,

$$X_T^n \mathbf{1}_{T < \infty} = \mathbf{E}[X^n \mathbf{1}_{T < \infty}|\mathcal{F}_T] \quad \text{a.s.} \tag{5.5}$$

The submartingale $(|X_t^n - X_t^m|)$, $0 < n < m$ satisfies the conditions of Lemma 5.2.1; hence

$$\mathbf{P}\left(\sup_t |X_t^n - X_t^m| > \epsilon\right) \leq \frac{1}{\epsilon}\mathbf{E}\left[|X|\mathbf{1}_{(n < |X| \leq m)}\right], \quad \epsilon > 0.$$

Therefore, there exists a subsequence (n') such that the sequence of the martingales $\{(X_t^{n'})\}$ converge uniformly a.s. on $[0, \infty)$. Let us denote by (X_t) the limit of the sequence $\{(X_t^{n'})\}$ as $n' \to \infty$. The process (X_t) is optional. Since the a.s. convergence is uniform on $[0, \infty)$, we see from (5.5) that, for any Markov time, T,

$$X_T \mathbf{1}_{(T < \infty)} = \lim_{n' \to \infty} X_T^{n'} \mathbf{1}_{(T < \infty)} = \lim_{n' \to \infty} \mathbf{E}[X^{n'} \mathbf{1}_{(T < \infty)}|\mathcal{F}_T] = \mathbf{E}[X\mathbf{1}_{(T < \infty)}|\mathcal{F}_T] \quad \text{a.s.}$$

We have thus proved the existence of the optional modification satisfying (5.4). ∎

Proof. Uniqueness. To prove uniqueness we assume two optional modifications (X_t) and $(\tilde{X}_t)$ are given satisfying (5.4). The set $A = \left\{(\omega, t) : X_t(\omega) > \tilde{X}_t(\omega)\right\}$ is $\mathcal{O}$-measurable. By the theorem on cross-sections, there exists a Markov time S such that (5.1) and (5.2) are satisfied. Assume that $\mathbf{P}(\pi(A)) > 0$. Then, by (5.1),

$$\mathbf{E}X_S \mathbf{1}_{(S < \infty)} > \mathbf{E}\tilde{X}_S \mathbf{1}_{(S < \infty)}.$$

On the other hand, by virtue of (5.4),

$$\mathbf{E}X_S \mathbf{1}_{(S < \infty)} = \mathbf{E}\tilde{X}_S \mathbf{1}_{(S < \infty)}.$$

The contradiction thus obtained implies that $\mathbf{P}(\pi(A)) = 0$. Similarly it can be shown that $\mathbf{P}\left\{\pi((\omega, t) : X_t(\omega) < \tilde{X}_t(\omega))\right\} = 0$. Therefore, the processes (X_t) and $(\tilde{X}_t)$ are indistinguishable, proving the theorem. ∎

Recall that under the usual conditions, or the σ-algebra $\mathcal{F}_t^*$ satisfying the usual conditions, an adapted process $X = (X_t)$, $t \in \mathbb{R}_+$, is a martingale (respectively, a supermartingale, submartingale) with respect to $\mathcal{F}_t^*$ if for any $t \in \mathbb{R}_+$, X_t is integrable and $X_s = \mathbf{E}[X_t|\mathcal{F}_s^*]$ (respectively, $X_s \geq \mathbf{E}[X_t|\mathcal{F}_s^*]$, $X_s \leq \mathbf{E}[X_t|\mathcal{F}_s^*]$) a.s. for any $s \leq t$. However, optional martingales (super and submartingales) as the case was for strong supermartingales in Chapter 4 are defined in terms of stopping times as:

Definition 5.2.3 We say that $X = (X_t)$, $t \in \mathbb{R}_+$ is an *optional martingale* (an *optional supermartingale*, an *optional submartingale*) if (a) X is an optional process (i.e., $X \in \mathcal{O}(\mathbf{F})$), (b) the random variable $X_T \mathbf{1}_{T < \infty}$ is integrable for any $T \in \mathcal{T}$, and (c) there exists an integrable r.v. ξ such that $X_T = \mathbf{E}[\xi|\mathcal{F}_T]$ (respectively, $X_T \geq \mathbf{E}[\xi|\mathcal{F}_T]$, $X_T \leq \mathbf{E}[\xi|\mathcal{F}_T]$) a.s. on $(T < \infty)$ for any $T \in \mathcal{T}$.

According to [30] an optional martingale X can be transformed by a continuous change of variable into a separable process to which we can apply the theorem on existence of a limit at infinity and right and left limits at each point. Consequently as $t \to \infty$ the martingale (X_t) has a limit (a.s. and in L_1) equal to ξ. This allows us to consider (X_t) on $\mathbb{R}_+ \cup \{\infty\}$, letting $X_\infty = \xi$. Hence we also conclude that an optional martingale has one-sided limits.

Separable supermartingale has left-hand and right-hand limits at each point $t \in \mathbb{R}_+$ with respect to a separability (see [55], Chapter VI, Theorem 3). Consequently, an optional supermartingale has left-hand and right-hand limits at each point we mention that the class of supermartingales given by the last definition is sufficiently broad. Indeed, let $X = (X_t)$, $t \in \mathbb{R}_+$, be an arbitrary supermartingale for which there is an integrable variable ξ such that $X \geq \mathbf{E}[\xi|\mathcal{F}_t]$ a.s. for any $t \in \mathbb{R}_+$. Then, according to [56], there exists a process X^* that is a modification of X and satisfies the conditions of the given definition. Moreover, the optional sampling theorem holds for X^*; namely,

$$X_S^* \geq \mathbf{E}[X_T^*|\mathcal{F}_S] \quad \text{a.s.} \tag{5.6}$$

for any S, $T \in \mathcal{T}$ with $S \leq T$. We remark that for $T = \infty$ it is assumed that $X_\infty^* = \xi$. We remark also that a supermartingale X^* has a limit in L_1 as $t \to \infty$ (see [55], Chapter VI, Theorem 6), and it can be assumed without loss of generality to coincide with ξ.

Definition 5.2.4 The optional process $X = (X_t)$, $t \in \mathbb{R}_+$, is called an (*optional*) *local* martingale and we write $X \in \mathcal{M}_{loc}$ ($X \in \mathcal{M}_{loc}^2$, respectively) if there exists a sequence $(R_n, X^{(n)})$, $n \in \mathbb{N}$, where $R_n \in \mathcal{T}_+$ is a s.t.b. $X^{(n)} \in \mathcal{M}$ ($X^{(n)} \in \mathcal{M}^2$, respectively) for all n and $R_n \uparrow \infty$ a.s., such that $X = X^{(n)}$ on the set $[\![0, R_n]\!]$ and the r.v. X_{R_n+} is integrable for all $n \in \mathbb{N}$.

5.3 Increasing and Finite Variation Processes

If there is any useful information to be gained from a random process is to know whether it is increasing, bounded, or finite variation. These notions, of finding deterministic facts, about stochastic processes is captured by the following definitions of increasing, integrable and finite variation processes.

Definition 5.3.1 We say that the process $A = (A_t)$, $t \in \mathbb{R}_+$, is *increasing* if it is non-negative, the trajectories do not decrease and for all t the r.v. A_t is $\mathcal{F}_t$-measurable. Let $\mathcal{V}^+$ be the collection of increasing processes that are $\mathcal{F}_t$-measurable.

An increasing process A is said to be *integrable* if $\mathbf{E}A_\infty < \infty$ and local integrable if there is a sequence of s.t.b. (R_n), $R_n \uparrow \infty$ a.s., such that $\mathbf{E}A_{R_n} < \infty$ for all n. Let $\mathcal{A}^+$ be the collection of increasing integrable processes and $\mathcal{A}_{loc}^+$ for locally integrable increasing processes that are $\mathcal{F}_t$-measurable.

Definition 5.3.2 The *variation* of an optional process $A = (A_t)$ is understood to be

$$\mathbf{Var}\,(A)_t = \sum_{0 \le s < t} |A_{s+} - A_s| + \int_{0+}^{t} |dA^r|.$$

Definition 5.3.3 A *finite variation process*, $A = (A_t)$, $t \in \mathbb{R}_+$, is a process with finite variation, $\mathbf{Var}(A)_t < \infty$, a.s. on any interval $[0, t]$, $t \in \mathbb{R}_+$. Such a process can be written in the form

$$A_t = \sum_{0 \le s < t} (A_{s+} - A_s) + A_t^r,$$

where A^r is a right-continuous finite-variation process and the series is absolutely convergent. Let $\mathcal{V}$ be the space of finite variation processes.

A process $A = (A_t)$, $t \in \mathbb{R}_+$, belongs to the space $\mathcal{V}_{loc}$ if there exists a sequence (R_n), $R_n \in \mathcal{T}_+$, $n \in \mathbb{N}$, $R_n \uparrow \infty$ a.s., such that $A\mathbf{1}_{[0,R_n]} \in \mathcal{V}$ for any n.

We say that a finite variation process $A = (A_t)$, $t \in \mathbb{R}_+$ belongs to the space $\mathcal{A}$ of integrable variation if

$$\mathbf{E}\,[\mathbf{Var}\,(A)_\infty] < \infty.$$

A process $A = (A_t)$, $t \in \mathbb{R}_+$, belongs to the space $\mathcal{A}_{loc}$ if there exists a sequence (R_n), $R_n \in \mathcal{T}_+$, $n \in \mathbb{N}$, $R_n \uparrow \infty$ a.s., such that $A\mathbf{1}_{[0,R_n]} \in \mathcal{A}$ for any n.

Evidently,

$$\mathcal{V}_{loc}^+ = \mathcal{V}^+, \quad \mathcal{V}_{loc} = \mathcal{V},$$
$$\mathcal{A}^+ \subseteq \mathcal{A}_{loc}^+ \subseteq \mathcal{V}^+, \quad \mathcal{A} \subseteq \mathcal{A}_{loc} \subseteq \mathcal{V}$$

5.3.1 Integration with respect to increasing and finite variation processes

The integral with respect to increasing right- or left-continuous processes is well defined in the *Lebesgue-Stieltjes* sense.

Definition 5.3.4 Suppose $H = (H(t, \omega))$ is a measurable non-negative function and $A = (A_t)$, $t \in \mathbb{R}_+$, is an increasing right- (left-) continuous process. By the integral

$$H \circ A_t = \int_{[0,t]} H_s dA_s \quad \left(= \int_{[0,t[} H_s dA_{s+} \right)$$

we mean the *Lebesgue-Stieltjes* integral defined trajectory-wise. In particular, If the function H is such that for all t,

$$\int_{[0,t]} |H_s| dA_s < \infty \quad \left(\int_{[0,t[} |H_s| dA_{s+} < \infty \right) \quad \text{a.s.}$$

then the process $H \circ A$ is continuous from the *right* (left) and has a limit from the *left* (right) and

$$\triangle(H \circ A)_t = H_t \triangle A_t, \quad \triangle(H \circ A)_0 = H_0 \triangle A_0 = H_0 \circ A_0,$$

$$(\triangle^+(H \circ A)_t = H_t \triangle^+ A_t), \quad t \in \mathbb{R}_+.$$

Further, if $A = (A_t)$, $t \in \mathbb{R}_+$, is an increasing process, then it can be represented in the form $A = A^c + A^d + A^g$, where A^c, A^d and A^g are increasing processes, A^c is a continuous component with $A_0^c = 0$, A^g is continuous from the left with $A^g = 0$ and A^d is continuous from the right with $\triangle A_0^d = A_0$. To extend the integrals above to any increasing optional process, it is desirable to obtain optional processes of two components, a continuous from the right component and a continuous from the left one. For the increasing process $A = A^r + A^g$ where $A^r = A^c + A^d$ and a Borel, $\mathcal{B}([0,\infty[) \times \mathcal{F}$-measurable function, $K(t,\omega)$, such that

$$\int_{[0,\infty[} |K_t||dA_t^r| + \sum_{0 \le t < \infty} |K_t|| \triangle^+ A_t| < \infty,$$

we would like to define an integral $Y := \int K dA$ having the following properties:

(a) The process $Y = (Y_t)$, $t \in \mathbb{R}_+$, has bounded variation, i.e., $Y = Y^r + Y^g$, where Y^g is a process continuous from the left, $Y^r = Y^c + Y^d$ is continuous from the right and

$$\sum_{0 \le t < \infty} | \triangle^+ Y_t| + \int_{[0,\infty[} |dY_t^r| < \infty.$$

(b)

$$Y_t^r = \int_{[0,t]} K_s dA_s^r, \quad Y_t^g = \int_{[0,t[} K_s dA_{s+}^g$$

$$\triangle Y_t = K_t \triangle A_t, \quad \triangle^+ Y_t = K_t \triangle^+ A_t \quad t \in \mathbb{R}_+$$

Starting from this, we introduce the integral with respect to an increasing optional process:

Definition 5.3.5 By the integral of the function $K = (K(t,\omega))$ with respect to the increasing process $A = (A_t)$, $t \in \mathbb{R}_+$, we mean the process $Y = \int K dA$, where

$$Y_t = \int_0^t K_s dA_s = \int_{[0,t]} K_{s-} dA_s^r + \int_{[0,t[} K_s dA_{s+}^g, \quad t \in \mathbb{R}_+,$$

where the summands on the right are defined.

It is easy to check that by properties (a) and (b), the definition is proper and yields a unique process Y. The first summand on the right side yields a process continuous from the right and the second yields a process continuous from the left.

5.3.2 Dual projections

We present several results on dual projections.

Theorem 5.3.6 Let A be a right continuous increasing process and X is a non-negative optional martingale. Then, for every s.t.b. T

$$\mathbf{E} \int_{[0,T]} X_{t+} dA_t = \mathbf{E} X_{T+} A_T,$$

where it is assumed that $X_{\infty+} = X_\infty$.

Proof. The assertion of this theorem coincides with Theorem 47 of [28], Chapter IV, where the process X was assumed to be right continuous and the usual conditions are satisfied. However, the proof in [28] also holds for our case. ■

Theorem 5.3.7 Suppose the process $Y = (Y_t) \in \mathcal{A}_{loc}$ and is left-continuous and increasing. Suppose the sequence (T_n), $n \in \mathbb{N}$, of s.t.b. absorbs the jumps of the process Y and has the property that $\mathbf{P}(T_n = T < \infty) = 0$ for every n and an arbitrary s.t. T. Then, there exists a unique continuous increasing process $A = (A_t) \in \mathcal{A}_{loc}$ such that for every non-negative measurable process X

$$\mathbf{E} \int_0^{\infty-} {}^o X_t dY_{t+} = \mathbf{E} \int_0^{\infty-} X_t dA_t,$$

where ${}^o X$ is the optional projection of the process X.

Proof. Recall, the optional projection ${}^o X$ of the process X is the $\mathcal{O}(\mathbf{F})$-measurable process such that $\mathbf{E} X_T 1_{T<\infty} = \mathbf{E} {}^o X_T 1_{T<\infty}$ for every s.t. T.

The existence of the optional projection is proved in [4] where the above theorem is a particular case of a result in [4] on optional dual projections for measures commutative with optional projections. ■

Theorem 5.3.8 Suppose $A = (A_t) \in \mathcal{A}_{loc}$ is a right-continuous increasing process. There exists a unique right-continuous $\mathcal{P}(\mathbf{F})$-predictable increasing process $B = (B_t) \in \mathcal{A}_{loc}$ such that for every positive $\mathcal{P}(\mathbf{F})$-predictable process X

$$\mathbf{E} \int_0^{\infty-} X_s dA_s = \mathbf{E} \int_0^{\infty-} X_s dB_s.$$

In particular, if S and T are s.t.b., $S \leq T$, then

$$\mathbf{E} \left[\int_{]S,T]} X_s dA_s | \mathcal{F}_S \right] = \mathbf{E} \left[\int_{]S,T]} X_s dB_s | \mathcal{F}_S \right]$$

a.s. on the set $(T < \infty)$.

Proof. For the usual conditions, a proof of this theorem is given in [28] (Chapter V, Theorems 28 and 30; see also [55], Chapters VII and VIII). For the case without the usual conditions, if we also make use of Theorem 5.1.6 we are able to obtain the assertions of the theorem. ∎

Remark 5.3.9 If in the preceding theorem the process A has jumps only at totally inaccessible s.t., then B is a continuous process (see [55], Chapter VII, Theorem 37).

Lemma 5.3.10 Let the process $A = (A_t) \in \mathcal{P}_s \cap \mathcal{V}$. Then there exists a sequence $(R_n) \subset \mathcal{T}$, $R_n \uparrow \infty$ a.s., such that for any n the stopped process A^{R_n} is bounded.

Proof. Note that in the decomposition $A = A^r + A^g$ the processes A^r, $A^g \in \mathcal{P}_s \cap \mathcal{V}$, by virtue of the definition of the process A^g. In [57] it has been established that for A^r there exists a sequence $(\tau_n) \subset \mathcal{T}$, $\tau_n \uparrow \infty$, a.s., such that, for the stopped process $A^{r\tau_n}$, $\sup_t |A_t^{r\tau_n}| \leq n$ for any $n \in \mathbb{N}$. Even though in [57] the proof was made under the assumption that the usual conditions hold, it is still valid in our case.

Further, set $\sigma_n = \inf\{t : |A_{t+}^g| > n\}$. Since $(\sigma_n > t) = \cap_{s \leq t}(A_{s+}^g \leq n)$ for any $t \in \mathbb{R}_+$, we know that $\sigma_n \in \mathcal{T}$ for any $n \in \mathbb{N}$. Moreover, $\sigma_n \uparrow \infty$ a.s., and $\sup_t |A_t^{g\sigma_n}| \leq n$. Set $R_n = \sigma_n \wedge \tau_n$, then (R_n) is the required sequence. ∎

Lemma 5.3.11 Let the process $A = (A_t) \in \mathcal{A}_{loc}$ and let it be **F**-adapted, and its trajectories be left-continuous.

(a) There exists a unique (to within indistinguishability) left-continuous strongly predictable process $A^o = (A_t^o) \in \mathcal{A}_{loc}$, such that, for any non-negative optional processes X,

$$\mathbf{E} \int_{[0,\infty[} X_s dA_{s+} = \mathbf{E} \int_{[0,\infty[} X_s dA_{s+}^o. \tag{5.7}$$

In particular, for any $S, T \in \mathcal{T}$, $S \leq T$,

$$\mathbf{E}\left[\int_{[S,T[} X_t dA_{t+} | \mathcal{F}_s\right] = \mathbf{E}\left[\int_{[S,T[} X_t dA_{t+}^o | \mathcal{F}_s\right]. \tag{5.8}$$

This means that $A - A^o \in \mathcal{M}_{loc}$.

(b) If the sequence $(R_n) \subset \mathcal{T}_+$, absorbing the jumps of the process A, possesses the property $\mathbf{P}(R_n = T < \infty) = 0$ for any $n \in \mathbb{N}$, $T \in \mathcal{T}$, then the process A^o is continuous.

Proof. Let us define on the σ-algebra $\mathcal{O}(\mathbf{F})$ the measure

$$\mu(X) = \mathbf{E} \int_{[0,\infty[} {}^o X_t dA_{t+},$$

where ${}^o X$ is the optional projection for the non-negative measurable process X (see [4]-[9]). By [4], the measure μ is generated by a unique (to within indistinguishability) right-continuous process $B = (B_t) \in \mathcal{A}_{loc}$. Set $A^o = B_- = (B_{t-})$. Equalities (5.7) and (5.8) are clearly fulfilled. Assertion (b) has been established in Theorem 5.3.7 (see also [54], Theorem 2.2). ∎

5.4 Decomposition Results

Here we study the properties of optional martingales and decomposition results. We begin with an important theorem pertaining to the jumps of local optional martingales.

Theorem 5.4.1 For $X \in \mathcal{M}_{loc}$ there exist sequences (S_n), (T_n) and (U_n), $n \in \mathbb{N}$, of predictable s.t., totally inaccessible s.t. and s.t.b., respectively, absorbing all stopping times of the process X and satisfying the conditions of the preceding theorem.

Proof. From [30] (see also [7]) the process X has a.s. finite limits from the left and right. Existence of the sequences of s.t. predictable, totally inaccessible and broad-sense follows from Theorem 5.1.10. ∎

Theorem 5.4.2 Suppose $X \in \mathcal{M}^2$. Then,

$$\mathbf{E}\left[\left(\triangle^+ X_t\right)^2 + \left(\triangle X_t\right)^2\right] \leq \mathbf{E}\left[X_\infty^2\right].$$

Proof. According to [30] (see also [5] and [58]), every optional process X is optionally separable. This presupposes existence of a sequence of s.t. (S_n), $n \in \mathbb{N}$, everywhere dense in $\mathbb{R}_+$, such that, the graph of the process (X_t), $t \in \mathbb{R}^+$, is contained in the closure of the graph of the sequence (X_{S_n}), $n \in \mathbb{N}$.

Suppose (S_n) is such a sequence of s.t. for the martingale X. From it we choose a finite collection $S_1, ..., S_n$. Let $S_1', ..., S_n'$ denote the s.t. obtained from $S_1, ..., S_n$ such that $S_1' \leq ... \leq S_n'$ and $\cup_1^n[S_k'] = \cup_1^n[S_k]$. The process $(X_{S_k'}, \mathcal{F}_{S_k'})$, $k \in (0, 1, ..., n, \infty)$, is a martingale, where $S_0' = 0$ and $S_\infty' = \infty$. Therefore,

$$\mathbf{E}(X_\infty^2 - X_0^2) = \mathbf{E}\left[\left(X_\infty - X_{S_n'}\right)^2 + \left(X_{S_n'} - X_{S_{n-1}'}\right)^2 + ... + \left(X_{S_1'} - X_0\right)^2\right].$$

If we sum back to the original s.t., the above equality asserts that for every finite collection $S_0, S_1, ..., S_n, S_\infty$ ($S_0 = 0, S_\infty = \infty$)

$$\mathbf{E}\left(X_\infty^2 - X_0^2\right) = \mathbf{E} \sum_{k,m \in (0,1,...,n,\infty)} \mathbf{E}\left(X_{S_k} - X_{S_m}\right)^2,$$

where on the right side we have differences of values of the process close in time. Now, letting $n \to \infty$, apply Fatou's lemma; from the last equality we obtain

$$\mathbf{E} \sum_k \left[\left(X_{S_k+} - X_{S_k}\right)^2 + \left(X_{S_k} - X_{S_k-}\right)^2\right] \leq \mathbf{E}\left(X_\infty^2 - X_0^2\right),$$

where $X_{S_k+} = \lim_{S_n \downarrow S_k} X_{S_n}$, $X_{S_k-} = \lim_{S_n \uparrow S_k} X_{S_n}$ and it is assumed that $X_{S_k+} - X_{S_k} = X_{S_k} - X_{S_k-} = 0$ if $S_k = \infty$. ∎

5.4.1   Decomposition of elementary processes

Here we study Doob-Meyer decomposition of elementary processes. Elementary processes are indicators of stopping times; a process of the form $\mathbf{1}_{T \leq t}$ where T is a random time. For the various types of stopping times, these processes play an important role in establishing various decomposition results of optional martingales and in the definition of stochastic integral with respect to optional semimartingales. Here we present various decomposition results of elementary processes.

Let us first define the notion of orthogonal local optional martingales:

Definition 5.4.3 Suppose $X, Y \in \mathcal{M}_{loc}$. We say that X is orthogonal to Y ($X \perp Y$) if $XY \in \mathcal{M}_{loc}$.

Theorem 5.4.4 Suppose T is a *totally inaccessible* s.t. Let ξ be an $\mathcal{F}_T$-measurable integrable ($\mathbf{E}\xi^2 < \infty$, respectively) r.v. and $Y_t = \xi \mathbf{1}_{T \leq t}$.
   (1) There exists a unique predictable process $A = (A_t) \in \mathcal{A}$ such that,

$$Z = Y - A \in \mathcal{M}.$$

A non-decreasing $B = (B_t) \in \mathcal{A}$ such that,

$$Z = Y - B \in \mathcal{M}^2,$$

and $Z^2 - B \in \mathcal{M}$. Moreover, the processes A and B are continuous.
   (2) In either case, Z is orthogonal to every martingale $X \in \mathcal{M}_{loc}$ whose trajectories are right- or left-continuous and do not have common jump times with Z. Moreover, in the case $Z \in \mathcal{M}^2$ the process Z is orthogonal to every martingale $X \in \mathcal{M}^2$ whose trajectories are left-continuous.

Proof. Assertion (1) follows from Theorem 5.3.8 and Remark 5.3.9. The case where $\mathbf{E}\xi^2 < \infty$ was investigated in [55] (Chapter VIII, Theorem 31).
   Assertion (2) is about orthogonality; for $X \in \mathcal{M}_{loc}$ we let $(U_n, X^{(n)})$ denote the corresponding localized sequence. We fix the index n. If the trajectories are right- (left-, respectively) continuous, then the process $X^{(n)}$ ($(X_{t+}^{(n)})$, respectively) is a local martingale with respect to the family $(\mathcal{F}_{t+}^{\mathbf{P}})$. According to [59], there is a sequence (V_k), $k \in \mathbb{N}$, of $\mathbf{F}_+^{\mathbf{P}}$-s.t. such that $V_k \downarrow \infty$ a.s. and for all k the jumps of the stopped martingales $(Z_{t \wedge V_k})$ and $(X_{t \wedge V_k}^{(n)})$ ($(X_{t \wedge V_k+}^{(n)})$, respectively) are uniformly integrable. By Theorem 5.1.5 there is a sequence of s.t.b. (V_k') such that $V_k' = V_k$ a.s. for all k.
   Let

$$R_n = (\inf(t \geq 0 : |X_t| \geq n \text{ or } |Z_t| \geq n)) \wedge V_n' \wedge U_n.$$

For all n, R_n is a s.t.b. and $R_n \uparrow \infty$ a.s. Since X does not have common jump times with Z, for every s.t. $S \leq R_n$ we have

$$
\begin{aligned}
\mathbf{E}|X_{S+}Z_S| &\leq \mathbf{E}\left[|X_{S-}Z_{S-}| + |X_{S-}\triangle Z_S| + |Z_{S-}(X_{S+} - X_{S-})|\right] \\
&\leq n^2 + n\mathbf{E}\left[|\triangle Z_S| + |X_{S+} - X_{S-}|\right] < \infty.
\end{aligned}
$$

We choose (R_n) as the localizing sequence for XZ. According to Definitions 5.2.4 and 5.4.3, the orthogonality of Z and X will be established if we construct a sequence of processes $(Y^{(n)})$ such that $Y^{(n)} \in \mathcal{M}$, $XZ = Y^{(n)}$ on the set $[\![0, R_n]\!]$ and the r.v. $X_{R_n} + Z_R$, is integrable for all n. Clearly for the last equality it is sufficient that $\mathbf{E}X_S Z_S = \mathbf{E}Y_S^{(n)}$ for all s.t. $S \leq R_n$ and arbitrary n.

For $(Y^{(n)})$ we choose the optional modification of the martingale $(\mathbf{E}[X_{R_n} + Z_{R_n}|\mathcal{F}_t])$. We note that integrability of the r.v. $X_{R_n} + Z_{R_n}$ can be verified analogously to that of the r.v. $X_{S} + Z_S$, $S \leq R_n$. Let $Z_t - Z_0 = \tilde{Z}_t$. Considering the equality $\mathbf{E}X_S^{(n)} = \mathbf{E}X_{S+}^{(n)}$, Theorem 5.3.6, and the fact that X does not have common jump times with Z, we obtain

$$\mathbf{E}X_S \tilde{Z}_S = \mathbf{E}X_{S+} \tilde{Z}_S = \mathbf{E}\int_{[0,S]} X_{t+}^{(n)} d\tilde{Z}_t = \mathbf{E}\int_{[0,S]} X_{t-}^{(n)} d\tilde{Z}_t.$$

The set $]S, R_n]$ is predictable; therefore, on the basis of Theorem 5.3.8

$$\mathbf{E}\int_{]S,R_n]} X_{t-}^{(n)} d\tilde{Z}_t = 0.$$

Considering this and continuing the chain of equalities, we have

$$\mathbf{E}X_S \tilde{Z}_S = \mathbf{E}\int_{[0,R_n]} X_{t-}^{(n)} d\tilde{Z}_t = \mathbf{E}\int_{[0,R_n]} X_{t+}^{(n)} d\tilde{Z}_t$$

Hence on the strength of Theorem 5.3.6

$$\mathbf{E}\left[X_S \tilde{Z}_S\right] = \mathbf{E}\left[X_{R_n+}^{\langle n \rangle} \tilde{Z}_{R_n}\right] = \mathbf{E}\left[X_{R_n} + \tilde{Z}_{R_n}|\mathcal{F}_S\right] = \mathbf{E}Y_S^{(n)} - \mathbf{E}X_S Z_0.$$

Comparing the left and right sides, we obtain $\mathbf{E}X_S Z_S = \mathbf{E}Y_S^{\langle n \rangle}$ for all s.t. $S < R_n$.

Now suppose $Z \in \mathcal{M}^2$, $X \in \mathcal{M}^2$ and the trajectories of X are left-continuous. We represent X in the form

$$X = (X_t - \triangle^+ X_T \mathbf{1}_{T<t}) + (\triangle^+ X_T \mathbf{1}_{T<t}).$$

Both summands on the right side belong to $\mathcal{M}^2$, where the first does not have jumps in common with Z and according to the above is orthogonal to Z. It remains to be shown that the second summand is also orthogonal to Z. For every s.t. S we have

$$\mathbf{E}Z_\infty \triangle^+ X_T \mathbf{1}_{T<\infty} = \mathbf{E}Z_\infty \triangle^+ X_T \mathbf{1}_{T<S} + \mathbf{E}Z_\infty \triangle^+ X_T \mathbf{1}_{S\leq T<\infty}.$$

The first summand on the right is equal to $\mathbf{E}Z_S \triangle^+ X_T \mathbf{1}_{T<S}$, and the second, to zero. Indeed, since Z is continuous from the right and $\mathbf{E}[\triangle^+ X_T|\mathcal{F}_T] = 0$, we have

$$\begin{aligned}
\mathbf{E}\left[Z_\infty \triangle^+ X_T \mathbf{1}_{S\leq T<\infty}\right] &= \mathbf{E}\left[\triangle^+ X_T \mathbf{1}_{S\leq T<\infty} \mathbf{E}[Z_\infty|\mathcal{F}_{T+}]\right] \\
&= \mathbf{E}\left[\triangle^+ X_T \mathbf{1}_{S\leq T<\infty} Z_T\right] = \mathbf{E}\left[Z_T \mathbf{1}_{S\leq T<\infty} \mathbf{E}[\triangle^+ X_T|\mathcal{F}_{T+}]\right] \\
&= 0.
\end{aligned}$$

Thus $\mathbf{E}[Z_\infty \triangle^+ X_T \mathbf{1}_{T<\infty}] = \mathbf{E}[Z_S \triangle^+ X_T \mathbf{1}_{T<S}]$. ∎

Remark 5.4.5 If $\mathbf{E}\xi^2 < \infty$, then, by the remark after Theorem 31 in Chapter VIII of [55], for all s.t. $S \leq U$ we have

$$\mathbf{E}[Z_U^2 - Z_S^2|\mathcal{F}_S] = \mathbf{E}[B_U - B_S|\mathcal{F}_S] = \mathbf{E}[\xi^2 \mathbf{1}_{S<T\leq U}|\mathcal{F}_S] \quad \text{a.s.}$$

Theorem 5.4.6 Suppose T is a *predictable* or *totaly inaccessible* s.t., and the r.v. ξ is integrable ($\mathbf{E}\xi^2 < \infty$, respectively) and $\mathcal{F}_{T+}$-measurable. Let $Y_t = \xi 1_{T<t}$ for $t > 0$ and $Y_0 = 0$.

(1) There exists a unique right-continuous strongly predictable process $A = (A_t) \in \mathcal{A}$, $A_0 = 0$ (respectively increasing $B \in \mathcal{A}$, $B_0 = 0$) such that $Z = Y - Z \in \mathcal{M}$, ($Z = Y - A \in \mathcal{M}^2$ and $Z^2 - B \in \mathcal{M}$, respectively).

(2) Z is orthogonal to every martingale $X \in \mathcal{M}_{loc}$ whose trajectories are continuous from the left or the right and which do not have common discontinuity times with Z. Moreover, in the case $Z = Y - B \in \mathcal{M}^2$, Z is orthogonal to every martingale $X \in \mathcal{M}_{loc}^2$ whose trajectories are right-continuous.

Proof. (1) Let $\tilde{\xi} = \xi - \mathbf{E}[\xi|\mathcal{F}_T]$, $A_t = \mathbf{E}[\xi|\mathcal{F}_T]1_{T<t}$, $B_t = \mathbf{E}\left[|\tilde{\xi}|^2|\mathcal{F}_T\right]1_{T<t}$ for $t > 0$, and $A_0 = B_0 = 0$. The processes $A = (A_t)$ and $B = (B_t)$ are left-continuous, strongly predictable, and belong to the space $\mathcal{A}$.

We establish that the process $Z = (\tilde{\xi}1_{T<t})$, $t > 0$, $Z_0 = 0$, belongs to $\mathcal{M}$. We have

$$\mathbf{E}Z_\infty = \mathbf{E}\tilde{\xi}1_{T<S} + \mathbf{E}\tilde{\xi}1_{S\le T<\infty}$$

for every s.t. S. The second summand on the right side is equal to zero, since $\{S \le T < \infty\} \in \mathcal{F}_T$ and $\mathbf{E}\left[\tilde{\xi}|\mathcal{F}_T\right] = 0$ a.s. Consequently $\mathbf{E}[Z_\infty|F_S] = Z_S$ a.s. for all s.t. S, i.e., $Z \in \mathcal{M}$. In the case $\mathbf{E}\xi^2 < \infty$, it is easy to verify that $\mathbf{E}Z_\infty^2 < \infty$ and hence $Z \in \mathcal{M}^2$.

Uniqueness. Suppose there is another process $C = (C_t)$, $C_0 = 0$, satisfying the hypothesis of the theorem. Then the process $A - C \in \mathcal{M}$. For every s.t. σ and sequence of s.t. (τ_n), $\tau_n \downarrow \sigma$, we obtain

$$\mathbf{E}\left[A_{\tau_n} - A_\sigma|\mathcal{F}_\sigma\right] = \mathbf{E}[C_{\tau_n} - C_\sigma|\mathcal{F}_\sigma] \quad \text{a.s.}$$

Passing to the limit in this equality and considering that $A_{\sigma+}$ and $C_{\sigma+}$ are $\mathcal{F}_0$-measurable, we find that $\triangle^+ A_\sigma = \triangle^+ C_\sigma$ a.s. Consequently the jumps of the processes A and C occur at one and the same times and are identical in size. From this it follows that $A - C$ is a continuous martingale of bounded variation exiting from zero. Such a martingale is not different from zero.

(2) Suppose $(U_n, X^{(n)})$ is a localizing sequence for $X \in \mathcal{M}_{loc}$. We fix the index n as in the preceding theorem; according to [59] and Theorem 5.1.5 there exists a sequence (S_k) of s.t.b., $S_k \uparrow \infty$ a.s., such that for all k the processes $\left(X_{S_k \wedge t+}^{(n)}\right)$ and $(Z_{S_k \wedge t+})$, $t \in \mathbb{R}_+$, have uniformly integrable jumps. Let

$$R_n = (\inf(t > 0 : |X_{t+}| \ge n \text{ or } |Z_{t+}| \ge n)) \wedge S_n \wedge U_n.$$

The sequence (R_n) consists of s.t.b., $R_n \uparrow \infty$ a.s. It is easy to see that for all n the r.v. $Z_{R_n+}X_{R_n+}$ is integrable. As $Y^{(n)}$ we choose the optional modification of the martingale $(\mathbf{E}[X_{R_n+}Z_{R_n+}|\mathcal{F}_t])$. Then for every s.t. $S \le R_n$ we have

$$\begin{aligned} \mathbf{E}Y_S^{(n)} &= \mathbf{E}X_{R_n+}Z_{R_n+} = \mathbf{E}[Z_S + \tilde{\xi}1_{S\le T\le R_n}]X_{R_n+} \\ &= \mathbf{E}Z_S X_S + \mathbf{E}\tilde{\xi}1_{S\le T\le R_n}X_{R_n+}. \end{aligned}$$

Since the set $\{S \le T \le R_n\} \in \mathcal{F}_T$ and $\mathbf{E}[X_{R_n+}|\mathcal{F}_T] = X_{T+}$ a.s. the second summand on the right is equal to the variable $\mathbf{E}\tilde{\xi}1_{S\le T\le R_n}X_{T+}$. The process X is continuous at the moment of time T; therefore, the indicated variable is equal to $\mathbf{E}\tilde{\xi}1_{S\le T\le R_n}X_T$ which in turn is equal to zero since $\mathbf{E}[\tilde{\xi}|\mathcal{F}_T] = 0$, and the remaining variables under the expectation sign are $\mathcal{F}_T$-measurable.

Thus $\mathbf{E}Y_S^{(n)} = \mathbf{E}X_S Z_S$ for all s.t. $S \leq R_n$. This means that $XZ \in \mathcal{M}_{loc}$ with localizing sequence $(R_n, Y^{(n)})$, i.e., $X \perp Z$.

In the case $Z \in \mathcal{M}^2$ and $X \in \mathcal{M}^2$, when the trajectories of X are right continuous, for every s.t. S we have

$$\mathbf{E}X_\infty Z_\infty = \mathbf{E}X_\infty \tilde{\xi}1_{T<\infty} = \mathbf{E}X_\infty \tilde{\xi}1_{T<S} + \mathbf{E}X_\infty \tilde{\xi}1_{S\leq T<\infty}.$$

The first summand on the right is equal to $\mathbf{E}X_S Z_S$, since the variable $Z_S = \tilde{\xi}1_{T\leq S}$ is $\mathcal{F}_S$-measurable and $\mathbf{E}[X_\infty | \mathcal{F}_S] = X_S$ a.s. The second quantity on the right is equal to $\mathbf{E}X_{T+}\tilde{\xi}1_{S\leq T<\infty}$, since the variable $\tilde{\xi}1_{S\leq T<\infty}$ is $\mathcal{F}_{T+}$-measurable. Since the process X is right-continuous, $X_{T+} = X_T$. Hence

$$\mathbf{E}X_\infty \tilde{\xi}1_{S\leq T<\infty} = \mathbf{E}X_T \tilde{\xi}1_{S\leq T<\infty}.$$

The last variable is equal to zero because $X_T 1_{S\leq T<\infty}$ is $\mathcal{F}_T$-measurable and $\mathbf{E}\left[\tilde{\xi}|\mathcal{F}_T\right] = 0$ a.s. Thus $\mathbf{E}X_\infty Z_\infty = \mathbf{E}X_S Z_S$ for every s.t. ∎

Theorem 5.4.7 Suppose T is a *totally inaccessible* s.t.b., where $\mathbf{P}(T = S < \infty) = 0$ for all s.t. S. Suppose the r.v. ξ is $\mathcal{F}_{T+}$-measurable and integrable (respectively, $\mathbf{E}\xi^2 < \infty$). Let $Y_t = \xi 1_{T<t}$, $t > 0$ and $Y_0 = 0$.

(1) There exists a continuous process $A = (A_t) \in \mathcal{A}$ (respectively, nondecreasing $B = (B_t) \in \mathcal{A}$) such that $Z = Y - A \in \mathcal{M}$ (respectively, $Z = Y - A \in \mathcal{M}^2$ and $Z^2 - B \in \mathcal{M}$). The process A (respectively, B) is unique in the class of strongly predictable processes.

(2) Z is orthogonal to every martingale $X \in \mathcal{M}_{loc}$ whose trajectories are continuous from the left or the right and which do not have common jump times with Z. If $Z \in \mathcal{M}^2$, then Z is orthogonal to every martingale $X \in \mathcal{M}^2$ whose trajectories are continuous from the right.

Proof. On the strength of Theorem 5.3.7, for the process Y there exists a unique continuous process A such that

$$\mathbf{E}\int_{[0,\infty[} {}^o X_t dY_{t+} = \mathbf{E}\int_{[0.\infty[} X_t dA_t,$$

where X is an arbitrary positive measurable process and ${}^o X$ is its optional projection. Setting $X = 1_H 1_{[S,\infty[}$, where S is an arbitrary s.t. and $H \in \mathcal{F}_S$, we obtain

$$\mathbf{E}1_H[\xi 1_{T<\infty} - \xi 1_{T\leq S<\infty}] = \mathbf{E}1_H[A_\infty - A_S].$$

Since $\mathbf{P}(T = S < \infty) = 0$, it now follows that

$$\mathbf{E}1_H Z_\infty = \mathbf{E}1_H[\xi 1_{T<\infty} - A_\infty] = \mathbf{E}1_H[\xi 1_{T<S} - A_S] = \mathbf{E}1_H Z_S,$$

i.e., $Z \in \mathcal{M}$.

In the case $\mathbf{E}\xi^2 < \infty$ the process $Z \in \mathcal{M}^2$. This follows from the fact that the r.v. $Z_\infty = \xi 1_{T<\infty} - A_\infty$ is the limit at infinity of the right-continuous martingale $(Z_{t+}, \mathcal{F}_{t+}^{\mathbf{P}})$ and is square integrable (see [55], Chapter VIII, Theorem 31).

The existence and uniqueness of a continuous process B such that $Z^2 - B \in \mathcal{M}$, as in the case of the process A, is a consequence of Theorem 5.3.7.

Orthogonality is established as in Theorem 5.4.4. We write down only the basic chain of equalities. Defining the times R_n and the martingales $Y^{(n)} = (\mathbf{E}Z_{R_n+}X_{R_n+}|\mathcal{F}_t])$ as in Theorem 5.4.4, for every s.t. $S \le R_n$ we have

$$\mathbf{E}Y_S^{(n)} = \mathbf{E}Z_{R_n+}X_{R_n+} = \mathbf{E}\int_{[0,R_n]} X_{t+}dZ_{t+}.$$

Since X and Z do not have common jump times, we can replace X_{t+} by X_t, on the right. In view of Theorem 5.3.7, we obtain

$$
\begin{aligned}
\mathbf{E}Y_S^{(n)} &= \mathbf{E}\int_{[0,R_n]} X_t dZ_{t+} = \mathbf{E}\int_{[0.S]} X_t dZ_{t+} \qquad (5.9) \\
&= \mathbf{E}\int_{[0,S]} X_{t+}dZ_{t+} = \mathbf{E}X_{S+}Z_{S+}.
\end{aligned}
$$

Now we consider two cases. If X is continuous from the right, then we can replace X_{S+} by X_S on the right side of (5.9), and considering that $\mathbf{E}[Z_{S+}|\mathcal{F}_S] = Z_s$, we obtain the required equality

$$\mathbf{E}Y_S^{(n)} = \mathbf{E}X_S Z_S.$$

If X is continuous from the left, then

$$\mathbf{E}X_{S+}X_{S+} = \mathbf{E}\left[X_S Z_S + X_S \triangle^+ Z_S + Z_S \triangle^+ X_S\right] = \mathbf{E}X_S Z_S.$$

Substituting this in (5.9), we again obtain the required equality. Orthogonality for the case $Z \in \mathcal{M}^2$ and a right-continuous martingale $X \in \mathcal{M}^2$ follows from the above, since all jump times of the process X are found among some sequence of s.t. and consequently X and Z do not have common jump times. ∎

Theorem 5.4.8 Suppose T is a predictable s.t. and the events $H \in \mathcal{F}_\infty$. Then $H1_{T=\infty} \in \mathcal{F}_{T-}$.

Proof. See [28] (Chapter III, Theorem 31). ∎

Theorem 5.4.9 Suppose T is a predictable s.t. and ξ is a r.v. which is $\mathcal{F}_T$-measurable and integrable (respectively, $E\xi^2 < \infty$). Let $Y_t = \xi 1_{T \le t}$.

(1) There exists a unique right-continuous predictable process $A \in \mathcal{A}$ (respectively, $B \in \mathcal{A}$) such that $Z = Y - A \in \mathcal{M}$ (respectively, $Z = Y - A \in \mathcal{M}^2$ and $Z^2 - B \in \mathcal{M}^2$).

(2) Z is orthogonal to every martingale $X \in \mathcal{M}$ whose trajectories are continuous from the left or the right and which do not have discontinuities in common with Z. If $Z \in \mathcal{M}^2$, then Z is orthogonal to every martingale $X \in \mathcal{M}^2$ whose trajectories are continuous from the left.

Proof. (1) Let $\tilde{\xi} = \xi - E[\xi|\mathcal{F}_{T-}]$, $A_t = E[\xi|\mathcal{F}_{T-}]1_{T \le t}$ and $B_t = \mathbf{E}[|\tilde{\xi}|^2|\mathcal{F}_{T-}]1_{T \le t}$. The processes A and B are continuous from the right, predictable, and belong to $\mathcal{A}$. We shall establish that $Z = Y - A \in \mathcal{M}$. We have $\tilde{\xi}1_{T=\infty} = 0$ a.s. Indeed, by the preceding lemma, $\tilde{\xi}1_{T=\infty} \in \mathcal{F}_{T-}$. Therefore $\tilde{\xi}1_{T=\infty} = \mathbf{E}[\tilde{\xi}|\mathcal{F}_{T-}]1_{T=\infty} = 0$ a.s. Consequently $Z_\infty = \tilde{\xi}1_{T<\infty}$ a.s. Further, for every s.t. S

$$\mathbf{E}Z_\infty = \mathbf{E}\tilde{\xi}1_{T \le S} + \mathbf{E}\tilde{\xi}1_{S<T<\infty}.$$

The second summand on the right is equal to zero since the variable $\mathbf{1}_{S<T<\infty}$ is $\mathcal{F}_{T-}$-measurable, and $\mathbf{E}[\tilde{\xi}|\mathcal{F}_{T-}] = 0$ a.s. Consequently

$$\mathbf{E}Z_\infty = \mathbf{E}\tilde{\xi}\mathbf{1}_{T\leq S} = \mathbf{E}Z_S,$$

i.e., $Z \in \mathcal{M}$. It is easy to verify that $Z \in \mathcal{M}^2$ for $\mathbf{E}\xi^2 < \infty$, and $Z^2 - B \in \mathcal{M}$ can be verified analogously to the above.

Uniqueness of the processes A and B follows from the uniqueness of the Doob-Meyer decomposition for right-continuous supermartingales.

(2) For the proof of orthogonality, we give only the basic chain of equalities. As in Theorem 5.4.6, we define the sequence of s.t.b. (R_n), and as $Y^{(n)}$ we take the optional modification of the martingale $(\mathbf{E}[Z_{R_n}X_{R_n+}|\mathcal{F}_t])$. Then for every s.t. $S \leq R_n$

$$\begin{aligned}\mathbf{E}Y_S^{(n)} &= \mathbf{E}Z_{R_n+}X_{R_n+} = \mathbf{E}\int_{[0,R_n]} X_{t+}dZ_t = \mathbf{E}\int_{[0,R_n]} X_{t-}dZ_t \\ &= \mathbf{E}\int_{[0,S]} X_{t+}dZ_t = \mathbf{E}X_{S+}Z_S = \mathbf{E}X_SZ_S,\end{aligned}$$

i.e., XZ coincides with the martingale $Y^{(n)}$ on $[0, R_n]$.

Orthogonality for the case $Z \in \mathcal{M}^2$ is established analogously to Theorem 5.4.4. ∎

Remark 5.4.10 Once again, we note that we have proved the Doob decomposition for elementary processes here. The Doob decomposition for supermartingales of the class D will be dealt with in Chapter 6.

5.4.2 Decomposition of optional martingales

Let us introduce a norm $||X||^2 = \mathbf{E}X_\infty^2$ for any $X \in \mathcal{M}^2$. Then, $\mathcal{M}^2$ is turned into a complete normed separable space. The separability follows from that of the space L^2 of variables X_∞. The completeness follows from Doob's inequality

$$\mathbf{P}(\sup_t |X_t^n - X_t^m| > \epsilon) \leq \frac{1}{\epsilon}||X^n - X^m||^2$$

for any optional martingales X^n and X^m (see Lemma 5.2.1 or [9]). Indeed, if the sequence (X_n) is Cauchy in $\mathcal{M}^2$, then from it we can choose a subsequence converging a.s. uniformly to an optional modification of the martingale $(\mathbf{E}[X_\infty|\mathcal{F}_t]) \in \mathcal{M}^2$ where $X_\infty = \lim_{n\to\infty} X_\infty^n$.

Theorem 5.4.11 Suppose $X \in \mathcal{M}^2$ then

$$X = X^c + X^d + X^g,$$

where X^c is a continuous process with $X_0^c = 0$, X^d and X^g have one-sided limits, X^d is continuous from the right, X^g from the left with $X_0^g = 0$ and $X^g, X^c, X^d \in \mathcal{M}^2$. This decomposition is unique.

In addition, X^g (X^d, respectively) is orthogonal to every martingale $Y \in \mathcal{M}_{loc}$ whose trajectories are one-sidedly continuous and do not have common discontinuity times with X^g (X^d, respectively). Moreover, X^g (X^d, respectively) is orthogonal to every martingale $Y \in \mathcal{M}^2$ whose trajectories are continuous from the right (from the left, respectively).

Proof. Suppose (S_n), (T_n) and (U_n) are sequences of predictable, totally inaccessible s.t. and totally inaccessible s.t.b., respectively, whose graphs do not intersect, $\mathbf{P}(U_n = T < \infty) = 0$ for all $n \in \mathbb{N}$ and s.t. T and these sequences absorb all the jumps of the martingale X. We let

$$X_t^g(n) = \sum_{k=1}^{n} [\triangle^+ X_{S_k} \mathbf{1}_{S_\kappa < t} + \triangle^+ X_{T_k} \mathbf{1}_{T_k < t} + (\triangle^+ X_{U_k} \mathbf{1}_{U_k < t} - B_t^k)], \tag{5.10}$$

$$X_t^d(n) = \sum_{k=1}^{n} \left[\triangle X_{S_k} \mathbf{1}_{S_k \le t} + (\triangle X_{T_k} \mathbf{1}_{T_k \le t} - A_t^k) \right], \tag{5.11}$$

where A^k and B^k are the strongly predictable processes from Theorems 5.4.4 and 5.4.7, respectively, such that the last summands on the right sides of (5.10) and (5.11) belong to $\mathcal{M}^2$. The processes $X^g(n)$ ($X^d(n)$, respectively) are continuous from the left (right, respectively) for all $n \in \mathbb{N}$.

On the strength of Theorems 5.4.6 and 5.4.7 the summands on the right side of (5.10) belong to $\mathcal{M}^2$ and are orthogonal to every one-sidedly continuous martingale not having common discontinuity times with them. On the strength of Theorems 5.4.4 and 5.4.9 the analogous fact is valid for the summands on the right side of (5.11). The sequence $(X^g(n))$, $n \in \mathbb{N}$ $((X^d(n))$, respectively) is Cauchy in $\mathcal{M}^2$:

$$\|X^g(n) - X^g(n+m)\|^2 = \mathbf{E} \sum_{k=n+1}^{n+m} \left[(\triangle^+ X_{S_k})^2 + (\triangle^+ X_{T_k})^2 + (\triangle^+ X_{U_k})^2 \right]$$

$$\left(\text{respectively } \|X^d(n) - X^d(n+m)\|^2 = \mathbf{E} \sum_{k=n+1}^{n+m} \left[(\triangle X_{S_k})^2 + (\triangle X_{T_k})^2 \right] \right),$$

which vanishes as $n, m \to \infty$ on the strength of Theorem 5.4.2.

We let $X^g = \lim X^g(n)$ and $X^d = \lim X^d(n)$, where the limit is taken in the space $\mathcal{M}^2$. As already noted, with respect to some subsequence, $X^g(n') \to X^g$, $X^d(n') \to X^d$ a.s., uniformly on $\mathbb{R}_+$. The processes X^g and X^d include all the jumps of X. Therefore, the process $X^c = X - X^d - X^g$ is a continuous martingale.

Orthogonality follows from that for the limit processes.

Now we prove uniqueness. Suppose we have the decomposition $X = \tilde{X}^c + \tilde{X}^d + \tilde{X}^g$ having the same orthogonality properties as in the theorem. Then

$$0 = \mathbf{E}[(\tilde{X}_T^g - X_T^g)^2 + (\tilde{X}_T^d - X_T^d)^2 + (\tilde{X}_T^c - X_T^c)^2]$$

for every s.t. T. Consequently, the decomposition is unique to within indistinguishability. ∎

Remark 5.4.12 If $X \in \mathcal{M}_{loc}^2$, then $X = X^c + X^d + X^g$, with $X^c, X^d, X^g \in \mathcal{M}_{loc}^2$ and the same orthogonality conditions on $Y \in \mathcal{M}_{loc}$ and $Y \in \mathcal{M}^2$ as in Theorem 5.4.11.

Remark 5.4.13 From the proof of the theorem above, it is evident that the processes X^g and X^d have the form

$$X_t^g = \sum_n [\triangle^+ X_{S_n} \mathbf{1}_{S_n < t} + \triangle^+ X_{T_n} \mathbf{1}_{T_n < t} + (\triangle^+ X_{U_n} \mathbf{1}_{U_n < t} - B_t^n)],$$

$$X_t^d = \sum_n [\triangle X_{S_n} \mathbf{1}_{S_n \leq t} + (\triangle X_{T_n} \mathbf{1}_{T_n \leq t} - A_t^n)],$$

where the sums converge in mean square.

Corollary 5.4.14 Since the summands on the right sides of the preceding equalities are mutually orthogonal, then the series $\sum_n [\triangle^+ X_{S_n} + \triangle^+ X_{T_n}]$ and $\sum_n \triangle X_{S_n}$ converge a.s. where $X \in \mathcal{M}^2$. If $X \in \mathcal{M}_{loc}^2$ then for all $t < \infty$ the series

$$\sum_n \triangle^+ X_{S_n} \mathbf{1}_{S_n < t} + \triangle^+ X_{T_n} \mathbf{1}_{T_n < t}, \quad \sum_n \triangle X_{S_n} \mathbf{1}_{S_n \leq t}$$

converge a.s.

Next we show that the conditional infinite sum of elementary processes is a.s. finite if it is locally integrable and a.s. finite.

Lemma 5.4.15 Suppose (T_k) and (ξ_k), $k \in \mathbb{N}$, are sequences of s.t. and r.v., respectively.
(1) Suppose for all $k \in \mathbb{N}$ the r.v. ξ_k is $\mathcal{F}_{T_k}$-measurable and $\hat{\xi}_k = \mathbf{E}[\xi_k | \mathcal{F}_{T_k-}]$ is defined. Suppose there exists a sequence of s.t.b. (σ_n) such that $\sigma_n \uparrow \infty$ a.s. and $\mathbf{E} \sum_n |\xi_k| \mathbf{1}_{T_k \leq \sigma_n} < \infty$ for all n. Then, for all $t \in \mathbb{R}_+$

$$\sum_k |\xi_k| \mathbf{1}_{T_k \leq t} < \infty \quad \text{a.s.} \quad \Rightarrow \quad \sum_k |\hat{\xi}_k| \mathbf{1}_{T_k \leq t} < \infty \quad \text{a.s.}$$

(2) Suppose the r.v. ξ_k is $\mathcal{F}_{T_k+}$-measurable and $\hat{\xi}_k = \mathbf{E}[\xi_k | \mathcal{F}_{T_k}]$ is defined for all $k \in \mathbb{N}$. Assume that there exists a sequence of s.t.b. (σ_n) such that $\sigma_n \uparrow \infty$ a.s. and $\mathbf{E} \sum_k |\xi_k| \mathbf{1}_{T_k \leq \sigma_n} < \infty$ for all n. Then for every $t \in \mathbb{R}_+$

$$\sum_k |\xi_k| \mathbf{1}_{T_k < t} < \infty \quad \text{a.s.} \quad \Rightarrow \quad \sum_k |\hat{\xi}_k| \mathbf{1}_{T_k < t} < \infty \quad \text{a.s.}$$

Proof. (1) Let $Z = \sum_k |\xi_k| \mathbf{1}_{T_k \leq t}$ and $\hat{Z}_t = \sum_k |\hat{\xi}_k| \mathbf{1}_{T_k \leq t}$; let $D = \left\{ (\omega, t) : \hat{Z}_t(\omega) = \infty \right\}$. We have

$$D = \bigcap_N \bigcup_{n_0} \bigcap_{n > n_0} \left\{ (\omega, t) : \sum_{k=1}^n |\hat{\xi}_k(\omega)| \mathbf{1}_{T_k \leq t} > N \right\} \in \mathcal{O}(\mathcal{F}).$$

From the theorem on sections (Theorem 2.4.48) it follows that for given $\epsilon > 0$ there exists a s.t. T such that $[T] \subseteq D$ and $\mathbf{P}(T < \infty) > \mathbf{P}(\pi(D)) - \epsilon$. Assume that $\mathbf{P}(T < \infty) > 0$ for some ϵ. Let

$$R_n = (\inf(t : Z_t \geq n)) \wedge \sigma_n.$$

The variable R_n is a s.t.b. and $R_n \uparrow \infty$ a.s. Choose n so large that $\mathbf{P}(\infty > R_n > T) > 0$. Then,

$$\infty = \mathbf{E}\hat{Z}_{T \wedge R_n} \leq \mathbf{E}\sum_k \mathbf{E}[|\xi_k||\mathcal{F}_{T_k-}]\mathbf{1}_{T_k \leq T \wedge R_n}.$$

The set $\{T_k \leq T \wedge R_n\} \in \mathcal{F}_{T_k-}$ (see [28], Chapter III, Theorem 29); therefore, on the right side of the last equality the indicators of the sets $\{T_k \leq T \wedge R_n\}$ can be put under the conditional expectation sign, after which we can remove the conditional expectation sign. As a result we have

$$\infty = \mathbf{E}\hat{Z}_{T \wedge R_n} \leq \mathbf{E}Z_{T \wedge R_n} \leq n + \mathbf{E}\sum_k |\xi_k|\mathbf{1}_{T_k = T \wedge R_n}.$$

We have arrived at a contradiction, since the right side is finite by the hypothesis of the lemma. Consequently the set D is negligible, which implies assertion (1).

Assertion (2) is proved analogously. Therefore, we write down only the basic inequality, which, as in part (1), leads to a contradiction:

$$
\begin{aligned}
\infty = \mathbf{E}\hat{Z}_{T \wedge R_n} &= \mathbf{E}\sum_k |\mathbf{E}[\xi_k|\mathcal{F}_{T_k}]|\mathbf{1}_{T_k < T \wedge R_n} \\
&\leq \mathbf{E}\sum_k \mathbf{E}[|\xi_k||\mathcal{F}_{T_k}]\mathbf{1}_{T_k < T \wedge R_n} \leq \mathbf{E}\sum_k \mathbf{E}[|\xi_k||\mathcal{F}_{T_k}]\mathbf{1}_{T_k \leq T \wedge R_n} \\
&= \mathbf{E}\sum_k \mathbf{E}\left[|\xi_k|\mathbf{1}_{T_k \leq T \wedge R_n}|\mathcal{F}_{T_k}\right] = \mathbf{E}Z_{T \wedge R_n+} \\
&\leq n + \mathbf{E}\sum_k |\xi_k|\mathbf{1}_{T_k = T \wedge R_n}.
\end{aligned}
$$

∎

Next, we present auxiliary results before proceeding to investigate integration with respect to local martingales. The following result is basic; it tells us that a local martingale is integrable over finite random intervals.

Lemma 5.4.16 For every $X \in \mathcal{M}_{loc}$ there exists a sequence of s.t.b. (σ_n), $n \in \mathbb{N}$, $\sigma_n \uparrow \infty$, such that the jumps of X on $[0, \sigma_n]$ are integrable for all n.

Proof. Suppose (S_k), (T_k) and (U_k), $k \in \mathbb{N}$, are sequences of predictable, totally inaccessible s.t. and totally inaccessible s.t.b., respectively, absorbing the jumps of X. We let $(R_n, X^{(n)})$, $n \in \mathbb{N}$, denote the localizing sequence for X. We fix the index n.

For the predictable times (S_k), for given k suppose the sequence of s.t.b. (W_m) predicts S_k, i.e., $W_m \uparrow S_k$ and $W_m < S_k$ a.s. on $\{S_k > 0\}$. Then,

$$\mathbf{E}|\triangle X_{S_k}|\mathbf{1}_{S_k \leq R_n} = \lim_{m \to \infty} \mathbf{E}|X_{S_k}^{(n)} - X_{W_m}^{(n)}|\mathbf{1}_{S_k \leq R_n} \leq 2\mathbf{E}|X_\infty^{(n)}|,$$

i.e., the jumps $\triangle X_{S_k}$ for all k are integrable on $[0, R_n]$. Further,

$$\mathbf{E}|\triangle^+ X_{S_k}|\mathbf{1}_{S_k < R_n} = \lim_{\epsilon \downarrow 0} \mathbf{E}|X_{S_k+\epsilon}^{(n)} - X_{S_k}^{(n)}|\mathbf{1}_{S_k < R_n} \leq 2\mathbf{E}|X_\infty^{(n)}|,$$

i.e., the jumps $\triangle^+ X_{S_k} 1_{S_k < R_n}$ are integrable on $[0, R_n]$. Analogously, it can be established that $\mathbf{E}| \triangle^+ X_{T_k}|1_{T_k < R_k} < \infty \ \forall k$.

For s.t.b. (U_k), the right-continuous martingale $(Y_t(n), \mathcal{F}^{\mathbf{P}}_{t+})$ where $Y_t^{(n)} = X_{t+}^{(n)}$, according to [59] there exists a sequence of $\mathcal{F}^{\mathbf{P}}_+$-s.t. (τ_k), $k \in \mathbb{N}$, such that the jumps of the stopped martingale $(Y_{\tau_k \wedge t}^{(n)})$ are uniformly integrable. Suppose the sequence of s.t.b. (τ'_k) is such that $\tau'_k = \tau_k$ a.s. for all k. Let $\sigma_n = R_n \wedge \tau'_n$. Then, since the process X is continuous from the left at the times (U_k),

$$\triangle^+ X_{U_k} 1_{U_k < \sigma_n} = \triangle Y_{U_k}^{(n)} 1_{U_k < \sigma_n}.$$

Hence for every k the variables on the left are integrable simultaneously with the variables on the right.

Finally for totally inaccessible s.t. (T_k), for all k

$$\triangle X_{T_k} 1_{T_k \leq \sigma_n} = \triangle X_{T_k}^{(n)} 1_{T_k \leq \sigma_n} = [\triangle Y_{T_k}^{(n)} - (X_{T_k+}^{(n)} + X_{T_k}^{(n)})]1_{\tau_k \leq \sigma_n}.$$

The variables on the right are integrable. Consequently the jumps $\triangle X_{T_k} 1_{T_k < \sigma_n}$ are integrable for every k on $[0, \sigma_n]$. ∎

The next lemmas will establish that the compensated sum of jumps of absolute sizes larger than 1 of local optional martingales are local optional martingales.

For the sequences (S_n) and (T_n), $n \in \mathbb{N}$, of predictable and totally inaccessible s.t., let us introduce the following notation:

$$\begin{aligned}
p_n &= \triangle X_{S_n} 1(| \triangle X_{S_n}| > 1, S_n < \infty), & \hat{p}_n &= \mathbf{E}[p_n | \mathcal{F}_{S_n-}], \\
y_n &= \triangle^+ X_{S_n} 1(| \triangle^+ X_{S_n}| > 1, S_n < \infty), & \hat{y}_n &= \mathbf{E}[y_n | \mathcal{F}_{S_n}], \\
z_n &= \triangle^+ X_{T_n} 1(| \triangle^+ X_{T_n}| > 1, T_n < \infty), & \hat{z}_n &= \mathbf{E}[z_n | \mathcal{F}_{T_n}].
\end{aligned}$$

Lemma 5.4.17 Suppose $X \in \mathcal{M}_{loc}$ then
(1) The conditional expectations $\hat{p}_n$, $\hat{y}_n$ and $\hat{z}_n$, $n \in \mathbb{N}$, are defined, and the processes

$$P = \left(\sum_{S_n \leq t} p_n \right), \quad \hat{P} = \left(\sum_{S_n \leq t} \hat{p}_n \right), \quad Y = \left(\sum_{S_n < t} y_n \right), \quad \hat{Y} = \left(\sum_{S_n < t} \hat{y}_n \right),$$

$$Z = \left(\sum_{T_n < t} z_n \right), \quad \hat{Z} = \left(\sum_{T_n < t} \hat{z}_n \right)$$

belong to $\mathcal{A}_{loc}$.
(2) The processes $\hat{P}$, $\hat{Y}$ and $\hat{Z}$ are strongly predictable and there exists a sequence of s.t. (V_n), $V_n \uparrow \infty$, a.s. such that

$$\sum_k |\hat{p}_k| 1_{S_k < t \wedge V_n} \leq n, \quad \sum_k |\hat{y}_k| 1_{S_k < t \wedge V_n} \leq n, \quad \sum_k |\hat{z}_k| 1_{T_k < t \wedge V_n} \leq n$$

for all $t \in \mathbb{R}_+$ and $n \in \mathbb{N}$.
(3) $\tilde{P} = P - \hat{P} \in \mathcal{M}_{loc}$, $\tilde{Y} = Y - \hat{Y} \in \mathcal{M}_{loc}$ and $\tilde{Z} = Z - \hat{Z} \in \mathcal{M}_{loc}$, and the processes $\hat{P}$, $\hat{Y}$ and $\hat{Z}$ are unique among the stongly predictable ones having this property.

Proof. (1) Suppose (R_n), $n \in \mathbb{N}$, is a localizing sequence of s.t.b. for X. For fixed k we obtain

$$\Omega_n = \{\omega : S_k(\omega) \le R_n(\omega)\} \cup \{S_k(\omega) = \infty\}.$$

We have $\Omega_n \uparrow \Omega$, $\Omega_n \in \mathcal{F}_{S_k-}$ and by Lemma 5.4.16, $\mathbf{E}|p_k|\mathbf{1}_{\Omega_n} < \infty$ for every n. Let $\mu(Y) = \mathbf{E}[p_k Y]$, where Y is a bounded $\mathcal{F}_{S_k-}$-measurable variable. Then we can determine the expectation $\hat{p}_k = \mathbf{E}[p_k | \mathcal{F}_{S_k-}]$ as a modification of the Radon-Nikodým derivative of the σ-finite measure μ with respect to the measure $\mathbf{P}$.

Analogously, we determine $\hat{y}_k$ ($\hat{z}_k$ respectively) by choosing

$$\Omega_n = \{(0 : S_k(\omega) < R_n(\omega)\} \cup \{\omega : S_k(\omega) = \infty\}$$

$$(\text{respectively } \Omega_n = \{\omega : T_k(\omega) \le R_n(\omega))\} \cup \{\omega : T_k(\omega) = \infty\}).$$

Further, since at each point the local martingale X has limits from the left and right, on any finite interval it has a finite number of jumps exceeding unity in absolute value. Therefore, the series

$$\sum_k p_k \mathbf{1}_{S_k \le t}, \quad \sum_k y_k \mathbf{1}_{S_k < t}, \quad \sum_k z_k \mathbf{1}_{T_k < t}$$

for each $t \in \mathbb{R}_+$ have only a finite number of summands, and consequently converge a.s. absolutely. From Lemma 5.4.16 it follows that P, Y, $Z \in \mathcal{A}_{loc}$. Convergence of the analogous series for $\hat{P}$, $\hat{Y}$, $\hat{Z}$ follows from Lemma 5.4.15. Then, from Lemma 5.4.16 it follows that $\hat{P}$, $\hat{Y}$, $\hat{Z} \in \mathcal{A}_{loc}$.

(2) The process $\hat{P}$ is continuous from the right, while the processes $\hat{Y}$ and $\hat{Z}$ are continuous from the left and they are all strongly predictable. The process $(\sum_k |\hat{p}_k| \mathbf{1}_{S_k < t})$ is increasing, continuous from the right and predictable. It is well known (see [57]) that for this process there exists a sequence of s.t. (σ_n), $\sigma_n \uparrow \infty$ a.s., for which

$$\sum_k |\hat{p}_k| \mathbf{1}_{S_k \le t \wedge \sigma_n} \le n$$

for all $n \in \mathbb{N}$. We set

$$u_t = \sum_k |\hat{y}_k| \mathbf{1}_{S_k < t}, \quad p_n = \inf(t > 0 : u_t \ge n);$$

then the p_n are stopping times. Indeed,

$$\{p_n < t\} = \cup_{r<t}\{u_r > n\} \in \mathcal{F}_t,$$

$$\{p_n = t\} = \cap_{r<t}\{u_r < n\} \cup \{u_t = n\} \cup \{u_{t+} \ge n\} \in \mathcal{F}_t$$

for all $t \in \mathbb{R}_+$, where r are rational numbers. We have $\rho_n \uparrow \infty$ a.s. and $u_{\rho_n} \le n$ for all $n \in \mathbb{N}$.

Analogously, for the process $v_t = \sum |\hat{z}_k| \mathbf{1}_{T_k < t}$ we determine a sequence of s.t. (δ_n) increasing to $+\infty$ such that $v_{\delta_n} \le n$ for all $n \in \mathbb{N}$. It remains to let $V_n = \sigma_n \wedge \rho_n \wedge \delta_n$.

(3) Suppose (σ_n) is the sequence of s.t.b. for X from Lemma 5.4.16, and the sequence of s.t. (V_n) is defined above. Let

$$v_n = \inf\left(t > 0 : \sum_{S_k \le t} |p_k| \ge n \text{ or } \sum_{S_k < t} |y_n| \ge n \text{ or } \sum_{T_k < t} |z_k| \ge n\right);$$

the ν_n are s.t.b. for all $n \in \mathbb{N}$; $\nu_n \uparrow \infty$ a.s. As localizing sequences $(R_n, Y^{(n)})$ for the processes $\tilde{P}$, $\tilde{Y}$ and $\tilde{Z}$ we take one and the same sequence of s.t.b. (R_n), where $R_n = \sigma_n \wedge V_n \wedge \nu_n$, and for $Y^{(n)}$ we take the optional modification of the martingales

$$\left(\mathbf{E} \left[\sum_{S_k \leq R_n} (p_h - \hat{p}_k) | \mathcal{F}_t \right] \right) \quad \text{for } \tilde{P},$$

$$\left(\mathbf{E} \left[\sum_{S_k < R_n} (y_k - \hat{y}_k) | \mathcal{F}_t \right] \right) \quad \text{for } \tilde{Y},$$

$$\left(\mathbf{E} \left[\sum_{T_k \leq R_n} (z_k - \hat{z}_k) | \mathcal{F}_t \right] \right) \quad \text{for } \tilde{Z}.$$

It is easy to verify that $\tilde{P}, \tilde{Y}, \tilde{Z} \in \mathcal{M}_{loc}$ for the localizing sequences thus introduced.

Uniqueness of the predictable process $\hat{P}$ follows from the uniqueness of the Doob-Meyer decomposition. Uniqueness of the strongly predictable processes is established analogously to that of Theorem 5.4.6. ∎

Suppose still that $X \in \mathcal{M}_{loc}$ and for X according to Theorem 5.4.1, let (T_k) and (U_k) be sequences of totally inaccessible s.t. and s.t.b., respectively. Let $v_k = \triangle X_{T_k} \mathbf{1}(|\triangle X_{T_k}| > 1, T_k < \infty)$ and $w_k = \triangle^+ X_{U_k} \mathbf{1}(|\triangle^+ X_{u_k}| > 1, U_k < \infty)$; then it must be that,

Lemma 5.4.18 For all $t \in \mathbb{R}_+$ the processes $V = (\sum_k v_k \mathbf{1}_{T_k < t})$ and $W = (\sum_k w_k \mathbf{1}_{U_k < t})$ belong to $\mathcal{A}_{loc}$. There exist a unique predictable process $A \in \mathcal{A}_{loc}$ and a unique strongly predictable process $B \in \mathcal{A}_{loc}$ such that $\tilde{V} = V - A \in \mathcal{M}_{loc}$ and $\tilde{W} = W - B \in \mathcal{M}_{loc}$. Moreover, the processes A and B are continuous.

Proof. The processes V and W have locally bounded variation. Let

$$\tau_n = \inf \left(t : \int_{[0,t]} |dV_s| \geq n \right)$$

For all $n \in \mathbb{N}$, τ_n is a s.t. Further, on the strength of Lemma 5.4.16 there exists a sequence of s.t.b. (σ_n), $\sigma_n \uparrow \infty$ a.s., such that the jumps of the process X are integrable on $[0, \sigma_n]$ for all $n \in \mathbb{N}$. Let $R_n = \tau_n \wedge \sigma_n$; R_n is a s.t.b. and $R_n \uparrow \infty$ a.s. We have

$$\mathbf{E} \left(\int_{[0,R_n]} |dV_s| \right) \leq \mathbf{E} \left(\int_{[0,R_n[} |dV_s| + |\triangle V_{R_n}| \right) \leq n + \mathbf{E}|\triangle X_{R_n}| < \infty,$$

i.e., $V \in \mathcal{A}_{loc}$. It is proved analogously that $W \in \mathcal{A}_{loc}$. From Theorem 5.3.8 and Remark 5.3.9 follows the existence of a unique predictable continuous process $A \in \mathcal{A}_{loc}$ such that $\hat{V} = V - A \in \mathcal{M}_{loc}$. From Theorem 5.3.7 follow the existence and uniqueness of the continuous strongly predictable process $B \in \mathcal{A}_{loc}$ such that $\tilde{W} = W - B \in \mathcal{M}_{loc}$. ∎

The next lemma tells us that any local optional martingale is the sum of a locally square integrable optional martingale and a local optional martingale of integrable finite variation.

Lemma 5.4.19 Any local optional martingale $X \in \mathcal{M}_{loc}$ can be decomposed a $X = \tilde{X} + \check{X}$, where $\tilde{X} \in \mathcal{M}^2_{loc}$ and $\check{X} \in \mathcal{M}_{loc} \cap \mathcal{A}_{loc}$.

Proof. Using the notation of Lemmas 5.4.17 and 5.4.18, we let $\check{X} = \hat{P} + \tilde{Y} + \tilde{Z} + \tilde{V} + \tilde{W}$. According to these lemmas, $\check{X} \in \mathcal{M}_{loc} \cap \mathcal{A}_{loc}$. For the process $\tilde{X} = X - \check{X}$ we have

$$\triangle \tilde{X}_{S_n} = \triangle X_{S_n} - \triangle \tilde{P}_{S_n} = \triangle X_{S_n} \mathbf{1}(|\triangle X_{S_n}| \leq 1) + \triangle \hat{P}_{S_n} \text{ on } \{S_n < \infty\},$$

$$\triangle^+ \tilde{X}_{S_n} = \triangle^+ X_{S_n} - \triangle^+ \tilde{Y}_{S_n} = \triangle^+ X_{S_n} \mathbf{1}(|\triangle^+ X_{S_n}| \leq 1) + \triangle^+ \hat{Y}_{S_n} \text{ on } \{S_n < \infty\},$$

$$\triangle \tilde{X}_{T_n} = \triangle X_{T_n} - \triangle \tilde{V}_{T_n} = \triangle X_{T_n} \mathbf{1}(|\triangle X_{T_n}| \leq 1) \text{ on } \{T_n < \infty\},$$

$$\triangle^+ \tilde{X}_{T_n} = \triangle^+ X_{T_n} - \triangle^+ \tilde{Z}_{T_n} = \triangle^+ X_{T_n} \mathbf{1}(|\triangle^+ X_{T_n}| \leq 1) + \triangle^+ \hat{Z}_{T_n} \text{ on } \{T_n < \infty\},$$

$$\triangle^+ \tilde{X}_{U_n} = \triangle^+ X_{U_n} - \triangle^+ \tilde{W}_{U_n} = \triangle^+ X_{U_n} \mathbf{1}(|\triangle^+ X_{U_n}| \leq 1) \text{ on } \{U_n < \infty\},$$

where the sequences (S_n), (T_n) and (U_n) are from Theorem 5.4.1. On the strength of Lemma 5.4.17 the processes $\hat{P}$, $\hat{Y}$ and $\hat{Z}$ are locally bounded. Consequently, their jumps are also locally bounded. Hence the jumps of the process $\tilde{X}$ are also locally bounded. From this it is easy to derive that $\tilde{X} \in \mathcal{M}^2_{loc}$. ∎

Finally, we proceed to the proof of the main theorem on decomposition of local optional martingales.

Theorem 5.4.20 Every process $X \in \mathcal{M}_{loc}$ has a unique representation $X = X^c + X^d + X^g$, where X^c, X^d, $X^g \in \mathcal{M}_{loc}$, X^c is continuous, $X^c_0 = 0$, X^g and X^d have one-sided limits, X^g is continuous from the left, $X^g_0 = 0$, X^d is continuous from the right, and $X^g(X^d)$ is orthogonal to every martingale $Y \in \mathcal{M}_{loc}$ whose trajectories are one-sidedly continuous and do not have common discontinuity times with $X^g(X^d)$.

Proof. According to the preceding lemma, we have $X = \tilde{X} + \check{X}$, where $\tilde{X} \in \mathcal{M}^2_{loc}$ and $\check{X} = \tilde{R} + \tilde{Y} + \tilde{Z} + \tilde{V} + \tilde{W} \in \mathcal{M}_{loc} \cap \mathcal{A}_{loc}$. By Remark 5.4.12, $\tilde{X} = \tilde{X}^g + \tilde{X}^c + \tilde{X}^d$, where the processes $\tilde{X}^g$, $\tilde{X}^c$, $\tilde{X}^d \in \mathcal{M}^2_{loc}$ and satisfy the required orthogonality conditions. Let $X^g = \tilde{X}^g + \tilde{Y} + \tilde{Z} + \tilde{W}$ and $X^c = \tilde{X}^c$, $X^d = \tilde{X}^d + \tilde{R} + \tilde{V}$. The processes X^g, X^c and X^d satisfy the hypotheses of the theorem.
Uniqueness follows from Remark 5.4.12 and Lemmas 5.4.17 and 5.4.18. ∎

5.5 Quadratic Variation

5.5.1 Predictable and optional

A fundamental role is played by the increasing quadratic variation processes $\langle X \rangle$ and $[X]$ in constructing the stochastic integral from a square integrable martingale X whose trajectories are continuous from the right. The process $\langle X \rangle$ is unique, continuous and predictable having the property $\mathbf{E} X_T^2 = \mathbf{E} \langle X \rangle_T$ for every s.t. T or that the process $X^2 - \langle X \rangle \in \mathcal{M}$.

In order to determine the process $\langle X \rangle$ for our case, $X \in \mathcal{M}^2$ a square integrable optional martingale, it suffices to consider the case where the trajectories of X are continuous from the left. Indeed, by Theorem 5.4.11, $X = X^c + X^d + X^g$ and

$$\mathbf{E} X_T^2 = \mathbf{E}[(X_T^g)^2 + (X_T^c)^2 + (X_T^d)^2] = \mathbf{E}[(X_T^g)^2 + \langle X^c \rangle_T + \langle X^d \rangle_T]$$

for every s.t. T. So, $\langle X^c \rangle_T$ and $\langle X^d \rangle_T$ are defined, and, we are left with $\mathbf{E}[(X_T^g)^2]$ to define.

On account of Remark 5.4.13 where the form of the process X^g is given and Theorems 5.4.7 and 5.4.6, we conclude that the process $\langle X^g \rangle$ should be sought for in the class of increasing strongly predictable processes. From the orthogonality of the summands comprising X^g, for every s.t. T we obtain

$$\mathbf{E}\,(X_T^g)^2 = \mathbf{E} \sum_n \left[\left(\triangle^+ X_{S_n} \right)^2 \mathbf{1}_{S_n < T} + \left(\triangle^+ X_{T_n} \right)^2 \mathbf{1}_{T_n < t} + \left(\triangle^+ X_{U_n} \mathbf{1}_{U_n < T} - B_T^{(n)} \right)^2 \right],$$

where $B^{(n)}$ is a continuous process from T such that $\left(\triangle^+ X_{U_n} \mathbf{1}_{U_n < t} - B_t^{(n)} \right) \in \mathcal{M}^2$. For the second summand on the right side we have

$$\mathbf{E} \sum_n \left(\triangle^+ X_{U_n} \mathbf{1}_{U_n < T} - B_T^{(n)} \right)^2 = \mathbf{E} \sum_n (\triangle^+ X_{U_n})^2 \mathbf{1}_{U_n < T}$$

(this property is analogous to the one presented in Remark 5.4.5 and is established by the same means).

Thus, for every s.t. T

$$\mathbf{E}(X_T^g)^2 = \mathbf{E} \sum_n \left[(\triangle^+ X_{S_n})^2 \mathbf{1}_{S_n < T} + (\triangle^+ X_{T_n})^2 \mathbf{1}_{S_n < T} + (\triangle^+ X_{U_n})^2 \mathbf{1}_{U_n < T} \right]. \tag{5.12}$$

From Theorem 5.3.7 follows the existence and uniqueness of the increasing continuous process $B = (B_t)$, $t \in \mathbb{R}_+$, such that for all s.t. T

$$\mathbf{E} \sum_n (\triangle^+ X_{U_n})^2 \mathbf{1}_{U_n < T} = \mathbf{E} B_T.$$

Further, on the basis of Theorem 5.4.6, the first two summands on the right side of (5.12) are equal to

$$\mathbf{E} \sum_n \mathbf{E} \left((\triangle^+ X_{S_n})^2 | \mathcal{F}_{S_n} \right) \mathbf{1}_{S_n < T} + \mathbf{E}((\triangle^+ X_{T_n})^2 | \mathcal{F}_{T_n}) \mathbf{1}_{T_n} \right) \mathbf{1}_{T_n < T}.$$

Hence we conclude that the process $\langle X^g \rangle$ has the form

$$\langle X^g \rangle_t = B_t + \sum_n [\mathbf{E}((\triangle^+ X_{S_n})^2 | \mathcal{F}_{S_n}) \mathbf{1}_{S_n < t} + \mathbf{E}((\triangle^+ X_{T_n})^2 | \mathcal{F}_{T_n}) \mathbf{1}_{T_n < t}]. \tag{5.13}$$

As for the component X^d, from Remarks 5.4.13 and 5.4.5 we have

$$\mathbf{E}(X_T^d)^2 = \mathbf{E}\sum_n[(\triangle X_{S_n})^2 1_{S_n \leq T} + (\triangle X_{T_n})^2 1_{T_n \leq T}]. \tag{5.14}$$

From Theorem 5.3.8 and Remark 5.3.9 follows the existence and uniqueness of a continuous increasing process A such that for every s.t. T

$$\mathbf{E}\sum_n(\triangle X_{T_n})^2 1_{T_n \leq T} = \mathbf{E}A_T.$$

According to Theorem 5.4.4 the first summand on the right side of (5.14) is equal to

$$\mathbf{E}\sum_n \mathbf{E}[(\triangle X_{S_n})^2|\mathcal{F}_{S_n-}]1_{S_n \leq T}.$$

From the above, it follows that

$$\langle X^d \rangle_t = \sum_n \mathbf{E}[(\triangle X_{S_n})^2|\mathcal{F}_{S_n-}]1_{S_n \leq t} + A_t. \tag{5.15}$$

We restate the above derivation in the form of the following assertion.

Lemma 5.5.1 (1) Suppose $X \in \mathcal{M}^2$. There exists a unique increasing strongly predictable process $\langle X \rangle \in \mathcal{A}$ such that $EX_T^2 = E\langle X \rangle_T$ for every s.t. T, or, equivalently, $X^2 - \langle X \rangle \in \mathcal{M}$, where $\langle X \rangle = \langle X^g \rangle + \langle X^c \rangle + \langle X^d \rangle$, where $\langle X^g \rangle$ and $\langle X^d \rangle$ are given by (5.13) and (5.15).

(2) If $X, Y \in \mathcal{M}^2$, then there exists a unique strongly predictable process $\langle X, Y \rangle \in \mathcal{A}$ such that $XY - \langle X, Y \rangle \in \mathcal{M}$, where

$$\langle X, Y \rangle = \frac{1}{2}[\langle X + Y \rangle - \langle X \rangle - \langle Y \rangle].$$

Proof. The proof of part (2) is carried out with the use of a so-called polarization identity. ∎

From (5.12) and (5.14) we have the following result.

Lemma 5.5.2 Suppose the process $X \in \mathcal{M}^2$ and the continuous component $X^c = 0$. Then, for every s.t. T

$$EX_T^2 = \mathbf{E}\left[\sum_{s \leq T}(\triangle X_s)^2 + \sum_{s < T}(\triangle^+ X_s)\right] \quad (\triangle X_0 = X_0).$$

Lemma 5.5.3 Suppose $X, Y \in \mathcal{M}^2$ and one of the processes does not have a continuous component. Then for every s.t. T

$$EX_TY_T = \mathbf{E}\left[\sum_{s \leq T}\triangle X_s \triangle Y_s + \sum_{s < T}\triangle^+ X_s \triangle^+ Y_s\right].$$

This means that the process

$$\left(X_t Y_t - \sum_{\leq st} \triangle X_s \triangle Y_s - \sum_{s < t} \triangle^+ X_s \triangle^+ Y_s\right) \in \mathcal{M}.$$

This process exits from zero, since $\triangle X_0 = X_0$ and $\triangle Y_0 = Y_0$.

Proof. If both processes do not have a continuous component, then the conclusion follows from Lemma 5.5.2 applied to the processes X, Y and $X + Y$. Now, suppose $Y^c = 0$ and X^c is a continuous component of X. Then by Theorem 5.4.11, $\mathbf{E} Y_T X_T^c = 0$ for every s.t. T. ∎

The previous lemmas lead to the intuitive definition of the increasing process $[X, X]$.

Definition 5.5.4 Suppose $X \in \mathcal{M}^2$ and X^c is its continuous component. Let

$$[X, X]_t = \langle X^c \rangle_t + \sum_{s \leq t} (\triangle X_s)^2 + \sum_{s < t} (\triangle^+ X_s)^2, \quad t \in \mathbb{R}_+.$$

The process $[X, X]$ is increasing, $\mathbf{F}$-adapted and by Theorem 5.5.2 is integrable.

If $X^c = 0$, then Lemma 5.5.3 tells us that the process $X^2 - [X, X] \in \mathcal{M}$. On the other hand, $(X^c)^2 - \langle X^c \rangle \in \mathcal{M}$. Since $X^c \perp X^g + X^d$, also in the general case we have $X^2 - [X, X] \in \mathcal{M}$. Now, setting

$$[X, Y] = \frac{1}{2}([X + Y, X + Y] - [X, X] - [Y, Y])$$

for $X, Y \in \mathcal{M}^2$, we obtain that $XY - [X, Y] \in \mathcal{M}$. Taking Definition 5.5.4 into account, we find that

$$[X, Y]_t = \langle X^c, Y^c \rangle_t + \sum_{s \leq t} \triangle X_s \triangle Y_s + \sum_{s < t} \triangle^+ X_s \triangle^+ Y_s.$$

Occasionally the process $[X, X]$ is written as $[X]$ and the process $\langle X \rangle$ is written $\langle X, X \rangle$.

5.5.2 Kunita-Watanabe inequalities

The increasing processes $\langle X, X \rangle$ and $[X, X]$ introduced above consist of parts continuous from the left and parts continuous from the right. We present and prove important inequalities, the Kunita-Watanabe inequalities for optional processes which relate quadratic variation of two processes to their individual quadratic variations.

Theorem 5.5.5 Kunita–Watanabe Inequalities. Suppose X, $Y \in \mathcal{M}^2$, and H and K are measurable random processes. Then, a.s. for every stopping time T in the broad sense,

$$\int_0^T |H||K||d\langle X, Y\rangle| \le \left(\int_0^T H^2 d\langle X\rangle\right)^{1/2} \left(\int_0^T K^2 d\langle Y\rangle\right)^{1/2},$$

and

$$\int_0^T |H||K||d[X, Y]| \le \left(\int_0^T H^2 d[X, X]\right)^{1/2} \left(\int_0^T K^2 d[Y, Y]\right)^{1/2}.$$

Proof. We establish only the first of the inequalities, since the second can be derived analogously. We follow the method proposed by P.A. Meyer [60].

The processes $\langle X, Y\rangle$, $\langle X\rangle$ and $\langle Y\rangle$ have a.s. not more than a countable number of jumps. We let Q denote the set consisting of the rational numbers, the jump times of the indicated processes not exceeding T and the time T.

(1) First we consider the components $\langle X^g, Y^g\rangle$, $\langle X^g\rangle$ and $\langle Y^g\rangle$. For $s < u$, $s, u \in Q$, and arbitrary λ the increment of the process $\langle X^g + \lambda Y^g\rangle$ on the interval $[s, u[$ satisfies $\langle X^g + \lambda Y^g\rangle_u - \langle X^g + \lambda Y^g\rangle_s \ge 0$ a.s. Hence, making the obvious simplification, we obtain

$$|\langle X^g, Y^g\rangle_s^u| \le (\langle X^g\rangle_s^u)^{1/2}(\langle Y\rangle_s^u)^{1/2} \quad \text{a.s.}$$

We choose an arbitrary collection of numbers from $Q : 0 = t_0 < t_1 < ... < t_n < t_{n+1} = T$, and bounded r.v. H_{t_i} and K_{t_i}, $i \le n$, and let

$$H_t = \sum_{i=0}^n H_{t_i} \mathbf{1}_{[t_i, t_{i+1}[}(t).$$

We define K_t analogously. We rewrite the preceding equality for $s = t_i$ and $u = t_{i+1}$, multiply by $|H_{t_i} K_{t_i}|$, sum over i and apply the Cauchy–Schwarz–Bunyakovsky inequality. We find

$$\left|\int_{[0,T[} H_t K_t d\langle X^g, Y^g\rangle_{t+}\right| \le \sum_{i=0}^n |H_{t_i} K_{t_j}||\langle X^g, Y^g\rangle_{t_i}^{t_{i+1}}| \tag{5.16}$$

$$\le \left(\int_{[0,T[} H_t^2 d\langle X^g\rangle_{t+}\right)\left(\int_{[0,T]} K_t^2 d\langle Y^g\rangle_{t+}\right)$$

For step processes we have obtained an auxiliary inequality.

Since such step processes form a σ-algebra of measurable sets on the space $\Omega \times \mathbb{R}_+$, with the help of the theorem on monotone classes, inequality (5.16) can be extended to all bounded measurable processes H and K.

(2) Suppose (R_t) is a measurable process with values in the set $\{-1, 1\}$ such that $|d\langle X^g, Y^g\rangle_{t+}| = R_t d\langle X^g, Y^g\rangle_{t+}$. Applying (5.16) to the processes (H_t) and $(K_t R_t)$, we introduce $|d\langle X^g, Y^g\rangle|$ on the left side in place of $d\langle X^g, Y^g\rangle$. Further, since H and K are arbitrary, we can substitute $|H|$ and $|K|$. Finally, if H and K are unbounded, then using the theorem on monotone limit transition in the inequality for truncated processes, we arrive at the inequality

$$\int_{[0,T[} |H_t||K_t||d\langle X^g, Y^g\rangle_{t+}| \le \left(\int_{[0,T[} H_t^2 dX^g\right)^{1/2}\left(\int_{[0,T[} K_t^2 d\langle Y^g\rangle_{t+}\right)^{1/2}. \tag{5.17}$$

(3) Now *we* need to repeat this reasoning for the processes $\langle X^r, Y^r \rangle$, $\langle X^r \rangle$ and $\langle Y^g \rangle$ ($X^r = X^c + X^d$; Y^g is defined analogously). We note that the increment of the process $\langle X^r + \lambda Y^r \rangle$ has to be considered on the interval $]s, u]$, and it is equal to

$$\langle X^r + \lambda Y^r \rangle_s^u = (X^r + \lambda Y^r)_u - \langle X^r + \lambda Y^r \rangle_s \geq 0.$$

Hence

$$|\langle X^r, Y^r \rangle_s^u| \leq (\langle X^r \rangle_s^u)^{1/2} (\langle Y_s^u \rangle)^{1/2},$$

and at zero $|\triangle \langle X^r, Y^r \rangle_0| = |X_0^r||Y_0^r|$. Here the step process H should be chosen in the form

$$H_t = H_0 1_{t=0} + \sum_{i=1}^{n} H_{t_i} 1_{]t_i, t_{t+1}]}(t),$$

and K_t is derived analogously. Proceeding with the reasoning we arrive at the inequality

$$\int_{[0,T]} |H_t||K_t| \, |d\langle X^r, Y^r \rangle_t| \leq \left(\int_{[0,T]} H_t^2 dX_t^r \right)^{1/2} \left(\int_{[0,T]} K_t^2 dY_t \right)^{1/2} \tag{5.18}$$

(4) Combining (5.17) and (5.18), applying the Cauchy-Schwarz-Bunyakovsky inequality on the right side and taking account of the notation introduced before the theorem, we obtain the first of the required inequalities. ∎

5.6 Optional Stochastic Integral

Here we study the properties of optional martingales, decomposition results, stochastic integrals with respect to optional processes and related results.

5.6.1 Integral with respect to square integrable martingales

The stochastic integral with respect to martingales plays an important role in applications of martingale theory. If the usual conditions are satisfied and if a martingale X has trajectories that are continuous from the right where $X \in \mathcal{M}^2$ and if the integrated process h is predictable such that $\mathbf{E}h^2 \circ \langle X \rangle_\infty < \infty$ then there exists a unique process $Y \in \mathcal{M}^2$ with trajectories continuous from the right such that $\langle Y, Z \rangle_T = h \circ \langle X, Z \rangle_T$ for all right-continuous martingales $Z \in \mathcal{M}^2$ for all s.t. T (see [60]). The process Y is called a stochastic integral and denoted as

$$Y = h \circ X = (h \circ X_t) = \left(\int_{[0,t]} h_s dX_s \right), \quad t \in \mathbb{R}_+.$$

In the case without the usual conditions, we like to define the stochastic integral with respect to an optional martingale X so that as a result we would also obtain an optional martingale. Suppose $X \in \mathcal{M}^2$ and taking into account the decomposition $X = X^g + X^c + X^d$ and the fact that according to what was said above the integral with respect to the right-continuous martingale $X^c + X^d$ is defined, we need only to define the integral for the component X^g.

Definition 5.6.1 Suppose the simple function h has the form

$$h_i = \sum_{j=0} h_{t_i} 1_{[t_i.t_{i+1}[}(t),$$

where h_{t_i} is a bounded $\mathcal{F}_{t_i}$-measurable r.v., $0 \leq i \leq n$, and the s.t. $0 = t_0 < t_1 < \ ... \ < t_n < t_{n+1} = \infty$ yield a partition of the line. Let

$$h \circ X_\infty^g = \sum_{i=0}^n h_{t_i} \left(X_{t_{t+1}}^g - X_{t_i}^g \right) = \int_{[0,\infty[} h_t dX_{t+}^g.$$

In particular, for every s.t. T

$$h \circ X_T^g = \sum_{i=0}^n h_{t_i} (X_{t_{i+1} \wedge T}^g - X_{t_i \wedge T}^g) = \int_{[0.T[} h_t dX_{t+}^g.$$

We call $h \circ X^g$ the stochastic integral of the function h with respect to X^g.

Lemma 5.6.2 Properties of the stochastic integral:
    (a) The process $h \circ X^g$ is continuous from the left and has limits from the right.
    (b) $h \circ X^g \in \mathcal{M}^2$.
    For every s.t.b. T the following assertions hold:
    (c) $\triangle^+ h \circ X_T^g = h_T \triangle^\div X_T^g$ a.s. on $(T < \infty)$.
    (d) $\mathbf{E} Z_T h \circ X_T^g = \mathbf{E} h \circ \langle Z^g, X^g \rangle_T = \mathbf{E} h \circ [Z^g, X^g]_T$ for $Z \in \mathcal{M}^2$.
    (e) $\langle Z, h \circ X^g \rangle = h \circ \langle Z^g, X^g \rangle$ and $[Z, h \circ X^g] = h \circ [Z^g, X^g]$, $Z \in \mathcal{M}^2$.

Proof. Properties (a) and (b) are obvious. Property (c), For $t_i \leq T < t_{i+1}$ we have

$$\triangle^+ h \circ X_T = h \circ X_{T+} - h \circ X_T = h_{t_i} (X_{T+}^g - X_T^g) = h_{t_i} \triangle^+ X_T^g.$$

Multiplying both sides by $1_{[t_i,t_{i+1}[}(T)$ summing and taking account of the form of the function h we obtain property (c).

Property (d), Given the martingale $Z = Z^c + Z^d + Z^g$; since the process $h \circ X^g \in \mathcal{M}^2$, it is orthogonal to Z^c and Z^d. Since $h \circ X_0^g = 0$, we have $\mathbf{E} Z_t h \circ X_T^g = \mathbf{E} Z_T^g h \circ X_T^g$ for all s.t. T. Further, omitting the index g from X^g and Z^g for the sake of simplicity, we have

$$\mathbf{E} Z_T h \circ X_T = \mathbf{E} \left[\sum_{k=0}^n (Z_{t_{k+1} \wedge T} - Z_{t_k \wedge T}) \right] \left[\sum_{i=0}^n h_{t_i} (X_{t_{i+1} \wedge T} - X_{t_i \wedge T}) \right]$$

For $s < u \leq t < v$

$$\mathbf{E}(Z_{v \wedge T} - Z_{t \wedge T})h_s(X_{u \wedge T} - X_{s \wedge T}) = \mathbf{E}(Z_{u \wedge T} - Z_{s \wedge T})h_t(X_{v \wedge T} - X_{t \wedge T}) = 0;$$

therefore

$$\mathbf{E}Z_T h \circ X_T = \mathbf{E} \sum_{i=0}^{n} h_{t_j}(Z_{t_{i+1} \wedge T-} - Z_{t_i \wedge T})(X_{t_{i+1} \wedge T} - X_{t_i \wedge T}) \qquad (5.19)$$

$$= \mathbf{E} \sum_{i=0}^{n} h_{t_t} \mathbf{E}[Z_{t_{i+1} \wedge T} X_{t_{i+1} \wedge T} - Z_{t_i \wedge T} X_{t_i \wedge T} | \mathcal{F}_{t_i}].$$

From (5.19) on the strength of Lemma 5.5.1 it follows that

$$\mathbf{E}Z_T h \circ X_T = \mathbf{E} \sum_{i=0}^{n} h_{t_i} \left(\langle Z, X \rangle_{t_{i+1} \wedge T} - \langle Z, X \rangle_{t_i \wedge T} \right) = \mathbf{E}h \circ \langle Z, X \rangle_T \qquad (5.20)$$

i.e., the first equality in (d). On the strength of Lemma 5.5.3 and Definition (5.5.4) from (5.19) we have

$$\mathbf{E}Z_T h \circ X_T = \mathbf{E} \sum_{i=0}^{n} h_{t_i}([Z, X]_{t_{i+1} \wedge T} - [Z, X]_{t_i \wedge T}) = \mathbf{E}h \circ [Z, X]_T \qquad (5.21)$$

i.e., the second equality in (d).

Property (e) follows from equalities (5.20) and (5.21), since the latter are satisfied for all s.t. T.
∎

We now extend the above definition of the stochastic integral from simple integrators to functions $\mathrm{L}^2(\langle X^g \rangle)$. Suppose Λ is the collection of optional bounded simple functions $h = (h_i)$ such that $h_i = h_{t_i}$ for $t \in [t_i, t_{i+1}[$, where the s.t.

$$0 = t_0 < t_1 < \ ... \ < t_n < t_{n+1} = \infty$$

assign a partition in $\mathbb{R}_+$. We let $\mathrm{L}^2(\langle X^g \rangle)$ denote the set of optional processes $h = (h_t)$ such that $\|h\|_{\mathrm{L}^2(\langle X^g \rangle)} = (\mathbf{E}h^2 \circ \langle X \rangle_\infty)^{1/2} < \infty$.

Theorem 5.6.3 For every function $h \in \mathrm{L}^2(\langle X^g \rangle)$ there exists a unique process $h \circ X^g$ having properties (a)-(e) of Lemma 5.6.2.

Proof. (1) *Existence and uniqueness.* The space $\mathrm{L}^2(\langle X^g \rangle)$ is the space L^2 on the product $\Omega \times \mathbb{R}_+$ with σ-family $\mathcal{O}(\mathbf{F})$ and measure $d\langle X^g \rangle d\mathbf{P}$. The set Λ is everywhere dense in $\mathrm{L}^2(\langle X^g \rangle)$. The definition of the stochastic integral $h \circ X^g$ for a simple function h establishes an isometric mapping of the subspace Λ into a subspace of $\mathcal{M}^2$. This mapping can be extended in a unique manner up to isometry between $\mathrm{L}^2(\langle X^g \rangle)$ and some subspace of $\mathcal{M}^2$. This latter means that to every function $h \in \mathrm{L}^2(\langle X^g \rangle)$ there corresponds a unique process $h \circ X^g \in \mathcal{M}^2$.

(2) *Properties.* Suppose the sequence of simple functions $(h(n)) \in \Lambda$ converges to $h \in \mathrm{L}^2(\langle X^g \rangle)$ and the sequence of martingales $(Y(n))$, $Y(n) = h(n) \circ X^g$, converges to $Y = h \circ X^g$ in $\mathcal{M}^2$. On

the basis of Doob's inequality (see Lemma 5.2.1 or [9]), for every $\epsilon > 0$

$$\mathbf{P}\left(\sup_t |Y_t(n) - Y_t| > \epsilon\right) \leq \epsilon^{-1} \|Y(n) - Y\|_{\mathcal{M}^2}$$
$$= \epsilon^{-1} \left(\mathbf{E}|h(n) - h) \circ X_\infty^g|^2\right)^{1/2}$$
$$= \epsilon^{-1} \|h(n) - h\|_{\mathrm{L}^2(\langle X^g \rangle)}.$$

The variable on the right vanishes. There exists a subsequence of indices (n') such that $Y(n') \to Y$ a.s. uniformly on $\mathbb{R}_+$. This implies property (a), and also for every s.t. T the convergence $\triangle^+ Y_T(n) \to \triangle^+ Y_T$ a.s. on $(T < \infty)$. On the other hand, we have the convergence $h_T(n) \triangle^+ X_T^g \to h_T \triangle^+ X_T^g$ a.s. on $(T < \infty)$. On the basis of property 5.6.2(c) for simple functions the left sides are equal in the last two relations, and therefore the rigt sides are also equal, which yields property 5.6.2(c) in the general case. We shall show why $h_T(n) \triangle^+ X_T^g \to h_T \triangle^+ X_T^g$. We have

$$\mathbf{E}|h_T(n) \triangle^+ X_T^g - h_T \triangle^+ X_T^g|^2 \mathbf{1}_{T<\infty} = \mathbf{E}|h_T(n) - h_T|^2 (\triangle^+ X)^2 \mathbf{1}_{T<\infty}$$
$$= \mathbf{E}|h_T(n) - h_T|^2 \mathbf{1}_{T<\infty} \triangle^+ \langle X^g \rangle_T$$
$$\leq \mathbf{E}|h(n) - h|^2 \circ \langle X^g \rangle_\infty \to 0.$$

Hence, passing to the subsequence, we obtain the required convergence.

We establish property (d) first for the s.t. $T = \infty$. The linear functional

$$h \to \mathbf{E}\left[Z_\infty h \circ X_\infty - h \circ \langle Z^g, X^g \rangle_\infty\right]$$

on $\mathrm{L}^2(\langle X_\infty \rangle)$ is continuous on account of the Kunita-Watanabe inequality of (Theorem 5.5.5); since it is equal to zero on Λ, it is equal to zero on all of $\mathrm{L}^2(\langle X_\infty \rangle)$.

For an arbitrary s.t. T we let $X^{gT} = (X_{t \wedge T}^g)$. Taking account of the equalities $\langle X^{gT} \rangle = \langle X^g \rangle^T$ and $\langle X^g \rangle_\infty^T = \langle X^g \rangle_T$ and substituting X^{gT} for X^g in the preceding functional, we obtain the first of the properties (d) for an arbitrary s.t. T. The second property (d) is a direct consequence of Lemma 5.5.3 and property 5.6.2(c).

The first equality of property 5.6.2(e), on the strength of Lemma 5.5.1, follows from the fact that $\langle Z, h \circ X^g \rangle = \langle Z^g, h \circ X^g \rangle$, the first of properties 5.6.2(d), and the fact that the process $h \circ \langle Z^g, X^g \rangle$ is strongly predictable. The second equality is easily derived from Definition 5.5.4 and property 5.6.2(c). ∎

5.6.2   Integral with respect to martingales with integrable variation

Now we consider the case of the martingale X having an integrable variation.

Theorem 5.6.4 Let $X = X^g \in \mathcal{M}^2 \cap \mathcal{A}$. Then for every $Y \in \mathcal{M}^2$

$$\mathbf{E}X_\infty Y_\infty = \mathbf{E}\sum_{t<\infty} \triangle^+ X_t \triangle^+ Y_t.$$

Proof. Applying Theorem 5.3.6 with the process of integrable variation $A = (X_{t+})$ and s.t. T, we find

$$\mathbf{E} \int_{]0.\infty[} Y_{t+} dX_{t+} = \mathbf{E} Y_\infty X_\infty.$$

On the other hand, on the strength of Theorem 5.3.7 and the definition of the process X^g in Remark 5.4.13 the martingale X generates a null measure on the optional σ-algebra $\mathcal{O}(\mathbf{F})$; therefore

$$\mathbf{E} \int_{[0.\infty[} Y_t dX_{t+} = 0.$$

Taking the difference of these two equalities, we obtain

$$\mathbf{E} Y_\infty X_\infty = \mathbf{E} \int_{[0.\infty[} \triangle^+ Y_t dX_{t+} = \mathbf{E} \sum_{t < \infty} \triangle^+ Y_t \triangle^+ X_t.$$

∎

If $X = X^g \in \mathcal{M} \cap \mathcal{A}$, the process $h = (h_t)$ is optional and $\mathbf{E} \int_{[0,\infty[} |h_t||dX_{t+}| < \infty$, then the stochastic *Stieltjes* integral $h^{(S)} \circ X$ is defined. On the strength of theorem 5.3.7 and by the definition of the martingale X^g in Remark 5.4.13 the process $h^{(S)} \circ X^g \in \mathcal{M} \cap \mathcal{A}$; its trajectories are continuous from the left and $\triangle^+ h^{(S)} \circ X_T^g = h_T \triangle^+ X_T^g$ a.s. on $(T < \infty)$ for all s.t.b. T.

We shall show that for the function $h \in L^2(\langle X^g \rangle)$ the Stieltjes integral coincides with the stochastic integral introduced earlier.

Theorem 5.6.5 Suppose $h \in L^2(\langle X^g \rangle)$ and $X = X^g \in \mathcal{M}^2 \cap \mathcal{A}$. Then the Stieltjes integral $h^{(S)} \circ X^g$ coincides with the stochastic integral $h \circ X^g$.

Proof. Suppose N_∞ is a bounded r.v. and $N = (N_t)$ is an optional modification of the martingale $(\mathbf{E}[N_\infty|\mathcal{F}_t])$. By Theorem 5.6.4

$$\mathbf{E} N_\infty h^{(S)} \circ X_\infty^g = \mathbf{E} \sum_{t < \infty} h_t \triangle^+ N_t \triangle^+ X_t^g.$$

On the other hand, according to Theorem 5.5.3,

$$\mathbf{E} N_\infty h \circ X_\infty^g = \mathbf{E} \sum_{t < \infty} h_t \triangle^+ N_t \triangle^+ X_t^g.$$

Comparing these two equalities, we obtain that for every bounded r.v. N_∞

$$\mathbf{E} N_\infty h^{(S)} \circ X_\infty^g = \mathbf{E} N_\infty h \circ X_\infty^g.$$

Consequently $h^{(S)} \circ X_\infty^g = h \circ X_\infty^g$ a.s. ∎

5.6.3 Integration with respect to local optional martingales

For the process $X \in \mathcal{M}_{loc}$ there exists a unique decomposition $X = X^c + X^d + X^g$, where $X^c + X^d$ is its r.c.l.l. component.

Definition 5.6.6 With the process $X \in \mathcal{M}_{loc}$ we associate the process $[X, X]$, setting

$$[X, X]_t = \langle X^c, X^c \rangle_t + \sum_{s \le t} (\triangle X_s)^2 + \sum_{s < t} (\triangle^+ X_s)^2, \quad t \in \mathbb{R}_+.$$

The sums on the right side are finite a.s. for all $t \in \mathbb{R}_+$. Indeed, on the basis of Lemma 5.4.18, $X = \tilde{X} + \check{X}$, where $\tilde{X} \in \mathcal{M}^2_{loc}$ and $\check{X} \in \mathcal{M}_{loc} \cap \mathcal{A}_{loc}$. Therefore,

$$\sum_{s \le t} (\triangle X_s)^2 + \sum_{s < t} (\triangle^+ X_s)^2 \le 2 \left[\sum_{s \le t} (\triangle \tilde{X}_s)^2 + (\triangle \check{X}_s)^2 + \sum_{s < t} (\triangle^+ \tilde{X}_s)^2 + (\triangle^+ \check{X}_s)^2 \right]$$

$$\le 2 \left[\sum_{s \le t} (\triangle \tilde{X}_s)^2 + \left(\sum_{s \le t} |\triangle \check{X}_s| \right)^2 + \sum_{s < t} (\triangle^+ \tilde{X}_s)^2 + \left(\sum_{s < t} |\triangle^+ \check{X}_s| \right)^2 \right] < \infty.$$

Thus the process $[X, X]$ is increasing and finite a.s. for all $t \in \mathbb{R}_+$. If $X, Y \in \mathcal{M}_{loc}$, we let

$$[X, Y]_t = \frac{1}{2}([X + Y, X + Y]_i - [X, Y]_t - [Y, Y]_t).$$

Taking consideration of Definition 5.6.6, we find that

$$[X, Y]_t = \langle X^c, Y^c \rangle_t + \sum_{s \le t} \triangle X_s \triangle Y_s + \sum_{s < t} \triangle^+ X_s \triangle^+ Y_s.$$

The process $[X, Y]$ has bounded variation a.s. up to every $t \in \mathbb{R}_+$.

Remark 5.6.7 The Kunita-Watanabe inequality (Theorem 5.5.5) for processes $[\cdot]$ extends which

$$\|X\|_{\mathrm{H}^{1,g}} = \mathbf{E}([X, X]^g_\infty)^{1/2} < \infty.$$

Theorem 5.6.8 The space $\mathrm{H}^{1,g}$ with the seminorm $\| \cdot \|_{\mathrm{H}^{1,g}}$ is a complete normed space.

Proof. The function $\| \cdot \|_{\mathrm{H}^{1,g}}$ is a seminorm; homogeneity is obvious, and the triangle inequality follows from the inequality

$$[X + Y, X + Y]^{1/2} < [X, X]^{1/2} + [Y, Y]^{1/2}$$

which in turn is derived with the help of the Kunita-Watanabe inequality, Theorem 5.5.5.

Completeness. Let II^1 denote the space of martingales Y such that $Y = (X_{t+}^g)$, $t \in \mathbb{R}_+$, $Y_{0-} = X_0^g$. Note that Y is r.c.l.l. martingale. Setting

$$\|Y\|_{\text{H}^1} = \mathbf{E}([Y,Y]_\infty)^{1/2},$$

we obtain that $\|Y\|_{\text{H}^1} = \|X\|_{\text{H}^{1,g}}$. In the spaces H^1 and $\text{H}^{1,g}$, we can introduce another norm, setting

$$\|Y\| = \mathbf{E}(\sup_{t<\infty} |Y_t|), \quad Y \in \text{H}^1 \quad \text{and}$$

$$\|X^g\| = \mathbf{E}(\sup_{t<\infty} |X_t^g|), \quad X^g \in \text{H}^{1,g}.$$

It is known that in H^1 the norms $\|\cdot\|$ and $\|\cdot\|_{\text{H}^1}$ are equivalent according to Meyer [59], Chapter V. Hence the space H^1 is complete and with the second norm ($\|\cdot\|$) it is the space $\text{L}^1(\Omega, \mathcal{F}_\infty^{\mathbf{P}}, \mathbf{P})$ of the r.v. $Y^* = \sup_t |Y_t|$. Hence we conclude that both norms ($\|\cdot\|$ and $\|\cdot\|_{\text{H}^{1,g}}$) are also equivalent in the space $\text{H}^{1,g}$, and it is also complete. ∎

Lemma 5.6.9 If $X = X^g \in \mathcal{M}_{loc}$ then $X \in \text{H}_{loc}^{1,g}$.

Proof. Suppose $(S_n, Y^{(n)})$ is a localizing sequence for X, where S_n is a s.t.b. and $Y^{(n)} \in \mathcal{M}$ for all n. Let

$$R_n = \inf(t \geq 0 : [X,X]_t > n).$$

For all n, R_n is a s.t.b., $R_n \uparrow \infty$ a.s. Let $T_n = S_n \wedge R_n$. The sequence $(T_n, Y^{(n)})$ is localizing for X. Since the process $[X,X]$ is continuous from the left, then $\mathbf{E}([X,X]_{R_n})^{1/2} \leq \sqrt{n}$. ∎

We proceed to the definition of the stochastic integral with respect to a local martingale. As in the class of square integrable martingales, it is sufficient to define the integral with respect to the process $X = X^g$.

Theorem 5.6.10 Suppose $X = X^g \in \mathcal{M}$, $X_0 = 0$, and the function h is optional and has the property that the process $(h^2 \circ [X,X]_t)^{1/2} \in \mathcal{A}_{loc}$. Then the following assertions are true:

(a) For every local martingale N the increasing process $\left(\int_{[0,t[} |h_s||d[X,N^g]_{s+}|\right)$ is finite for $t \in \mathbb{R}_+$.

(b) There exists a unique local martingale $Y = h \circ X (= Y^g)$ such that for every local martingale N

$$[Y,N] = h \circ [X, N^g]. \tag{5.22}$$

(c) $\wedge^+ Y_T = h_T \triangle^+ X_t$ a.s. on $(T < \infty)$ for all s.t.b. T.

Proof. First we establish *uniqueness*. Suppose Y^1 and Y^2 are two local martingales satisfying the theorem, and $D = Y^1 - Y^2$. Without resorting to localization, we assume that $D \in \mathcal{M}$. For every s.t. T we have $\triangle^+ D_T = 0$, and on the strength of (b) we have $[D,N]_T = 0$ for every local martingale N. Suppose that $D_T \neq 0$ for some s.t. T. We let N denote an optional modification of the martingale $(\mathbf{E}[\mathbf{1}_{D_T>0}|\mathcal{F}_t])$. For the bounded martingale $N, D \in \mathcal{M}$ and every s.t. T, on the strength of Theorem 5.6.4 we have $\mathbf{E}D_T N_T = \mathbf{E}[D, N]_T$. This equality leads to a contradiction, since the right side is equal to zero while the left is positive. Consequently, $D = 0$.

Assertion (a) follows from the Kunita-Watanabe inequality (see Remark 5.6.7)

$$\int_{[0,t[} |h_s| |d[X, N^g]_{s+}| \leq \left(\int_{[0,t[} h_s^2 d[X, X]_{s+} \right)^{1/2} [N^g, N^g]_t^{1/2}.$$

(b) Now we prove *existence*. Suppose that $\mathbf{E}(h^2 \circ [X, X]_\infty)^{1/2} < \infty$, and let $h(n) = h\mathbf{1}(|h| \leq n)$. We can define the stochastic integral for the function $h(n)$, letting

$$h(n) \circ X = h(n) \circ \tilde{X} + h(n) \circ \check{X},$$

where $X = \tilde{X} + \check{X}$, $\tilde{X} \in \mathcal{M}_{loc}^2$ and $\check{X} \in \mathcal{M}_{loc} \cap \mathcal{A}_{loc}$. From the properties of stochastic integrals we have

$$\mathbf{E}[(h(n) - h(m)) \circ X, (h(n) - h(m)) \circ X]_\infty = \mathbf{E}((h(n) - h(m))^2 \circ [X, X]_\infty)^{1/2}$$

By the *Lebesgue dominated convergence theorem* the last variable vanishes as $n, m \to \infty$. Thus the sequence of martingales $Y(n) = h(n) \circ X$ is Cauchy in $\mathrm{H}^{1,g}$. On account of Theorem 5.6.8, the space $\mathrm{H}^{1,g}$ is complete. Consequently, the limit $Y = h \circ X$ of the indicated sequence exists in it.

We shall show that

$$[Y, Y] = h^2 \circ [X, X]. \tag{5.23}$$

For $Y(n)$ and every s.t. T we have

$$[Y(n), Y(n)]_T = h^2(n) \circ [X, X]_T. \tag{5.24}$$

By the *theorem on monotone convergence*, $h^2(n) \circ [X, X] \uparrow h^2 \circ [X, X]$. On the other hand, by the triangle inequality, for every s.t. T

$$\begin{aligned} |[Y(n), Y(n)]_T^{1/2} - [Y, Y]_T^{1/2}| &\leq [Y(n) - Y, Y(n) - Y]_T^{1/2} \\ &\leq [Y(n) - Y, Y(n) - Y]_\infty^{1/2}. \end{aligned}$$

Since the right side vanishes in L^1, so does the left. Passing to subsequences (n') we obtain that $[Y(n'), Y(n')]_T \to [Y, Y]_T$ a.s. for every s.t. T. Therefore, passing to the limit in (5.24) with respect to the subsequence (n'), we obtain (5.23).

We now establish (5.22). Since $[Y(n), N]_T = h(n) \circ [X, N^g]_T$, on the set $\{T < \infty\}$ we have

$$[Y(n) - Y, N]_T + [Y, N]_T = (h(n) - h) \circ [X, N^g]_T + h \circ [X, N^g]_T. \tag{5.25}$$

By the Kunita-Watanabe inequality on the set $\{T < \infty\}$ we have

$$|(h(n) - h) \circ [X, N^g]_T| \leq ((h(n) - h)^2 \circ [X, X]_T)^{1/2} [N^g, N^g]_T^{1/2},$$

$$|[Y(n) - Y, N]_T| \leq ([Y(n) - Y, Y(n) - Y]_T)^{1/2} [N, N]_T^{1/2}.$$

The right sides of these inequalities vanish. Therefore, passing to the limit in (5.25), we obtain (5.22).

(c) For every s.t.b. T we have a.s.

$$\triangle^+ Y_T(n) = h_T(n) \triangle^+ X_T \tag{5.26}$$

on the set $\{T < \infty\}$. There exists a subsequence of indices (n') such that, for every s.t.b.T, $\triangle^+ Y_T(n') \to \triangle^+ Y_T$ and $h_T(n') \triangle^+ X_T \to h_T \triangle^+ X_T$ a.s. on the set $\{T < \infty\}$, since

$$\mathbf{E}\left[\sum_t (\triangle^+ Y_t(n) - \triangle^+ Y_t)^2\right]^{1/2} \to 0, \quad \mathbf{E}\left[\sum_t (h_t(n) - h_t)^2 (\triangle^+ X_t)^2\right]^{1/2} \to 0.$$

Therefore, passing to the limit in (5.26) we obtain assertion (c). ∎

As already noted, a local optional martingale $X = (X_t)$, $t \in \mathbb{R}_+$, decomposes uniquely into a sum $X = X^c + X^d + X^g$. The stochastic integral $f \cdot X^r$ with respect to the process $X^r = X^c + X^d$ is well defined for a predictable locally bounded process f. Furthermore, the stochastic integral with respect to X^g has also been defined for an optional $h \in \mathrm{H}^{1,g}$. Here we bring the two together in the form of the stochastic integral with respect to the local optional martingale X after we present some auxiliary results.

The next result from [61] and [6] has proved useful in order to define the stochastic integral $h \cdot X^d \in \mathcal{M}^d_{loc}$ with the property $\triangle h \cdot X^d = h \triangle X^d$.

Theorem 5.6.11 Let Y be an optional process. It is necessary and sufficient that (a) the predictable projection ${}^p Y$ be zero, and (b)

$$\left(\sum_{s \leq \cdot} Y_s^2\right)^{1/2} \in \mathcal{A}_{loc},$$

for the existence of a unique to within indistinguishability process $Z \in \mathcal{M}^d_{loc}$ with the property $\triangle Z = Y$.

Let us give an analogous result for the stochastic integral $h \cdot X^g \in \mathcal{M}^g_{loc}$ having the property $\triangle^+ h \cdot X^g = h \triangle^+ X^g$, where h is an optional process.

Theorem 5.6.12 Let Y be an $\mathcal{O}(\mathbf{F}_+)$-measurable process. For the existence of a unique to within indistinguishability process $Z \in \mathcal{M}^g_{loc}$ with the property $\triangle^+ Z = Y$ it is necessary and sufficient that (a) the $\mathcal{O}(F)$ projection ${}^o Y$ be equal to zero and (b)

$$\left(\sum_{s <} Y_s^2\right)^{1/2} \in \mathcal{A}_{loc}.$$

Proof. First we deal with *necessity*. The process $\triangle^+ Z$ is $\mathcal{O}(\mathbf{F}_+)$-measurable. If Z is an optional martingale, then $\mathbf{E}[\triangle^+ Z_T \mathbf{1}_{T<\infty}|\mathcal{F}_T] = 0$ a.s. for any $T \in \mathcal{T}$. If Z is a local optional martingale, then we denote by $(R_n, X^{(n)})$ its localizing sequence. We have

$$\mathbf{E}[\triangle^+ Z_T \mathbf{1}_{T \leq R_n}|F_T] = E[\triangle^+ X_T^{(n)} \mathbf{1}_{T \leq R_n}|F_T] = 0 \quad \text{a.s.}$$

for any $n \in \mathbb{N}$, $T \in \mathcal{T}$, since the set $(T \leq R_n) \in \mathcal{F}_T$. Hence, $\mathbf{E}[\triangle^+ Z_T \mathbf{1}_{\tau < \infty} | F_T] = 0$ a.s. for any $T \in \mathcal{T}$. But this means that $^{\circ}Y$ (see Section 5.2 or [4], [9]).

Condition (b) follows from the fact that, for $Z \in \mathcal{M}_{loc}^g$,

$$\left(\sum_{s < \cdot} (\triangle^+ Z_s)^2\right)^{1/2} = [Z, Z]^{1/2} \in \mathcal{A}_{loc}$$

(see Lemma 5.6.9 or [54], Lemma 7.10).

Now for *sufficiency*; from (b) it follows that the set $\{Y \neq 0\}$ is a rare event. By Theorem 5.1.12 there exist sequences $(S_n) \subset \mathcal{T}^p$, $(T_n) \subset \mathcal{T}^i$, $(U_n) \subset \mathcal{T}$, with disjoint graphs such that $\mathbf{P}(U_n = T < \infty) = 0$ for any n and $T \in \mathcal{T}$ and such that $\{Y \neq 0\} \subseteq \cup_n ([S_n] \cup [T_n] \cup [U_n])$.

Suppose that

$$\mathbf{E} \left(\sum_{s < \infty} Y_s^2\right)^{1/2} < \infty. \tag{5.27}$$

Introduce the processes

$$X_t^{1,m} = Y_{S_m} \mathbf{1}_{S_m < t}, \quad X_t^{2,m} = Y_{T_m} \mathbf{1}_{T_m < t}, \quad X_t^{3,m} = Y_{U_m} \mathbf{1}_{U_m < t} - {}^c X_t^m,$$

where $^c X_t^m$ is a compensating process, i.e., a strongly predictable process such that the difference in the right side of the last equality is a martingale (see [54], Theorem 3.5). The process $\left(\sum_{m=1}^n [X_t^{1,m} + X_t^{2,m} + X_t^{3,m}]\right)$ is a martingale in

$$\mathrm{H}^{1,g} = \left\{ X \in \mathcal{M}_{loc} : \mathbf{E} \left(\sum_{s < \infty} \triangle^+ X_s^2\right)^{1/2} < \infty \right\},$$

converging by (5.27) in $\mathrm{H}^{1,g}$ to some martingale $Z \in \mathrm{H}^{1,g}$ (see [54], Theorem 7.9). Passage from (5.27) to the general case (b) is done with the help of localization. ∎

Let us now derive a theorem giving definitions of the stochastic integral with respect to the *components* of an optional martingale.

Theorem 5.6.13 Let $X \in \mathcal{M}_{loc}$ and $X = X^c + X^d + X^g$ be its decomposition into continuous, right-continuous and left-continuous components. Let f be a predictable, h an optional function, and suppose they satisfy the conditions

$$(f^2 \cdot [X^r, X^r])^{1/2} \in \mathcal{A}_{loc}, \quad (h^2 \circ [X^g, X^g])^{1/2} \in \mathcal{A}_{loc},$$

where $X^r = X^c + X^d$, $[X^r, X^r]_t = \langle X^c, X^c \rangle_t + \sum_{s \leq t} (\triangle X_s)^2$, $[X^g, X^g]_t = \sum_{s < t} (\triangle^+ X_s)^2$. Then there exists a unique (to within indistinguishability) process $Y \in \mathcal{M}_{loc}$, which we can write as $Y = f \circ X^r + h \circ X^g$, possessing the properties $\triangle Y = f \triangle X$, $\triangle^+ Y = h \triangle^+ X$, $[Y, Z] = f \circ [X^r, Z^r] + h \circ [X^g, Z^g]$ for any $Z \in \mathcal{M}_{loc}$.

Proof. The existence and uniqueness of the processes $f \cdot X^d \in \mathcal{M}^d_{loc}$, $h \cdot X^g \in \mathcal{M}^g_{loc}$ follow from Theorems 5.6.11 and 5.6.12. The existence and uniqueness of the process $f \cdot X^c$ has been established in [62] (see, for example, [20, 63]). We assume that $Y = f \cdot X^c + f \cdot X^d + h \cdot X^g$. ∎

Let's summarize some results. A random process X is an **F**-martingale if it is **F**-adapted and, for any $s \leq t$, $\mathbf{E}|X_s| < \infty$, $\mathbf{E}[X_t|\mathcal{F}_s] = X_s$ a.s. A random process X is an optional **F**-martingale, $X \in \mathcal{M}$ (respectively, $X \in \mathcal{M}^2$) if X is an optional process and there exists an $\mathcal{F}_\infty$-measurable r.v. ξ with $\mathbf{E}|\xi| < \infty$ (respectively, $\mathbf{E}|\xi|^2 < \infty$) such that $X_T = \mathbf{E}[\xi|\mathcal{F}_T]$ a.s. for any bounded $T \in \mathcal{T}$. For optional martingales, **P**-almost all trajectories have left and right limits at each point t. An optional martingale X has a limit $\xi \in \mathrm{L}_1$ a.s. as $t \to \infty$.

We shall call the optional process X an optional local martingale and write $X \in \mathcal{M}_{loc}$ (respectively, $\mathcal{M}^2_{loc}$) if when X is localized it is an optional martingale (respectively, $\mathcal{M}^2$). A local optional martingale $X \in \mathcal{M}_{loc}$ can be decomposed into a sum $X = X^c + X^d + X^g$, where X^c, X^d, $X^g \in \mathcal{M}_{loc}$ the trajectories of X^c are continuous, X^d are right-continuous, and X^g are left-continuous. The processes X^d and X^g are orthogonal and to any continuous $Y \in \mathcal{M}_{loc}$. We denote by $\mathcal{M}^c_{loc}$ ($\mathcal{M}^d_{loc}$, $\mathcal{M}^g_{loc}$) the set of continuous (right-continuous, left-continuous) local martingales similarly $\mathcal{M}^{2,c}_{loc}$, $\mathcal{M}^{2,d}_{loc}$, $\mathcal{M}^{2,g}_{loc}$ for subsets of $\mathcal{M}^2_{loc}$.

We shall say that the process A is increasing if it is non-negative, its trajectories do not decrease and it is **F**-measurable. An increasing process A is integrable if $\mathbf{E}A_\infty < \infty$ and locally integrable if when localized it is integrable. A process of finite variation is a process having a.s. finite variation on any finite time segment. Increasing or a finite variation process A can be written in the form $A = A^c + A^d + A^g_t$, where A^c is a continuous process, A^d is right continuous and A^g is left continuous and $A^d = \sum_{s \leq .} \triangle A_s$ and $A^g = \sum_{s < .} \triangle^+ A_s$ where the series converge absolutely.

Chapter 6

Optional Supermartingales Decomposition

6.1 Introduction

The Doob decomposition for a supermartingale (X_n) is well known: $X_n = m_n - A_n$, where (m_n) is a martingale and (A_n) is an increasing process for which the variable A_n is $\mathcal{F}_{n-1}$-measurable for all n. This decomposition is unique to within a modification. Under the usual conditions for a r.c.l.l. supermartingale $X = (X_t)$, $t \in \mathbb{R}_+$ of class D (uniformly integrable) Meyer proved (see [55]) that the Doob decomposition is also valid: $X_t = m_t - A_t$, where (m_t) is a martingale and (A_t) is an increasing predictable process. Again, the decomposition is unique to within indistiguishability in the class of predictable processes A.

Here we will generalize the result of Doob-Meyer to the case where $X = (X_t, \mathcal{F}_t)$ is an optional supermartingale of the class D where we do not assume that the family $(\mathcal{F}_t)$ satisfies the usual conditions. We prove that there exist an optional martingale $m = (m_t, \mathcal{F}_t)$ and a strongly predictable increasing process $A = (A_t, \mathcal{F}_t)$ such that the Doob-Meyer decomposition $X_t = m_t - A_t$ is valid.

We begin by stating auxiliary definitions and proving supporting assertions. Recall that a process X is a *supermartingale* if X is integrable for every t and $X_s \geq \mathbf{E}[X_t|\mathcal{F}_s]$ ($X_s \leq \mathbf{E}[X_t|\mathcal{F}_s]$ for submartingale) a.s. for all $s \leq t$. On the other hand, we say that X is an *optional supermartingale* if $X \in \mathcal{O}(\mathbf{F})$, the random variables $X_T \mathbf{1}_{T<\infty}$ are integrable for all $T \in \mathcal{T}$ and there exists an integrable r.v. ξ such that $X_T \geq \mathbf{E}[\xi|\mathcal{F}_T]$ ($X_T \leq \mathbf{E}[\xi|\mathcal{F}_T]$ for submartingale) a.s. on $\{T < \infty\}$ for any $T \in \mathcal{T}$.

According to the definition of supermartingale, the class of supermartingales is sufficiently broad. Indeed, if X is an arbitrary supermartingale for which there is an integrable variable ξ such that $X_t \geq \mathbf{E}[\xi|\mathcal{F}_t]$ a.s. for any $t \in \mathbb{R}_+$. Then, according to [56], there exists a process X^* that is a modification of X and satisfies the conditions of the definition. Moreover, the optional sampling theorem holds for X^*; namely,

$$X_S^* \geq \mathbf{E}[X_T^*|\mathcal{F}_S] \quad \text{a.s.} \tag{6.1}$$

for any $S, T \in \mathcal{T}$ with $S \leq T$. We remark that for $T = \infty$ it is assumed that $X_\infty^* = \xi$. We remark also that a supermartingale X^* has a limit in L_1 as $t \to \infty$ (see [55], Chapter VI, Theorem 6), and it can be assumed without loss of generality to coincide with ξ.

Definition 6.1.1 A non-negative optional supermartingale $X = (X_t)$, $t \in \mathbb{R}_+$, will be called a *potential* if $\lim_{t \to \infty} \mathbf{E} X_t = 0$.

Definition 6.1.2 Let $X = (X_t)$, $t \in \mathbb{R}_+$, be an optional supermartingale. We say that X belongs to the class D if the family of r.v.s X_T, $T \in \mathcal{T}_+$, is uniformly integrable. Also, we say that X belongs to the class DL if the family of variables X_T, $T \in \mathcal{T}_+$, $T \leq a$, is uniformly integrable for any a, $0 \leq a < \infty$.

It follows easily from the definition of uniform integrability that the family X_{T+}, $T \in \mathcal{T}_+$ is uniformly integrable together with the family X_T, $T \in \mathcal{T}_+$.

6.2   Riesz Decomposition

Let $A = (A_t)$ and $B = (B_t)$, $t \in \mathbb{R}_+$, be two random processes. We say that B *majorizes* A (and write $A \leq B$) if outside some set of **P**-measure zero we have $A_t \leq B_t$ for any $t \in \mathbb{R}_+$.

We now give the Riesz decomposition theorem for supermartingales. Under the usual conditions, its proof can be found in [55]. Although the scheme of the proof in [55] is used here, the technical details that arise for optional supermartingales compel us to reproduce the proof in its entirety.

Theorem 6.2.1 Let $X = (X_t)$, $t \in \mathbb{R}_+$ be an optional supermartingale. Then, the following two properties are equivalent:

(a) X majorizes some optional submartingale.

(b) There exist an optional martingale Y and a potential Z such that $X = Y + Z$. This decomposition is unique (to within indistinguishability). The martingale Y majorizes each submartingale Y' satisfying the condition $Y' \leq X$.

Proof. Obviously, (b)$\Rightarrow$(a). We establish that (a)$\Rightarrow$(b). Suppose that X majorizes an optional submartingale Y'. For a fixed $S \in \mathcal{T}$ we introduce a family $\mathbf{G} = (\mathcal{G}_t)$, $t \in \mathbb{R}_+$, of σ-algebras by setting $\mathcal{G}_t = \mathcal{F}_S$ for any $t \in \mathbb{R}_+$. Let $(X_{S,t})$, $t \in \mathbb{R}_+$, be an $\mathcal{O}(\mathbf{G})$-measurable modification of the process $(\mathbf{E}[X_{S+t}|\mathcal{F}_S])$, $t \in \mathbb{R}_+$; such a process exists according to [4] and [9] or in section 5.2.

For a fixed $\mathbf{G}$-s.t. T we set

$$D = \{(\omega, u) : X_{S,T}(\omega) < X_{S,T+u}(\omega)\}.$$

by the section theorem (Theorem 37) in Chapter 1 of [4] and Theorem 59 in Chapter IV of [19], there exists a $\mathbf{G}$-s.t. U such that the graph $[\![U]\!]$ is contained in D (to within a negligible set) and the set $\{\omega : U(\omega) < \infty\}$ a.s. coincides with the projection $\pi(D)$ of D on Ω. We have

$$X_{S,T+U} = \mathbf{E}[X_{S+T+U}|\mathcal{F}_S] = \mathbf{E}\left[\mathbf{E}[X_{S+T+U}|\mathcal{F}_{S+T}]|\mathcal{F}_S\right].$$

By (6.1), the inner expectation on the right-hand side does not exceed X_{S+T}. Therefore,

$$X_{S,T+U} \leq \mathbf{E}[X_{S+T}|\mathcal{F}_S] = X_{S,T} \quad \text{a.s.}$$

which is possible only if the set D is negligible. Consequently, outside some set of **P**-measure zero the process $(X_{S,t})$, $t \in \mathbb{R}_+$, decreases with respect to t.

Suppose now that $D = \{(\omega, u) : Y'_{S,u}(\omega) > X_{S,u}(\omega)\}$. Then $D \in \mathcal{O}(\mathbf{G})$. Let T be an arbitrary **G**-s.t., again constructed by the section theorem. We have

$$X_{S,T} = \mathbf{E}[X_{S+T}|\mathcal{F}_S] \geq \mathbf{E}[Y'_{S+T}|\mathcal{F}_S] \geq Y'_S \quad \text{a.s.,}$$

which is possible only if D is a negligible set. Consequently, outside some set of **P**-measure zero we have $X_{S,T} \geq Y'_S$ for any **G**-s.t. T. Suppose that $\tilde{Y}_S = \lim_{t\to\infty} X_{S,t}$ a.s. Then $\tilde{Y}_S \geq Y'_S$ a.s. Further, for $S, T \in \mathcal{T}$

$$
\begin{aligned}
\mathbf{E}[\tilde{Y}_{S+T}|\mathcal{F}_S] &= \lim_{p\to\infty} \mathbf{E}[X_{S+T,p}|\mathcal{F}_S] = \lim_{p\to\infty} \mathbf{E}[\mathbf{E}[X_{S+T+p}|\mathcal{F}_{S+T}]|\mathcal{F}_S] \\
&= \lim_{p\to\infty} \mathbf{E}[X_{S+T+p}|\mathcal{F}_s].
\end{aligned}
\tag{6.2}
$$

On the one hand, continuing the right-hand side of (6.2) we get

$$\mathbf{E}[\tilde{Y}_{S+T}|\mathcal{F}_S] = \lim_{p\to\infty} X_{S,T+p} = \tilde{Y}_s \quad \text{a.s.}$$

The last relation shows that the process (Y_t), $t \in \mathbb{R}_+$, is a martingale with respect to $(\mathcal{F}_t)$. On the other hand, taking the limit on the right-hand side of (6.2) under the sign for the conditional expectation and considering that $X_t \to \tilde{X}$ in L_1 as $t \to \infty$, we have

$$E[\tilde{Y}_{S+T}|\mathcal{F}_S] = E[\tilde{X}|\mathcal{F}_S] \quad \text{a.s.}$$

Consequently, $\tilde{Y}_s = \mathbf{E}[\tilde{X}|\mathcal{F}_S]$ a.s.

Let us now take the martingale $Y = (Y_t)$ to be an optional modification of the process $(\mathbf{E}[\tilde{X}|\mathcal{F}_t])$, $t \in \mathbb{R}_+$, which exists, according to [4] and [9]. We have that $Y_T \geq Y'_T$ a.s. for any $T \in \mathcal{T}$. Since the processes Y and Y' are optional, the section theorem gives us that outside a set of **P**-measure zero $Y_t \geq Y'_t$ for all $t \in \mathbb{R}_+$.

Let $Z = X - Y$. Then $Z \geq 0$. We show that Z is a potential. From the definition of Z it follows that for any $T \in \mathcal{T}$

$$\lim_{p\to\infty} E[Z_{T+p}|\mathcal{F}_T] = 0 \text{ a.s.}$$

By Lebesgue's theorem, this implies that $\lim_{p\to\infty} \mathbf{E}Z_{T+p} = 0$. Consequently, Z is a potential.

Uniqueness. Suppose that there is another decomposition $X = \tilde{Y} + \tilde{Z}$ with $\tilde{Y}$ an optional martingale and $\tilde{Z}$ a potential. We have that $Y_{T+h} + Z_{T+h} = \tilde{Y}_{T+h} + \tilde{Z}_{T+h}$ for any $T \in \mathcal{T}$ and $h \in \mathbb{R}_+$. Taking the conditional expectation with respect to $\mathcal{F}_T$ in this equality and passing to the limit as $h \to \infty$, we get that $Y_T = \tilde{Y}_T$ a.s. for any $T \in \mathcal{T}$. Since both processes are optional, this implies, by section theorem, that $Y = \tilde{Y}$ outside a set of **P**-measure zero; and, consequently, $Z = \tilde{Z}$ outside some set of **P**-measure zero. ∎

6.3   Doob-Meyer-Galchuk Decomposition

We present an auxiliary result on variation of discrete supermartingales and extend it to continuous time optional supermartingale.

Definition 6.3.1 Let (X_n), $n \in \mathbb{N}$, be a sequence of random variables. The square variation of it is defined to be the r.v.

$$U = X_0^2 + \sum_n (X_{n+1} - X_n)^2.$$

Theorem 6.3.2 Let $(X_n, \mathcal{F}_n)$, $n \in \mathbb{N}$, be a non-negative finite supermartingale. Then its square variation is finite and:

(a) if the supermartingale (X_n) is bounded by a constant C, then $\mathbf{E}[U|\mathcal{F}_0] \leq 2CX_0$;

(b) in the general case $\mathbf{P}\left(U > a^2|\mathcal{F}_0\right) \leq 3X_0 a^{-1}$ for any $a \in \mathbb{R}_+$.

Proof. A proof of this theorem is given in [64]. ∎

We need a generalization of this theorem to the case of supermartingales with continuous time. Let

$$U = \sum_t [(\triangle X_t)^2 + (\triangle^+ X_t)^2],$$

where $\triangle X_t = X_t - X_{t-}$, $\triangle X_0 = X_0$ and $\triangle^+ X_t = X_{t+} - X_t$.

Theorem 6.3.3 Suppose that $X = (X_t)$, $t \in \mathbb{R}_+$, is a positive optional process such that $\mathbf{E}[X_t|\mathcal{F}_s] \leq X_s$ a.s. for any $s \leq t < \infty$.

(a) If $X \leq C$, then

$$\mathbf{E}[U|\mathcal{F}_0] \leq 2CX_0.$$

(b) In the general case $\mathbf{P}\left(U > a^2|\mathcal{F}_0\right) \leq 3X_0 a^{-1}$.

Proof. (a) Let (U_m), $m \in \mathbb{N}$, be a sequence of s.t.'s that make up a separability set for the process X (see [30]). Let $U_1, \ldots, U_n$ be the first n s.t.'s and $V_1 \leq \cdots \leq V_n$ the ordered s.t.'s constructed from $U_1, \ldots, U_n$, i.e.,

$$\bigcup_{m=1}^n [\![U_m]\!] = \bigcap_{m=1}^n [\![V_m]\!].$$

Let $V_0 = 0$. The sequence $(X_{V_m}, \mathcal{F}_{V_m})$, $m \in (0, \ldots, n)$, is a supermartingale. By Theorem 6.3.2, we have

$$\mathbf{E}\left[X_0^2 + \sum_{m=0}^{n-1} (X_{V_{m+1}} - X_{V_m})^2 | \mathcal{F}_0\right] \leq 2CX_0,$$

or, returning to the original s.t.,

$$\mathbf{E}\left[X_0^2 + \sum_{m,l}(X_{U_m} - X_{U_l})^2|\mathcal{F}_0\right] \leq 2CX_0,$$

where the differences on the left-hand side are taken over adjacent values of U_m and U_l, m, $l \in (0,\dots,n)$. Applying Fatou's lemma, we get the assertion (a).

(b) For arbitrary $c > 0$ we introduce the process $X \wedge c$, which is a supermartingale. Let U^c be the corresponding quantity analogous to U for X. Observe now that $U^c = U$ on the set $A = \{\sup_t X_t \leq c\}$. The set A is in $\mathcal{F}$. Indeed, $A = \Omega \setminus \pi(B)$, where $B = \{(\omega, t) : X_t(\omega) > c\} \in \mathcal{O}(\mathbf{F})$ and $\pi(B)$ is the projection of B on Ω; but $\pi(B) \in \mathbf{F}$ because the original probability space is complete (see [28]). Considering the foregoing, we have

$$\begin{aligned}
\mathbf{P}\left(U > a^2|\mathcal{F}_0\right) &\leq \mathbf{P}\left(\sup_t X_t > c|\mathcal{F}_0\right) + \mathbf{P}\left(A, U^c > a^2|\mathcal{F}_0\right) \\
&\leq \mathbf{P}\left(\sup_t X_t > c|\mathcal{F}_0\right) + \mathbf{P}\left(U^c > a^2|\mathcal{F}_0\right) \\
&\leq \mathbf{P}\left(\sup_t X_t > c|\mathcal{F}_0\right) + \frac{1}{a^2}\mathbf{E}\left[U^c|\mathcal{F}_0\right].
\end{aligned}$$

Observe now that

$$\mathbf{P}\left(\sup_t X_t > c|F\right) \leq \frac{1}{c}X_0$$

for non-negative supermartingales, as in the case of the "usual" conditions. Considering this inequality and (a), we get from the preceding chains of inequalities

$$\mathbf{P}\left(U > a^2|\mathcal{F}_0\right) \leq \frac{1}{c}X_0 + \frac{1}{a^2}2cX_0.$$

Setting $c = a$, we arrive at the required inequality. ∎

Theorem 6.3.4 Uniqueness. Let $X = (X_t)$, $t \in \mathbb{R}_+$, be an optional supermartingale of class D that admits a decomposition $X = Y - A$, where Y is an optional martingale, A is an increasing strongly predictable integrable process and $A_0 = 0$. Then this decomposition is unique to within negligibility.

Proof. Suppose that there is another decomposition $X = Y' - A'$ with a strongly predictable increasing process A', $A_0' = 0$, and an optional martingale Y'. For any finite s.t. S we have $\triangle^+(Y - Y')_S = \triangle^+(A - A')_S$. By strong predictability, the quantity on the right-hand side is $\mathcal{F}_S$-measurable; therefore,

$$\triangle^+(A - A')_S = \mathbf{E}[\triangle^+(Y - Y')_S|\mathcal{F}_S] = 0 \quad \text{a.s.}$$

for any finite s.t. S. Consequently, outside some set of $\mathbf{P}$-measure zero the process $A - A'$ is right-continuous. Since it is predictable, its jumps occur only at predictable s.t.'s. Suppose now that $S \in \mathcal{T}$ is a predictable s.t. We have

$$\triangle(A - A')_S = \mathbf{E}[\triangle(Y - Y')_S|\mathcal{F}_{S-}] = 0 \quad \text{a.s.}$$

Consequently, outside a set of **P**-measure zero the process $A - A'$ is left-continuous. Thus, the jumps of the processes A and A' coincide. Further, $Y - Y' = A - A'$. The process on the right-hand side is a continuous process of bounded variation that starts out from zero. Since the martingale $Y - Y'$ is on the left-hand side, $Y - Y' = const$ a.s. Considering that $Y_0 - Y_0' = 0$, we get $Y - Y' = A - A' = 0$ to within a set of **P**-measure zero. ∎

Theorem 6.3.5 Existence. Let $X = (X_t)$, $t \in R_+$, be an optional supermartingale of class D. There exist an increasing strongly predictable integrable process A, $A_0 = 0$, and an optional martingale Y such that $X = Y - A$.

The proof of the theorem consists of the sequence of lemmas given below. We remark that because of Riesz decomposition, it suffices for us to restrict ourselves to the case when X is a potential.

With a process X and a w.s.s.t. $T \in \mathcal{T}_+$ we associate the variable $U_T(X)$ by setting

$$U_T(X) = \sum_{0 \le t \le T} (\triangle X_t)^2 + \sum_{0 \le t < T} (\triangle^+ X_t)^2,$$

where $\triangle X_0 = X_0$. It is clear that $U_{T+}(X) = U_T(X) + (\triangle^+ X_T)^2$.

Lemma 6.3.6 Let X be an optional supermartingale of class DL for which there exists a sequence (R_n), $R_n \in \mathcal{T}_+$, $n \in N$, $R_n \uparrow \infty$ a.s., such that for any n

$$|\triangle X| \le C(n) \quad on \quad [\![0, R_n]\!], \quad |\triangle^+ X| \le C(n) \quad on \quad [\![0, R_n[\![,$$
$$U_{R_n}(X) \le C(n), \quad U_{R_n+}(X) \le C(n),$$

where $C(n)$ is a constant. Then there exist an optional local martingale $m \in \mathcal{M}_{loc}$ and a strongly predictable increasing process $A \in \mathcal{A}_{loc}$ such that $X = m - A$.

Proof. According to [54], there exist sequences (S_n), (T_n) and (U_n), with $S_n, T_n \in \mathcal{T}$ and $U_n \in \mathcal{T}_+$, that absorb the jumps of the process X. Moreover, S_n is a predictable s.t., T_n and U_n are totally inaccessible, and $\mathbf{P}(U_n = T < \infty) = 0$ for any $T \in \mathcal{T}$ and any n. The graphs of the s.t.'s in the sequences are disjoint. It will be assumed that $S_0 = 0$. According to Theorems 5.4.4-5.4.9 or 3.2-3.7 in [54], for any fixed N and any n we have for the processes

$$(\triangle X_{S_n} \mathbf{1}(S_n \le R_N \wedge t)), \quad (\triangle^+ X_{S_n} \mathbf{1}(S_n < R_N \wedge t)), \quad (\triangle^+ X_{T_n} \mathbf{1}(T_n < R_N \wedge t)),$$

$$(\triangle X_{T_n} \mathbf{1}(T_n \le R_N \wedge t)), \quad (\triangle^+ X_{U_n} \mathbf{1}(U_n < R_N \wedge t)), \quad t \in \mathbb{R}_+,$$

the Doob decompositions

$$\triangle X_{S_n} \mathbf{1}(S_n \le R_N \wedge t) = K_t^n(1, N) + a_t^n(1, N), \quad K_0^n(1, N) = 0,$$

$$\triangle^+ X_{S_n} \mathbf{1}(S_n < R_N \wedge t) = K_t^n(2, N) + a_t^n(2, N), \quad K_0^n(2, N) = 0,$$

$$\triangle^+ X_{T_n} \mathbf{1}(T_n < R_N \wedge t) = K_t^n(3, N) + a_t^n(3, N), \quad K_0^n(3, N) = 0, \quad (6.3)$$

$$\triangle X_{T_n}\mathbf{1}(T_n \leq R_N \wedge t) = K_t^n(4, N) + a_t^n(4, N), \quad K_0^n(4, N) = 0,$$

$$\triangle^+ X_{U_n}\mathbf{1}(U_n < R_N \wedge t) = K_t^n(5, N) + a_t^n(5, N), \quad K_0^n(5, N) = 0,$$

where $K^n(i, N)$, $i = 1, \ldots, 5$, are mutually orthogonal optional martingales, and the processes $a^n(i, N)$, $i = 1, \ldots, 5$, are strongly predictable and belong to the space $\mathcal{A}$ and $a^n(4, N)$ and $a^n(5, N)$ are continuous, while $a_0^n(2, N) = a_0^n(3, N) = 0$.

It is assumed that

$$\triangle X_{S_0}\mathbf{1}_{S_0 \leq t} = K_t^0(1, N), \quad K_t^0(1, N) = X_0.$$

These Doob decompositions are valid for all $t \in \mathbb{R}_+$ outside some set of **P**-measure zero. We remark that

$$K_t^n(1, N) = (\triangle X_{S_n} - \mathbf{E}[\triangle X_{S_n}|\mathcal{F}_{S_n-}])\mathbf{1}_{S_n \leq R_N \wedge t},$$

$$a_t^n(1, N) = \mathbf{E}[\triangle X_{S_n}|\mathcal{F}_{S_n-}]\mathbf{1}_{S_n \leq R_N \wedge t},$$

$$K_t^n(2, N) = (\triangle^+ X_{S_n} - \mathbf{E}[\triangle^+ X_{S_n}|\mathcal{F}_{S_n}])\mathbf{1}_{S_n < R_N \wedge t},$$

$$a_t^n(2, N) = \mathbf{E}[\triangle^+ X_{S_n}|\mathcal{F}_{S_n}]\mathbf{1}_{S_n < R_N \wedge t}$$

$$K_t^n(3, N) = (\triangle^+ X_{T_n} - \mathbf{E}[\triangle^+ X_{T_n}|\mathcal{F}_{T_n}])\mathbf{1}_{T_n < R_N \wedge t}$$

$$a_t^n(3, N) = \mathbf{E}[\triangle^+ X_{T_n}|\mathcal{F}_{T_n}]\mathbf{1}_{T_n < R_N \wedge t}.$$

By Remark 5.4.6 also ([54], Remark 3.4),

$$\mathbf{E}\,|K_\infty^n(4, N)|^2 = \mathbf{E}|\triangle X_{T_n}|^2\mathbf{1}_{T_n \leq R_N}.$$

Similarly, it can be shown that

$$\mathbf{E}|K_\infty^n(5, N)|^2 = \mathbf{E}|\triangle^+ X_{U_n}|^2\mathbf{1}_{U_n < R_N}.$$

From the form of the processes $K^n(i, N)$, $i = 1, 2, 3, \ldots$ it follows that

$$\mathbf{E}|K_\infty^n(1, N)|^2 \leq 2\mathbf{E}|\triangle X_{S_n}|^2\mathbf{1}_{S_n \leq R_N},$$

$$\mathbf{E}|K_\infty^n(2, N)|^2 \leq 2\mathbf{E}|\triangle^+ X_{S_n}|^2\mathbf{1}_{S_n < R_N},$$

$$\mathbf{E}|K_\infty^n(3, N)|^2 \leq 2\mathbf{E}|\triangle^+ X_{T_n}|^2\mathbf{1}_{T_n < R_N}.$$

From this, considering the hypothesis of the theorem, we get

$$\mathbf{E}\sum_n \sum_{t=1}^5 |K_\infty^n(i, N)|^2 \leq 2\mathbf{E}\sum_n \left[|\triangle X_{T_n}|^2\mathbf{1}_{T_n \leq R_N} + |\triangle^+ X_{U_n}|^2\mathbf{1}_{U_n < R_N} \right. \tag{6.4}$$

$$\left. +|\triangle X_{S_n}|^2\mathbf{1}_{S_n \leq R_N} + |\triangle^+ X_{S_n}|^2\mathbf{1}_{S_n < R_N} + |\triangle^+ X_{T_n}|^2\mathbf{1}_{T_n < R_N}\right]$$

$$= 2\mathbf{E}\left[U_{R_N}(X)\right] < 2C(N) < \infty.$$

Let H^2 be the space of optional martingales $Z = (\mathbf{E}[\xi|\mathcal{F}_t])$, $t \in \mathbb{R}_+$, $\mathbf{E}\xi^2 < \infty$. A norm $||\cdot||$ is introduced in H^2 by setting $||Z||^2 = \mathbf{E}\sup_t Z_t^2$. By Doob's inequality, $||Z||^2 \leq 4\mathbf{E}Z_\infty^2 = 4\mathbf{E}\xi^2$. The space H^2 is a complete normed space. Let

$$Y(k, N) = \sum_{n=0}^k \sum_{i=1}^5 K^n(i, N).$$

Then for $k < l$

$$\|Y(k,N) - Y(l,N)\| \le 2 \left(\mathbf{E} \sum_{n=k+1}^{l} \sum_{i=1}^{5} |K_\infty^n(i,N)|^2 \right)^{1/2},$$

which tends to zero as k, $l \to \infty$, by (6.4). Consequently, H^2 contains an optional martingale $m(N)$ to which the sequence $(Y(k,N))$, $k \in \mathbb{N}$, converges. It is easy to see that $m(N)$ coincides (to within indistinguishability) with an optional modification of the martingale

$$\mathbf{E} \left[\sum_n \sum_{i=1}^{5} K_\infty^n(i,N) | \mathcal{F}_i \right], \quad t \in \mathbb{R}_+,$$

where the series $\sum_n \sum_{i=1}^{5} K_\infty^n(i,N)$ converges in $\mathrm{L}^2(\Omega, \mathcal{F}_\infty, \mathbf{P})$. From the convergence of the sequence $(Y(k,N))$, $k \in \mathbb{N}$, to $m(N)$ in H^2 it follows that there exists a subsequence $(Y(k',N))$, $k' \in \mathbb{N}$, that converges to $m(N)$ a.s. uniformly on $\mathbb{R}_+$. From this it follows that $m(N)$ has jumps at the same times as X, and also

$$\triangle X_{T_n} \mathbf{1}_{T_n \le R_n} = \triangle m_{T_n}(N), \quad \triangle^+ X_{U_n} \mathbf{1}_{U_n < R_N} = \triangle^+ m_{U_n}$$

a.s. for any $n \in \mathbb{N}$ (the last two equalities hold because the jumps of the martingales $Y(k,N)$ at T_n and U_n, $n \le k$, coincide with the jumps of X at the same times).

By the uniqueness of the Doob decomposition (6.3), the processes $K^n(i,N)$ and $K^n(i,N+r)$, $i = 1, \ldots, 5$, coincide on $[\![0, R_N]\!]$ for any $r \ge 1$ and any $n \in \mathbb{N}$. But then the limit processes $m(N)$ and $m(N+r)$ also coincide on $[\![0, R_N]\!]$ for any $r \ge 1$. This implies that there is an optional local martingale $\tilde{m}$ such that $m(N) = \tilde{m}$ on $[\![0, R_N]\!]$ for any N. Moreover, the process $\tilde{m}$ has jumps at the same times as X, and

$$\triangle \tilde{m}_{T_n} = \triangle X_{T_n}, \quad \triangle^+ \tilde{m}_{U_n} = \triangle^+ X_{U_n} \qquad (6.5)$$

a.s. for any $n \in \mathbb{N}$.

Let $B = \tilde{m} - X$. Note that $B_0 = 0$. Indeed, for any N we have

$$\begin{aligned} B_0 &= \tilde{m}_0 - X_0 = \mathbf{E} \left[\sum_n \sum_{t=1}^{5} K_\infty^n(i,N) | \mathcal{F}_0 \right] - X_0 \\ &= \sum_n \sum_{t=1}^{5} K_0^n(i,N) - X_0 = K_0^0(1,N) - X_0 = X_0 - X_0 = 0. \end{aligned}$$

By (6.5), the jumps of the process B occur only at the times S_n and T_n, $n \in \mathbb{N}$, and for any n

$$\triangle B_{S_n} = -\mathbf{E}[\triangle X_{S_n} | \mathcal{F}_{S_n -}] \text{ a.s. on } \{S_n < \infty\},$$

$$\triangle^+ B_{S_n} = -\mathbf{E}[\triangle^+ X_{S_n} | \mathcal{F}_{S_n}] \text{ a.s. on } \{S_n < \infty\}, \qquad (6.6)$$

$$\triangle^+ B_{T_n} = -\mathbf{E}[\triangle^+ X_{T_n} | \mathcal{F}_{T_n}] \text{ a.s. on } \{T_n < \infty\}.$$

Since X is a supermartingale of the class DL, the jumps of B are non-negative and finite a.s. The process B is optional and its sample paths have one-sided limits a.s.

Let $\tau_N = \inf(t \geq 0 : |B_t| > N)$ with $\tau_N = \infty$ if the set $(\cdot)$ is empty. Then $\tau_N \in \mathcal{T}$ for any $N > 0$. Indeed, for any $t \in \mathbb{R}_+$.

$$\{\tau_N > t\} = \left\{ \bigcap_{r \leq t} \{|B_r| \leq N\} \right\} \bigcap_n \{|B_{S_n}| \leq N, S_n \leq t\} \bigcap_n \{|B_{S_n+}| \leq N, S_n \leq t\}$$

$$\bigcap_n \{|B_{T_n}| \leq N, T_n \leq t\} \bigcap_n \{|B_{T_n+}| \leq N, T_n \leq t\},$$

where r is a rational number. All the sets on the right-hand side are $\mathcal{F}_t$-measurable; therefore, $\{\tau_N > t\} \in \mathcal{F}_t$ and $\{\tau_N \leq t\} = \Omega \backslash \{\tau_N > t\} \in \mathcal{F}_t$. Moreover, $\tau_N \uparrow \infty$ a.s. as $N \to \infty$.

Suppose that the sequence (W_n), $W_n \in \mathcal{T}$, $n \in \mathbb{N}$, forms an optional separant set for the process B (see [30]). Fix some N and let $W^{nN} = W_n \wedge \tau_N$. For a fixed n let $V_1^{nN} \leq \cdots \leq V_n^{nN}$ be s.t.'s such that

$$\bigcup_{m=1}^{n} [\![W^{mN}]\!] = \bigcup_{m=1}^{n} [\![V_m^{nN}]\!].$$

In addition, let $V_0^{nN} = 0$ and $V_\infty^{nN} = \tau_N$. Then

$$\sum_i \mathbf{E}\left(B_{V_{i+1}^{nN}}^{\tau_N} - B_{V_i^{nN}}^{\tau_N}\right) \mathbf{1}(\tau_N \leq R_{N'}) = \mathbf{E} B_{\tau_N} \mathbf{1}(\tau_N \leq R_{N'}), \tag{6.7}$$

where (R_N) is the sequence in the lemma, N' is an index such that $\mathbf{P}(\tau_N > R_{N'}) < 2^{-N}$ and B^T is the process obtained from B by stopping at the s.t. T. We have

$$\mathbf{E} B_{\tau_N} \mathbf{1}(\tau_N \leq R_{N'}) = \mathbf{E} B_{\tau_N -} \mathbf{1}(\tau_N \leq R_{N'}) + \mathbf{E} \triangle B_{\tau_N} \mathbf{1}(\tau_N \leq R_{N'})$$
$$\leq N + C(N') < \infty.$$

For fixed N and N' the quantities $B_V^{\tau_N} \mathbf{1}(\tau_N \leq R_{N'})$, $V \in \mathcal{T}$ are uniformly integrable. Moreover, $\mathbf{P}$-a.s. the difference $B_{W_k} - B_W$ tends to 0, or to $\triangle^+ B_{T_r}$, or to $\triangle^+ B_{s_m}$, or to $\triangle B_{s_n}$, or to $B_{S_u+} - B_{S_u-} = \triangle^+ B_{s_u} + \triangle B_{s_u}$ for some r, m, n and u when $W_k - W_l \to 0$, $k, l \to \infty$.

Let us use Fatou's lemma in (6.7). We have

$$\sum_n \mathbf{E}\left[\triangle^+ B_{T_n}^{\tau_N} + \triangle^+ B_{S_N}^{\tau_N} + \triangle B_{S_N}^{\tau_N}\right] \mathbf{1}(\tau_N \leq R_{N'})$$

$$\leq \sum_i \liminf_{n \to \infty} \mathbf{E}\left(B_{V_{i+1}^{nM}}^{\tau_N} - B_{V_i^{nN}}^{\tau_N}\right) \mathbf{1}(\tau_N \leq R_{N'})$$

$$\leq \mathbf{E} B_{\tau_N} \mathbf{1}(\tau_N \leq R_{N'}) < \infty, n \to \infty.$$

Since N is arbitrary and $\tau_N \uparrow \infty$, $R_{N'} \uparrow \infty$ as $N, N' \to \infty$, the series jumps of the process B converges $\mathbf{P}$-a.s. to a finite quantity. Similarly, it can be shown that for any $S, T \in \mathcal{T}$, $S \leq T$,

$$\mathbf{E}\left[\sum_{S \leq T_n < T} \triangle^+ B_{T_n}^{\tau_N} + \sum_{S \leq S_n < T} \triangle^+ B_{S_n}^{\tau_N} + \sum_{S < S_n \leq T} \triangle B_{S_n}^{\tau_N}\right] \mathbf{1}(\tau_N \leq R_{N'})$$

$$\leq \mathbf{E}(B^\tau - B_s) \mathbf{1}(\tau_N \leq R_{N'}).$$

Let

$$\tilde{B}_t = B_t - \sum_n \left[\triangle^+ B_{T_n} \mathbf{1}_{T_n < t} + \triangle^+ B_{S_n} \mathbf{1}_{S_n < t} + \triangle B_{S_n} \mathbf{1}_{S_n \le t} \right].$$

The process $\tilde{B}$ is continuous **P**-a.s. Let $\sigma_n = \inf(t \ge 0 : \tilde{B}_t > n)$ with $\sigma_n = \infty$ if the set $(\cdot)$ is empty. We have $\sigma_n \uparrow \infty$ and $\sigma_n \in \mathcal{T}$ for any $n \in \mathbb{N}$. By the last inequality,

$$\mathbf{E} \left(\tilde{B}^{\sigma_n}_{T \wedge \tau_N} - \tilde{B}^{\sigma_n}_{S \wedge \tau_N} \right) \mathbf{1} \left(\tau_N \le R_{N'} \right) \ge 0$$

for any N and any $S, T \in \mathcal{T}$, $s \le T$. Since the process B^{σ_n} is bounded for fixed n, we can pass to the limit as N' and $N \to \infty$ in the preceding inequality, and this gives

$$\mathbf{E}(\tilde{B}^{\sigma_n}_T - \tilde{B}^{\sigma_n}_S) \ge 0$$

for any $S, T \in \mathcal{T}$, $S \le T$, and fixed $n \in \mathbb{N}$. Consequently, the process $\tilde{B}$ is a continuous local submartingale. We fix n and define a set function μ_n on the algebra of stochastic intervals of the form $]\!]S, T]\!]$, $S, T \in \mathcal{T}$, by letting

$$\mu_n(]\!]S, T]\!]) = \mathbf{E}(\tilde{B}^{\sigma_n}_T - \tilde{B}^{\sigma_n}_S).$$

Repeating the arguments of Section 5 in Chapter V of [28], we extend μ_n to a measure (with preservation of the notation) on the predictable σ-algebra $\mathcal{P}(\mathbf{F})$. To conclude the proof we need the following result. ∎

Lemma 6.3.7 For any $n \in \mathbb{N}$
  (a) $\mu_n([\![0]\!]) = 0$, $\mu_n(]\!][0, \infty]\!][) < \infty$, and
  (b) $\mu_n(H) = 0$ for any negligible $\mathcal{P}(\mathbf{F})$-measurable set H.

Proof. (a) $\mu_n([\![0]\!]) = 0$ from the definition of μ_n and

$$\mu_n(]\!][0, \infty]\!][) = \mathbf{E}(\tilde{B}^{\sigma_n}_\infty - \tilde{B}^{\sigma_n}_0) \le \mathbf{E}\tilde{B}_{\sigma_n} < \infty.$$

(b) We extend the measure μ_n to the σ-algebra $\mathcal{P}(\mathbf{F}_+)$ of predictable sets corresponding to the family $(\mathcal{F}^{\mathbf{P}}_{t+})$, where $\mathcal{F}^{\mathbf{P}}_{0+}$ is completed by including all **P**-null sets. It is known that for every set $H' \in \mathcal{P}(\mathbf{F}_+)$ there is a set $H \in \mathcal{P}(\mathbf{F})$ such that these sets are indistinguishable (see [19]). Therefore, we assume that $\mu(H') = \mu(H)$, $H' \in \mathcal{P}(\mathbf{F}_+)$, $H \in \mathcal{P}(\mathbf{F})$. It has been shown that $\mu_n(H') = 0$ for negligible sets $H' \in \mathcal{P}(\mathbf{F}_+)$ (see [28], p. 117). Consequently, $\mu_n(H) = 0$ also for negligible $H \in \mathcal{P}(\mathbf{F})$, since the sets in $\mathcal{P}(\mathbf{F})$ belong to $\mathcal{P}(\mathbf{F}_+)$. ∎

It can next be proved, as in Theorem 49 of Chapter V in [28], that the measure μ_n is generated by a unique $\mathcal{P}(\mathbf{F})$-measurable increasing process $B(n)$. In a standard way (for example, look at the dual projections) it can be verified that the process $B(n)$ is continuous. Thus, for $S \le T$, $S, T \in \mathcal{T}$,

$$\mathbf{E}[B_T(n) - B_S(n)] = \mu_n(]\!]S, T]\!]) = \mathbf{E}[\tilde{B}^{\sigma_n}_{T^n} - \tilde{B}^{\sigma_n}_S].$$

Since for $S \le T \le \sigma_n$ the measures μ_n and μ_{n+k} coincide on $]\!]S, T]\!]$, the uniqueness of the processes $B(n)$ and $B(n+k)$ implies that there is a continuous increasing process B' such that $B'_S = B_S(n)$ for $S \le \sigma_n$ and for any n.

Letting $S = t_H$, $H \in \mathcal{F}_t$ for any $t \in \mathbb{R}_+$, and $T = \infty$, and fixing n, we get from the preceding equality that

$$\mathbf{E}\mathbf{1}_H \cdot (B_\infty'^{\sigma_n} - B_t'^{\sigma_n}) = \mathbf{E}\mathbf{1}_H \cdot (\tilde{B}_\infty^{\sigma_n} - \tilde{B}_t^{\sigma_n}).$$

This means that

$$\mathbf{E}(\tilde{B}_\infty^{\sigma_n} - B_\infty'^{\sigma_n} | \mathcal{F}_t) = \tilde{B}_t^{\sigma_n} - B_t'^{\sigma_n}$$

a.s. or that $\tilde{B}^{\sigma_n} - B'^{\sigma_n}$, is an optional martingale. Since $\sigma_n \uparrow \infty$ a.s., the process $\tilde{B} - B'$ is a continuous local martingale. Recalling now the definitions of the processes B, $\tilde{B}$, and B', we have

$$\tilde{B} - B' + B' + \sum_l \left[\triangle^+ B_{T_n} \mathbf{1}_{T_n <\cdot} + \triangle^+ B_{S_n} \mathbf{1}_{S_n <\cdot} + \triangle B_{S_n} \mathbf{1}_{S_n \leq \cdot}\right] = \tilde{Y} - X.$$

From this, setting

$$A = B' + \sum_n \left[\triangle^+ B_{Tn} \mathbf{1}_{T_n <\cdot} + \triangle^+ B_{S_n} \mathbf{1}_{S_n <\cdot} + \triangle B_{S_n} \mathbf{1}_{S_n \leq \cdot}\right], \quad Y = \tilde{Y} + B' - \tilde{B},$$

we get the decomposition $X = Y - A$, where Y is an optional local martingale and A is a strongly predictable increasing process for which there is a sequence (R_N), $R_N \in \mathcal{T}_+$, $R_N \uparrow \infty$ a.s. such that $A\mathbf{1}_{\cdot \leq R_N}$ is an integrable process.

Let (S_n), (T_n) and (U_n) be the sequences of s.t.'s in Lemma 6.3.6 that absorb the jumps of the process X. We set

$$p_n = \triangle X_{S_n} \mathbf{1}(|\triangle X_{S_n}| > 1, S_n < \infty), \quad \hat{p}_n = \mathbf{E}[p_n | \mathcal{F}_{S_n -}],$$

$$y_n = \triangle^+ X_{S_n} \mathbf{1}(|\triangle^+ X_{S_n}| > 1, S_n < \infty), \quad \hat{y}_n = \mathbf{E}[y_n | \mathcal{F}_{S_n}],$$

$$z_n = \triangle^+ X_{T_n} \mathbf{1}(|\triangle^+ X_{T_n},| > 1, T_n < \infty), \quad \hat{z}_n = \mathbf{E}[z_n | \mathcal{F}_{T_n}],$$

$$v_n = \triangle X_{T_n} \mathbf{1}(|\triangle X_{T_n}| > 1, T_n < \infty), \quad w_n = \triangle^+ X_{U_n} \mathbf{1}(|\triangle^+ X_{U_n},| > 1, U_n < \infty).$$

Lemma 6.3.8 Let X a supermartingale of the class DL.

(1) The conditional expectations $\hat{p}_n$, $\hat{y}_n$, $\hat{z}_n$, $n \in \mathbb{N}$, are defined, and the processes

$$P = \sum_n p_n \mathbf{1}_{S_n \leq t}, \quad \hat{P} = \sum_n \hat{p}_n \mathbf{1}_{S_n \leq t}, \quad Y = \sum_n y_n \mathbf{1}_{S_n < t},$$

$$\hat{Y} = \sum_n \hat{y}_n \mathbf{1}_{S_n < t}, \quad Z = \sum_n z_n \mathbf{1}_{T_n < t}, \quad \hat{z} = \sum_n \hat{z}_n \mathbf{1}_{T_n < t}$$

belong to $\mathcal{V}_{loc}$.

(2) The processes $\hat{P}$, $\hat{Y}$, and $\hat{Z}$ are strongly predictable, and there exists a sequence (V_n), $V_n \in \mathcal{T}$, $n \in \mathbb{N}$, $V_n \uparrow \infty$ a.s., such that

$$\sum_k |p_k| \mathbf{1}(S_k \leq t \wedge V_n) \leq n, \quad \sum_k |\hat{y}_k| \mathbf{1}(S_k < t \wedge V_n) \leq n,$$

$$\sum_k |\hat{z}_k| \mathbf{1}(T_k < t \wedge V_n) \leq n$$

for any $t \in \mathbb{R}_+$ and $n \in \mathbb{N}$.

(3) $\tilde{P} = P - \hat{P} \in \mathcal{M}_{loc}$, $Y = Y - \hat{Y} \in \mathcal{M}_{loc}$ and $\tilde{Z} = Z - \hat{Z} \in \mathcal{M}_{loc}$, and the processes $\hat{P}$, $\hat{Y}$ and $\hat{Z}$ are unique (to within indistinguishability) among the strongly predictable processes having these properties.

Proof. These lemmas are proved in [54] (see Lemmas 5.4.17 and 5.4.18). ∎

Lemma 6.3.9 Let X be a supermartingale of the class DL. Then the processes

$$V = \left(\sum_k v_k \mathbf{1}(T_k \le t) \right), \quad W = \left(\sum_k w_k \mathbf{1}(U_k < t) \right), \quad t \in \mathbb{R}_+,$$

belong to $\mathcal{V}_{loc}$. There exist unique strongly predictable processes $\hat{V} \in \mathcal{V}_{loc}$ and $\hat{W} \in \mathcal{V}_{loc}$ such that $\tilde{V} = V - \hat{V} \in \mathcal{M}_{loc}$ and $\tilde{W} = W - \hat{W} \in \mathcal{M}_{loc}$. Moreover, $\hat{V}$ and W are continuous processes.

Proof. These lemmas are proved in Lemmas 5.4.17 and 5.4.18 (see also, [54] Lemmas 4.7 and 4.8). ∎

We remark that it is assumed in [54] that the process X is in $\mathcal{M}_{loc}$, but the proofs use only the fact that the jumps of X are locally integrable. The last condition is satisfied under our assumptions.

Lemma 6.3.10 Let X be a supermartingale of the class DL. Let

$$Q = X - (\tilde{P} + \tilde{Y} + \tilde{Z} + \tilde{V} + \tilde{W}),$$

where the optional local martingales $\tilde{P}$, $\tilde{Y}$, $\tilde{Z}$, $\tilde{V}$ and $\tilde{W}$ are from Lemmas 6.3.8 and 6.3.9. There exists a sequence (V_n), $V_n \in \mathcal{T}$, $V_n \uparrow \infty$ a.s., such that the stopped supermartingale Q^{V_n} has bounded jumps, i.e., $|\triangle Q^{V_n}| \le n+1$ and $|\triangle^+ Q^{V_n}| \le n+1$ for any $n \in \mathbb{N}$.

Proof. By the definitions of the processes $\tilde{P}$, $\tilde{Y}$, $\tilde{Z}$, $\tilde{V}$ and $\tilde{W}$, the jumps of Q have the form

$$\triangle Q_{S_n} = \triangle X_{S_n} \mathbf{1}(|\triangle X_{S_n}| \le 1) + \hat{p}_n \quad on \quad \{S_n < \infty\},$$

$$\triangle^+ Q_{S_n} = \triangle^+ X_{S_n} \mathbf{1}(|\triangle^+ X_{S_n}| \le 1) + \hat{y}_n \quad on \quad \{S_n < \infty\},$$

$$\triangle^+ Q_{T_n} = \triangle^+ X_{T_n} \mathbf{1}(|\triangle^+ X_{T_n}| \le 1) + \hat{z}_n \quad on \quad \{T_n < \infty\}, \tag{6.8}$$

$$\triangle Q_{T_n} = \triangle X_{T_n} \mathbf{1}(|\triangle X_{T_n}| \le 1) \quad on \quad \{T_n < \infty\},$$

$$\triangle^+ Q_{U_n} = \triangle^+ X_{U_n} \mathbf{1}(|\triangle^+ X_{U_n}| \le 1) \quad on \quad \{U_n < \infty\}.$$

By Lemma 6.3.8, there exists a sequence (V_n), $V_n \in \mathcal{T}$, $V_n \uparrow \infty$ a.s., such that for any fixed $n \in \mathbb{N}$

$$|\hat{p}_k| \mathbf{1}(S_k \le V_n) \le n, \quad |\hat{y}_k| \mathbf{1}(S_k < V_n) \le n, \quad |\hat{z}_k| \mathbf{1}(T_k < V_n) \le n$$

for arbitrary $k \in \mathbb{N}$. From these inequalities and (6.8) it follows that $|\triangle Q^{V_n}| \le n+1$ and $|\triangle^+ Q^{V_n}| \le n+1$ for any $n \in \mathbb{N}$. ∎

Lemma 6.3.11 Let X be a potential in the class DL and (V_n) a sequence of s.t.'s, and suppose that Q is the process in Lemma 6.3.10. There exists a sequence (R_n), $R_n \in \mathcal{T}_+$, $R_n \uparrow \infty$ a.s., such that

$$U_{R_n}(Q^{V_n}) \le C(n), \quad U_{R_n+}(Q^{V_n}) \le C(n),$$

where $C(n)$ is a constant.

Proof. By Theorem 6.3.3, the variable $U(X) = U_\infty(X)$ is finite a.s. Further, the variable $U(\tilde{P} + \tilde{Y} + \tilde{Z} + \tilde{V} + \tilde{W})$ is also finite a.s. Consequently, $U(Q) < \infty$ a.s. Let

$$R_n = \inf(t \ge 0 : U_t(Q) > n),$$

with $R_n = \infty$ if the set $(\cdot)$ is empty. Then $R_n \in \mathcal{T}_+$ for any $n \in \mathbb{N}$, and $R_n \uparrow \infty$ a.s. We have

$$U_{R_n}(Q^{V_n}) \le n + (\triangle Q^{V_n}_{R_n})^2 \le n + (n+1)^2,$$

$$U_{R_n+}(Q^{V_n}) \le n + (\triangle Q^{V_n}_{R_n})^2 + (\triangle^+ Q^{V_n}_{R_n})^2 \le n + 2(n+1)^2 = C(n).$$

∎

Lemma 6.3.12 Let X be a potential of the class D. Then there exist an optional martingale m and a strongly predictable increasing integrable process A, $A_0 = 0$, such that $X = m - A$.

Proof. By Lemmas 6.3.10 and 6.3.11, the process $Q = X - \tilde{P} - \tilde{Y} - \tilde{Z} - \tilde{V} - \tilde{W}$ is a supermartingale of class DL for which there exist sequences (V_n), $V_n \in \mathcal{T}$, and (R_n), $R_n \in \mathcal{T}_+$, $V_n \uparrow \infty$, $R_n \uparrow \infty$ a.s. as $n \to \infty$, such that $|\triangle Q^{V_n}| \le n + 1$, $|\triangle^+ Q^{V_n}| \le n + 1$ on $[\![0, R_n]\!]$ and $U_{R_n}(Q) \le C(n)$, $U_{R_n+}(Q) \le C(n)$. By Lemma 6.3.6, $Q = Y - A$, where $Y \in \mathcal{M}_{loc}$ and A is a strongly predictable increasing process with $A \in \mathcal{V}_{loc}$, $A_0 = 0$. Consequently

$$X = m - A \tag{6.9}$$

where $m = \tilde{Y} + \tilde{P} + \tilde{Y} + \tilde{Z} + \tilde{V} + \tilde{W}$.

We show that m is an optional martingale, while the process A is integrable. Let $\tilde{A}_t = \sum_n \triangle^+ A_{T_n} \mathbf{1}_{T_n < t}$, $t > 0$, and $\tilde{A}_0 = 0$. The process $\tilde{A}$ increases and is left-continuous. Let $\sigma_n = \inf(t > 0 : \tilde{A}_t > n)$, with $\sigma_n = \infty$ if the set $(\cdot)$ is empty. It is easy to see that $\sigma_n \in \mathcal{T}$ and $\sigma_n \uparrow \infty$ a.s. Let (R_n), $R_n \in \mathcal{T}_+$ be the sequence indicated above. Without changing the notation we assume that (R_n) simultaneously localizes the m in class D. Considering that $\mathbf{E}[\triangle^+ m_{\sigma_n} | \mathcal{F}_{\sigma_n}] \mathbf{1}_{\sigma_n \le R_{n'}} = 0$ for any n, $n' \in \mathbb{N}$. we get from (6.9) that

$$\mathbf{E} \triangle^+ X_{\sigma_n} \mathbf{1}_{\sigma_n \le R_{n'}} = -\mathbf{E} \triangle^+ A_{\sigma_n} \mathbf{1}_{\sigma_n \le R_{n'}}.$$

Since the variables $\triangle^+ X_{\sigma_n} \mathbf{1}_{\sigma_n \le R_{n'}}$ are integrable uniformly with respect to n', and $\triangle^+ A_{\sigma_n} \mathbf{1}_{\sigma_n \le R_{n'}}$, $\triangle^+ A_{\sigma_n} \mathbf{1}_{\sigma_n < \infty}$ as $n' \to \infty$ the Lebesgue theorem on passage to the limit as $n' \to \infty$ can be applied in the last equality. We have

$$\mathbf{E} \triangle^+ X_{\sigma_n} = -\mathbf{E} \triangle^+ A_{a_n},$$

i.e., the jump is integrable. Consequently, the increasing process $\tilde{A}^{\sigma_n}$ is integrable for any $n \in \mathbb{N}$.

Suppose now that $\check{A} = A - \tilde{A}$. The process $\check{A}$ is increasing and predictable, with sample paths that are right-continuous and have limits from the left. According to [57], there exists a sequence

(τ_n), $\tau_n \in \mathcal{T}$, $\tau_n \uparrow \infty$ a.s., such that the stopped process $\check{A}^{\tau_n}$ is bounded for any $n \in \mathbb{N}$. Let $\delta_n = \tau_n \wedge \sigma_n$. Then the stopped process A^{δ_n} is integrable (the variable A_{δ_n+} is also integrable) for any $n \in \mathbb{N}$.

Since X is in the class D and the process A^{δ_n} is integrable, it is not hard to see that the stopped process $m^{\delta_n} = X^{\delta_n} + A^{\delta_n}$ is an optional martingale for any $n \in \mathbb{N}$. We now have that

$$\mathbf{E}X_{\delta_n} = \mathbf{E}m_{\delta_n} - \mathbf{E}A_{\delta_n} = \mathbf{E}X_0 - \mathbf{E}A_{\delta_n},$$

because $\mathbf{E}m_{\delta_n} = \mathbf{E}m_0 = \mathbf{E}X_0$ (by the definition of m). Let us take the limits of the left-hand and right-hand sides of this equality as $n \to \infty$. Since the process (X_t) has a limit X_∞ in L_1 as $t \to \infty$, we get $\mathbf{E}X_\infty = \mathbf{E}X_0 - \mathbf{E}A_\infty \to \infty$. Consequently, the process A is integrable. Then the process m is an optional martingale. $\blacksquare$

6.3.1   Decomposition of DL class

Theorem 6.3.13 An optional supermartingale $X = (X_t)$, $t \in \mathbb{R}_+$, admits a decomposition $X = Y - A$ where $Y \in \mathcal{M}_{loc}$ and A is an increasing strongly predictable locally integrable process with $A_0 = 0$ if and only if X belongs to the class DL. This decomposition is unique to within indistinguishability.

Proof. The proof repeats almost word-for-word the proof of the analogous statement in [55] (see Chapter VII, Theorem 31). $\blacksquare$

For optional submartingale decomposition, results can be obtained by the same line of reasoning presented above.

Chapter 7

Calculus of Optional Semimartingales

In this chapter we shall consider integration with respect to optional semimartingales and formula of change of variables. Further, we study random measures generated by jumps of the form $\triangle X_t = X_t - X_{t-}$, $\triangle^+ X_t = X_{t+} - X_t$ of the process X, define the compensators of these measures, and construct stochastic integrals over semimartingales random measures. As an application of stochastic integrals of optional semimartingales, we study an analogue of Dolean's equation.

7.1 Integral with Respect to Optional Semimartingales

Definition 7.1.1 The random process $X = (X_t)$, $t \in \mathbb{R}_+$, is called an (optional) semimartingale if it is representable in the form

$$X = M + A, \quad M \in \mathcal{M}_{loc}, \quad A \in \mathcal{V}, \quad M_0 = 0. \tag{7.1}$$

A semimartingale X is called special if there exists a representation (7.1) with a strongly predictable process $A \in \mathcal{A}_{loc}$.

Let us agree to denote by $\mathcal{S}$ (respectively, $\mathcal{S}_s$) the set of all (respectively, special) semimartingales. For an n-dimensional process $X = (X^1, ..., X^n)$ the notation $X \in y$ means that $X^i \in \mathcal{S}$, $i = 1, ..., n$.

The next result is an analogue of an assertion in [63] when the usual conditions are fulfilled.

Lemma 7.1.2 Let $X \in \mathcal{S}$. Then the following assertions are equivalent:

(a) There exists a decomposition (7.1) with strongly predictable process $A \in \mathcal{A}_{loc}$, and this decomposition is unique.

(b) There exists a decomposition (7.1) with $A \in \mathcal{A}_{loc}$.

(c) For any decomposition (7.1), $A \in \mathcal{A}_{loc}$.

(d) The increasing process $X^* \in \mathcal{A}_{loc}$, where $X_t^* = \sup_{s \le t} |X_s|$.

Proof. (c)⇒(b) is obvious.

(b)⇒(a): Let $A \in \mathcal{A}_{loc}$. We have $A = A^r + A^g$, A^r, $A^g \in \mathcal{A}_{loc}$. For A^r there exists a unique process $A^{rp} \in \mathcal{P} \cap \mathcal{A}_{loc}$ for which $A^r - A^{rp} \in \mathcal{M}_{loc}$ (see [63]). According to Lemma 5.3.11, there exists for A^g a unique strongly predictable process $A^{go} \in \mathcal{A}_{loc}$ such that $A^g - A^{go} \in \mathcal{M}^g_{loc}$. Thus we have arrived at a new decomposition

$$X = M' + A' \text{ with } M' = M + (A^r - A^{rp}) + (A^g - A^{go}) \in \mathcal{M}_{loc},$$

$$A' = A^{rp} + A^{go} \in \mathcal{A}_{loc}.$$

Let us establish the uniqueness. Assume one has the decompositions $X = M' + A' = M + A$, $M'_0 = M_0 = A'_0 - A_0 = 0$, with strongly predictable processes A', $A \in \mathcal{A}_{loc}$. Put $m = M - M'$, $a = A' - A$. Then $m = a$. Since a is strongly predictable, its jumps have the form $\triangle a_S$, $\triangle^+ a_S$, $\triangle^+ a_T$, where $S \in \mathcal{T}^p$, $T \in \mathcal{T}$. We have

$$\triangle m_S = \triangle a_S, \quad \triangle^+ m_S = \triangle^+ a_S \text{ on } (S < \infty)$$

$$\triangle^+ m_T = \triangle^+ a_T \text{ on } (T < \infty) \text{ for } S \in \mathcal{T}^p, T \in \mathcal{T}.$$

In these equalities we take the conditional expectations with respect to $\mathcal{F}_{S-}$, $\mathcal{F}_S$, $\mathcal{F}_T$, respectively. Noting that

$$\mathbf{E}[\triangle m_S | \mathcal{F}_{S-}] = \mathbf{E}[\triangle^+ m_S | \mathcal{F}_S] = 0 \text{ a.s.on } (S < \infty) \text{ and}$$

$$\mathbf{E}[\triangle^+ m_T | \mathcal{F}_T] = 0 \text{ a.s. on } (T < \infty) \text{ for } S \in \mathcal{T}^p, T \in \mathcal{T},$$

we get $\triangle a_S = \triangle^+ a_S = 0$ a.s. on $(S < \infty)$, $\triangle^+ a_T = 0$ a.s. on $(T < \infty)$, for any $S \in \mathcal{T}^p$, $T \in \mathcal{T}$. With the aid of the theorem on sections, it is deduced that a is a continuous process. From the equality $m = a$ it follows that m is a local martingale of bounded variation. So $m = $ const. But m goes out from zero, so $m = 0$.

(a)⇒(d). We have $X = M + A$, where M, M^r, $M^g \in \mathcal{M}_{loc}$, A, A^r, $A^g \in \mathcal{A}_{loc}$. From elementary inequalities we get that

$$X^*_t = \sup_{s \leq t} |M_s + A_s| \leq \sup_{s \leq t} |M_s| + \sup_{s \leq t} |A_s| \leq \sup_{s \leq \iota} |M^r_s| + \sup_{s \leq t} |M^g_s|$$

$$+ \sup_{s \leq t} |A^r_s| + \sup_{s \leq t} |A^g_s| = M^{r*}_t + M^{g*}_t + A^{r*}_t + A^{g*}_t$$

The processes $M^{r*} \in \mathcal{A}_{loc}$ (see [63]), $M^{g*} \in \mathcal{A}_{loc}$ (see [54], Theorem 7.9). Since A^r, $A^g \in \mathcal{A}_{loc}$, we know that A^{r*}, $A^{g*} \in \mathcal{A}_{loc}$. Hence, $X^* \in \mathcal{A}_{loc}$.

(d)⇒(c). Let $X = M + A$ be an arbitrary decomposition. Set $T_n = \inf\{t > 0 : \mathbf{Var}_{[0,t]}(A) > n\}$. Let $(R_n) \in T_+$ be the localizing sequence for M and X^* Set $\tau_n = T_n \wedge R_n$. We have $\tau_n \in \mathcal{T}_+$, $\tau_n \uparrow \infty$ a.s. Further,

$$\mathbf{Var}_{[0,\tau_n]}(A) \leq n + |\triangle A_{\tau_n}| + |\triangle^+ A_{\tau_n}|.$$

From the triangle inequality we have, for any n,

$$|\triangle A_{\tau_n}| \leq |\triangle M_{\tau_n}| + |\triangle M_{\tau_n} + \triangle A_{\tau_n}| = |\triangle M_{\tau_n}| + \triangle X^*_{\tau_n},$$

where the quantity on the right is integrable. It is shown similarly that $\triangle^+ A_\tau$ is integrable. Consequently, $\mathbf{EVar}_{[0,\tau_n]}(A) < \infty$ for any n. ∎

The optional stochastic integral with respect to optional semimartingales X is defined in terms of the stochastic integrals with respect to its components A and m as

$$
\begin{aligned}
Y_t &= \int_0^t h_s dX_s = h \circ X_t = \int_0^t h_s dA_s + \int_0^t h_s dm_s \\
&= h \circ A_t + h \circ m_t = h \circ A_t^r + h \circ A_t^g + h \circ m_t^r + h \circ m_t^g,
\end{aligned}
$$

where

$$
h \circ A_t^r = \int_{0+}^t h_{s-} dA_s^r, \quad h \circ A_t^g = \int_0^{t-} h_s dA_{s+}^g,
$$

$$
h \circ m_t^r = \int_{0+}^t h_{s-} dm_s^r, \quad h \circ m_t^g = \int_0^{t-} h_s dm_{s+}^g.
$$

Also, since $X = X^r + X^g = X^c + X^d + X^g$ we can write the integral with respect to X as

$$
h \circ X = h \circ X^r + h \circ X^g = h \circ X^c + h \circ X^d + h \circ X^g
$$

where $X^r = A^r + m^r = A^c + m^c + A^d + m^d$ and $X^g = A^g + m^g$.

Extensions of the above integral to a larger class of integrands is given by the bilinear form $(f, g) \circ X_t$,

$$
Y_t = (f, g) \circ X_t = f \circ X_t^r + g \circ X_t^g,
$$

where Y_t is again an optional semimartingale $f_- \in \mathcal{P}(\mathbf{F})$, and $g \in \mathcal{O}(\mathbf{F})$. Hence, for the stochastic integral with respect to optional semimartingales, the space of integrands is the product space of predictable and optional processes, $\mathcal{P}(\mathbf{F}) \times \mathcal{O}(\mathbf{F})$. Further, extension is possible in terms of all the components of the

$$
Y_t = \phi \circ A_t^c + \varphi \circ A_t^d + \psi \circ A_t^g + f \circ m_t^c + g \circ m_t^d + h \circ m_t^g,
$$

where $(f, g, h) \in \mathcal{P}(\mathbf{F}) \times \mathcal{P}(\mathbf{F}) \times \mathcal{O}(\mathbf{F})$ and the processes ϕ, φ, and ψ are $\mathcal{B}(\mathbb{R}_+) \times \mathcal{F}$ measurable.

7.2 Formula for Change of Variables

Theorem 7.2.1 Suppose X is an n-dimensional semimartingale, i.e., $X = (X^1, ..., X^n)$, $X^k = X_0^k + A^k + m^k \in \mathcal{S}$, $k = 1, ..., n$, and $F(x) = F(x^1, ..., x^n)$ is a twice continuously differentiable function on $\mathbb{R}^n$. Then the process $F(x) \in \mathcal{S}$, and for all $t \in \mathbb{R}_+$

$$
\begin{aligned}
F(X_t) \;=\; & F(X_0) + \sum_{k=1}^{n} \int_{]0,t]} D^k F(X_{s-}) d(A^{kr} + m^{kr})_s \\
& + \frac{1}{2} \sum_{k,l=1}^{n} \int_{]0,t]} D^k D^l F(X_{s-}) d\langle m^{kc}, m^{lc}\rangle_s \qquad (7.2) \\
& + \sum_{0<s\leq t} \left[F(X_s) - F(X_{s-}) - \sum_{k=1}^{n} D^k F(X_{s-}) \triangle X_s^k \right] \qquad (7.3) \\
& + \sum_{k=1}^{n} \int_{[0,t[} D^k F(X_s) d(A^{kg} + m^{kg})_{s+} \\
& + \sum_{0\leq s<t} \left[F(X_{s+}) - F(X_s) - \sum_{k=1}^{n} D^k F(X_s) \triangle^+ X_s^k \right],
\end{aligned}
$$

where D^k is the differentiation operator with respect to the kth coordinate, and the process $m^r = m^c + m^d$ (A^r is defined analogously).

Proof. We carry out the proof by the method of Meyer (see [62]). It consists of several lemmas, and we restrict consideration to the case $n = 1$. First we explain why the summands on the right side of (7.3) are meaningful.

The first two integrals are defined because the functions under the integral are locally bounded, and as localizing sequence we can take the sequence of s.t.b. (σ_n), where $\sigma_n = \inf(t : |X_t| \geq n)$, $n \in \mathbb{N}$.

Consider the summand $\sum_{s\leq t}[\cdot]$. Outside some set of **P**-measure zero the process X is finite and has one-sided limits for all $t \in \mathbb{R}_+$. Therefore, for every interval $[0, T]$ there is a variable $K_t(\omega) > 0$ such that $|X_s(\omega)| \leq K_t(\omega)$ for $s \leq t$. Then, by the continuity of the second derivative F'', some variable $C_t(\omega) > 0$ is uniformly bounded on $[-K(\omega), K(\omega)]$. Now applying Taylor's formula, we obtain

$$ |F(X_s) - F(X_{s-}) - F'(X_{s-}) \triangle X_s| \leq \frac{1}{2} C_t(\omega) [\triangle X_s]^2 \text{ for all } s \leq t. $$

Thus

$$ \sum_{s\leq t} |[\cdot]| \leq \frac{1}{2} C_t \sum_{s\leq t} (\triangle X_s)^2 \leq C_t \left[\sum_{s\leq t} (\triangle m_s)^2 + \left(\sum_{s\leq t} |\triangle A_s| \right)^2 \right] < \infty $$

for all $t \in \mathbb{R}_+$. From the absolute convergence of $\sum_{s\leq t}[\cdot]$ it follows that the corresponding process $\left(\sum_{s\leq t}[\cdot] \right)$, $t \in \mathbb{R}_+$, is continuous from the right and has limits from the left.

It can be established analogously that the second sum of the form $\sum_{s\leq t}[\cdot]$ in (7.3) a.s. converges absolutely for every $t \in \mathbb{R}_+$ and determines a process continuous from the left and having limits from the right.

Finally, we consider the summand

$$ \int_{[0,t[} F'(X_s) d(A^g + m^g)_{s+} = \int_{[0,t[} F'(X_s) dA^g_{s+} + \int_{[0,t[} F'(X_s) dm^g_{s+}. $$

Let $\tau_n = \inf(t > 0 : [m^g, m^g]_t > n)$ and $R_n = \sigma_n \wedge \tau_n$, where σ_n was introduced above. Then

$$\mathbf{E} \left[\int_{[0, R_n[} |F'(X_s)|^2 d[m^g, m^g]_{s+} \right]^{1/2} \leq Cn,$$

where $C = \max |F'(x)|$ for $x \in [-n, n]$. Consequently, by Theorem 5.6.10 the stochastic integral $F' \circ m^g \in \mathcal{M}_{loc}$ is defined. On the strength of the local boundedness of F' the Lebesgue-Stieltjes integral $F' \circ A^g$ is defined. Thus the summand $F \circ (A^g + m^g)$ is meaningful.

Using the properties of the integrals involved on the right side of (7.3), it is easy to verify that the jumps $\triangle F$ and $\triangle^+ F$ of the left side and the corresponding jumps of the right side coincide. ∎

The next lemma establishes that if a change of variable formula is true for a semimartingale that is the sum of a square integrable martingale and integrable finite variation process, then it is true for the sum of local optional martingale and finite variation process of Theorem 7.2.1.

Lemma 7.2.2 If Equation (7.3) holds for the semimartingale $X = X_0 + A + m$, where X_0 is an integrable $\mathcal{F}_0$-measurable r.v., $A \in \mathcal{A}$, $m \in \mathcal{M}^2$ and $m_0 = A_0 = 0$, then it also holds under the conditions of Theorem 7.2.1.

Proof. Suppose $X = X_0 + A + m$ in Theorem 7.2.1. On the basis of Lemma 5.4.19, $m = \tilde{m} + \check{m}$, where $\tilde{m} \in \mathcal{M}_{loc}^2$ and $\check{m} \in \mathcal{M} \cap \mathcal{A}_{loc}$. Therefore, combining $\check{m}$ and A and not changing notation, we can assume that $X = X_0 + A + m$, where $A \in \mathcal{V}$ and $m \in \mathcal{M}_{loc}^2$. Let $\delta_n = 0$ if $|X_0| > n$, $\delta_n = \infty$ if $|X_0| \leq n$, and

$$\rho_n = \inf \left(t : \int_{[0,t]} |dA_s^r| \geq n \right), \quad \sigma_n = \inf \left(t : \int_{[0,t[} |dA_{s+}^g| > n \right).$$

For all $n \in \mathbb{N}$ the variables δ_n and ρ_n are s.t., and σ_n is a s.t.b. Let

$$A^{g,n} = \begin{cases} A^g & \text{on } (t \leq \sigma_n), \\ A_{\sigma_n -}^g & \text{on } (t > \sigma_n), \end{cases} \quad A^{r,n} = \begin{cases} A^r & \text{on } (t < \rho_n), \\ A_{\rho_n -}^r & \text{on } (t \geq \rho_n). \end{cases}$$

Then the process $A^{(n)} = A^{g,n} + A^{r,n}$ is $\mathcal{F}$-adapted, belongs to $\mathcal{A}$, and coincides with A on $[0, \rho_n \wedge \sigma_n]$.

Suppose $(\tau_n, Y^{(n)})$ is a localizing sequence for m. Let $R_n = \tau_n \wedge \sigma_n \wedge \rho_n \wedge \delta_n$ and $X^{(n)} = X_0 \mathbf{1}_{R_n > 0} + A^{(n)} + Y^{(n)}$. We have that $R_n \uparrow \infty$ a.s., the process $X^{(n)}$ satisfies the conditions of the lemma and $X^{(n)} = X$ on $[0, R_n]$. By hypothesis, (7.3) holds for $X^{(n)}$ for all $n \in \mathbb{N}$. Consequently it holds for X for $t < R_n$. Since the jumps $\triangle F$ and $\triangle^+ F$ of the left side at time R_n coincide with the analogous jumps of the right side, (7.3) extends to $t = R_n$. We repeat the reason: for $R_n < t < R_{n+1}$. Taking into account that $R_n \uparrow \infty$, a.s. we obtain (7.3) on the entire line. ∎

For a function of bounded first and second derivatives, the following lemma proves the change of variable formula for a sum of integrable finite variation process and square integrable martingale. This lemma is a precursor to the previous lemma.

Lemma 7.2.3 If Equation (7.3) holds for the semimartingale $X = X_0 + A + m$, where $A \in \mathcal{A}$, $m \in \mathcal{M}^2$ and X_0 is an integrable r.v., and a function F having two continuous bounded derivatives, then it also holds for a function F is twice continuously differentiable.

Proof. Let $\sigma_n = \inf(t : |X_t| \geq n)$. For every $n \in \mathbb{N}$, σ_n is a s.t.b. and $\sigma_n \uparrow \infty$ a.s. for $0 \leq t < \sigma_n$ we have $|X_t| \leq n$. Let G denote a function having two continuous bounded derivatives and coincide with F on $[-n, n]$.

Formula (7.3) holds for G. For $t < \sigma_n$ we substitute F for G in (7.3). This can be done for the following reason: in the stochastic integrals with respect to m^{kr} the functions under the integrals depend on X_-, and consequently $F'(X_-) = G'(X_-)$ on every predictable set $[0, \sigma_n]$; the integrals with respect to A^{kr}, $\langle \cdot, \cdot \rangle$ and A^{kg} are Lebesgue-Stieltjes integrals and we can make the indicated substitution in them; the stochastic integrals with respect to m^{kg} are defined for $F'(X)\mathbf{1}_{[0,\sigma_n[}$. Further, since the jumps of the right and left sides coincide, formula (7.3) for F extends to $t = \sigma_n$. Since $\sigma_n \uparrow \infty$ a.s., we obtain (7.3) for every $t \in \mathbb{R}_+$. ∎

Suppose (R_n) consists of the s.t.b. S_n, T_n and U_n from Theorem 5.1.10 renumbered into one sequence. The sequence (R_n) absorbs all the jumps of the process $X = A + m$, where $A \in \mathcal{A}$ and $m \in \mathcal{M}^2$. Let

$$\begin{aligned} K_t(n) &= \triangle m_{R_n}\mathbf{1}_{R_n \leq t}, \quad H(n) = \triangle^+ m_{R_n}\mathbf{1}_{R_n < t}, \\ L_t(n) &= \triangle A_{R_n}\mathbf{1}_{R_n \leq t}, \quad M_t(n) = \triangle^+ A_{R_n}\mathbf{1}_{R_n < t}. \end{aligned}$$

For every n, we let $(\,^c K_t(n)) \in \mathcal{M}^2$ and $(\,^c H_t(n)) \in \mathcal{M}^2$, $t \in \mathbb{R}_+$, denote the martingales obtained from $(K_t(n))$ and $(H_t(n))$ by compensating in accordance with Theorems 5.4.4-5.4.9.

Lemma 7.2.4 Suppose the function F has two continuous bounded derivatives. If (7.3) holds for this function and the semimartingale $X = X_0 + A + m$, where $A \in \mathcal{A}$ and $m \in \mathcal{M}^2$ having a finite number of jumps, then it also holds in the case where there is no restriction on the number of jumps.

Proof. Suppose (R_n) is a s.t.b. absorbing the jumps of X. Let

$$X^n = X_0 + A^c + m^c + \sum_{i=1}^{n} [\,^c K(i) + \,^c H(i) + L(i) + M(i)].$$

On the strength of Theorem 5.4.11, the processes $\sum_{i=1}^{n} {}^c K(i)$ and $\sum_{i=1}^{n} {}^c H(i)$ converge in probability uniformly on $\mathbb{R}_+$ to m^d and m^g, respectively. The process $\sum_{i=1}^{n} L(i)$ and $\sum_{i=1}^{n} M(i)$ converge a.s. in variation to A^d and A^g, respectively. Consequently, the process X^n converges in probability to X uniformly on $\mathbb{R}_+$.

By hypothesis, Equation (7.3) holds for the semimartingale X^n. From our reasoning, it follows that $F(X^n) \to F(X)$ in probability uniformly on $\mathbb{R}_+$. Consider the difference of the right sides of (7.3) written for X and X^n. Let

$$I_1(t, n) = \int_{]0,t]} F'(X_{s-})d(A_s^r + m_s^r) - \int_{]0,t]} F'(X_{s-}^n)d(A_s^{r,n} + m_s^{r,n}).$$

Then $I_1(t, n) = \sum_{i=1}^{4} J_i(t, n)$, where

$$J_1(t, n) = \int_{]0,t]} [F'(X_{s-}) - F'(X^n_{s-})] dm^r_s,$$

$$J_2(t, n) = \int_{]0,t]} F'(X^n_{s-}) d(m^r_s - m^{r,n}_s),$$

$$J_3(t, n) = \int_{]0,t]} [F'(X_{s-}) - F'(X^n_{s-})] dA^r_s,$$

$$J_4(t, n) = \int_{]0.t]} F'(X^n_{s-}) d(A^r_s - A^{r,n}_s).$$

On the basis of Doob's inequality and the properties of stochastic integrals we have

$$\mathbf{E}[\sup_t |J_1(t, n)|]^2 \leq 2 \sup_t \mathbf{E} |J_1(t, n)|^2$$

$$= 2 \sup_t \mathbf{E} \int_{]0,t]} |F'(X_{s-}) - F'(X^n_{s-})|^2 d\langle m^r \rangle_s$$

$$\leq 2\mathbf{E} \int_{]0,\infty[} |F'(X_{s-}) - F'(X^n_{s-})|^2 d\langle n\iota^r \rangle_s.$$

The variable on the right side is finite and vanishes by the Lebesgue dominated convergence theorem. Consequently, $J_1(t, n) \to 0$ as $n \to \infty$ in probability uniformly on $\mathbb{R}_+$.

Analogously,

$$\mathbf{E}[\sup_t |J_2(t, n)|]^2 \leq 2 \sup_t \mathbf{E} |J_2(t, n)|^2$$

$$= \sup_t \mathbf{E} \int_{]0.t]} |F'(X^n_{s-})|^2 d[m^r - m^{r,n}, m^r - m^{r,n}]_s$$

$$\leq C\mathbf{E} [m^r - m^{r,n}, m^r - m^{r,n}]_\infty$$

$$= C\mathbf{E} \sum_{k=n+1}^{\infty} (\triangle m^r_{R_k})^2,$$

where $|F'| \leq C$. The last variable vanishes as $n \to \infty$. Thus $J_2(t, n) \to 0$ as $n \to \infty$ in probability uniformly on $\mathbb{R}_+$. Further,

$$\mathbf{E} \sup_t |J_3(t, n)| \leq \mathbf{E} \sup_t \int_{]0,t]} |F'(X_{s-}) - F'(X^n_{s-})| |dA^r_s|$$

$$\leq \mathbf{E} \int_{]0,\infty[} |F'(X_{s-}) - F'(X^n_{s-})| |dA^r_s|.$$

Since $F'(X^n_t) \to F'(X_t)$ a.s. uniformly on $\mathbb{R}_+$, by the Lebesgue dominated convergence theorem the variable on the right vanishes. Consequently, $J_3(t, n) \to 0$ as $n \to \infty$ in probability uniformly on $\mathbb{R}_+$. For J_4 we have

$$\mathbf{E} \sup_t |J_4(t, n)| \leq \mathbf{E} \int_{]0,\infty[} |F'(X^n_{s-})| |d(A^r_s - A^{r,n}_s)|$$

$$\leq C\mathbf{E} \int_{]0,\infty[} |d(A^r_s - A^{r,n}_s)| \to 0 \text{ as } n \to \infty,$$

since the variation $(A^r - A^{r,n}) \to 0$. Therefore, $J_4(t,n) \to 0$ in probability uniformly on $\mathbb{R}_+$. From the reasoning for $J_1, ..., J_4$ it follows that $I_1(t,n) \to 0$ as $n \to \infty$ in probability uniformly on $\mathbb{R}_+$.

By analogous reasoning, it can be shown that such convergence is satisfied for

$$I_2(t,n) = \int_{]0,t[} F'(X_s)d(A^g + m^g)_{s+} - \int_{]0,t[} F'(X_s^n)d(A^{g,n} + m^{g,n})_{s+}.$$

Let

$$\begin{aligned}
I_3(t,n) &= \int_{]0,t]} F''(X_{s-})d\langle m^c\rangle_s - \int_{]0,t]} F''(X_{s-}^n)d\langle m^{c,n}\rangle_s \\
&= \int_{]0,t]} [F''(X_{s-}) - F''(X_{s-}^n)]d\langle m^c\rangle_s.
\end{aligned}$$

Similar to the case of J_3, it can be shown that $I_3(t,n) \to 0$ as $n \to \infty$ in probability uniformly on $\mathbb{R}_+$.

Let

$$I_4(t,n) = \sum_{s \leq t}[F(X_s) - F(X_{s-}) - F'(X_{s-}) \triangle X_s - F(X_s^n) + F(X_s^n) + F'(X_{s-}^n) \triangle X_s^n].$$

On the strength of the uniform convergence $X^n \to X$ on $\mathbb{R}_+$ every term in this sum vanishes as $n \to \infty$. In order to pass to the limit under the summation, we have to show that the series converges uniformly with respect to n.

Applying Taylor's formula, we can represent every term $[\cdot]_s$ of the sum as

$$\frac{1}{2}[F''(X_{s-} + \theta \triangle X_s)(\triangle X_s)^2 - F''(X_{s-}^n + \theta_n \triangle X_s^n)(\triangle X_s^n)^2 = \frac{1}{2}\epsilon_s(n)(\triangle X_s)^2 + \qquad (7.4)$$

$$+\frac{1}{2}F''(X_{s-}^n + \theta_n \triangle X_s^n)[(\triangle X_s)^2 - (\triangle X_s^n)^2],$$

where $\epsilon_s(n) = F''(X_{s-} + \theta \triangle X_s) - F''(X_{s-}^n + \theta_n \triangle X_s^n)$. From this we draw two conclusions.

(1) Since every term $[\cdot]_s$ of the sum tends to zero and $X^n \to X$ uniformly on $\mathbb{R}_+$, it follows that $\epsilon_s(n) \to 0$ for all $s \leq t$. Therefore $\sum_{s \leq t}\epsilon_s(n)(\triangle X_s)^2 \to 0$ as $n \to \infty$ for all $t \in \mathbb{R}_+$ by the Lebesgue dominated convergence theorem.

(2) Since

$$I_4(t,n) = \frac{1}{2}\sum_{s \leq t}\epsilon_s(n)(\triangle X_s)^2 + \frac{1}{2}\sum_{s \leq t}F''(X_{s-}^n + \theta_n \triangle X_s^n)[(\triangle X_s)^2 - (\triangle X_s^n)^2]$$

and $|F''| \leq C$, for uniform convergence of the original sum it suffices to establish uniform convergence for $\sum_{s \leq t}(\triangle X_s^n)^2$. But for all n such sum are majorized by one convergent sum $\sum_{s \leq t}(\triangle X_s)^2$ because X^n has only n jumps, coinciding with the analogous jumps of the process X. Thus, in the last representation for $I_4(t,n)$ we can pass to the limit under the summation signs. Hence as $n \to \infty$ the sum $I_4(t,n)$ converges to zero absolutely for all $t \in \mathbb{R}_+$.

Analogously it can be established that as $n \to \infty$ the sum

$$I_5(t,n) = \sum_{s < t}[F(X_{s+}) - F(X_s) - F'(X_s) \triangle^+ X_s - F(X_{s+}^n) + F(X_s^n) + F'(X_s^n) \triangle^+ X_s^n]$$

also a.s. converges to zero absolutely for all $t \in \mathbb{R}_+$.

Thus, having formula (7.3) for the semimartingale X^n in our chosen form and passing to the limit in it as $n \to \infty$, we obtain (7.3) for X in the hypothesis of the lemma. ∎

Since we have reduced the problem to the case of a semimartingale $X = X_0 + A + m$, $A \in \mathcal{A}$, $m \in \mathcal{M}^2$, having a finite number of jumps, we can assume that m is a continuous martingale including truncated martingales (of which there are a finite number) in the process A.

Lemma 7.2.5 Suppose the function F has two continuous bounded derivatives, and suppose that the semimartingale $X = X_0 + A + m$, where X_0 is an integrable $\mathcal{F}_0$-measurable variable, the process $A \in \mathcal{A}$ has a finite number of jumps and the martingale $m \in \mathcal{M}^2$ is continuous. Then (7.3) holds.

Proof. Suppose the s.t.b. $0 = R_0 \le R_1 \le ...R_n < \infty$ absorb all the jumps of the process A. Formula (7.3) is obviously satisfied for $t = 0$. Further, since

$$F(X_{0+}) = F(X_0) + \int_{[0]} F'(X_s)d(A^g + m^g)_s + \left[F(X_{0+}) - F(X_0) - F'(X_0) \triangle^+ X_0 \right],$$

where the second term on the right is $F'(X_0) \triangle^+ X_0$, it follows that (7.3) is valid for $t = 0+$. For $0 < t < R_1$ the semimartingale $\tilde{X}_t = X_{0+} + (X_t - X_{0+})$ is continuous, and for this case (7.3) was proved in [62]. We have

$$F(\tilde{X}_t) = F(X_{0+}) + \int_{]0,t]} F'(\tilde{X}_{s-})d(A^c + m)_s + \frac{1}{2} \int_{]0,t]} F''(\tilde{X}_{s-})d \langle m \rangle_s .$$

Replacing $\tilde{X}$ by X and taking account of the previous representation for $F(X_{0+})$, we obtain (7.3) on $[0, R_1[$. We take (7.3) at the time R_{1-}. Adding the variable $F(X_{R_1}) - F(X_{R_1-})$ to both sides and making some transformations on the right side, we obtain (7.3) on $[0, R_1]$. Repeating this procedure with the variable $F(X_{R_1+}) - F(X_{R_1})$, we get (7.3) at the time R_1+. Next, on $]R_1, R_2[$ we consider the continuous semimartingale $\tilde{X}_t = X_{R_1+} + (X_{R_1+t} - X_{R_1+})$. The formula from [62] is again applicable for it. Then at time R_2 we repeat the reasoning for time R_1. After n steps we arrive at the interval $]R_n, \infty[$, where again the process X is continuous. This completes the proof of (7.3). ∎

7.3 Stochastic Integrals of Random Measures

Let E be a Lusin space (i.e., a space homeomorphic to a Borel subset of a compact metric space) with σ-algebra of Borel subsets $\mathcal{E}$. Put

$$\tilde{\Omega} = \Omega \times \mathbb{R}_+ \times E, \quad \tilde{E} = \mathbb{R}_+ \times E, \quad \tilde{\mathcal{E}} = \mathcal{B}(\mathbb{R}_+) \times \mathcal{E},$$

$$\tilde{\mathcal{O}}(\mathbf{F}) = \mathcal{O}(\mathbf{F}) \times \mathcal{E}, \quad \tilde{\mathcal{O}}(\mathbf{F}_+) = \mathcal{O}(\mathbf{F}_+) \times \mathcal{E}, \quad \tilde{\mathcal{P}} = \mathcal{P}(\mathbf{F}) \times \mathcal{E}.$$

A non-negative random set function $\mu(\omega, \Gamma)$, $\omega \in \Omega$, $\Gamma \in \tilde{\mathcal{E}}$, is called a random measure on $\tilde{\mathcal{E}}$ if $\mu(\cdot, \Gamma) \in \mathcal{F}$ for any $\Gamma \in \tilde{\mathcal{E}}$ and $\mu(\omega, \cdot)$ is a σ-finite measure on $(\tilde{E}, \tilde{\mathcal{E}})$ for each $\omega \in \Omega$.

A random measure is called integer-valued if

$$\mu(\omega, \Gamma) \in \{0, 1, ..., +\infty\} \text{ and } 0 \leq \mu(\omega, \{t\} \times E) \leq 1$$

for any $\omega \in \Omega$, $t \in R_+$, $\Gamma \in \tilde{\mathcal{E}}$.

For a non-negative function $f \in \mathcal{F} \times \tilde{\mathcal{E}}$ and measure μ let us form the process $f * \mu$, where

$$f * \mu_t = \int_{[0,t] \times E} f(\omega, s, x) \mu(\omega, ds, dx), \quad t < \infty.$$

We shall say that the random measure μ is $\mathcal{O}(\mathbf{F}_+)$-optional if the process $f * \mu \in \mathcal{O}(\mathbf{F}_+)$ for any non-negative $f \in \tilde{\mathcal{O}}(\mathbf{F})$. Similarly, the random measure μ is $\mathcal{O}(\mathbf{F})$-optional (for brevity, optional) if $f * \mu \in \mathcal{O}(F)$ for any $f \in \hat{\mathcal{O}}(\mathbf{F})$, $f \geq 0$. The random measure μ is predictable if $f * \mu \in \varphi$ for any $f \in \tilde{\varphi}$, $f \geq 0$.

The next important result was shown in [57] and [63]. Let μ be an optional measure. On $(\tilde{\Omega}, \tilde{\mathcal{O}}(\mathbf{F}))$ define the measure M_μ^P by setting

$$\mathsf{M}_\mu^P(f) = \mathbf{E} f * \mu_\infty, \quad f \in \tilde{G}(F), \quad f \geq 0.$$

Lemma 7.3.1 Let the optional measure μ be such that the measure M_μ^P is $\tilde{\mathcal{P}}$-σ-finite (i.e., the restriction of M_μ^P to $(\tilde{\Omega}, \tilde{\mathcal{P}})$ is σ-finite). Then there exists a unique (to within a set of $\mathcal{P}$-null measure) predictable measure $v = \nu(\omega, dt, dx)$ such that for any function $f \in \tilde{\mathcal{P}}$, $f \geq 0$, one has $\mathsf{M}_\mu^P(f) = \mathsf{M}_\nu^P(f)$.

The measure v can be written in the form

$$\nu(\omega, dt, dx) = dA_t(\omega) K(\omega, t, dx),$$

where A is an increasing predictable right-continuous process, $K(\omega, t, dx)$ is the kernel of the space $(\Omega \times \mathbb{R}_+, \mathcal{P}(\mathbf{F}))$ into $(E, \mathcal{E})$.

If the measure μ does not load any predictable stopping times whatsoever, then the same is true of v, and the process $f * v$ is continuous for any function $f \in \tilde{\mathcal{P}}$, $f \geq 0$. Moreover, for any $S \in \mathcal{T}^p$ and any $f \in \tilde{\mathcal{P}}$, $f \geq 0$, on $(S < \infty)$,

$$\mathbf{E}\left[\int_E f(S, x) \mu(\{S\}, dx) | \mathcal{F}_{S-}\right] = \int_E f(S, x) \nu(\{S\}, dx).$$

The process $f * v$ is the dual predictable projection for the process $f * \mu$, $f \in \tilde{\mathcal{P}}$, $f \geq 0$. If $f * \mu \in \mathcal{A}_{loc}$, then the process $f * \mu - f * v \in \mathcal{M}_{loc}$.

In the case of integer-valued μ, outside a set of $\mathbf{P}$-null measure, $0 \leq \nu(\omega, \{t\} \times E) \leq 1$ for all $t \in \mathbb{R}_+$. The measure v is called the compensator of the measure μ.

Remark 7.3.2 The proof in [57] and [63] was carried out under the assumption that the usual conditions hold. Without essential modifications it is applicable in our case. Note that there one can select a modification of the measure v with the property $0 \le \nu(\omega, \{t\} \times E) \le 1$ for all $\omega \in \Omega$, $t \in \mathbb{R}_+$. In our case this is false.

Let us derive an analogue of this result for $\mathcal{O}(\mathbf{F}_+)$-optional measures. Let η be an $\mathcal{O}(\mathbf{F}_+)$-optional measure. On the space $(\tilde{\Omega}, \tilde{\mathcal{O}}(\mathbf{F}_+))$ define the measure M_η^P by setting $M_\eta^P(f) = \mathbf{E} f * \eta_\infty$ for any $f \in \tilde{\mathcal{O}}(F_+)$, $f \ge 0$.

Lemma 7.3.3 Let the $\mathcal{O}(\mathbf{F}_+)$-optional measure η be such that the measure M_η^P is $\tilde{\mathcal{O}}(\mathbf{F})$-$\sigma$-finite. Then:

(a) There exists a unique (to within a set of $\mathbf{P}$-null measure) optional measure $\lambda = \lambda(\omega, dt, dx)$ such that, for any $f \in \tilde{\mathcal{O}}(\mathbf{F})$, $f \ge 0$, one has $\mathsf{M}_\eta^P(f) = \mathsf{M}_\lambda^P(f)$; the measure λ can be written in the form

$$\lambda(\omega, dt, dx) = dA_t(\omega)K(\omega, t, dx),$$

where A is an increasing right-continuous $\mathcal{O}(\mathbf{F})$-measurable process, $K(\omega, t, dx)$ is the kernel from the space $(\Omega \times \mathbb{R}_+, \mathcal{O}(\mathbf{F}))$ into $(E, \mathcal{E})$.

(b) If the measure η does not load any stopping time $T \in \mathcal{T}$ whatsoever, then the same is true for the measure λ, and the process $f * \lambda$ is continuous for any $f \in \tilde{\mathcal{O}}(\mathbf{F})$, $f \ge 0$.

(c) The process

$$f * \lambda_+ = \left(\int_{[0,t] \times E} f\lambda(ds, dx) \right)$$

is the dual optional projection for the process

$$f * \eta_+ = \left(\int_{[0,t] \times E} f\eta(ds, dx) \right)$$

for $f \in \tilde{\mathcal{O}}(\mathbf{F})$, $f \ge 0$. In particular, for $f \in O(\mathbf{F})$, $f \ge 0$ and $T \in \mathcal{T}$.

$$\mathbf{E}\left[\int_E f(T, x)\eta(\{T\}, dx) | \mathcal{F}_T \right] = \mathbf{E}\left[\int_E f(T, x)\lambda(\{T\}, dx) | \mathcal{F}_T \right] \text{ a.s. on } (T < \infty);$$

if $f * \eta_+ \in \mathcal{A}_{loc}$, then the process $f * \eta - f * \lambda \in \mathcal{M}_{loc}$, where

$$f * \eta_t - f * \lambda_t = \int_{[0,t] \times E} f\eta(ds, dx) - \int_{[0,t] \times E} f\lambda(ds, dx). \tag{7.5}$$

(d) If the measure η is integer-valued, then there exists a modification of the measure λ such that outside a set of $\mathbf{P}$-null measure, $0 \le \lambda(\omega, \{t\}, E) \le 1$ for any $t \in \mathbb{R}_+$. We shall call the measure λ the compensator of the $(\mathcal{O}(\mathbf{F}_+)$-optional measure η.

Proof. Suppose that $M_\eta^P(\tilde{\Omega}) < \infty$.

(a) The relation $m(\cdot) = \mathsf{M}_\eta^P(\cdot, E)$ defines a finite measure on the space $(\Omega \times \mathbb{R}_+, \mathcal{O}(F))$. For a bounded $\mathcal{E}$-measurable function f define the measure m^f by setting $m^f(\cdot) = M_\eta^P(\cdot, f)$. We have

$m^f << m$. Denote by Tf the derivative dm^f/dm. T stands for a non-negative linear mapping of bounded $\mathcal{E}$-measurable functions into the space of $\mathcal{O}(\mathbf{F})$-measurable functions. Since E is a Lusin space, there exists a kernel $K(\omega, t, dx)$ from $(\Omega \times \mathbb{R}_+, \mathcal{O}(\mathbf{F}))$ into the space $(E, \mathcal{E})$ such that $Tf(\cdot) = K(\cdot, f)$, i.e., $Tf(\cdot) = \int_E K(\cdot, dx)f(x)$ (see, for example, [65]). Consequently, the measure M_η^P can be represented in the form

$$M_\eta^P(d\omega, dt, dx) = m(d\omega, dt)K(\omega, t, dx).$$

According to [4], there exists an increasing right-continuous $\mathcal{O}(\mathbf{F})$-measurable process $A = (A_t)$, $t \in \mathbb{R}_+$, such that for any $f \in \mathcal{O}(\mathbf{F})$, $f \geq 0$, one has $m(f) = \mathbf{E}\int_{[0,\infty[} f dA$. Set

$$\lambda(\omega, dt, dx) = dA_t(\omega)K(\omega, t, dx).$$

The measure λ is optional, as is easily checked with the help of the theorem on monotonic classes. Further, for fixed set $C \subset \mathcal{E}$, the process $\lambda_t - \lambda([0, t], C)$ is the dual $\mathcal{O}(\mathbf{F})$-projection of the process $\eta_t = \eta([0, t], C)$ defined uniquely to within a set of $\mathbf{P}$-null measure. Since the σ-algebra $\mathcal{E}$ is separable, it follows that λ is uniquely defined to within a set of P-null measure.

(b) If the measure η does not load any stopping time $T \in \mathcal{T}$ whatsoever, then for any set $C \in \mathcal{E}$,

$$0 = \mathbf{E}\mathbf{1}_{[T] \times C} * \eta_\infty = \mathbf{E}\mathbf{1}_{[T] \times C} * \lambda_\infty = \mathbf{E}K(T, C) \triangle A_T \mathbf{1}_{T < \infty}.$$

From the second equality it follows that $\lambda(\{T\}, C) = 0$ a.s., i.e., λ does not load any s.t. $T \in \mathcal{T}$ whatsoever. Since C is arbitrary, it follows from the third equality that $\triangle A_T = 0$ a.s. on $(T < \infty)$ for any $T \in \mathcal{T}$. With the aid of the theorem on monotone class, we get that the process $f * \lambda$ is continuous for any function $f \in \tilde{\mathcal{O}}(\mathbf{F})$, $f \geq 0$.

(c) From the fact that

$$\lambda(\omega, dt, dx) = dA_t(\omega)K(\omega, t, dx),$$

at once it is deduced using the theorem on monotonic classes that the process $f * \lambda_+$ is the dual optional projection for the process $f * \eta_+$ for $f \in \tilde{\mathcal{O}}(F)$, $f \geq 0$. Then the process (7.5) belongs to $\mathcal{M}_{loc}$ according to Lemma 5.3.11.

(d) For an integer-valued measure η we have $0 \leq \eta(\omega, \{t\}, E) \leq 1$ for all $\omega \in \Omega$, $t \in \mathbb{R}_+$. Since, by point (c), $\lambda(\{T\}, E) = \mathbf{E}[\eta(\{T\}, E)|F_T]$ a.s. on $(T < \infty)$ for any $T \in \mathcal{T}$, we have $0 \leq \lambda(\omega, \{T\}, E) \leq 1$ a.s. for any $T \in \mathcal{T}$. Put $D = \{(\omega, t) : \lambda(\omega, \{t\}, E) > 1\}$. By a section theorem in [4] there exists an s.t. T for which $[T] \subseteq D$. If $\mathbf{P}(T < \infty) > 0$, then $\lambda(\{T\}, E) > 1$ with positive probability, which leads to a contradiction. ∎

Example 7.3.4 Let X be an optional semimartingale, i.e., $X = m + A$, $m \in \mathcal{M}_{loc}$, $A \in \mathcal{V}$. We shall suppose that X has values in E, and E is an additive group. Set

$$\mu(\omega, [0, t] \times \Gamma) = \sum_{0 < s \leq t} \mathbf{1}(\triangle X_s \neq 0)\mathbf{1}(\triangle X_s \in F), \quad \Gamma \in \mathcal{E},$$

$$\eta(\omega, [0, t] \times \Gamma) = \sum_{0 \leq s \leq t} \mathbf{1}(\triangle^+ X_s \neq 0)\mathbf{1}(\triangle^+ X_s \in \Gamma).$$

Then μ is an optional integer-valued measure, and M_μ^P is $\tilde{\mathcal{P}}$-σ-finite. Indeed, for fixed $m > 0$ define recursively: $T(0, m) = 0$,

$$T(n+1, m) = \inf\{t > T(n, m) : \rho(X_{t-}, X_t) \in [1/m, 1/(m-1)[\},$$

where $\rho(\cdot, \cdot)$ is the distance in E. We have $T(n, m) \in \mathcal{T}$. Then M_μ^P is finite on each set

$$\{(\omega, t, x) : t \le T(n, m), \; \rho(X_{t-}, x) \in [1/m, 1/(m-1)[\} \in \tilde{\mathcal{P}}(\mathbf{F}).$$

The union of these sets over n and m yields $\tilde{\Omega}$.

The measure η is $\mathcal{O}(\mathbf{F}_+)$-optional and integer-valued. The measure M_η^P is $\tilde{\mathcal{O}}(\mathbf{F})$-$\sigma$-finite; again define recursively:

$$T(0, m) = 0, \; T(n+1, m) = \inf\{t > T(n, m) : \rho(X_t, X_{t+}) \in [1/m, 1/(m-1)[\},$$

and the measure M_η^P is finite on each set

$$\{(\omega, t, x) : t \le T(n, m), \; \rho(X_t, x) \in [1/m, 1/(m-1)[\} \in \tilde{\mathcal{O}}(\mathbf{F}),$$

the union of which equals $\tilde{\Omega}$.

Let us turn to the definition of stochastic integrals over random measures. Let μ be an optional, η an $\mathcal{O}(\mathbf{F}_+)$-optional measure, and let ν and λ be their respective compensators. Let us agree to write

$$a_t = \nu(\{t\}, E), \quad b_t = \lambda(\{t\}, E),$$

$$\hat{f}_t(\nu) = \int_E f(t, x)\nu(\{t\}, dx), \quad \hat{h}_t(\lambda) = \int_E h(t, x)\lambda(\{t\}, dx),$$

$$\tilde{f}_t(\mu, \nu) = \int_E f(t, x)\mu(\{t\}, dx) - \hat{f}_t(\nu), \quad \tilde{h}_t(\eta, \lambda) = \int_E h(t, x)\eta(\{t\}, dx) - \hat{h}_t(\lambda),$$

if all these integrals are defined; $\tilde{f} = \infty$ if the corresponding integral is undefined (similarly for h). We assume that the indeterminate form $\infty - \infty = \infty$.

Our goal is to define the stochastic integrals

$$f * (\mu - \nu) \in \mathcal{M}_{loc}^d, \quad h * (\eta - \lambda) \in \mathcal{M}_{loc}^g,$$

possessing the properties

$$\triangle f * (\mu - \nu)_t = \tilde{f}_t(\mu, \nu), \quad \triangle^+ h * (\eta - \lambda)_t = \tilde{h}_t(\eta, \lambda). \tag{7.6}$$

It is implicit that

$$f * (\mu - \nu)_t = \int_{]0,t] \times E} f(s, x)(\mu - \nu)(ds, dx),$$

$$h * (\eta - \lambda)_t = \int_{[0,t[\times E} h(s, x)(\eta - \lambda)(ds, dx).$$

Introduce the spaces

$$G^i(\mu) = \left\{ f \in \tilde{\mathcal{P}} : \sum_{0 < s \leq \cdot} |\tilde{f}_s(\mu, \nu)|^2)^{1/2} \in \mathcal{A}_{loc} \right\},$$

$$G^i(\eta) = \left\{ h \in \tilde{\mathcal{O}}(F) : \left(\sum_{0 \leq s < \cdot} |\tilde{h}_s(\eta, \lambda)|^2 \right)^{i/2} \in \mathcal{A}_{loc} \right\}.$$

Theorem 7.3.5 For the existence of unique (to within indistinguishability) elements $f * (\mu - \lambda) \in \mathcal{M}_{loc}^{i,d}$ and $h * (\eta - \lambda) \in \mathcal{M}_{loc}^{i,g}$ with the properties (7.6) it is necessary and sufficient that $f \in G^i(\mu)$, $h \in G^i(\eta)$, $i = 1, 2$.

Theorem 7.3.6 Let $f \in \tilde{\mathcal{P}}$ and let the constant $c > 0$. Then the following assertions are equivalent:
  (a) $f \in G^1(\mu)$,
  (b)

$$\frac{|f - \hat{f}(v)|^2}{1 + |f - \hat{f}(v)|} * v + \sum_{0 < s \leq \cdot} \frac{|\hat{f}_s(v)|^2}{1 + |\hat{f}_s(v)|}(1 - a_s) \in \mathcal{A}_{loc},$$

  (c)

$$\left[|f - \hat{f}(v)|^2 \mathbf{1}_{\{|f - \hat{f}(v)| \leq c\}} + |f - \hat{f}(v)| \mathbf{1}_{\{|f - \hat{f}(v)| > c\}} \right] * v$$

$$+ \sum_{0 < s \leq} (|\hat{f}_s(v)|^2 \mathbf{1}_{\{|\hat{f}(v)| \leq c\}} + |v| \mathbf{1}_{\{|\hat{f}(v)| > c\}}(1 - a_s)) \in \mathcal{A}_{loc}.$$

  (d) If $\hat{f} \geq -1$, then these conditions are equivalent to the following:

$$(1 - (1 + f - \hat{f}(v))^{1/2})^2 * v + \sum_{0 < s \leq \cdot} (1 - a_s)(1 - (1 - \hat{f}_s(v))^{1/2})^2 \in \mathcal{A}_{loc}.$$

The analogous assertion is true when $f \in \tilde{\mathcal{P}}$, $\hat{f}(v)$, μ, ν, a, $\sum_{0 < s \leq \cdot}$ are replaced by $h \in \tilde{\mathcal{O}}(\mathbf{F})$, $\hat{h}(\lambda)$, η, λ, b, $\sum_{0 \leq s < \cdot}$, respectively.

The proofs of these theorems for the function f are derived in [63], [66] and [67]. The case of the function h is considered similarly.

7.4   Semimartingales and Their Characteristics

The *canonical* and *component* representation of semimartingales is of fundamental importance in stochastic analysis. It is also essential to our development of stochastic integral equations driven by optional semimartingales. The canonical and component representation of optional semimartingales can be seen as a natural consequence of the decomposition

$$X = X_0 + X^c + X^d + X^g.$$

where X^c is a continuous optional semimartingale with decomposition, $X^c = a + m$, where a is continuous strongly predictable with locally integrable variation ($a \in \mathcal{P}_s \cap \mathcal{A}_{0,loc}$), and m a continuous local martingale ($m \in \mathcal{M}^c_{0,loc}$). The discrete optional semimartingale parts, $X^d = a^d + m^d$ and $X^g = a^g + m^g$, (a^d, $a^g \in \mathcal{A}_{loc}$, $m^d \in \mathcal{M}^d_{loc}$, $m^g \in \mathcal{M}^g_{loc}$) are representable in terms of some underlying measures of right and left jumps, respectively.

7.4.1   Canonical representation

Let us go over to the definition of the characteristics of an optional semimartingale. Let $X = (X^1, ..., X^n)$ be an n-dimensional semimartingale with decomposition $X^i = A^i + M^i$, $M^i \in \mathcal{M}_{loc}$, $A^i \in \mathcal{V}$, $M^i_0 = 0$, $i = 1, ..., n$. For $(E, \mathcal{E})$ we take the space $(\mathbb{R}^n \backslash \{0\}, \mathcal{B}(\mathbb{R}^n \backslash \{0\}))$. On the σ-algebra $\tilde{\mathcal{E}} = \mathcal{B}(\mathbb{R}_+) \times \mathcal{E}$ define the integer-valued random measures μ and η by setting

$$\mu(dt, dx) = \sum_{0 < s} \mathbf{1}_{\{\triangle X_s \neq 0\}} \varepsilon(s, \triangle X_s)(dt, dx),$$

$$\eta(dt, dx) = \sum_{0 \leq s} \mathbf{1}_{\{\triangle^+ X_s \neq 0\}} \varepsilon(s, \triangle^+ X_s)(dt, dx),$$

where $\varepsilon_{(s,y)}(dt, dx)$ is the Dirac measure. Let v and λ denote the compensators of μ and η, respectively.

Lemma 7.4.1 The function $f(\omega, s, x) = x\mathbf{1}_{(|x| \leq b)}$, $b > 0$, belongs to the spaces $\mathrm{G}^2(\mu)$ and $\mathrm{G}^2(\eta)$.

Proof. Set

$$\tau_N = \inf \left\{ t > 0 : \sum_{0 < s \leq t} |\triangle X_s|^2 \mathbf{1}_{\{|\triangle X_s| \leq b\}} > N \right\}.$$

We have $\tau_N \in \mathcal{T}_+$ for any $N \in \mathbb{N}$, $\tau_N \uparrow \infty$ a.s. Set $B =]0, \tau_N] \times \{|x| \leq b\} \in \tilde{\mathcal{P}}$. Noting that μ is an integer-valued measure, we get

$$\mathbf{E} \sum_{0 < s \leq \tau_N} \left| \tilde{f}_s(\mu, \nu) \right|^2 \leq 2\mathbf{E} \sum_{0 < s \leq \tau_N} \left[\left(\int_{|x| \leq b} x\mu(\{s\}, dx) \right)^2 + \left(\int_{|x| \leq b} x\nu(\{s\}, dx) \right)^2 \right]$$

$$\leq 2\mathbf{E} \sum_{0 < s \leq \tau_N} \left[\int_{|x| \leq b} |x|^2 \mu(\{s\}, dx) + \int_{|x| \leq b} |x|^2 \nu(\{s\}, dx) \right]$$

$$\leq 2\mathbf{E} \left[\int_B |x|^2 \mu(ds, dx) + \int_B |x|^2 \nu(ds, dx) \right]$$

$$= 4\mathbf{E} \int_B |x|^2 \mu(ds, dx),$$

where the last equality is written because ν is the compensator of μ. From this and the definition of τ_N, we find that

$$\mathbf{E} \sum_{0 < s \leq \tau_N} \left| \tilde{f}_s(\mu, \nu) \right|^2 \leq 4\mathbf{E} \int_B |x|^2 \mu(ds, dx) = 4\mathbf{E} \sum_{0 < s \leq \tau_N} |\triangle X_s|^2 \mathbf{1}_{\{|\triangle X_s| \leq b\}} \leq 4(N + b^2).$$

It is shown similarly that $f \in G^2(\eta)$. We need only define τ_N in the form

$$\tau_N = \inf \left\{ t \geq 0 : \sum_{0 \leq s < t} |\triangle^+ X_s|^2 \mathbf{1}_{\{|\triangle^+ X_s| \leq b\}} > N \right\}$$

and take the set $B = [0, \tau_N] \times \{|x| \leq b\} \in \tilde{\mathcal{P}}$. ∎

Theorem 7.4.2 The semimartingale $X = (X_t)$ with values in $\mathbb{R}^n$ is representable in the form

$$X_t = X_0 + a_t + m_t^c + m_t^d + m_t^g + \int_{|x| > 1} x\mu(]0, t], dx) + \int_{|x| > 1} x\eta([0, t[, dx), \qquad (7.7)$$

where $a = (a^1, ..., a^n)$ is a strongly predictable process,

$$m_t^d = \int_{|x| \leq 1} x(\mu - \nu)(]0, t], dx) \in \mathcal{M}_{loc}^d, \quad m_t^g = \int_{|x| \leq 1} x(\eta - \lambda)([0, t[, dx) \in \mathcal{M}_{loc}^g.$$

We shall call the collection $(a, \langle m^c \rangle, v, \lambda)$ the collection of local characteristics of the semimartingale X.

Proof. The last two summands in (7.7), equal to

$$\sum_{0 < s \leq 1} |\triangle X_s| \mathbf{1}_{\{|\triangle X_s| > 1\}}, \quad \sum_{0 \leq s < t} |\triangle^+ X_s| \mathbf{1}_{\{|\triangle^+ X_s| > 1\}},$$

have a.s. finite variation on any finite segment. By Lemma 7.4.1 and Theorem 7.3.5, the stochastic integrals m_t^d, m_t^g are defined, and $m^d \in \mathcal{M}_{loc}^{2,d}$, $m^g \in \mathcal{M}_{loc}^{2,g}$.

Set

$$\tilde{X}_t = X_t - X_0 - m_t^d - m_t^g - \int_{|x| > 1} x\mu(]0, t], dx) - \int_{|x| > 1} x\eta([0, t[, dx). \qquad (7.8)$$

For any $T \in \mathcal{T}$ and $S \in \mathcal{T}_+^-$ on the sets $(T < \infty)$ and $(S < \infty)$ we have, by virtue of the properties of stochastic integrals,

$$\triangle \tilde{X}_T = \triangle X_T - \triangle X_T \mathbf{1}_{\{|\triangle X_T| \leq 1\}} + \int_{|x| \leq 1} x \nu(\{T\}, dx) - \triangle X_T \mathbf{1}_{\{|\triangle X_T| > 1\}},$$

$$\triangle^+ \tilde{X}_s = \triangle^+ X_s - \triangle^+ X_s \mathbf{1}_{\{|\triangle^+ X_s| \leq 1\}} + \int_{|x| \leq 1} x \lambda(\{S\}, dx) - \triangle^+ X_s \mathbf{1}_{\{|\triangle^+ X_s| < 1\}}.$$

Whence, if $T \in \mathcal{T}^i$, then $\triangle \tilde{X}_T = 0$, and if $T \in \mathcal{T}^p$, then

$$\triangle \tilde{X}_T = \int_{|x| \leq 1} x \nu(\{T\}, dx). \tag{7.9}$$

Further, if $S \in \mathcal{T}_+$ and $\mathbf{P}(S = T < \infty) = 0$ for any $T \in \mathcal{T}$, then $\triangle^+ \tilde{X}_s = 0$. But if $S \in \mathcal{T}$, then

$$\triangle^+ \tilde{X}_s = \int_{|x| \leq 1} x \lambda(\{S\}, dx). \tag{7.10}$$

The process $\tilde{X}$ has bounded jumps. Hence, by point (d) of Lemma 4.2, $\tilde{X}$ is a special semimartingale and representable in the form

$$\tilde{X} = a + M, \quad a \in \mathcal{A}_{loc}, \quad M \in \mathcal{M}_{loc}, \quad M_0 = 0, \tag{7.11}$$

where a is strongly predictable. Let us show that M is continuous. The jumps of a and $\tilde{X}$ of the form $\triangle a_T$ and $\triangle \tilde{X}_T$ occur only at predictable times (see (7.9)) and are $\mathcal{F}_T$-measurable. Since $M = \tilde{X} - a$, $\triangle M_T$ is an $\mathcal{F}_T$-measurable quantity for any $T \in \mathcal{T}^p$ and

$$\triangle M_T = \mathbf{E}[\triangle M_T | \mathcal{F}_{T-}] = 0 \quad a.s. \quad on \ (T < \infty).$$

Similarly (see (7.9)), the jumps of a and $\tilde{X}$ of the form $\triangle^+ a_T$ and $\triangle^+ \tilde{X}_T$ occur only at times $T \in \mathcal{T}$ and are $\mathcal{F}_T$-measurable. Hence $\triangle^+ M_T$ is $\mathcal{F}_T$-measurable and $\triangle^+ M_T = \mathbf{E}[\triangle^+ M_T | \mathcal{F}_T] = 0$ a.s. on $(T < \infty)$ for any $T \in \mathcal{T}$.

With the aid of the theorem on sections (see [4]) we see that almost all trajectories of M are continuous. Setting $m^c = M$ from (7.8) and (7.11) we get (7.7). ∎

Corollary 7.4.3 From (7.11) and the continuity of M it follows that the jumps of $\tilde{X}$ and a coincide. Since $a \in \mathcal{A}_{loc}$, we get, noting (7.9) and (7.10), that the series

$$\sum_{0 < s \leq t} \left| \int_{|x| \leq 1} x \nu(\{s\}, dx) \right|, \quad \sum_{0 \leq s < t} \left| \int_{|x| \leq 1} x \lambda(\{s\}, dx) \right|$$

converge a.s. for any t.

7.4.2   Component representation

Here we are going to consider the Lusin space $(\mathbb{E}, \mathscr{E})$ where $\mathbb{E} = (\mathbb{R}^d \backslash \{0\}) \cup \{\delta^d\} \cup \{\delta^g\}$; δ^d and δ^g are some supplementary points or are the set of processes with finite variation on any segment $[0, t]$, **P**-a.s.; $\mathscr{E} = \mathcal{B}(\mathbb{E})$ is the Borel σ-algebra in $\mathbb{E}$. Also, define the spaces

$$\widetilde{\Omega} = \Omega \times \mathbb{R}_+ \times \mathbb{E}, \quad \widetilde{\mathbb{E}} = \mathbb{R}_+ \times \mathbb{E}, \quad \widetilde{\mathscr{e}} = \mathcal{B}(\mathbb{R}_+) \times \mathscr{E}, \quad \widetilde{\mathcal{G}} = \mathcal{G} \times \mathcal{B}(\mathbb{E}), \tag{7.12}$$

$$\widetilde{\mathcal{O}}(\mathbf{F}) = \mathcal{O}(\mathbf{F}) \times \mathscr{E}, \quad \widetilde{\mathcal{O}}(\mathbf{F}_+) = \mathcal{O}(\mathbf{F}_+) \times \mathscr{E}, \ and \ \widetilde{\mathcal{P}}(\mathbf{F}) = \mathcal{P}(\mathbf{F}) \times \mathscr{E}.$$

It was shown in [11] that there exist sequences $\{S_n\}$, $\{T_n\}$, and $\{U_n\}$ for $n \in \mathbb{N}$ of predictable stopping time (s.t.), totally inaccessible stopping time and totally inaccessible stopping time in the broad sense (s.t.b.), respectively, absorbing all jumps of the process X such that the graphs of these stopping times do not intersect within each sequence. On $\widetilde{\Omega}$ let $\mu^i(\omega, \cdot, \cdot)$, $p^i(\omega, \cdot, \cdot)$ and $\eta^g(\omega, \cdot, \cdot)$ where $i \in (d, g)$ be integer valued measures defined on the σ-algebra $\mathcal{B}(\mathbb{R}_+) \times \mathcal{B}(\mathbb{E})$ that are associated with the sequences of stopping times that are associated with X. On the σ-algebra $\mathcal{B}(\mathbb{R}_+) \times \mathcal{B}(\mathbb{E})$ we define the random integer-valued measures by the relations,

$$p^d(B \times \Gamma) = \sum_n \mathbf{1}_{B \times \Gamma}(S_n, \beta^d_{S_n}), \quad p^g(B \times \Gamma) = \sum_n \mathbf{1}_{B \times \Gamma}(S_n, \beta^g_{S_n}),$$

$$\mu^d(B \times \Gamma) = \sum_n \mathbf{1}_{B \times \Gamma}(T_n, \beta^d_{T_n}), \quad \mu^g(B \times \Gamma) = \sum_n \mathbf{1}_{B \times \Gamma}(T_n, \beta^g_{U_n}),$$

$$\eta^g(B \times \Gamma) = \sum_n \mathbf{1}_{B \times \Gamma}(T_n, \beta^g_{T_n}),$$

where $B \in \mathcal{B}(\mathbb{R}_+)$, $\Gamma \in \mathcal{B}(\mathbb{E})$, $\beta^d_t = \triangle X_t$ if $\triangle X_t \neq 0$ and $\beta^d_t = \delta^d$ if $\triangle X_t = 0$, $\beta^g_t = \triangle^+ X_t$ if $\triangle^+ X_t \neq 0$, $\beta^g_t = \delta^g$ if $\triangle^+ X_t = 0$, $t > 0$, $\mathbf{1}_A(x)$ is the indicator of the set A. For the measures $\mu^d(\omega, \cdot)$ and $\mu^g(\omega, \cdot)$ there exist unique random measures $\nu^d(\omega, \cdot)$ and $\nu^g(\omega, \cdot)$, respectively, on $\mathcal{B}(\mathbb{R}_+) \times \mathcal{B}(\mathbb{E})$ such that, for any non-negative functions $\varphi_d \in \widetilde{\mathcal{P}}(\mathbf{F})$ and $\varphi_g \in \widetilde{\mathcal{O}}(\mathbf{F})$
(i) The process

$$\int_{0+}^t \int_{\mathbb{E}} \varphi_d(s, u) \nu^d(ds, du), \quad \int_0^{t-} \int_{\mathbb{E}} \varphi_g(s, u) \nu^g(ds, du)$$

is $\mathcal{P}(\mathbf{F})$-measurable and $\mathcal{O}(\mathbf{F})$-measurable, respectively.
(ii) The equalities

$$\mathbf{E} \int_{0+}^\infty \int_{\mathbb{E}} \varphi_d(s, u) \mu^d(ds, du) \ = \ \mathbf{E} \int_{0+}^\infty \int_{\mathbb{E}} \varphi_d(s, u) \nu^d(ds, du)$$

$$\mathbf{E} \int_0^\infty \int_{\mathbb{E}} \varphi_g(s, u) \mu^g(ds, du) \ = \ \mathbf{E} \int_0^\infty \int_{\mathbb{E}} \varphi_g(s, u) \nu^g(ds, du)$$

are valid.

The measures ν^i, $i \in (d, g)$ possess the property $0 \leq \nu^i(\omega, \{t\} \times \mathbb{E}) \leq 1$ for all ω and t except for some set of **P** measure zero. We denote by $\lambda^i(\omega, \cdot)$, $i \in (d, g)$ the analogous measures for $p^i(\omega, \cdot)$, and $\theta^g(\omega, \cdot)$ that of $\eta^g(\omega, \cdot)$. The measures ν^i, λ^i and θ^g are called the (dual) predictable projections (compensator) for the measures μ^i, p^i and η^g, respectively. Note, μ^g, p^g, and η^g are $\mathcal{O}(\mathbf{F}_+)$-optional with their compensator being $\mathcal{O}(\mathbf{F})$-optional. On the other hand μ^d and p^d is $\mathcal{O}(\mathbf{F})$-optional with their compensator being $\mathcal{P}(\mathbf{F})$-predictable.

Having defined integer valued measures and stochastic integrals, with respect to them one can write a representation of discrete optional semimartingales,

$$
X_t^d = \int_{0+}^t \int_{\mathbb{E}} u\mathbf{1}_{|u|\leq 1}(\mu^d - \nu^d)(ds, du) + \int_{0+}^t \int_{\mathbb{E}} u\mathbf{1}_{|u|>1}\mu^d(ds, du) + \int_{0+}^t \int_{\mathbb{E}} up^d(ds, du),
$$

$$
X_t^g = \int_0^{t-} \int_{\mathbb{E}} u\mathbf{1}_{|u|\leq 1}(\mu^g - \nu^g)(ds, du) + \int_0^{t-} \int_{\mathbb{E}} u\mathbf{1}_{|u|>1}\mu^g(ds, du) + \int_0^{t-} \int_{\mathbb{E}} up^g(ds, du)
$$

$$
+ \int_0^{t-} \int_{\mathbb{E}} u\eta^g(ds, du).
$$

With $X^c = a + m$, we can write the component decomposition of X as

$$
\begin{aligned}
X = {} & X_0 + a + m \\
& + \int_{0+}^t \int_{\mathbb{E}} u\mathbf{1}_{|u|\leq 1}(\mu^d - \nu^d)(ds, du) + \int_{0+}^t \int_{\mathbb{E}} u\mathbf{1}_{|u|>1}\mu^d(ds, du) + \int_{0+}^t \int_{\mathbb{E}} up^d(ds, du), \\
& + \int_0^{t-} \int_{\mathbb{E}} u\mathbf{1}_{|u|\leq 1}(\mu^g - \nu^g)(ds, du) + \int_0^{t-} \int_{\mathbb{E}} u\mathbf{1}_{|u|>1}\mu^g(ds, du) + \int_0^{t-} \int_{\mathbb{E}} up^g(ds, du) \\
& + \int_0^{t-} \int_{\mathbb{E}} u\eta^g(ds, du).
\end{aligned}
$$

Note that the local martingales of the process X are,

$$
m^d = \int_{0+}^t \int_{\mathbb{E}} u\mathbf{1}_{|u|\leq 1}(\mu^d - \nu^d)(ds, du),
$$

$$
m^g = \int_0^{t-} \int_{\mathbb{E}} u\mathbf{1}_{|u|\leq 1}(\mu^g - \nu^g)(ds, du).
$$

and the characteristic of the process X is $\left(a, \langle m, m \rangle, \nu^d, \lambda^d, \nu^g, \lambda^g, \theta^g\right)$. For further details on the construction of the component decomposition of optional semimartingales, see [11].

Now, let's consider integrals with respect to the components of X. The process, a, is a continuous locally finite variation process that is strongly predictable, ($a \in \mathcal{P}_s \cap \mathcal{A}_{0,loc}$). An integral of a function, f, with respect, a, is well defined in the Lebesgue-Stieltjes sense, $f \cdot a_t \in \mathcal{A}_{loc}$, where the integral is over the interval $[0, t]$ and f is $\mathcal{P}(\mathbf{F})$-measurable. For the continuous local martingale $m \in \mathcal{M}_{0,loc}^c$, if the function g is $\mathcal{P}(\mathbf{F})$-measurable and $\left[|g|^2 \cdot \langle m, m \rangle\right]^{j/2} \in \mathcal{A}_{loc}$, then the stochastic integral, $g \cdot m \in \mathcal{M}_{loc}^{j,c}$, $j = 1, 2$, is well defined; again, the integral is over the interval $[0, t]$.

Integrals with respect to random measures are

$$k_d * \mu_t^d = \int_{0+}^{t} \int_{\mathbb{E}} k_d(s,u) \mu^d(ds,du), \quad k_g * \mu_t^g = \int_{0}^{t-} \int_{\mathbb{E}} k_g(s,u) \mu^g(ds,du),$$

$$h_d * (\mu^d - \nu^d)_t = \int_{0+}^{t} \int_{\mathbb{E}} h_d(\omega,s,u)(\mu^d - \nu^d)(ds,du),$$

$$h_g * (\mu^g - \nu^g)_t = \int_{0}^{t-} \int_{\mathbb{E}} h_g(\omega,s,u)(\mu^g - \nu^g)(ds,du),$$

$$r_d * p^d = \int_{0+}^{t} \int_{\mathbb{E}} r_d(\omega,s,u) p^d(ds,du), \quad r_g * p^g = \int_{0}^{t-} \int_{\mathbb{E}} r_g(\omega,s,u) p^g(ds,du),$$

$$w_g * \eta^g = \int_{0}^{t-} \int_{\mathbb{E}} w_g(\omega,s,u) \eta^g(ds,du),$$

where "$*$" means integral with respect to random measures for any type of jump, and differences are recognized by the symbols d and g for right and left jumps, respectively.

If the function h_d is $\widetilde{\mathcal{P}}(\mathbf{F})$-measurable and $\left[|h_d|^2 * \nu^d\right]^{j/2} \in \mathcal{A}_{1oc}$, then the stochastic integral $h_d * (\mu^d - \nu^d) \in \mathcal{M}_{1oc}^{j,d}$, $j = 1,2$, is well defined. If the function h_g is $\widetilde{\mathcal{O}}(\mathbf{F})$-measurable and $\left[|h_g|^2 * \nu^g\right]^{j/2} \in \mathcal{A}_{1oc}$, then the stochastic integral $h_g * (\mu^g - \nu^g) \in \mathcal{M}_{1oc}^{j,g}$, $j = 1,2$, is also well defined. If k_d is $\widetilde{\mathcal{G}}(\mathbf{F})$-measurable and $|k_d| * \mu^d \in \mathcal{V}$, then the integral $f * \mu^d \in \mathcal{V}$ is defined (see [63]). And, if k_g is $\widetilde{\mathcal{G}}(\mathbf{F})$-measurable and $|k_g| * \mu^g \in \mathcal{V}$, then the integral $k_g * \mu^g \in \mathcal{V}$ is defined.

If r_d is $\widetilde{\mathcal{P}}(\mathbf{F})$-measurable, $\left[|r_d|^2 * p^d\right]^{j/2} \in \mathcal{A}_{1oc}$ and for any predictable stopping time S, we have that $\mathbf{E}\left[r_d(S, \beta_S^d) | \mathcal{F}_{S-}\right] = 0$ a.s., then the stochastic integral $r_d * p^d \in \mathcal{M}_{1oc}^{j,d}$, $j = 1,2$, is defined. And, if $r_d \in \widetilde{\mathcal{G}}(\mathbf{F})$ and $|r_d| * p^d \in \mathcal{V}$ then the integral $r_d * p^d \in \mathcal{V}$ is defined (see [68]). Note that the facts used below in the theory of martingales can be found in [59, 63, 66]. If r_g is $\widetilde{\mathcal{O}}(\mathbf{F})$-measurable, $\left[|r_g|^2 * p^g\right]^{j/2} \in \mathcal{A}_{1oc}$ and for any totally inaccessible stopping time T, $\mathbf{E}\left[r_g(T, \beta_T^g) | \mathcal{F}_T\right] = 0$ a.s., then the stochastic integral $r_g * p^g \in \mathcal{M}_{1oc}^{j,g}$, $j = 1,2$, is defined. And, if $r_g \in \widetilde{\mathcal{G}}(\mathbf{F})$ and $|r_g| * p^g \in \mathcal{V}$ then the integral $r_g * p^g \in \mathcal{V}$ is defined. If w_g is $\widetilde{\mathcal{O}}(\mathbf{F})$-measurable, $\left[|w_g|^2 * \eta^g\right]^{j/2} \in \mathcal{A}_{1oc}$ and for any totally inaccessible stopping time in the broad sense U, $\mathbf{E}\left[w_g(U, \beta_U^g) | \mathcal{F}_U\right] = 0$ a.s., then the stochastic integral $w_g * \eta^g \in \mathcal{M}_{1oc}^{j,g}$, $j = 1,2$, is defined. And, if $w_g \in \widetilde{\mathcal{G}}(\mathbf{F})$ and $|w_g| * \eta^g \in \mathcal{V}$, then the integral $w_g * \eta^g \in \mathcal{V}$ is defined.

Notation 7.4.4 We have used $i \in (d,g)$ to clearly identify the different types of optional semimartingales. However, from now on, for convenience, we are going to identify "d" by 1 the right-continuous discrete component of the semimartingale and "g" by 2 the left-continuous discrete part of the semimartingale to give a concise description (i.e., $i \in (1,2)$).

With new notations, the optional semimartingale X has the following component representation,

$$X = X_0 + a + m$$
$$+ \int_{0+}^{t} \int_{\mathbb{E}} U(\mu^1 - \nu^1)(ds, du) + \int_{0+}^{t} \int_{\mathbb{E}} V\mu^1(ds, du) + \int_{0+}^{t} \int_{\mathbb{E}} up^1(ds, du)$$
$$+ \int_{0}^{t-} \int_{\mathbb{E}} U(\mu^2 - \nu^2)(ds, du) + \int_{0}^{t-} \int_{\mathbb{E}} V\mu^2(ds, du) + \int_{0}^{t-} \int_{\mathbb{E}} up^2(ds, du)$$
$$+ \int_{0}^{t-} \int_{\mathbb{E}} u\eta(ds, du),$$

where $U = u1_{|u| \leq 1}$, $V = u1_{|u| > 1}$ and $\eta = \eta^g$. The component representation of optional semimartingale will be the representation form that we will use to construct the comparison lemma.

Before we get to the main theorem, we need to extend the change of variables formula of the component representation of semimartingales in the usual conditions (cf. [68]) to optional semimartingales in the *un*usual case.

Lemma 7.4.5 Suppose an optional semimartingale $Y = (Y^1, Y^2, ..., Y^k)$ is defined by the relation

$$Y_t = Y_0 + f \circ a_t + g \circ m_t + (r + w) * \eta_t,$$
$$+ \sum_j U H_j * (\mu^j - \nu^j)_t + V h_j * \mu_t^j + (k_j + l_j) * p_t^j,$$

where all the integrals are well defined. Consider the function $F(y) = F(y^1, y^2, ..., y^k)$ to be twice continuously differentiable on $\mathbb{R}^k$.
Then the process $F(Y) = (F(Y_t))_{t \geq 0}$ is an optional semimartingale and has the representation

$$F(Y_t) = F(Y_0) + F'(Y)f \circ a_t + F'(Y)g \circ m_t + \frac{1}{2}F''(Y)g^2 \circ \langle m, m \rangle_t$$
$$+ \sum_j U\left[F\left(Y + H_j\right) - F(Y)\right] * (\mu^j - \nu^j)_t$$
$$+ \sum_j V\left[F\left(Y + h_j\right) - F(Y)\right] * \mu_t^j$$
$$+ \sum_j U\left[F\left(Y + H_j\right) - F(Y) + F'(Y)H_j\right] * \nu_t^j$$
$$+ \sum_j \left[F\left(Y + (k_j + l_j)\right) - F(Y)\right] * p_t^j$$
$$+ \left[F\left(Y + (r + w)\right) - F(Y)\right] * \eta_t.$$

Proof. Galchuk [68] proved the change of variable formula for semimartingales under the usual conditions. Extending the proof to optional semimartingales is straightforward. ∎

7.5   Uniform Doob-Meyer Decompositions

Under the *usual conditions*, the Doob-Meyer decomposition (DMD) and the Uniform Doob-Meyer decomposition (UDMD), commonly known as the *optional decomposition*, are fundamental results in stochastic analysis with many applications. For example, in the area of mathematical finance, UDMD allows for the construction of superhedging strategies in incomplete markets. It asserts that if, for a non-empty set, $\mathbb{Q}(X)$, of equivalent local martingale measures for the semi-martingale, X, such that, the stochastic process Y is a local supermartingale with respect to all measures in $\mathbb{Q}(X)$, then there exists a unique predictable stochastic integrand φ such that the difference $Y - \varphi \cdot X$ is a decreasing optional process for all measures in $\mathbb{Q}(X)$. In contrast to the classical Doob-Meyer decomposition, the process $Y - \varphi \cdot X$ is optional and is not uniquely determined. Moreover, UDMD is *universal*, meaning that it holds simultaneously for all probability measures in $\mathbb{Q}(X)$.

In incomplete markets, the value process Y of an $\mathcal{F}_T$-measurable contingent claim $H \geq 0$ can be defined as the essential supremum of the conditional expectations over the class of all equivalent martingale measures in $\mathbb{Q}(X)$. In this case, Y is the right-continuous version of the process given by

$$Y_t = ess \sup_{\mathbf{Q} \in \mathbb{Q}(X)} \mathbf{E_Q}[H|\mathcal{F}_t].$$

It follows that Y is a supermartingale with respect to all $\mathbf{Q} \in \mathbb{Q}(X)$. Then, UDMD is used to identify Y as the value process of a superhedging strategy (φ, C) satisfying

$$Y = Y_0 + \varphi \cdot X - C,$$

where the integrand φ specifies the number of units of the underlying asset X, and generates an increasing optional process $-C = Y - Y_0 - \varphi \cdot X$ of cumulative side payments with $C_0 = 0$. This strategy induces a perfect hedge in the class of investment strategies with terminal capital $Y_T = H$.

The existence of UDMD for supermartingales was first presented by El Karoui and Quenez [69] for diffusion processes. Kramkov [70] proved the existence of UDMD for locally bounded super-martingales. Follmer and Kabanov [71] proved UDMD without the local boundedness assumption thus permitting the inclusion of models with unbounded jumps. The authors also provided an interpretation of the integrand values as Lagrange multipliers for an optimization problem with constraints. Stricker and Yan [72] proved UDMD for super- and submartingales in a general context. Recently, Jacka [73] presented a simple proof of UDMD without integral representation of the local martingale in terms of the underlying assets and with the set of local martingale measures satisfying some closure property. Karatzas *et al.* [74] presented a specific treatment of the UDMD for continuous semimartingales and general filtrations where their method does not assume the existence of equivalent local martingale measure(s), only that of strictly positive local martingale deflator(s). All these results assumed that the stochastic basis satisfies the *usual conditions* and all the processes are r.c.l.l. Therefore, the problem of finding an adequate form of the Uniform

Doob-Meyer decomposition for *optional local supermartingales* on *un*usual probability spaces becomes an important question to be answered. In this section, we extend the works of Kabanov and Follmer, and, Stricker and Yan to derive a uniform Doob-Meyer decomposition theorem for *optional* super/submartingales on *un*usual probability spaces.

Next we prove the existence of UDMD on *un*usual probability spaces. In Chapter 11 we show how to apply this decomposition to the *filtering problem* for optional supermartingales.

Theorem 7.5.1 Let X be an $\mathbb{R}^d$-valued r.l.l.l. optional semimartingale on the *un*usual stochastic basis $(\Omega, \mathcal{F}, \mathbf{F} = (\mathcal{F}_t), \mathbf{P})$. Let $\mathbb{Q}(X)$ be the set of all probability measures $\mathbf{Q} \sim \mathbf{P}$ such that X is a $\mathbf{Q}$-local optional martingale. Suppose that $\mathbb{Q}(X) \neq \emptyset$ and Y is a non-negative process. Then Y is a $\mathbf{Q}$ local optional supermartingale for each $\mathbf{Q} \in \mathbb{Q}(X)$ if and only if there exist a unique optional process φ integrable with respect to X and an adapted optional increasing process C with $C_0 = 0$ such that $Y = Y_0 + \varphi \circ X - C$ for each $\mathbf{Q} \in \mathbb{Q}(X)$.

Proof. In the proof, we will be using auxiliary facts formulated as lemmas right after the formal proof. Let $\mathbb{Z}(X)$ be the set of all strictly positive local optional martingales Z with $Z_0 = 1$ such that ZX is a local optional martingale. Corresponding to $\mathbb{Z}(X)$ is $\mathbb{Q}(X)$ the set of all probability measures $\mathbf{Q}$ such that $\mathbf{Q} \sim \mathbf{P}$ and X is a local optional martingale with respect to $\mathbf{Q}$.

Suppose $\mathbb{Z}(X) \neq \varnothing$ and consider the right-continuous version $\mathcal{F}_{t+}$ of the filtration $\mathcal{F}_t$ where X_{t+}, Y_{t+}, $Z_{t+}Y_{t+}$ and $Z_{t+}X_{t+}$ are r.c.l.l. $\mathcal{F}_{t+}$-measurable processes. ZX is a local optional martingale and ZY is a local optional supermartingale. As well, $Z_{t+}X_{t+}$ is a local martingale and $Z_{t+}Y_{t+}$ a local supermartingale (see Lemma 7.5.3).

Let $\mathbb{Z}^+(X)$ be the collection of all Z_{t+} r.c.l.l. version of the local martingale deflator Z of X. With the collection $\mathbb{Z}^+(X)$, X_{t+} is the r.c.l.l. $\mathcal{F}_{t+}$-measurable local martingale and Y_+ is r.c.l.l. local supermartingale (see Lemma 7.5.4). Let us re-label processes and filtrations that are subscripted with "$t+$", with the superscript "$+$" to avoid any possible confusion as to the meaning of the plus sign. So, let $X_t^+ := X_{t+}$, $Y_t^+ := Y_{t+}$, $Z_t^+ := Z_{t+}$ and $\mathcal{F}_t^+ := \mathcal{F}_{t+}$.

Since $(\Omega, \mathcal{F}, \mathbf{P})$ is a complete probability space and $\mathbf{F}_+ = (\mathcal{F}_{t+})$ is right-continuous filtration, then by way of Remark 7.5.5 and Lemma 7.5.6 the probability space $(\Omega, \mathcal{F}, \mathbf{F}_+, \mathbf{P})$ satisfies the *usual* conditions. Therefore, by uniform Doob-Meyer decomposition in the *usual* case, we find (φ, C^+), where, in this case, φ is $\mathcal{F}_t$-measurable and C^+ is $\mathcal{F}_{t+}$-measurable increasing process such that

$$Y_t^+ = Y_0 + \varphi \cdot X_t^+ - C_t^+, \tag{7.13}$$

with $C_0^+ = 0$ for any Z^+ in $\mathbb{Z}^+(X)$. By taking optional projection of Equation (7.13) on $\mathcal{F}_t$ with the expectation operator "$\mathbf{E}$", we arrive at the result we desire. In what follows, we show the steps in detail.

Z^+Y^+ is a local supermartingale for any $Z^+ \in \mathbb{Z}^+(X)$, whereas Z^+X^+ is a local martingale under the right-continuous filtration $\mathcal{F}_t^+$ and complete space $(\Omega, \mathcal{F}, \mathbf{P})$. By UDMD in the usual case we get

$$Z^+Y^+ = Y_0 + \varphi \cdot Z^+X^+ - C^+ \Rightarrow$$
$$Z^+Y^+ + C^+ = Y_0 + \varphi \cdot Z^+X^+. \tag{7.14}$$

The right hand side of Equation (7.14) is a local martingale, therefore, the left-hand side is a also a local martingale. Taking projections on the *un*usual σ-algebra $\mathcal{F}_t$ we get

$$\mathbf{E}\left[Z_t^+ Y_t^+ + C_t^+ | \mathcal{F}_t\right] = Y_0 + \mathbf{E}\left[\varphi \cdot (Z^+ X^+)_t | \mathcal{F}_t\right] \tag{7.15}$$

which, by the argument we will present next, implies

$$Z_t Y_t + C_t = Y_0 + \mathbf{E}\left[\varphi \cdot (Z^+ X^+)_t | \mathcal{F}_t\right] \Rightarrow \tag{7.16}$$

$$Z_t Y_t = Y_0 + \varphi \circ (ZX)_t - C_t, \tag{7.17}$$

is true for any Z, where $C_t = \mathbf{E}\left[C_t^+ | \mathcal{F}_t\right]$.

Equation (7.16) leads to Equation (7.17) in which the usual stochastic integral $\varphi \cdot Z^+ X^+$ under projection on $\mathcal{F}_t$ has been transformed to the optional stochastic integral with respect to an r.l.l.l. process, that is $\mathbf{E}\left[\varphi \cdot Z^+ X^+ | \mathcal{F}_t\right] = \varphi \circ ZX$. To show how we arrived to this result, let $M^+ := Z^+ X^+$ and consider the statement

$$\mathbf{E}\left[\varphi \cdot M_t^+ - \varphi \cdot M_s^+ | \mathcal{F}_s\right] = \mathbf{E}\left[\mathbf{E}\left[\varphi \cdot M_t^+ - \varphi \cdot M_s^+ | \mathcal{F}_s^+\right] | \mathcal{F}_s\right] = 0,$$

since $\mathcal{F}_s^+ \supseteq \mathcal{F}_s$ for all s. Then,

$$
\begin{aligned}
\mathbf{E}\left[\varphi \cdot M_t^+ - \varphi \cdot M_s^+ | \mathcal{F}_s\right] &= \mathbf{E}\left[\mathbf{E}\left[\varphi \cdot M_t^+ - \varphi \cdot M_s^+ | \mathcal{F}_s^+\right] | \mathcal{F}_s\right] \\
&= \mathbf{E}\left[\int_{(0,t]} \varphi_{u-} dM_u^+ - \int_{(0,s]} \varphi_{u-} dM_u^+ | \mathcal{F}_s\right] \\
&= \mathbf{E}\left[\int_{(0,s]} \varphi_{u-} dM_u^+ + \int_{(s,t]} \varphi_{u-} dM_u^+ - \int_{(0,s]} \varphi_{u-} dM_u^+ | \mathcal{F}_s\right] \\
&= \mathbf{E}\left[\int_{(s,t]} \varphi_{u-} dM_u^+ | \mathcal{F}_s\right] \\
&= \mathbf{E}\left[\int_{(s,t]} \varphi_{u-} \left(dM_u^{r,+} + \triangle M_u^{g,+}\right) | \mathcal{F}_s\right],
\end{aligned}
$$

since $M_{u+} = M_{u+}^r + M_{u+}^g$,

$$= \mathbf{E}\left[\int_{(s,t]} \varphi_{u-} dM_u^{r,+} | \mathcal{F}_s\right] + \mathbf{E}\left[\int_{(s,t]} \varphi_{u-} \triangle M_u^{g,+} | \mathcal{F}_s\right]; \tag{7.18}$$

But since $M_u^{g,+}$ (i.e., M_{u+}^g) is evolving in the interval $(s,t]$, it follows that M_u^g is evolving in the interval $[s,t)$ and (7.18) can be written as

$$\mathbf{E}\left[\varphi \cdot M_t^+ - \varphi \cdot M_s^+ | \mathcal{F}_s\right] = \mathbf{E}\left[\int_{(s,t]} \varphi_{u-} dM_u^{r,+} | \mathcal{F}_s\right] + \mathbf{E}\left[\int_{[s,t)} \varphi_v \triangle M_v^{g,+} | \mathcal{F}_s\right].$$

But since $\triangle M_v^{g,+} = M_{v+}^g - M_v^g = \triangle^+ M_v^g$ and $M_u^{r,+} = M_u^r$ then,

$$= \mathbf{E}\left[\int_{(s,t]} \varphi_{u-} dM_u^r + \int_{[s,t)} \varphi_v \triangle^+ M_v^g | \mathcal{F}_s\right]$$

$$= \mathbf{E}\left[\int_{(s,t]} \varphi_{u-} dM_u^r + \int_{[s,t)} \varphi_v dM_{v+}^g | \mathcal{F}_s\right]$$

$$= \mathbf{E}\left[\int_{[s,t]} \varphi_u dM_u | \mathcal{F}_s\right]$$

$$= \mathbf{E}\left[\int_{[0,t]} \varphi_u dM_u | \mathcal{F}_s\right] + \mathbf{E}\left[\int_{[0,s)} \varphi_u dM_u | \mathcal{F}_s\right]$$

$$= \mathbf{E}\left[\varphi \cdot M_t^+ | \mathcal{F}_s\right] - \mathbf{E}\left[\varphi \circ M_s | \mathcal{F}_s\right]$$

$$= \mathbf{E}\left[\varphi \cdot M_t^+ - \varphi \circ M_s | \mathcal{F}_s\right]$$

$$= \mathbf{E}\left[\varphi \cdot M_t^+ | \mathcal{F}_s\right] - \varphi \circ M_s = 0 \Rightarrow \mathbf{E}\left[\varphi \cdot Z^+ X_t^+ | \mathcal{F}_s\right] = \varphi \circ Z X_s. \tag{7.19}$$

Alternatively, the same result (7.19) can be proved by using the definition of the integral by limits of sums. We will demonstrate this results as follows. In the limit of the partitions Π_n of $[s,t]$ for any s and t and for any $u \in [s,t]$ the integrand φ is the limit of simple functions,

$$\varphi_t \longleftarrow \varphi_s + \sum \varphi_{t_{i-1}} 1_{]t_{i-1},t_{i+1}]}(u).$$

Note that we have chosen the intervals $]t_{i-1}, t_{i+1}]$ where $\varphi_{t_{i-1}} = \varphi_{t_i}$. Since the stochastic integral is well defined under any refinement, we are justified in using this interval.

Consider a partition Π_n of $[s,t]$. The projection of the integral $\int_s^t \phi_{u-} dM_u^+$ on $\mathcal{F}_s$ is

$$\mathbf{E}\left[\varphi \cdot M_t^+ - \varphi \cdot M_s^+ | \mathcal{F}_s\right] = \mathbf{E}\left[\int_s^t \varphi_{u-} dM_u^+ | \mathcal{F}_s\right]$$

$$= \mathbf{E}\left[\lim \sum \varphi_{t_{i-1}} \left(M_{t_{i+1}}^+ - M_{t_{i-1}}^+\right) | \mathcal{F}_s\right],$$

by localization and integrability,

$$= \lim \mathbf{E}\left[\sum \varphi_{t_{i-1}} \left(M_{t_{i+1}}^+ - M_{t_{i-1}}^+\right) | \mathcal{F}_s\right]$$

$$= \lim \mathbf{E}\left[\sum \varphi_{t_{i-1}} \left(M_{t_{i+1}}^+ - M_{t_i}^+\right) + \varphi_{t_{i-1}} \left(M_{t_i}^+ - M_{t_{i-1}}^+\right) | \mathcal{F}_s\right],$$

by $\varphi_{t_{i-1}} = \varphi_{t_i}$,

$$= \lim \mathbf{E}\left[\sum \varphi_{t_i} \left(M_{t_{i+1}}^+ - M_{t_i}^+\right) + \varphi_{t_{i-1}} \left(M_{t_i}^+ - M_{t_{i-1}}^+\right) | \mathcal{F}_s\right]. \tag{7.20}$$

Using the properties of conditional expectations and using the fact that $\mathcal{F}_{t_i} \supseteq \mathcal{F}_s$ for any t_i and that φ_t is $\mathcal{F}_t$ measurable, the chain of equalities (7.20) continues as

$$= \mathbf{E}\left[\lim \sum \begin{array}{c} \mathbf{E}\left[\varphi_{t_i}M_{t_{i+1}}^+|\mathcal{F}_{t_{i+1}}\right] - \mathbf{E}\left[\varphi_{t_i}M_{t_i}^+|\mathcal{F}_{t_i}\right] \\ +\mathbf{E}\left[\varphi_{t_{i-1}}M_{t_i}^+|\mathcal{F}_{t_i}\right] - \mathbf{E}\left[\varphi_{t_{i-1}}M_{t_{i-1}}^+|\mathcal{F}_{t_{i-1}}\right] \end{array} |\mathcal{F}_s\right]$$

$$= \mathbf{E}\left[\lim \sum \begin{array}{c} \varphi_{t_i}\mathbf{E}\left[M_{t_{i+1}}^+|\mathcal{F}_{t_{i+1}}\right] - \varphi_{t_i}\mathbf{E}\left[M_{t_i}^+|\mathcal{F}_{t_i}\right] \\ +\varphi_{t_{i-1}}\mathbf{E}\left[M_{t_i}^+|\mathcal{F}_{t_i}\right] - \varphi_{t_{i-1}}\mathbf{E}\left[M_{t_{i-1}}^+|\mathcal{F}_{t_{i-1}}\right] \end{array} |\mathcal{F}_s\right],$$

by $M_u^+ = M_{u+} = M_u + \triangle^+ M_u$

$$= \mathbf{E}\left[\lim \sum \begin{array}{c} \varphi_{t_{i-1}}\left(M_{t_i} - M_{t_{i-1}}\right) + \varphi_{t_i}\left(M_{t_{i+1}} - M_{t_i}\right) \\ +\varphi_{t_{i-1}}\left(\triangle^+ M_{t_i} - \triangle^+ M_{t_{i-1}}\right) + \varphi_{t_i}\left(\triangle^+ M_{t_{i+1}} - \triangle^+ M_{t_i}\right) \end{array} |\mathcal{F}_s\right]$$

$$- \mathbf{E}\left[\lim \sum \begin{array}{c} \varphi_{t_{i-1}}\left(M_{t_i} - M_{t_{i-1}}\right) + \varphi_{t_i}\left(M_{t_{i+1}} - M_{t_i}\right) \\ +\varphi_{t_{i-1}}\left(\triangle^+ M_{t_i} - \triangle^+ M_{t_{i-1}}\right) + \varphi_{t_i}\left(\triangle^+ M_{t_{i+1}} - \triangle^+ M_{t_i}\right) \end{array} |\mathcal{F}_s\right]$$

$$= \mathbf{E}\left[\int_{s+}^t \varphi_{u-}dM_u^r + \int_s^{t-} \varphi_u dM_{u+}^g|\mathcal{F}_s\right]$$

$$= \mathbf{E}\left[\int_{0+}^t \varphi_{u-}dM_u^r + \int_0^{t-} \varphi_u dM_{u+}^g - \int_{0+}^s \varphi_{u-}dM_u^r - \int_0^{s-} \varphi_u dM_{u+}^g|\mathcal{F}_s\right]$$

$$= \mathbf{E}\left[\int_0^t \varphi_{u-}dM_u^+ - \int_{0+}^s \varphi_{u-}dM_u^r - \int_0^{s-} \varphi_u dM_{u+}^g|\mathcal{F}_s\right]$$

$$= \mathbf{E}\left[\varphi \cdot M_t^+ - \varphi_- \cdot M_s^r - \varphi \odot M_{s+}^g|\mathcal{F}_s\right] \; = \; \mathbf{E}\left[\varphi \cdot M_t^+ - \varphi \circ M_s|\mathcal{F}_s\right]$$

$$= \mathbf{E}\left[\varphi \cdot M_t^+|\mathcal{F}_s\right] - \varphi \circ M_s \; = \; 0$$

Therefore, $\mathbf{E}\left[\varphi \cdot M_t^+|\mathcal{F}_s\right] = \varphi \circ M_s = \varphi \cdot M_s^r + \varphi \odot M_{s+}^g$ for any $s \le t$. Note that in the limit $\triangle^+ M_{t_i} - \triangle^+ M_{t_{i-1}} = \triangle^+ M_{t_{i+1}} - \triangle^+ M_{t_i} = 0$.

Having carried optional projection on the right-hand side of Equation (7.14) we consider next the left-hand side, $Z^+ Y^+ + C^+$. Let $N_t^+ := Z_{t+}Y_{t+} + C_t^+ = Z_t^+ Y_t^+ + C_t^+$ which is an $\mathbf{F}_+$ r.c.l.l. local martingale. Using optional projection of N^+ on $\mathbf{F}$ we find the local optional martingale $N_u := \mathbf{E}\left[N_t^+|\mathcal{F}_u\right]$ for all $u \le t$ by Lemma 7.5.2 or simply by equality (7.15).

The process $Z^+ Y^+$ is a local supermartingale for which the projection on $\mathbf{F}$ gives $\mathbf{E}\left[Z_{t+}Y_{t+}|\mathcal{F}_t\right] \le Z_t Y_t$. Since $N_t = \mathbf{E}\left[N_t^+|\mathcal{F}_t\right]$, we have

$$N_{t+} = \mathbf{E}\left[N_t^+|\mathcal{F}_{t+}\right] = \mathbf{E}\left[Z_{t+}Y_{t+} + C_t^+|\mathcal{F}_{t+}\right] = Z_{t+}Y_{t+} + C_t^+$$

and $N_{t+} = N_t^+$ is r.c.l.l. Consequently, $N_t = Z_t Y_t + C_{t-}^+$ where C_{t-}^+ is optional $\mathcal{F}_t$ adapted increasing process.

Alternatively, one can consider the difference $N_{t+} - Z_{t+}Y_{t+}$, which is an increasing process. Hence, $N_t - Z_t Y_t$ is a local optional submartingale, i.e., $N_t - Z_t Y_t = \mathbf{E}\left[N_{t+} - Z_{t+}Y_{t+}|\mathcal{F}_t\right]$ having the decomposition, $N_t - Z_t Y_t = \tilde{N}_t + A_t$, where A is an increasing optional process and $\tilde{N}_t$ is a local optional martingale. Therefore, $N_t - \tilde{N}_t = Z_t Y_t + A_t$ and $N_{t+} - \tilde{N}_{t+} = Z_{t+}Y_{t+} + A_{t+}$ which implies that $C_t^+ = \tilde{N}_{t+} + A_{t+}$. This leads us to conclude that since C_t^+ is increasing, then it must be that $\tilde{N}_{t+} = 0$ and $\tilde{N}_t = \mathbf{E}\left[\tilde{N}_{t+}|\mathcal{F}_t\right] = 0$ for all t. ∎

7.5.1 Supporting lemmas

Lemma 7.5.2 Suppose that N_{t+} is an $\mathbf{F}_+$ local martingale, then N_t is an $\mathbf{F}$ local optional martingale.

Proof. Let $N_t^+ := N_{t+}$. For any $\mathbf{F}_+$ stopping time $\tau \geq t$, $\mathbf{E}\left[N_\tau^+ | \mathcal{F}_t^+\right] = N_t^+$, since N_{t+} is an $\mathbf{F}_+$ local martingale. Then, N_t is an $\mathbf{F}$ local optional martingale, for that

$$N_t := \mathbf{E}\left[N_\tau^+ | \mathcal{F}_t\right] \Rightarrow$$

$$\mathbf{E}\left[N_t | \mathcal{F}_s\right] = \mathbf{E}\left[\mathbf{E}\left[N_\tau^+ | \mathcal{F}_t\right] | \mathcal{F}_s\right] = \mathbf{E}\left[N_\tau^+ | \mathcal{F}_s\right] = N_s$$

for any $s \leq t$. ∎

Lemma 7.5.3 Given that $Z_t X_t$ is a local optional martingale and $Z_t Y_t$ is a local optional supermartingale on $\mathbf{F}$, then $Z_{t+} X_{t+}$ is a local martingale and $Z_{t+} Y_{t+}$ a local supermartingale on $\mathbf{F}_+$.

Proof. First we show that, if ZY is a local optional supermartingale under $\mathbf{F}$, then $Z^+ Y^+$ is a local supermartingale under $\mathbf{F}_+$, that is $\mathbf{E}\left[Z_t^+ Y_t^+ | \mathcal{F}_s^+\right] \leq Z_s^+ Y_s^+$ for any $s \leq t$.
Let $S_t = Z_t Y_t$ then $\mathbf{E}[S_t | \mathcal{F}_s] \leq S_s$ and $\mathbf{E}[S_{t+\epsilon} | \mathcal{F}_{s+\delta}] \leq S_{s+\delta}$ for any $s + \delta \leq t + \epsilon$ and $\delta > 0$ and $\epsilon > 0$. Then,

$$\lim_{\delta \downarrow 0} \lim_{\epsilon \downarrow 0} \mathbf{E}\left[S_{t+\epsilon} | \mathcal{F}_{s+\delta}\right] \leq \lim_{\delta \downarrow 0} \lim_{\epsilon \downarrow 0} S_{s+\delta} \Rightarrow$$

$$\lim_{\delta \downarrow 0} \mathbf{E}\left[S_{t+} | \mathcal{F}_{s+\delta}\right] \leq S_{s+} \Rightarrow \mathbf{E}\left[S_{t+} | \mathcal{F}_{s+}\right] \leq S_{s+} \Rightarrow$$

$$\mathbf{E}\left[S_t^+ | \mathcal{F}_s^+\right] \leq S_s^+ \Rightarrow \mathbf{E}\left[Z_t^+ Y_t^+ | \mathcal{F}_s^+\right] \leq Z_s^+ Y_s^+.$$

Similarly, ZX is a local optional martingale under $\mathbf{F}$, then $Z^+ X^+$ is a local martingale under $\mathbf{F}_+$. ∎

Corollary 7.5.4 Let the collection $\mathbb{Z}^+(X)$ be of all r.c.l.l. Z_{t+} versions of the local optional martingale deflators Z of X. With the collection $\mathbb{Z}^+(X)$, X_{t+} is r.c.l.l. local martingale and Y_+ is r.c.l.l. local supermartingale.

Proof. By way of the above lemma. ∎

Remark 7.5.5 Under the *un*usual conditions, the probability space $(\Omega, \mathcal{F}, \mathbf{P})$ is complete – $\mathcal{F}$ is a complete σ-algebra and $\mathbf{P}$ is a complete probability measure. Hence, the expectation operator $\mathbf{E}$ is also complete as a measure. However, under the *un*usual conditions the filtrations $\mathcal{F}_t$ and $\mathcal{F}_{t+}$ are not complete. In the proof of UDMD theorem, we have used the optional projection of stochastic process on $\mathcal{F}_t$ and $\mathcal{F}_{t+}$ where in certain cases the conditioning of an optional process on an *un*usual filtration leads to subsets of null sets that are not measurable in neither $\mathcal{F}_t$ nor $\mathcal{F}_{t+}$. But since $\mathbf{P}$ has been completed, these sets pose no problem as their probability values are zero, hence their expectation.

Lemma 7.5.6 A complete measure completes an incomplete sub-σ-algebra.

Proof. Given a complete probability space $(\Omega, \mathcal{F}, \mathbf{P})$ and a σ-algebra $\mathcal{G} \subset \mathcal{F}$ that is not complete, then, for any zero measure set $A \in \mathcal{F}$, i.e., $\mathbf{E}[1_A] = 0$, the expectation $\mathbf{E}[1_A|\mathcal{G}] = \mathbf{E}[\mathbf{E}[1_A|\mathcal{F}]|\mathcal{G}] = 0$ a.s. $\mathbf{P}$. Moreover, for any subset B of C such that $\mathbf{E}[1_C|\mathcal{G}] = 0$, the expectation $\mathbf{E}[1_B|\mathcal{G}] = \mathbf{E}[\mathbf{E}[1_B|\mathcal{F}]|\mathcal{G}] = 0$. Therefore, the measure $\mathbf{E}_\mathcal{G}[\cdot] := \mathbf{E}[\cdot|\mathcal{G}]$ maps subsets of zero measure sets in $\mathcal{G}$ to zero. Let

$$\mathcal{N}_\mathcal{G} := \{B : \mathbf{E}_\mathcal{G}[B] = 0\}.$$

If we augment $\mathcal{G}$ with $\mathcal{N}_\mathcal{G}$, we complete $\mathcal{G}$ and obtain a complete measurable space $(\Omega, \mathcal{G} \vee \mathcal{N}_\mathcal{G}, \mathbf{E}_\mathcal{G})$. ∎

Stricker and Yan [72] proved uniform Doob-Meyer decomposition of optional submartingales under the usual conditions. An extension to the *un*usual case is straightforward by employing the same methods we have developed in the proof of Theorem 7.5.1. Below, a version of UDMD for local optional submartingales under the *un*usual conditions is presented.

Theorem 7.5.7 Let X be $\mathbb{R}^d$-valued r.l.l.l. optional semimartingale on the *un*usual stochastic basis $(\Omega, \mathcal{F}, \mathbf{F} = (\mathcal{F}_t)_{t \geq 0}, \mathbf{P})$. Let $\mathbb{Q}(X)$ be the set of all probability measures $\mathbf{Q} \sim \mathbf{P}$ such that X is a $\mathbf{Q}$ local optional martingale. Suppose $\mathbb{Q}(X) \neq \emptyset$ and let U be a non-negative process. Then U admits the decomposition $U = U_0 + \psi \circ X + A$, where ψ is X-integrable optional process and A is an adapted increasing process with $A_0 = 0$, if and only if U is a local optional submartingale for each $\mathbf{Q} \in \mathbb{Q}(X)$.

Proof. By way of the procedure we have developed in Theorem 7.5.1 and submartingale uniform Doob-Meyer decomposition in the usual conditions proved in [72], the result of the theorem is established. ∎

As an application of the uniform Doob-Meyer decomposition, we will look at the filtering of optional semimartingales. Filtering theory has a vast literature investigating different aspects of the filtering problem and its applications, see [2, 75, 76, 77, 78, 79, 80, 81, 82].

Chapter 8

Optional Stochastic Equations

Stochastic equations have numerous applications in engineering, finance, and physics and are a central part of stochastic analysis. Moreover, and perhaps a fundamental aspect of stochastic equations, is that they provide a way of manufacturing complex processes from ones that are simpler. For example, a geometric Brownian motion is constructed from the simpler Weiner process. Progress in the study of stochastic equations came as a result of the development of semimartingale integration theory which gave the study of stochastic equations a strong theoretical basis. As the theory of stochastic integration advances beyond usual probability spaces and cadlag processes, stochastic equations will also be advanced in those directions too.

This is the main goal of this chapter, to study stochastic equations driven by optional semimartingales. We will cover the topics of stochastic linear equations, stochastic exponential and logarithms, solutions of the nonhomogeneous stochastic linear equation, the Gronwall lemma, existence and uniqueness of solution of stochastic equations under monotonicity conditions and comparison lemma under Yamada conditions.

8.1 Linear Equations, Exponentials and Logarithms

Stochastic exponentials and logarithms are indispensable tools of stochastic calculus, differential equations and financial mathematics. For example, they describe relative returns, link hedging with the calculation of minimal entropy and utility indifference. Moreover, they determine the structure of the Girsanov transformation.

8.1.1 Stochastic exponential

For optional semimartingales $X \in \mathcal{S}(\mathbf{F}, \mathbf{P})$, then there exists a unique semimartingale $Z \in \mathcal{S}(\mathbf{F}, \mathbf{P})$, the stochastic exponential of X, such that

$$Z_t = Z_0 + Z \circ X_t = Z_0 + \int_{0+}^{t} Z_{s-} dX_s^r + \int_{0}^{t-} Z_s dX_{s+}^g, = Z_0 \mathcal{E}(X)_t, \tag{8.1}$$

$$\mathcal{E}(X)_t = \exp\left(X_t - \frac{1}{2}\langle X^c, X^c \rangle\right) \prod_{0 \leq s \leq t} (1 + \triangle X_s)e^{-\triangle X_s} \prod_{0 \leq s < t} (1 + \triangle^+ X_s)e^{-\triangle^+ X_s}.$$

Here is the theorem that proves its existence and uniqueness:

Theorem 8.1.1 Let $X \in \mathcal{S}$. There exists a unique (to within indistinguishability) semimartingale $Z \in \mathcal{S}$ such that

$$Z_t = Z_0 + \int_{]0,t]} Z_{s-} dX_s^r + \int_{[0,t[} Z_s dX_{s+}^g. \tag{8.2}$$

The process Z is given by the formula

$$Z_t = Z_0 \exp\left(X_t - \frac{1}{2}\langle X^c, X^c \rangle\right) \prod_{0 < s \leq t} (1 + \triangle X_s)e^{-\triangle X_s} \prod_{0 \leq s < t} (1 + \triangle^+ X_s)e^{-\triangle^+ X_s}. \tag{8.3}$$

Proof. Set $U_0 = Z_0$, $V_0 = 1$,

$$U_t = Z_0 \prod_{0 < s \leq t} (1 + \triangle X_s)e^{-\triangle X_s}, \quad V_t = \prod_{0 \leq s < t} (1 + \triangle^+ X_s)e^{-\triangle^+ X_s}, \quad t > 0,$$

and let us show that these products converge absolutely a.s. for any $t \in \mathbb{R}_+$.

Since there are almost surely a finite number of jumps of the form $|\triangle X| > 1/2$, $|\triangle^+ X| > 1/2$ on any time segment, the processes

$$\prod_{0 < s \leq t} (1 + \triangle X_s \mathbf{1}(A_s^c))e^{-\triangle X_s \mathbf{1}(A_s^c)}, \quad \prod_{0 \leq s < 1} (1 + \triangle^+ X_s \mathbf{1}(B_s^c))e^{-\triangle^+ X_s \mathbf{1}(B_s^c)}$$

have a.s. finite variation on any segment, where A_s^c, B_s^c are the complements of the sets

$$A_s = \left\{|\triangle X_s| \leq \frac{1}{2}\right\}, \quad B_s = \left\{|\triangle^+ X_s| \leq \frac{1}{2}\right\}.$$

Set

$$U_t' = \prod_{0 < s \leq 1} (1 + \triangle X_s \mathbf{1}(A_s))e^{-\triangle X_s \mathbf{1}(A_s)}, \quad V_t' = \prod_{0 \leq s < t} (1 + \triangle^+ X_s \mathbf{1}(B_s))e^{-\triangle^+ X_s \mathbf{1}(B_s)}.$$

Since the products contain strictly positive quantities, one can take their logarithm:

$$\log U_t' = \sum_{0 < s \leq t} [\log(1 + \triangle X_s \mathbf{1}(A_s)) - \triangle X_s \mathbf{1}(A_s)],$$

$$\log V_t^{'} = \sum_{0 \leq s < t} [\log(1 + \triangle^+ X_s \mathbf{1}(B_s) - \triangle^+ X_s \mathbf{1}(B_s)].$$

These series a.s. converge absolutely since they are majorized by the series $c \sum_{0 < s \leq t} (\triangle X_s)^2$, $c \sum_{0 \leq s < t} (\triangle^+ X_s)^2$, see Theorems 5.4.2 and 5.4.19 or (see [54] Theorems 1.16 and 4.9).

Let us verify that (8.3) is a solution of (8.2). Set

$$K_t = X_t - \frac{1}{2} \langle X^c, X^c \rangle_t, \quad F(x,y,z) = e^x yz$$

and apply the change of variables formula to F (7.2) (see [54]). We have

$$
\begin{aligned}
Z_t &= F(K_t, U_t, V_t) = Z_0 + \int_{]0,t]} \left(Z_{s-} d(X_s^r - \frac{1}{2} \langle X^c, X^c \rangle_s) + e^{K_{s-}} V_{s-} dU_s \right) \\
&\quad + \frac{1}{2} \int_{]0,t]} Z_{s-} d\langle X^c, X^c \rangle_s + \int_{[0,t[} Z_s dX_{s+}^g + \int_{[0,t[} e^{K_s} U_s dV_{s+} \\
&\quad + \sum_{0 < s \leq t} \left[Z_s - Z_{s-} - Z_{s-} \triangle X_s - e^{K_{s-}} V_{s-} \triangle U_s \right] \\
&\quad + \sum_{0 \leq s < t} \left[Z_{s+} - Z_s - Z_s \triangle^+ X_s - e^{K_s} U_s \triangle^+ V_s \right].
\end{aligned}
\tag{8.4}
$$

Reducing similar terms and noting that

$$
\int_{]0,t]} e^{K_{s-}} V_{s-} dU_s = \sum_{0 < s \leq t} e^{K_{s-}} V_{s-} \triangle U_s,
$$

$$
\int_{[0,t[} e^{K_s} U_s dV_{s+} = \sum_{0 \leq s < t} e^{K_s} U_s \triangle^+ V_s,
$$

and also that $\triangle Z_t = Z_{t-} \triangle X_t$, $\triangle^+ Z_t = Z_t \triangle^+ X_t$, we see that (8.4) coincides with (8.2).

Uniqueness. Let $\tilde{Z} \in \mathcal{S}$ be yet another solution of (8.2), $\tilde{Z}_0 = Z_0$. Put $Y_t = e^{-K_t} \tilde{Z}_t$. From the change of variables formula, reducing similar terms, we get

$$Y_t = Z_0 + \sum_{0 < s \leq t} \left[e^{-K_s} \tilde{Z}_s - e^{-K_{s-}} - \tilde{Z}_{s-} \right] + \sum_{0 \leq s < t} \left[e^{-K_{s+}} \tilde{Z}_{s+} - e^{-K_s} \tilde{Z}_s \right],$$

or

$$Y_t = Z_0 + \sum_{0 < s \leq t} \triangle Y_s + \sum_{0 \leq s < t} \triangle^+ Y_s,$$

where the series converge absolutely a.s. It follows that Y has a.s. finite variation on any segment $[0, t]$.

Since

$$\triangle Y_s = Y_{s-} \left[e^{-\triangle X_s} (1 + \triangle X_s) - 1 \right], \quad \triangle^+ Y_s = Y_s \left[e^{-\triangle^+ X_s} (1 + \triangle^+ X_s) - 1 \right],$$

the process Y is a solution of the equation

$$Y_t = Z_0 + \int_{0,t} Y_{s-} dA_s^r + \int_{0,t} Y_s dA_{s+}^g,$$

where

$$A_t^r = \sum_{0<s\le t} [e^{-\triangle X_s}(1+\triangle X_s)-1], \quad A_t^g = \sum_{0\le s<t}[e^{-\triangle^+ X_s}(1+\triangle^+ X_s)-1] \quad (8.5)$$

are processes a.s. of bounded variation on any segment. By virtue of Lemma 24 in [15], the equations

$$Y = Y_0^r + \int_{0,t} Y_{s-}dA_s^r, \quad Y_t^g = Y_0^g + \int_{0,t} Y_s dA_{s+}^g$$

have at most one solution. Consequently, Equation (8.5) has at most one solution. It follows that Equation (8.2) also has at most one solution. ∎

Two useful properties of stochastic exponentials are the inverse and product formulas.

Lemma 8.1.2 The product of stochastic exponentials is

$$\mathcal{E}(X)\mathcal{E}(Y) = \mathcal{E}(X+Y+[X,Y]).$$

Proof. Using the change variable formula and definition of stochastic exponential,

$$\begin{aligned}
\mathcal{E}(X)\mathcal{E}(Y) &= 1+\mathcal{E}(X)\circ\mathcal{E}(Y)+\mathcal{E}(Y)\circ\mathcal{E}(X)+[\mathcal{E}(X),\mathcal{E}(Y)] \\
&= 1+\mathcal{E}(X)\mathcal{E}(Y)\circ Y+\mathcal{E}(X)\mathcal{E}(Y)\circ X+\mathcal{E}(X)\mathcal{E}(Y)\circ[X,Y] \\
&= 1+\mathcal{E}(X)\mathcal{E}(Y)\circ(X+Y+[X,Y]) \\
&= \mathcal{E}(X+Y+[X,Y]).
\end{aligned}$$

∎

Lemma 8.1.3 The inverse of stochastic exponential for the semimartingale X is $\mathcal{E}^{-1}(X) = \mathcal{E}(-X^*)$, where,

$$X_t^* = X_t - \langle X^c, X^c\rangle_t - \sum_{0<s\le t}\frac{(\triangle X_s)^2}{1+\triangle X_s} - \sum_{0\le s<t}\frac{(\triangle^+ X_s)^2}{1+\triangle^+ X_s}.$$

Proof. Let X^* be such that $\mathcal{E}(-X^*)\mathcal{E}(X) = \mathcal{E}(X-X^*-[X,X^*]) = 1$. Hence, $X-X^* - [X,X^*] = 0$ and

$$\begin{aligned}
X^* &= X-\langle X^c\rangle_t + \sum_{s\le t}(\triangle X_s)^2 + \sum_{s<t}(\triangle^+ X_s)^2 \\
&= X-\langle X^c\rangle_t + \sum_{0<s\le t}\frac{(\triangle X_s)^2+(\triangle X_s)^3}{1+\triangle X_s} + \sum_{0\le s<t}\frac{(\triangle^+ X_s)^2+(\triangle^+ X_s)^3}{1+\triangle^+ X_s} \\
&= X-\langle X^c\rangle_t + \sum_{0<s\le t}\frac{(\triangle X_s)^2}{1+\triangle X_s} + \sum_{0\le s<t}\frac{(\triangle^+ X_s)^2}{1+\triangle^+ X_s}.
\end{aligned}$$

The sum of $(\triangle X)^3/(1+\triangle X)$ is zero. To see this, consider

$$\sum \frac{(\triangle X)^3}{1+\triangle X} = \left[\sum \frac{(\triangle X)^2}{1+\triangle X}, \sum \triangle X\right] = 0$$

and as well, for $(\triangle^+ X)^3/(1+\triangle^+ X)$. Hence

$$X^* = X - \langle X^c \rangle_t + \sum_{0<s\leq t} \frac{(\triangle X_s)^2}{1+\triangle X_s} + \sum_{0\leq s<t} \frac{(\triangle^+ X_s)^2}{1+\triangle^+ X_s}.$$

∎

8.1.2   Stochastic logarithm

The stochastic logarithm is defined by the following theorem.

Theorem 8.1.4 Let Y be a real-valued optional semimartingale such that the processes Y_- and Y do not vanish, then the process

$$X_t = \frac{1}{Y} \circ Y_t = \int_{0+}^{t} \frac{1}{Y_{s-}} dY_s^r + \int_{0}^{t-} \frac{1}{Y_s} dY_{s+}^g, \quad X_0 = 0, \tag{8.6}$$

also denoted by $X = \mathcal{L}ogY$, called the stochastic logarithm of Y, is the unique semimartingale X such that $Y = Y_0 \mathcal{E}(X)$. Moreover, if $\triangle X \neq -1$ and $\triangle^+ X \neq -1$ we also have

$$\begin{aligned}
\mathcal{L}ogY_t &= \log\left|\frac{Y_t}{Y_0}\right| + \frac{1}{2Y^2} \circ \langle Y^c, Y^c \rangle_t - \sum_{0<s\leq t}\left(\log\left|1+\frac{\triangle Y_s}{Y_{s-}}\right| - \frac{\triangle Y_s}{Y_{s-}}\right) \\
&\quad - \sum_{0\leq s<t}\left(\log\left|1+\frac{\triangle^+ Y_s}{Y_s}\right| - \frac{\triangle^+ Y_s}{Y_s}\right).
\end{aligned} \tag{8.7}$$

It is important to note that the process Y need not be positive for $\mathcal{L}og(Y)$ to exist, in accordance with the fact that the stochastic exponential $\mathcal{E}(X)$ may take negative values.

Proof. The assumption that Y_- and Y don't vanish implies that $S_n = \inf\left(t : |Y_{t-}| \leq \frac{1}{n}\right) \uparrow \infty$; hence, $1/Y_-$ is locally bounded, likewise, $T_n = \inf\left(t : |Y_t| \leq \frac{1}{n}\right) \uparrow \infty$; hence, $1/Y$ is also locally bounded and the stochastic integral in (8.6) makes sense. Let $\tilde{Y} = Y/Y_0$, then $\tilde{Y}_0 = 1$ and by (8.6), $X = (1/\tilde{Y}) \circ \tilde{Y}$. Thus,

$$1 + \tilde{Y} \circ X = 1 + \tilde{Y} \circ \left(\frac{1}{\tilde{Y}} \circ \tilde{Y}\right) = 1 + \left(\tilde{Y}\frac{1}{\tilde{Y}}\right) \circ \tilde{Y} = \tilde{Y},$$

or $\tilde{Y} = \mathcal{E}(X)$. Furthermore, $\triangle X = \triangle Y/Y_- \neq -1$ and $\triangle X = \triangle^+ Y/Y \neq -1$. Let $\tilde{X}$ be any other semimartingale satisfying $Y = \mathcal{E}(\tilde{X})$. With $\tilde{Y} = Y/Y_0$, yields $\tilde{Y} = \mathcal{E}(\tilde{X})$. So, $\tilde{Y} = 1 + \tilde{Y} \circ \tilde{X}$ or $Y = Y_0 + \tilde{Y} \circ \tilde{X}$. Since $X_0 = 0$ we have,

$$\tilde{X} = \frac{\tilde{Y}}{\tilde{Y}} \circ \tilde{X} = \frac{1}{\tilde{Y}} \circ (Y - Y_0) = \frac{1}{\tilde{Y}} \circ Y = X,$$

and we get uniqueness.

To prove representation (8.7) we have to apply the change of variable formula to optional semimartingale $\log |Y|$. Since the log explodes at 0, we consider: for each n, the C^2 function f_n on $\mathbb{R}$, with $f_n(x) = \log |x|$ when $|x| \geq 1/n$. Then for all n and $t < T_n$ and $t < S_n$ we have

$$\log |Y_t| = \log |Y_0| + \frac{1}{Y} \circ Y_t - \frac{1}{2Y^2} \circ \langle Y^c, Y^c \rangle_t + \sum_{0 < s \leq t} \left(\triangle \log |Y_s| - \frac{\triangle Y_s}{Y_{s-}} \right)$$

$$+ \sum_{0 \leq s < t} \left(\triangle^+ \log |Y_s| - \frac{\triangle^+ Y_s}{Y_s} \right).$$

This, together with (8.6), yields (8.7) for $t < T_n$ and and $t < S_n$. Since $T_n \uparrow \infty$ and $S_n \uparrow \infty$, then we obtain (8.7) everywhere. $\blacksquare$

Now, we present some of the properties of stochastic logarithms.

Lemma 8.1.5 (a) If X is a semimartingale satisfying $\triangle X \neq -1$ and $\triangle^+ X \neq -1$, then $\mathcal{L}og(\mathcal{E}(X)) = X - X_0$. (b) If Y is a semimartingale such that Y and Y_- do not vanish, then $\mathcal{E}(\mathcal{L}og(Y)) = Y/Y_0$. (c) For any two optional semimartingales X and Z we get the following identities:

$$\mathcal{L}og(XZ) = \mathcal{L}ogX + \mathcal{L}ogZ + [\mathcal{L}ogX, \mathcal{L}ogZ];$$

and

$$\mathcal{L}og\left(\frac{1}{X}\right) = 1 - \mathcal{L}og(X) - \left[X, \frac{1}{X}\right].$$

Proof. (a)

$$\mathcal{L}og(\mathcal{E}(X)) = \mathcal{E}(X)^{-1} \circ \mathcal{E}(X) = \frac{\mathcal{E}(X)}{\mathcal{E}(X)} \circ X = X - X_0.$$

(b) Let $Z = \mathcal{E}(\mathcal{L}og(Y))$ then by (a) we find that

$$\mathcal{L}og(Z) = \mathcal{L}og(\mathcal{E}(\mathcal{L}og(Y))) = \mathcal{L}og(Y) - \mathcal{L}og(Y_0) = \mathcal{L}og(Y/Y_0).$$

Therefore, $Z = Y/Y_0$. (c) By the integral representation of the stochastic logarithm we find

$$\mathcal{L}og(XZ) = \frac{1}{XZ} \circ (X \circ Z + Z \circ X + [X, Z]) = \mathcal{L}ogX + \mathcal{L}ogZ + [\mathcal{L}ogX, \mathcal{L}ogZ].$$

Using integration by parts and the integral definition of the stochastic logarithm,

$$\mathcal{L}og\left(\frac{1}{X}\right) = X \circ \left(\frac{1}{X}\right) = 1 - \frac{1}{X} \circ X - \left[X, \frac{1}{X}\right] = 1 - \mathcal{L}og(X) - \left[X, \frac{1}{X}\right].$$

$\blacksquare$

8.1.3 Nonhomogeneous linear equation

A generalization of the stochastic exponential integral Equation (8.1) is the nonhomogeneous linear stochastic integral Equation [83], $X = G + X \circ H$. This equation has a natural application in finance: G is a stochastic cash flow, H is the interest rate of the money market account and X is the time value of the cash flow accumulated in the money market account. Here we will give the solution of the nonhomogeneous linear stochastic integral equation to the case where G and H are optional semimartingales.

Theorem 8.1.6 Consider the nonhomogeneous linear stochastic integral equation,

$$X_t = G_t + \int_0^t X_s dH_s \tag{8.8}$$

$$= G_t + \int_{0+}^t X_{s-} dH_s^r + \int_0^{t-} X_s dH_{s+}^g$$

$G = (G_t)_{t \geq 0} \in \mathcal{S}(\mathbf{F}, \mathbf{P})$ is an optional semimartingale. The solution is

$$X_t = \mathcal{E}_t(H) \left[G_0 + \int_0^t \mathcal{E}_s(H)^{-1} d\tilde{G}_s \right], \tag{8.9}$$

$$d\tilde{G}_t = dG_t - d\left[G, \tilde{H} \right]_t,$$

$$\tilde{H}_t = H_t^c + \sum_{0 < s \leq t} \frac{\triangle H_s}{1 + \triangle H_s} + \sum_{0 \leq s < t} \frac{\triangle^+ H_s}{1 + \triangle^+ H_s}.$$

Proof. Let's consider a solution of the following form $X_t = \mathcal{E}_t(H) Z_t$ where Z_t is related to G_t. The differential of X_t is

$$dX_t = d\left(\mathcal{E}_t(H) Z_t\right)$$
$$= X_t dH_t + \mathcal{E}_t(H) \{dZ_t + d[H, Z]_t\}. \tag{8.10}$$

Comparing Equation (8.8) to Equation (8.10), we find that $dG_t = \mathcal{E}_t(H) \{dZ_t + d[H, Z]_t\}$. We would like to solve this equation for Z_t. To do so, we choose

$$\tilde{H}_t = H_t^c + \sum_{0 < s \leq t} \frac{\triangle H_s}{1 + \triangle H_s} + \sum_{0 \leq s < t} \frac{\triangle^+ H_s}{1 + \triangle^+ H_s}$$

and compute the quadratic variation of G with $\tilde{H}$. Note that the quadratic variation for optional processes belongs to the space $\mathcal{A}_{loc}$ and is defined as

$$[G, \tilde{H}]_t = \langle G^c, \tilde{H}^c \rangle_t + \sum_{0 < s \leq t} \triangle G_s \triangle \tilde{H}_s + \sum_{0 \leq s < t} \triangle^+ G_s \triangle_s^+ \tilde{H}.$$

Hence,

$$d\left[G,\tilde{H}\right]_t = \mathcal{E}_t(H)\left\{d\left[Z,\tilde{H}\right]_t + d\left[[H,Z],\tilde{H}\right]_t\right\}$$

$$= \mathcal{E}_t(H)\left\{d\left[Z^c,H^c\right]_t + \frac{\triangle H_t \triangle Z_t}{1+\triangle H_t} + \frac{\triangle^+ H_t \triangle^+ Z_t}{1+\triangle^+ H_t}\right.$$

$$\left. + \frac{(\triangle H_t)^2}{1+\triangle H_t}\triangle Z_t + \frac{(\triangle^+ H_t)^2}{1+\triangle^+ H_t}\triangle^+ Z_t\right\}$$

$$= \mathcal{E}_t(H)\left\{d\left[Z^c,H^c\right]_t + \triangle H_t \triangle Z_t\left(\frac{1+\triangle H_t}{1+\triangle H_t}\right)\right.$$

$$\left. + \triangle^+ H_t \triangle^+ Z_t\left(\frac{1+\triangle^+ H_t}{1+\triangle^+ H_t}\right)\right\}$$

$$= \mathcal{E}_t(H)\left\{d\left[Z^c,H^c\right]_t + \triangle H \triangle Z + \triangle^+ H_t \triangle^+ Z_t\right\}$$

$$= \mathcal{E}_t(H)d\left[Z,H\right]_t.$$

Then we calculate Z in the following way,

$$dG_t = \mathcal{E}_t(H)\left\{dZ_t + d[H,Z]_t\right\}$$

$$\mathcal{E}_t(H)^{-1}dG_t = dZ_t + d[H,Z]_t$$

$$= dZ_t + \mathcal{E}_t(H)^{-1}d\left[G,\tilde{H}\right]_t$$

$$dZ_t = \mathcal{E}_t(H)^{-1}\left[dG_t - d\left[G,\tilde{H}\right]_t\right]$$

Note that,

$$\left[[H,Z],\tilde{H}\right] = \left[\langle H^c,Z^c\rangle + \sum_{s\leq\cdot}\triangle H\triangle Z + \sum_{s<\cdot}\triangle^+ H\triangle^+ Z,\tilde{H}\right]$$

$$= \left[\langle H^c,Z^c\rangle,\tilde{H}\right] + \left[\left(\sum_{s\leq\cdot}\triangle H\triangle Z\right),\tilde{H}\right] + \left[\left(\sum_{s<\cdot}\triangle^+ H\triangle^+ Z\right),\tilde{H}\right]$$

$\left[\langle H^c,Z^c\rangle,\tilde{H}\right] = 0$ because $\langle H^c,Z^c\rangle$ is continuous locally bounded variation processes and $\tilde{H}$ a semimartingale. Then,

$$\left[[H,Z],\tilde{H}\right] = \left[\left(\sum_{s\leq\cdot}\triangle H\triangle Z\right),\tilde{H}\right] + \left[\left(\sum_{s<\cdot}\triangle^+ H\triangle^+ Z\right),\tilde{H}\right]$$

$$= \sum_{s\leq t}(\triangle H_s)^2\triangle Z_s + \sum_{s<t}(\triangle^+ H_s)^2\triangle^+ Z_s.$$

■

Another important application of stochastic exponential is the Gronwall Lemma.

8.1.4 Gronwall lemma

The Gronwall Lemma is a fundamental inequality in analysis and has far-reaching consequences. For example, a fundamental problem in the study of differential or integral equations or their stochastic generalization is that of existence and uniqueness of solutions for which many variants of Gronwall's lemma were extensively used. Basically, Gronwall's lemma allows us to put bounds on functions that satisfy an integral or differential inequality by a solution of a supposed equality. In stochastic analysis, the Gronwall lemma is essential and many extensions have been proposed; see for example Metivier [84], Melnikov [85] and others [86, 87, 88]. It is used to study the stability of solutions of stochastic equations of semimartingales. Here we will extend the Gronwell lemma to optional semimartingales in *un*usual probability spaces.

Lemma 8.1.7 Let X be an optional process and H be an optional increasing process and C_t a constant such that

$$X_t \leq C_t + \int_0^t X_s dH_s$$

$$= C_t + \int_{0+}^t X_{s-} dH_s + \int_0^{t-} X_s dH_{s+}$$

for all $t \in [0, \infty)$. *Then,*

$$X_t \leq C_t \mathcal{E}_t(H).$$

Proof. Let

$$N_t = C_t + \int_0^t X_s dH_s - X_t$$

then $N_t \geq 0$ for all t. Therefore,

$$X_t = C_t - N_t + \int_0^t X_s dH_s$$

is a nonhomogeneous stochastic integral equation whose solution is given by

$$X_t = \mathcal{E}_t(H) \left[G_0 + \int_0^t \mathcal{E}_s(H)^{-1} d\tilde{G}_s \right], \qquad (*)$$

$$G_t = C_t - N_t$$

$$\tilde{G}_t = G_t - \left[G, \tilde{H} \right]_t,$$

$$\tilde{H}_t = H_t^c + \sum_{0 < s \leq t} \frac{\triangle H_s}{1 + \triangle H_s} + \sum_{0 \leq s < t} \frac{\triangle^+ H_s}{1 + \triangle^+ H_s}.$$

Since H is increasing, then $\triangle H_s \geq 0$ and $\triangle^+ H_s \geq 0$ and, therefore, $\tilde{H}$ is increasing. Hence, $\left[G, \tilde{H}\right] = 0$, $\tilde{G}_t = G_t = C_t - N_t \leq C_t$ for all t since $N_t \geq 0$. Knowing all this, we can write (*) as

$$X_t = \mathcal{E}_t(H)\left[G_0 + \int_0^t \mathcal{E}_s(H)^{-1}d\tilde{G}_s\right]$$

$$= \mathcal{E}_t(H)\left[G_0 + \int_0^t \mathcal{E}_s(H)^{-1}dG_s\right]$$

$$\leq C_t\mathcal{E}_t(H),$$

where $\int_0^t \mathcal{E}_s(H)^{-1}dG_s \leq 0$ and $G_0 \leq C_0$. ∎

Next we develop the theory of existence and uniqueness of solutions of stochastic equations of optional semimartingales given that the coefficients satisfy a version of the monotonicity conditions. Monotonicity conditions are less stringent assumptions than Lipschitz conditions.

8.2 Existence and Uniqueness of Solutions of Optional Stochastic Equations

A central problem in the theory of stochastic equations is the study of existence and uniqueness of solutions under certain conditions placed on the semimartingale driver and on coefficients of this semimartingale. A plethora of stochastic equations, models and proofs were proposed, see [89, 90] for a comprehensive review.

However, a brief history of existence and uniqueness theorems is as follows. Itô [91] proved that there is a unique, strong solution of the diffusion equation under Lipschitz conditions. Gihman and Skorokhod [92] proved existence and uniqueness of solution for equations driven by Wiener and Poisson measures, also under Lipschitz conditions. Kazamaki [93] extended Ito's result to the case of bounded coefficients of a continuous increasing process and *cadlag* square integrable martingale drivers. For a stochastic equation with respect to a semimartingale driver, Protter [94] proved existence and uniqueness of a strong solution given that the coefficient of the semimartingale is Lipschitz. Galchuk [95] proved existence and uniqueness of strong solution for a stochastic equation driven by the components of a semimartingale such that the components' integrands satisfy the Lipschitz conditions. While previous results were obtained for real-valued finite dimensional stochastic equations Gyongy and Krylov [96] proved existence and uniqueness of solution of stochastic equations of locally square integrable *cadlag* martingale taking values in a Hilbert space under monotonicity condition.

However, little was done in showing existence and uniqueness of solution of stochastic equations driven by optional semimartingales on *un*usual probability spaces, except for the work of Gasparyan

[12] on existence and uniqueness under Lipschitz conditions. In this paper, following Gyongy and Krylov [96] exposition for the proof of existence and uniqueness of solution of stochastic equations of *cadlag* semimartingales under monotonicity conditions, we will consider the question of existence and uniqueness of a strong solution, x, taking values in $\mathbb{R}^d$ of the following equation

$$x = \xi + a(\cdot, x) \circ A + b(\cdot, x) \circ M$$

for the underlying optional *ladlag* vector-semimartingales $X = (A, M)$ on *unusual* probability space. A is increasing or is of bounded variation optional process taking values in $\mathbb{R}$ and M is a locally square integrable optional martingale (i.e., $\mathcal{M}_{loc}^2(\mathbf{F}, \mathbf{P})$) taking values in the Euclidean space $\mathbb{R}^k$. ξ is a r.v. in the Euclidean space $\mathbb{R}^d$. Under additional conditions of monotonicity and restriction on growth of the random functions a and b, existence and uniqueness will be demonstrated.

The main difference between our approach and that of Gyongy and Krylov [96] is that we are working with optional *ladlag* semimartingale driver taking values in Euclidean space $\mathbb{R}^d$, not *cadlag* semimartingales in a general Hilbert space. The reason for not considering general Hilbert space valued processes is that the stochastic integral and calculus with respect to optional semimartingales taking values in Hilbert space has not been defined yet. Our motivation for studying the problem of existence and uniqueness of stochastic equations driven by *ladlag* optional semimartingales is for their potential applications to mathematical finance and physics. We present examples of such applications in Section 8.2.3.

8.2.1 Stochastic equation with monotonicity condition

Let $(\Omega, \mathcal{F}, \mathbf{F} = (\mathcal{F}_t)_{t\geq 0}, \mathbf{P})$ be the *unusual* probability space. Consider a Euclidean space $\mathbb{R}^d$ with Borel σ-algebra $\mathcal{B}(\mathbb{R}^d)$ and $x = (x_t)_{t\geq 0}$ an $\mathbb{R}^d$ valued optional process defined on $\mathbb{R}_+ \times \Omega$. Let $\mathbb{R}^k$ be a k-dimensional Euclidean space with Borel σ-algebra $\mathcal{B}(\mathbb{R}^k)$ and M is $\mathcal{M}_{loc}^2$ of locally square integrable optional martingales with respect to the family $(\mathcal{F}_t)_{t\geq 0}$ taking values in $\mathbb{R}^k$ with $M_0 = 0$. Let $\mathcal{V}^+$ be the set of all increasing *ladlag* real-valued processes and $\mathcal{V} = \mathcal{V}^+ - \mathcal{V}^+$ the set of finite variation processes. A is *ladlag* $\mathbb{R}$ valued process with $A_0 = 0$, that is either increasing, $\mathcal{V}^+$, or finite variation, $\mathcal{V}$, and either strongly predictable or optional.

The process x is given by the stochastic equation

$$x = \xi + a(\cdot, x) \circ A + b(\cdot, x) \circ M \tag{8.11}$$

where ξ is $\mathcal{F}_0$-measurable r.v. in $\mathbb{R}^d$. The integrals $a(\cdot, x) \circ A$ and $b(\cdot, x) \circ M$ are understood as

$$
\begin{aligned}
a(\cdot, x) \circ A_t &= \int_{0+}^{t} a(s-, x_{s-}) dA_s^r + \int_0^{t-} a(s, x_s) dA_s^g \\
&= a^1(\cdot, x) \cdot A_t^r + a^2(\cdot, x) \odot A_t^g \\
b(\cdot, x) \circ M_t &= \int_{0+}^{t} b(s-, x_{s-}) dM_s^r + \int_0^{t-} b(s, x_s) dM_s^g \\
&= b^1(\cdot, x) \cdot M_t^r + b^2(\cdot, x) \odot M_t^g,
\end{aligned}
$$

respectively. a^1 and a^2 are $\mathbb{R}^d$ valued processes integrable with respect to A such that $a^1 \in \mathcal{P}$ and $a^2 \in \mathcal{O}$. b^1 and b^2 are $\mathbb{R}^d \times \mathbb{R}^k$ valued processes integrable with respect to M^r and M^g, respectively, with $b^1 \in \mathcal{P}$ and $b^2 \in \mathcal{O}$. To further clarify the structure of the stochastic equation, we list all domains, ranges and mapping of the different objects of the stochastic equation:

$$
\begin{aligned}
x &: \quad (\mathbb{R}_+ \times \Omega, \mathcal{B}(\mathbb{R}_+) \times \mathbf{F}) \longrightarrow \left(\mathbb{R}^d, \mathcal{B}(\mathbb{R}^d)\right), \\
\xi &: \quad (\Omega, \mathcal{F}) \longrightarrow \left(\mathbb{R}^d, \mathcal{B}(\mathbb{R}^d)\right), \\
a^1 &: \quad \left(\mathbb{R}_+ \times \Omega \times \mathbb{R}^d, \mathcal{P}(\mathbb{R}_+) \times \mathcal{B}(\mathbb{R}^d)\right) \longrightarrow \left(\mathbb{R}^d, \mathcal{B}(\mathbb{R}^d)\right), \\
a^2 &: \quad \left(\mathbb{R}_+ \times \Omega \times \mathbb{R}^d, \mathcal{O}(\mathbb{R}_+) \times \mathcal{B}(\mathbb{R}^d)\right) \longrightarrow \left(\mathbb{R}^d, \mathcal{B}(\mathbb{R}^d)\right), \\
A &: \quad (\mathbb{R}_+ \times \Omega, \mathcal{B}(\mathbb{R}_+) \times \mathbf{F}) \longrightarrow (\mathbb{R}, \mathcal{B}(\mathbb{R})), \\
b^1 &: \quad \left(\mathbb{R}_+ \times \Omega \times \mathbb{R}^d, \mathcal{P}(\mathbb{R}_+) \times \mathcal{B}(\mathbb{R}^d)\right) \longrightarrow \left(\mathbb{R}^k \times \mathbb{R}^d, \mathcal{B}(\mathbb{R}^k) \times \mathcal{B}(\mathbb{R}^d)\right), \\
b^2 &: \quad \left(\mathbb{R}_+ \times \Omega \times \mathbb{R}^d, \mathcal{O}(\mathbb{R}_+) \times \mathcal{B}(\mathbb{R}^d)\right) \longrightarrow \left(\mathbb{R}^k \times \mathbb{R}^d, \mathcal{B}(\mathbb{R}^k) \times \mathcal{B}(\mathbb{R}^d)\right), \\
M &: \quad (\mathbb{R}_+ \times \Omega, \mathcal{B}(\mathbb{R}_+) \times \mathbf{F}) \longrightarrow \left(\mathbb{R}^k, \mathcal{B}(\mathbb{R}^k)\right).
\end{aligned}
$$

Let $\mathsf{L}_1(\mathbb{R}^k, \mathbb{R}^k)$ be the space of the *nuclear operators* on $\mathbb{R}^k$, and $\mathsf{L}_2(\mathbb{R}^k, \mathbb{R}^d)$ the space of the *Hilbert-Schmidt operators* on $\mathbb{R}^k$ into $\mathbb{R}^d$. The product $\mathbb{R}^k \otimes_1 \mathbb{R}^k$ is the *projective tensor product* of $\mathbb{R}^k$ by itself and $\mathbb{R}^k \otimes_2 \mathbb{R}^d$ the *Hilbertian tensor product* of $\mathbb{R}^k$ by $\mathbb{R}^d$. If $x, y \in \mathbb{R}^d$, then xy denotes the scalar product of x and y in $\mathbb{R}^d$ and $|x| = (xx)^{\frac{1}{2}}$ is the scalar norm of x. If $b \in \mathsf{L}_2(\mathbb{R}^k, \mathbb{R}^d)$ then $|b|$ denotes the Hilbert-Schmidt norm of b which is the same as the norm in $\mathbb{R}^k \otimes_2 \mathbb{R}^d$. If $\Theta \in \mathsf{L}_1(\mathbb{R}^k, \mathbb{R}^k)$ is a non-negative operator, then $\mathsf{L}_\Theta(\mathbb{R}^k, \mathbb{R}^d)$ denotes the set of all linear not necessarily bounded operators b mapping $\Theta^{\frac{1}{2}}(\mathbb{R}^k)$ into $\mathbb{R}^d$ such that $b\Theta^{\frac{1}{2}} \in \mathsf{L}_2(\mathbb{R}^k, \mathbb{R}^d)$.

Given that $M \in \mathcal{M}^2_{loc}$ then the sharp-bracket process $\langle M \rangle$ is $\mathbb{R}$ valued predictable and increasing $\mathcal{P} \cap \mathcal{V}^+$ with $\langle M \rangle_0 = 0$ and $M^2 - \langle M \rangle$ is $\mathbb{R}$ valued local optional martingale. On the other hand, the sharp tensor process $\langle\langle M \rangle\rangle \in \mathcal{P}(\mathbb{R}^k \otimes_1 \mathbb{R}^k)$ with $\langle\langle M \rangle\rangle_0 = 0$ and $M \otimes_1 M - \langle\langle M \rangle\rangle$ are local optional martingales taking values in $\mathbb{R}^k \otimes_1 \mathbb{R}^k$. Let $V \in \mathcal{V}^+$ be an optional increasing process such that $dA \leqq dV$, $d\langle M \rangle \leqq dV$, and $d\langle\langle M \rangle\rangle \leqq dV$. Let $\mathcal{L}$ denote the set of all real-valued, non-negative processes which are locally integrable with respect to dV_t. Then, the following processes can be defined

$$
\Lambda := \frac{dA}{dV} \in \mathcal{O}(\mathbb{R}), \quad \Theta := \frac{d\langle\langle M \rangle\rangle}{dV} \in \mathcal{O}(\mathbb{R}^k \otimes_1 \mathbb{R}^k) \quad and \quad \mathrm{tr}\Theta = \frac{d\langle M \rangle}{dV} \quad or
$$

$$
A_t = \Lambda \circ V_t = \Lambda^1 \circ V_t^r + \Lambda^2 \circ V_t^g \quad and
$$

$$
\langle\langle M \rangle\rangle_t = \Theta^1 \circ V_t^r + \Theta^2 \circ V_t^g
$$

for every $t \in \mathbb{R}_+$ and almost all $\omega \in \Omega$.

For us to carry on with the proof of existence and uniqueness of Equation (8.11) we, in addition to the constructs we have presented above, make some additional assumptions. We have to require the random functions a and b to satisfy the following conditions:

1. Function a is continuous in x, and for every fixed x, a is locally integrable with respect to dA_t for almost all ω. That is, there is a sequence $(T_n) \subset \mathcal{T}_+$, $n \in \mathbb{N}$, $T_n \uparrow \infty$ a.s. such that $|a|| \circ A|_{T_{n+}} < \infty$ for all $n \in \mathbb{N} < \infty$. From a and Λ we write $\alpha^1 := a^1 \Lambda^1$ and $\alpha^2 := a^2 \Lambda^2$ predictable and optional functions, respectively.

2. Function b is continuous in x, and for every fixed x, b is integrable with respect to M, i.e., $|b|^2 \circ \langle M \rangle < \infty$ a.s. Moreover, we require that the stochastic integral $b\Theta^{\frac{1}{2}} \circ V$ be defined which requires us to assume that the processes b^1 and b^2 are $\mathsf{L}_\Theta(\mathbb{R}^k, \mathbb{R}^d)$, $b^{1\frac{1}{2}} \in \mathcal{P}(\mathbb{R}^k \otimes_2 \mathbb{R}^d)$ and $b^{2\frac{1}{2}} \in \mathcal{O}(\mathbb{R}^k \otimes_2 \mathbb{R}^d)$ such that $\mathbf{E} \left| b\Theta^{\frac{1}{2}} \right|^2 \circ V_t < \infty$ and $|b|^2 |\Theta| \circ V < \infty$ a.s. From b and Λ we write $\beta^1 := b^1 \Theta^{1\frac{1}{2}}$ and $\beta^2 := b^2 \Theta^{2\frac{1}{2}}$ which are $\mathsf{L}_2(\mathbb{R}^k, \mathbb{R}^d)$ where β^1 is predictable and β^2 optional operators.

3. For each $\rho \geq 0$ there exists a process $K_t(\rho) \in \mathcal{L}$, such that, for every x, $y \in \mathbb{R}^d$ $|x| \leq \rho$, $|y| \leq \rho$ the following inequalities hold for almost all (t, ω) with respect to $d\mathbf{P}dV_t$:

I. Monotonicity Condition:

$$
\begin{aligned}
2(x_- - y_-)(\alpha^1(t, x) - \alpha^1(t, y)) + \triangle V_t \left| \alpha^1(t, x) - \alpha^1(t, y) \right|^2 \\
+ |\beta^1(t, x) - \beta^1(t, y)|^2 \leq K_t(\rho)|x_- - y_-|^2
\end{aligned}
\tag{8.12}
$$

$$
\begin{aligned}
2(x - y)(\alpha^2(t, x) - \alpha^2(t, y)) + \triangle^+ V_t \left| \alpha^2(t, x) - \alpha^2(t, y) \right|^2 \\
+ |\beta^2(t, x) - \beta^2(t, y)|^2 \leq K_t(\rho)|x - y|^2
\end{aligned}
\tag{8.13}
$$

II. Restriction on Growth:

$$
2x_- \alpha^1(t, x) + \triangle V_t |\alpha^1(t, x)|^2 + |\beta^1(t, x)|^2 \leq K_t(\rho)(1 + |x_-|^2),
\tag{8.14}
$$

$$
2x \alpha^2(t, x) + \triangle^+ V_t |\alpha^2(t, x)|^2 + |\beta^2(t, x)|^2 \leq K_t(\rho)(1 + |x|^2).
\tag{8.15}
$$

This brings us to the statement of existence and uniqueness theorem.

8.2.2 Existence and uniqueness results

Theorem 8.2.1 (Existence and Uniqueness) *If the conditions 1-3 hold, then there exists one and only one* $\mathbb{R}^d$*-valued adapted* *ladlag* *process* x *satisfying the equation*

$$
x_t = \xi + a(\cdot, x) \circ A_t + b(\cdot, x) \circ M_t.
\tag{8.16}
$$

The proof of existence and uniqueness of Theorem 8.2.1 will come as a result of a sequence of several facts and Lemmas presented below. We begin by stating the following observations. The functions $\alpha^1(t, x)$ and $\alpha^2(t, x)$ satisfy the Lipschitz conditions in x on the sets $\{(t, \omega) : \triangle V_t(\omega) > 0\}$ and $\{(t, \omega) : \triangle^+ V_t(\omega) > 0\}$, respectively. Moreover, for $|x| \leq \rho$ and $|y| \leq \rho$ we have

$$
2 \triangle V_t |\alpha^1(t, x_-) - \alpha^1(t, y_-)| \leq (4 + K_t(\rho) \triangle V_t)|x_- - y_-|,
\tag{8.17}
$$

which follows from (8.12) by multiplying both sides of the inequality by $\triangle V$ and using the inequality $a^2 \geq 4(a-1)$. Similarly, from (8.13) we get

$$2 \triangle^+ V_t |\alpha^2(t,x) - \alpha^2(t,y)| \leq (4 + K_t(\rho) \triangle^+ V_t)|x-y|. \tag{8.18}$$

Notation 8.2.2 We will use bold Delta "$\boldsymbol{\Delta}$" of a *ladlag* process X to mean $\boldsymbol{\Delta} X := (\triangle X, \triangle^+ X)$. For example, consider a pure jump process X and a process Y then we write the quadratic variation of X with Y as

$$[X,Y]_t = \int_{0+}^t \triangle X_{s-} dY_s + \int_0^{t-} \triangle^+ X_s dY_{s+} = \boldsymbol{\Delta} X \circ Y_t.$$

The following Lemma will be an important tool in the proof of Theorem 8.2.1.

Lemma 8.2.3 Let φ be a solution of the equation

$$\begin{aligned}
\varphi_t &= 1 + \int_{0+}^t \gamma_s \varphi_{s-} dV_s^r + \int_0^{t-} \gamma_s \varphi_s dV_{s+}^g \\
&= 1 + \gamma\varphi \circ V_t = 1 + \varphi \circ (\gamma \circ V)_t = \mathcal{E}(\gamma \circ V)_t,
\end{aligned} \tag{8.19}$$

where $\gamma \in \mathcal{L}$. Then

$$\varphi_t^{-1} = 1 - \int_{0+}^t \gamma_s \varphi_{s-}^{-1} dV_s^r - \int_0^{t-} \gamma_s \varphi_s^{-1} dV_s^g. \tag{8.20}$$

Also, if $y \in \mathcal{O}$ is an optional semimartingale with decomposition $y = B + N$ where $N \in \mathcal{M}_{loc}^2$ and B is either locally integrable monotone or finite variation, then the following equality holds

$$\varphi_t^{-1} |y_t|^2 = \varphi^{-1}(2y + \boldsymbol{\Delta}B) \circ B + \varphi^{-1} \circ \langle N, N \rangle - \varphi^{-1}|y|^2 \gamma \circ V + m_t', \tag{8.21}$$

where

$$m_t' = \varphi^{-1} 2(y + \boldsymbol{\Delta}B) \circ N + \varphi^{-1} \circ ([N,N] - \langle N,N \rangle)$$

consequently, m_t' is a local optional martingale.

Proof. Equation (8.20) follows from the integral of functions of increasing and finite variation processes. Since φ is either finite variation, increasing or decreasing depending on γ then,

$$\varphi_t^{-1} = 1 - \frac{1}{\varphi^2} \circ \varphi_t = 1 - \frac{\gamma}{\varphi} \circ V_t.$$

The second equality, Equation (8.21), can be derived as follows,

$$
\begin{aligned}
\varphi_t^{-1}|y|_t^2 &= \varphi^{-1} \circ |y|_t^2 + |y|^2 \circ \varphi_t^{-1} \\
&= \varphi^{-1} \circ (2y \circ y + [y, y])_t - |y|^2 \gamma \varphi^{-1} \circ V_t \\
&= \varphi^{-1} 2y \circ y + \varphi^{-1} \circ [y, y]_t - |y|^2 \gamma \varphi^{-1} \circ V_t \\
&= \varphi^{-1} 2y \circ y + \varphi^{-1} \Delta y \circ y_t - |y|^2 \gamma \varphi^{-1} \circ V_t \\
&= \varphi^{-1} (2y + \Delta y) \circ y - \varphi^{-1} |y|^2 \gamma \circ V \\
&= \varphi^{-1} (2y + \Delta y) \circ (B + N) - \varphi^{-1} |y|^2 \gamma \circ V \\
&= \varphi^{-1} 2y \circ B + \varphi^{-1} \Delta y \circ B + \varphi^{-1} 2y \circ N + \varphi^{-1} \Delta y \circ N - \varphi^{-1} |y|^2 \gamma \circ V \\
&= \varphi^{-1} 2y \circ B + \varphi^{-1} \circ [y, B] + \varphi^{-1} 2y \circ N + \varphi^{-1} \circ [y, N] - \varphi^{-1} |y|^2 \gamma \circ V \\
&= \varphi^{-1} 2y \circ B + \varphi^{-1} \circ [B + N, B] + \varphi^{-1} 2y \circ N + \varphi^{-1} \circ [B + N, N] - \varphi^{-1} |y|^2 \gamma \circ V \\
&= \varphi^{-1} 2y \circ B + \varphi^{-1} \circ [B, B] + \varphi^{-1} \circ [N, B] + \varphi^{-1} 2y \circ N + \varphi^{-1} \circ [B, N] \\
&\quad + \varphi^{-1} \circ [N, N] - \varphi^{-1} |y|^2 \gamma \circ V \\
&= \varphi^{-1} 2y \circ B + \varphi^{-1} \Delta B \circ B + \varphi^{-1} 2y \circ N + 2\varphi^{-1} \circ [B, N] \\
&\quad + \varphi^{-1} \circ [N, N] - \varphi^{-1} |y|^2 \gamma \circ V \\
&= \varphi^{-1} (2y + \Delta B) \circ B + \varphi^{-1} 2y \circ N + 2\varphi^{-1} \circ [B, N] + \varphi^{-1} \circ [N, N] - \varphi^{-1} |y|^2 \gamma \circ V \\
&= \varphi^{-1} (2y + \Delta B) \circ B + \varphi^{-1} \circ \langle N, N \rangle - \varphi^{-1} |y|^2 \gamma \circ V + \varphi^{-1} 2 (y + \Delta B) \circ N \\
&\quad + \varphi^{-1} \circ ([N, N] - \langle N, N \rangle)
\end{aligned}
$$

hence $m' = \varphi^{-1} 2 (y + \Delta B) \circ N + \varphi^{-1} \circ ([N, N] - \langle N, N \rangle)$ and

$$
\varphi_t^{-1} |y|_t^2 = \varphi^{-1} (2y + \Delta B) \circ B + \varphi^{-1} \circ \langle N, N \rangle - \varphi^{-1} |y|^2 \gamma \circ V + m'.
$$

■

8.2.2.1 Uniqueness

Now, we are ready to prove the *uniqueness of solution* part of Theorem 8.2.1.

Proof. Uniqueness of solution of Theorem 8.2.1. Let x and y be *ladlag* optional processes satisfying Equation (8.11). Let φ be the solution of the equation $\varphi = 1 + \gamma \varphi \circ V_t$. From Lemma 8.2.3 we get

$$
\begin{aligned}
\varphi^{-1} |x - y|^2 &= \varphi^{-1} \left(2 (x - y) (a(x) - a(y)) + (a(x) - a(y))^2 \Delta A \right) \circ A \\
&\quad + \varphi^{-1} (b(x) - b(y))^2 \circ \langle M, M \rangle - \varphi^{-1} \gamma |x - y|^2 \circ V \\
&\quad + \varphi^{-1} 2 ((x - y) + (a(x) - a(y)) \Delta A) (b(x) - b(y)) \circ M \\
&\quad + \varphi^{-1} (b(x) - b(y))^2 \circ ([M, M] - \langle M, M \rangle)
\end{aligned}
$$

where we omit the variable t in functions a, α, b and β to simplify notations. Since $\langle M, M \rangle = \Theta \circ V$ and $A = \Lambda \circ V$, then

$$
\begin{aligned}
\varphi^{-1} |x - y|^2 \;=\;& \varphi^{-1} 2 \, (x - y) \, (a(x) - a(y)) \, \Lambda \circ V \\
&+ \varphi^{-1} \, (a(x) - a(y))^2 \, \Lambda^2 \boldsymbol{\Delta} V \circ V \\
&+ \varphi^{-1} \, (b(x) - b(y))^2 \, Q \circ V - \varphi^{-1} \gamma |x - y|^2 \circ V \\
&+ \varphi^{-1} 2 \, (x - y) \, (b(x) - b(y)) \circ M \\
&+ \varphi^{-1} \, (b(x) - b(y))^2 \circ ([M, M] - \langle M, M \rangle) .
\end{aligned}
$$

Let $m'' = \varphi^{-1} 2 \, (x - y) \, (b(x) - b(y)) \circ M + \varphi^{-1} \, (b(x) - b(y))^2 \circ ([M, M] - \langle M, M \rangle)$, and, as in condition 2, set the predictable part $a_- \Lambda_- = a^1 \Lambda^1 := \alpha^1$ and the optional part $a\Lambda = a^2 \Lambda^2 := \alpha^2$, also, the predictable part of $b_- \Theta_-^{\frac{1}{2}} = b^1 \Theta^{1\frac{1}{2}} := \beta^1$ and the optional part is $b \Theta^{\frac{1}{2}} = b^2 \Theta^{2\frac{1}{2}} := \beta^2$ then

$$
\begin{aligned}
\varphi^{-1} |x - y|^2 \;=\;& \varphi^{-1} 2 \, (x - y) \, (\alpha^1(x) - \alpha^1(y)) \cdot V + \varphi^{-1} 2 \, (x - y) \, (\alpha^2(x) - \alpha^2(y)) \odot V \\
&+ \varphi^{-1} \, (\alpha^1(x) - \alpha^1(y))^2 \, \triangle V \cdot V + \varphi^{-1} \, (\alpha^1(x) - \alpha^1(y))^2 \, \triangle^+ V \odot V \\
&+ \varphi^{-1} \, (\beta^1(x) - \beta^1(y))^2 \cdot V + \varphi^{-1} \, (\beta^2(x) - \beta^2(y))^2 \odot V \\
&- \varphi^{-1} \gamma |x - y|^2 \cdot V - \varphi^{-1} \gamma |x - y|^2 \odot V + m'' \\
\;=\;& \varphi^{-1} \Big[2 \, (x - y) \, (\alpha^1(x) - \alpha^1(y)) + (\alpha^1(x) - \alpha^1(y))^2 \, \triangle V \\
&+ (\beta^1(x) - \beta^1(y))^2 - \gamma_- |x_- - y_-|^2 \Big] \cdot V \\
&+ \varphi^{-1} \Big[2 \, (x - y) \, (\alpha^2(x) - \alpha^2(y)) + (\alpha^2(x) - \alpha^2(y))^2 \, \triangle^+ V \\
&+ (\beta^2(x) - \beta^2(y))^2 - \gamma |x - y|^2 \Big] \odot V + m''.
\end{aligned}
$$

Since the first two terms

$$
\begin{aligned}
C^1 \; : \; &= 2 \, (x - y) \, (\alpha^1(x) - \alpha^1(y)) + (\alpha^1(x) - \alpha^1(y))^2 \, \triangle V + (\beta^1(x) - \beta^1(y))^2 - \gamma_- |x_- - y_-|^2 \le 0 \\
C^2 \; : \; &= 2 \, (x - y) \, (\alpha^2(x) - \alpha^2(y)) + (\alpha^2(x) - \alpha^2(y))^2 \, \triangle^+ V + (\beta^2(x) - \beta^2(y))^2 - \gamma |x - y|^2 \le 0
\end{aligned}
$$

are negative in accordance with the monotonicity conditions and that $\varphi^{-1} |x - y|^2 \ge 0$ we get,

$$
\begin{aligned}
0 \le \varphi^{-1} |x - y|^2 \le m'' \;&\Rightarrow \\
0 \le \varphi^{-1} C^1 \cdot V + \varphi^{-1} C^2 \odot V + m'' \le m'' \;&\Rightarrow \\
\varphi^{-1} C^1 \cdot V + \varphi^{-1} C^2 \odot V = 0. &
\end{aligned}
$$

Now, let $\gamma = K(\rho)$, for each $\rho \ge 0$ define the stopping time $\tau(\rho)$ in the broad sense

$$
\tau(\rho) = \inf \{ t : \max \, (|x_t|, |y_t|) \ge \rho \} .
$$

By stopping $\varphi^{-1}|x-y|^2$ with $\tau(\rho)$ we get,

$$
\begin{aligned}
\varphi^{-1}_{t\wedge\tau(\rho)}\left|x_{t\wedge\tau(\rho)} - y_{t\wedge\tau(\rho)}\right|^2 &= \int_{0+}^{t\wedge\tau(\rho)} \varphi^{-1}_{s-}\left[2(x_{s-} - y_{s-})(\alpha^1(x_{s-}) - \alpha^1(y_{s-}))\right. \\
&\quad + \triangle V_s\left|\alpha^1(x_{s-}) - \alpha^1(y_{s-})\right|^2 + \left|\beta^1(x_{s-}) - \beta^1(y_{s-})\right|^2 \\
&\quad \left. -K_s(\rho)|x_{s-} - y_{s-}|^2\right] dV_s^r \\
&\quad + \int_0^{t\wedge\tau(\rho)-} \varphi^{-1}_s\left[2(x_s - y_s)(\alpha^2(x_s) - \alpha^2(y_s))\right. \\
&\quad + \triangle^+ V_s\left|\alpha^2(x_s) - \alpha^2(y_s)\right|^2 + \left|\beta^2(x_s) - \beta^2(y_s)\right|^2 \\
&\quad \left. -K_s(\rho)|x_s - y_s|^2\right] dV_s^g \\
&\quad + m''_{t\wedge\tau(\rho)},
\end{aligned}
$$

where m'' is a local optional martingale with $(m''_0 = 0)$. Hence, by conditions (8.12) and (8.13) we have

$$
0 \le \varphi^{-1}_{t\wedge\tau(\rho)}\left|x_{t\wedge\tau(\rho)} - y_{t\wedge\tau(\rho)}\right|^2 \le m''_{t\wedge\tau(\rho)}.
$$

Therefore, $m''_{\tau(\rho)}$ is a non-negative local optional martingale and by Fatou's Lemma it is a non-negative optional supermartingale. Since, $m''_0 = 0$ it follows that $m''_t = 0$ for all t. Consequently, $\varphi^{-1}_t|x_t - y_t|^2 = 0$ for $t \in [0, \tau(\rho)]$ for almost all $\omega \in \Omega$. Finally, since $\varphi^{-1}_t > 0$ and $\tau(\rho) \uparrow \infty$ a.s., then $x_t = y_t$ for all $t \in \mathbb{R}_+$ almost surely. ∎

The existence of the solution will be proved next, after additional prerequisite lemmas.

8.2.2.2 Existence

The proof of existence is carried out by an iteration scheme similar to that of Picard's iteration. Given a sequence of processes $(x^n)_{n\ge 0}$ satisfying

$$
x^{n+1} = \xi + a(\cdot, x^n) \circ A + b(\cdot, x^n) \circ M, \quad x^0 \in \mathcal{F}_0,
$$

where a, b, A and M are as in Equation (8.11), then a solution exists if $\mathbf{E}\left|x^{n+1} - x^n\right|^2 \to 0$. Even though the idea behind the proof is simple, the steps to get to the result are many. So, the proof is broken down into several lemmas.

Lemma 8.2.8 proves existence in the case that the coefficients of Equation (8.11) are bounded in the way given in Lemma 8.2.6. Lemma 8.2.6 basically says that if the coefficients of Equation (8.11) are defined as in (8.22), then they satisfy the inequalities (8.23) and (8.24), a modified restriction on growth conditions, and (8.25) and (8.26) the monotonicity conditions. Lemma 8.2.7 proves existence in the case that the coefficients of Equation (8.11) satisfy the inequalities given in Lemma 8.2.6. Lemmas 8.2.7 and 8.2.8 are tied by Lemma 8.2.6. Finally, the proof of the existence theorem uses Lemma 8.2.8 to establish existence of solution for Equation (8.11) for coefficients satisfying conditions 1, 2 and 3.

We begin by two supporting Lemmas 8.2.4 and 8.2.5. Lemma 8.2.4 establishes a finiteness condition on the integral of the supremum of the coefficients of Equation (8.11).

Lemma 8.2.4 *For $T > 0$ and $\rho > 0$ we have*

$$\left[\sup_{|x| \le \rho} |a(\cdot, x)| \right] \circ A_T = \int_{0+}^{T} \sup_{|x_-| \le \rho} |a(t, x_-)| dA_t + \int_{0}^{T-} \sup_{|x| \le \rho} |a(t, x)| dA_{t+} < \infty,$$

$$\left[\sup_{|x| \le \rho} |\beta(\cdot, x)|^2 \right] \circ V_T = \int_{0+}^{T} \sup_{|x_-| \le \rho} |\beta(t, x_-)|^2 dV_t + \int_{0}^{T-} \sup_{|x| \le \rho} |\beta(t, x)|^2 dV_{t+} < \infty,$$

for almost all ω.

Proof. The right-continuous case was proved in [96]. Here, we proceed as follows. The functions a and β are continuous in x. So, over the finite interval $|x| \le \rho$, the functions attain their supremum at a point in the interval. Let this point be $x^*(\rho)$ and replace the sup over $|x| \le \rho$ by $x^*(\rho)$ (e.g., $\sup_{|x| \le \rho} |a(\cdot, x)| = |a(\cdot, x^*(\rho))|$). By integrability conditions imposed on a and β, the results of the lemma are established. ∎

The lemma states the results for a and β, but similar results follow for α and b. The next lemma is obvious. It allows us to construct bounded functions from locally bounded ones.

Lemma 8.2.5 Suppose f is locally bounded function on R^d, n an integer and $N = \sup \{|f(x)| : |x| \le n\}$. Then, there exists a function $\phi(x)$ such that $\phi(x) = f(x)$ for $|x| \le n$, $\phi(x) = 0$ for $|x| \ge n+1$ and $|\phi(x)| \le |f(x)|$, $|\phi(x) - \phi(y)|^2 \le |f(x) - f(y)|^2 + N^2|x-y|^2$ for every $x, y \in R^d$. Moreover, if f is continuous in x and a measurable function, then so is ϕ.

Given a bound on x, the next lemma tells us that the coefficients of the stochastic equation (8.11) will satisfy certain boundedness conditions. By way of these conditions, Lemma 8.2.8 proves existence of solution to Equation (8.11).

Lemma 8.2.6 For a fixed integer m define the functions $\tilde{a}$ and $\tilde{b}$ as

$$\tilde{a} = \begin{cases} a & |x| \le m \\ 0 & |x| \ge m+1 \end{cases} \quad and \quad \tilde{b} = \begin{cases} b & |x| \le m \\ 0 & |x| \ge m+1 \end{cases}. \tag{8.22}$$

Then, there exists a process $L \in L$ such that for all x, y the following inequalities hold

$$\left| \tilde{\alpha}^1(t, x_-) \right| + \left| \tilde{\beta}^1(t, x_-) \right|^2 \le L_t \tag{8.23}$$

$$\left| \tilde{\alpha}^2(t, x) \right| + \left| \tilde{\beta}^2(t, x) \right|^2 \le L_t \tag{8.24}$$

$$2(x_- - y_-)(\tilde{\alpha}^1(t, x_-) - \tilde{\alpha}^1(t, y_-)) + \triangle V_t |\tilde{\alpha}^1(t, x_-) - \tilde{\alpha}^1(t, y_-)|^2 \tag{8.25}$$
$$+ |\tilde{\beta}^1(t, x_-) - \tilde{\beta}^1(t, y_-)|^2 \le L_t |x_- - y_-|^2$$

$$2(x - y)(\tilde{\alpha}^2(t, x) - \tilde{\alpha}^2(t, y)) + \triangle^+ V_t |\tilde{\alpha}^2(t, x) - \tilde{\alpha}^2(t, y)|^2 \tag{8.26}$$
$$+ |\tilde{\beta}^2(t, x) - \tilde{\beta}^2(t, y)|^2 \le L_t |x - y|^2.$$

for almost all (t,ω) with respect to $dV\,dP$ where $\tilde{\alpha}^1 := \tilde{a}^1\Lambda^1 = \tilde{a}_-\Lambda_-$ and $\tilde{\beta}^1 := \tilde{b}^1\Theta^{1\frac{1}{2}} = \tilde{b}_-\Theta_-^{\frac{1}{2}}$ and $\tilde{\alpha}^2 := \tilde{a}^2\Lambda^2 = \tilde{a}\Lambda$ and $\tilde{\beta}^2 := \tilde{b}^2\Theta^{2\frac{1}{2}} = \tilde{b}\Theta^{\frac{1}{2}}$ are continuous in x.

Proof. For all (t,ω), $\Theta_t(\omega)$ is $\mathcal{O}$ measurable self-adjoint nuclear operator. Therefore, there exists a finite set $\{\theta_j(t,\omega)\}_{j=1}^k$ of $\mathcal{O}$ measurable eigenvectors that form an orthonormal basis in $\overline{\{\Theta^{\frac{1}{2}}\}}$ (the closure of the set $\{\Theta^{\frac{1}{2}}\}$). Let $b_j := b\theta_j$ and truncate each by an integer n as in Lemma 8.2.5. For every (t,ω,x) the truncated functions $\tilde{b}(t,x)$ are linear operators in $\overline{\{\Theta^{\frac{1}{2}}\}}$ defined as $\tilde{b}(t,x)\theta_j := \tilde{b}_j(t,x)$. Since $|\Theta|_1 \leq 1$ then by Lemma 8.2.5 there exists a process $L^1 \in \mathcal{L}$ such that

$$\left|\left(\tilde{b}(t,x_-) - \tilde{b}(t,y_-)\right)\Theta_{t-}^{\frac{1}{2}}\right|^2 \leq L_t^1 |x_- - y_-|^2 + \left|(b(t,x_-) - b(t,y_-))\Theta_{t-}^{\frac{1}{2}}\right|^2, \quad (8.27)$$

$$\left|\left(\tilde{b}(t,x) - \tilde{b}(t,y)\right)\Theta_t^{\frac{1}{2}}\right|^2 \leq L_t^1 |x - y|^2 + \left|(b(t,x) - b(t,y))\Theta_t^{\frac{1}{2}}\right|^2. \quad (8.28)$$

Let $0 \leq \eta \leq 1$ and $\eta \in C_0^\infty(\mathbb{R}^d)$. Define $\eta(x) = 1$ for $|x| \leq n+2$, for an integer n, and $\eta(x) = 0$ otherwise . Let $\tilde{a}(t,x) = a(t,x)\eta(x)$, $\tilde{\alpha}^1 := \tilde{a}^1\Lambda^1$ and $\tilde{\alpha}^2 := \tilde{a}^2\Lambda^2$. By Lemma 8.2.4 there exists a process $L^2 \in \mathcal{L}$ such that

$$|\alpha^1(t,x)| + |\beta^1(t,x)|^2 \leq L_t^2, \quad (8.29)$$
$$|\alpha^2(t,x)| + |\beta^2(t,x)|^2 \leq L_t^2. \quad (8.30)$$

where $\alpha^1 = a_-\Lambda_-$, $\alpha^2 = a\Lambda$, $\beta^1 = b_-\Theta_-^{\frac{1}{2}}$ and $\beta^2 = b\Theta^{\frac{1}{2}}$.

Next, we show that there exists a process $L^3 \in \mathcal{L}$ such that

$$2(x_- - y_-)(\alpha^1(t,x_-) - \alpha^1(t,y_-)) + \triangle V_t \left|\alpha^1(t,x_-) - \alpha^1(t,y_-)\right|^2 \leq L_t^3 |x_- - y_-|^2 \quad (8.31)$$
$$+ \eta(x_-)\eta(y_-)\left\{2(x_- - y_-)(\alpha^1(t,x_-) - \alpha^1(t,y_-)) + \triangle V_t|\alpha^1(t,x_-) - \alpha^1(t,y_-)|^2\right\}.$$

$$2(x - y)(\alpha^2(t,x) - \alpha^2(t,y)) + \triangle^+ V_t\left|\alpha^2(t,x) - \alpha^2(t,y)\right|^2 \leq L_t^3 |x - y|^2 \quad (8.32)$$
$$+ \eta(x)\eta(y)\left\{2(x - y)(\alpha^2(t,x) - \alpha^2(t,y)) + \triangle^+ V_t|\alpha^2(t,x) - \alpha^2(t,y)|^2\right\}.$$

Because Equations (8.31) and (8.32) are symmetric with respect to x, y, it is therefore sufficient to consider the case $0 \leq \eta(x) < \eta(y)$. Observe that for $|y| \leq n+3$ the left-hand side of (8.31) is not greater than

$$\eta(x_-)2(x_- - y_-)(\alpha^1(t,x_-) - \alpha^1(t,y_-)) + \triangle V_t|\alpha^1(t,x_-) - \alpha^1(t,y_-)|^2$$
$$+ 2(x_- - y_-)\alpha^1(t,y_-)(\eta(x_-) - \eta(y_-)) + |\alpha^1(t,y_-)|^2(\eta(x_-) - \eta(y_-))^2 \triangle V_t$$
$$+ 2\triangle V_t\eta(x_-)|\alpha^1(t,x_-) - \alpha^1(t,y_-)\|\alpha^1(t,y_-)\|\eta(x_-) - \eta(y_-)|.$$

Similarly, the left-hand side of (8.32) is not greater than

$$\eta(x)2(x - y)(\alpha^2(t,x) - \alpha^2(t,y)) + \triangle^+ V_t|\alpha^2(t,x) - \alpha^2(t,y)|^2$$
$$+ 2(x - y)\alpha^2(t,y)(\eta(x) - \eta(y)) + |\alpha^2(t,y)|^2(\eta(x) - \eta(y))^2 \triangle^+ V_t$$
$$+ 2\triangle^+ V_t\eta(x)|\alpha^2(t,x) - \alpha^2(t,y)\|\alpha^2(t,y)\|\eta(x) - \eta(y)|.$$

By (8.17) and (8.18), the last term can be estimated by

$$L_t^4 |x_- - y_-|^2 = 2C \sup_{|y_-| \leq n+3} |\alpha^1(t, y_-)|(4 + K_t(n+3) \triangle V_t)|x_- - y_-|^2,$$

$$L_t^4 |x - y|^2 = 2C \sup_{|y| \leq n+3} |\alpha^2(t, y)|(4 + K_t(n+3) \triangle^+ V_t)|x - y|^2,$$

where C is Lipschitz constant associated with η. From Lemma 8.2.4 and conditions (8.17), and (8.18) it follows that $L^4 \in \mathcal{L}$. Also, by using the trivial results

$$|\alpha^1(t, y_-)|^2 (\eta(x_-) - \eta(y_-))^2 \triangle V_t \leq C^2 \sup_{|y_-| \leq n+3} |\alpha^1(t, y_-)|^2 \triangle V_t |x_- - y_-|^2,$$

$$|\alpha^2(t, y)|^2 (\eta(x) - \eta(y))^2 \triangle^+ V_t \leq C^2 \sup_{|y| \leq n+3} |\alpha^2(t, y)|^2 \triangle^+ V_t |x - y|^2$$

and

$$2(x_- - y_-)\alpha^1(t, y_-)(\eta(x_-) - \eta(y_-)) \leq 2C \sup_{|y_-| \leq n+3} |\alpha^1(t, y_-)| \|x_- - y_-|^2,$$

$$2(x - y)\alpha^2(t, y)(\eta(x) - \eta(y)) \leq 2C \sup_{|y| \leq n+3} |\alpha^2(t, y)| \|x - y|^2,$$

and by Lemma 8.2.4 and conditions (8.17) and (8.18) we obtain the results (8.31) and (8.32).

If $|x| \leq n+2$ and $|y| \leq n+2$, then $\eta(x) = \eta(y) = 1$ and by (8.27), (8.28), (8.31) and (8.32) and (8.12) and (8.13) the inequalities (8.25) and (8.26) are valid with $L = L^{(1)} + K(n+3)$.

If $|x| \geq n+1$ and $|y| \geq n+1$, then $\beta^1(t, x_-) = \beta^1(t, y_-) = 0$ and $\beta^2(t, x) = \beta^2(t, y) = 0$ (8.25) and (8.26) holds with $L = L^3 + K(n+3)$.

If one of the values $|x|$, $|y|$ is smaller than $n+1$ and the other is greater than $n+2$, then $|x - y| \geq 1$ and one of the values of $\beta^1(t, x_-)$, $\beta^1(t, y_-)$ is zero similarly for $\beta^2(t, x)$ and $\beta^2(t, y)$. Therefore, in this case (8.29) and (8.30) imply that

$$\left|\beta^1(t, x_-) - \beta^1(t, y_-)\right|^2 \leq L_t^2 |x_- - y_-|, \tag{8.33}$$

$$\left|\beta^2(t, x) - \beta^2(t, y)\right|^2 \leq L_t^2 |x - y|. \tag{8.34}$$

Consequently, from (8.27-8.33) we get that the inequalities (8.23) and (8.28) and (8.25) and (8.26) hold with $L = L^1 + L^2 + L^3 + K(n+3)$, where $m = n+2$. ∎

Suppose the coefficients of Equation (8.11) satisfy the inequalities in Lemma 8.2.6, then does a solution to the stochastic differential equation exist? Lemma 8.2.7 proves that, indeed, a solution exists.

Lemma 8.2.7 *Let $L \in \mathcal{L}$ be such that the following inequalities*

$$|\alpha^1(t, x_-)| + |\beta^1(t, x_-)|^2 \leq L_t \tag{8.35}$$

$$|\alpha^2(t, x)| + |\beta^2(t, x)|^2 \leq L_t \tag{8.36}$$

$$2|x_- - y_-| |\alpha^1(t, x_-) - \alpha^1(t, y_-)| + |\beta^2(t, x_-) - \beta^2(t, y_-)|^2 \leq L_t |x_- - y_-|^2 \tag{8.37}$$

$$2|x - y| |\alpha^2(t, x) - \alpha^2(t, y)| + |\beta^2(t, x) - \beta^2(t, y)|^2 \leq L_t |x - y|^2 \tag{8.38}$$

for every x, $y \in \mathbb{R}^d$ a.s. $dV d\mathbf{P}$. Then, Equation (8.11) has a unique solution.

Proof. Define an *iteration* procedure as follows: $x_t^0 = \xi$ and for $n \geq 0$,

$$x_t^{n+1} = \xi + \int_{0+}^t a(s, x_{s-}^n)dA_s + \int_0^{t-} a(s, x_s^n)dA_{s+} + \int_{0+}^t b(s, x_{s-}^n)dM_s + \int_0^{t-} b(s, x_s^n)dM_{s+}. \quad (8.39)$$

Let $\gamma_t^1 = L_t(6 + 4 \bigtriangleup V_t)$, $\gamma_t^2 = L_t(6 + 4 \bigtriangleup^+ V_t)$ and φ a solution of Equation (8.19). Set $\psi_t = \exp(-|\xi|)\varphi_t^{-1}$ and apply Lemma 8.2.3 to $|x_t^{n+1} - x_t^n|^2 \psi_t$,

$$\begin{aligned}
|x_t^{n+1} - x_t^n|^2 \psi_t &= \int_{0+}^t \psi_s \left\{ 2(x_{s-}^{n+1} - x_{s-}^n)(\alpha^1(s, x_{s-}^n) - \alpha^1(s, x_{s-}^{n-1})) \right. \\
&\quad + \bigtriangleup V_s |\alpha^1(s, x_{s-}^n) - \alpha^1(s, x_{s-}^{n-1})|^2 + |\beta^1(s, x_{s-}^n) - \beta^1(s, x_{s-}^{n-1})|^2 \\
&\quad \left. -\gamma_s^1 |x_{s-}^{n+1} - x_{s-}^n|^2 \right\} dV_s \\
&\quad + \int_0^{t-} \psi_s \left\{ 2(x_s^{n+1} - x_s^n)(\alpha^2(s, x_s^n) - \alpha^2(s, x_s^{n-1})) \right. \\
&\quad + \bigtriangleup^+ V_s |\alpha^2(s, x_s^n) - \alpha^2(s, x_s^{n-1})|^2 + |\beta^2(s, x_s^n) - \beta^2(s, x_s^{n-1})|^2 \\
&\quad \left. -\gamma_s^2 |x_s^{n+1} - x_s^n|^2 \right\} dV_{s+} + m_t^n,
\end{aligned}$$

where m_t^n is a local optional martingale with $m_0^n = 0$. Using the inequality $2|pq| \leq |p|^2 + |q|^2$ and assumptions (8.37) and (8.38) we get

$$\begin{aligned}
|x_t^{n+1} - x_t^n|^2 \psi_t &\leq \int_{0+}^t \psi_s \left\{ \frac{1}{4}\gamma_s^1 |x_{s-}^n - x_{s-}^{n-1}|^2 - \frac{1}{2}\gamma_s^1 |x_{s-}^{n+1} - x_{s-}^n|^2 \right\} dV_s \quad (8.40) \\
&\quad + \int_0^{t-} \psi_s \left\{ \frac{1}{4}\gamma_s^2 |x_s^n - x_s^{n-1}|^2 - \frac{1}{2}\gamma_s^2 |x_s^{n+1} - x_s^n|^2 \right\} dV_{s+} \\
&\quad + m_t^n.
\end{aligned}$$

Let τ be a stopping time in the *broad sense* and (τ^i) an increasing sequence of stopping times in the *broad sense*, such that $\lim_{i \to \infty} \tau^i = \infty$ a.s. For each τ^i the stopped local optional martingale $m_{t \wedge \tau^i}^n$ is uniformly integrable. Replace t in (8.40) with $t \wedge \tau \wedge \tau^i$, rearrange terms, take expectation and let $t \to \infty$, to get

$$4\mathbf{E}|x_\tau^{n+1} - x_\tau^n|^2 \psi_\tau + 2\mathbf{E} \int_{0+}^\tau \gamma_s^1 \psi_s |x_{s-}^{n+1} - x_{s-}^n|^2 dV_s + 2\mathbf{E} \int_0^{\tau-} \gamma_s^2 \psi_s |x_s^{n+1} - x_s^n|^2 dV_{s+} \leq \quad (8.41)$$

$$\leq \mathbf{E} \int_{0+}^\tau \gamma_s^1 \psi_s |x_{s-}^n - x_{s-}^{n-1}|^2 dV_s + \mathbf{E} \int_0^{\tau-} \gamma_s^2 \psi_s |x_s^n - x_s^{n-1}|^2 dV_{s+}.$$

Similarly for $|x_\tau^1 - \xi|^2$ we find

$$4\mathbf{E}|x_\tau^1 - \xi|^2 \psi_\tau + 2\mathbf{E} \int_{0+}^\tau \gamma_s^1 \psi_s |x_{s-}^1 - \xi|^2 dV_s + 2\mathbf{E} \int_{0+}^\tau \gamma_s^2 \psi_s |x_s^1 - \xi|^2 dV_{s+} \leq$$

$$\leq \mathbf{E} \int_{0+}^\tau \gamma_s^1 \psi_s |\xi|^2 dV_s + \mathbf{E} \int_0^{\tau-} \gamma_s^2 \psi_s |\xi|^2 dV_s$$

$$\leq \mathbf{E}|\xi|^2 \left[\int_{0+}^\tau \gamma_s^1 \psi_s dV_s + \int_0^{\tau-} \gamma_s^2 \psi_s dV_s \right]$$

$$\leq \mathbf{E}|\xi|^2 \exp(-|\xi|)(1 - \varphi_\tau^{-1}) < \infty.$$

Repeat the iteration on inequality (8.41), for example, for $n = 1$

$$4\mathbf{E}|x_\tau^2 - x_\tau^1|^2 \psi_\tau + 2\mathbf{E}\left(\int_{0+}^\tau \gamma_s^1 \psi_s |x_{s-}^2 - x_{s-}^1|^2 dV_s + \int_0^{\tau-} \gamma_s^2 \psi_s |x_s^2 - x_s^1|^2 dV_{s+}\right) \leq$$

$$\leq \mathbf{E}\left(\int_{0+}^\tau \gamma_s^1 \psi_s |x_{s-}^1 - \xi|^2 dV_s + \int_0^{\tau-} \gamma_s^2 \psi_s |x_s^1 - \xi|^2 dV_{s+}\right)$$

$$\leq 2\mathbf{E}|x_\tau^1 - \xi|^2 \psi_\tau + \mathbf{E}\left(\int_{0+}^\tau \gamma_s^1 \psi_s |x_{s-}^1 - \xi|^2 dV_s + \int_0^{\tau-} \gamma_s^2 \psi_s |x_s^1 - \xi|^2 dV_{s+}\right)$$

$$\leq \frac{1}{2}\mathbf{E}|\xi|^2 \exp(-|\xi|)(1 - \varphi_\tau^{-1})$$

and for n we get

$$\mathbf{E}\left(\int_{0+}^\tau \gamma_s^1 \psi_s \left|x_{s-}^{n+1} - x_{s-}^n\right|^2 dV_s + \int_0^{\tau-} \gamma_s^2 \psi_s \left|x_s^{n+1} - x_s^n\right|^2 dV_{s+}\right) \leq 2^{-n} C$$

and $\mathbf{E}\left|x_\tau^{n+1} - x_\tau^n\right|^2 \psi_\tau \leq 2^{-n} C$. Note that on $(\tau = \infty)$ set $\left|x_\tau^{n+1} - x_\tau^n\right|^2 \psi_\tau = 0$.

So, for $\tau = \tau_n = \inf\left\{t : |x_t^{n+1} - x_t^n|^2 \psi_t \geq n^{-4}\right\}$ we get $\mathbf{P}(\tau_n < \infty) \leq n^4 C 2^{-n}$. Hence,

$$\mathbf{P}\left(\sup_{t\geq 0}\left|x_t^{n+1} - x_t^n\right| \psi_t^{1/2} \geq n^{-2}\right) \leq n^4 2^{-n} C.$$

By the Borel-Cantelli Lemma, the convergence of $\sum_{n=0}^\infty n^4 2^{-n}$ implies uniform convergence of

$$\sum_{n=0}^\infty \left|x_t^{n+1} - x_t^n\right| \psi_t^{1/2}$$

with probability 1. Therefore, the sequence of adapted *laglad* processes $x_t^n = \xi + \sum_{i=0}^{n-1}(x_t^{i+1} - x_t^i)$ converges uniformly on bounded intervals to an optional *ladlag* process x_t with probability 1.

Finally, we verify that x_t satisfies Equation (8.11). For a fixed $\rho > 0$ define the stopping time $\tau(\rho) = \inf(t : |x_t| \geq \rho)$. Since $x_t^n \to x_t$ uniformly with probability 1 for every t on all bounded intervals and by the Lebesgue Theorem, for $n \to \infty$

$$\int_{0+}^{t\wedge\tau(\rho)} a^1(s, x_{s-}^n) dA_s \quad \to \quad \int_{0+}^{t\wedge\tau(\rho)} a^1(s, x_{s-}) dA_s^r,$$

$$\int_0^{t\wedge\tau(\rho)-} a^2(s, x_s^n) dA_{s+} \quad \to \quad \int_0^{t\wedge\tau(\rho)-} a^2(s, x_s) dA_{s+}^g \quad a.s.$$

and

$$\left\langle \int_0^\cdot (b^1(s, x_{s-}^n) - b^1(s, x_{s-})) dM_s \right\rangle_{t\wedge\tau(\rho)} = \int_{t\wedge]0,\tau(\rho)]} |\beta^1(s, x_{s-}^n) - \beta^1(s, x_{s-})|^2 dV_s^r$$

$$+ \int_{t\wedge[0,\tau(\rho)[} |\beta^2(s, x_s^n) - \beta^2(s, x_s)|^2 dV_{s+}^g$$

$$\to 0 \quad a.s.$$

for that a and b are continuous in x and by Lemma 8.2.4. Therefore, as $n \to \infty$ in (8.39) x_t solves Equation (8.11). ∎

Lemma 8.2.8 Let $\tilde{a}$ and $\tilde{b}$ be the random functions defined in Lemma 8.2.6. *If we replace a with $\tilde{a}$ and b with $\tilde{b}$ in Equation (8.11), then this equation admits a unique solution.*

Proof. Let

$$J(z) = \begin{cases} C\exp\left(\frac{-1}{1-|z|^2}\right) & if \;\; |z| < 1 \\ 0 & if \;\; |z| \geq 1 \end{cases},$$

in $\mathbb{R}^d$ where $C \in \mathbb{R}$ such that $\int_{\mathbb{R}^d} J(z)dz = 1$. Approximate the functions a and b by smooth functions with respect to x using $J(z)$ as follows

$$\tilde{a}^k(t, x) := \int_{\mathbb{R}^d} \tilde{a}\left(t, x - k^{-1}z\right) J(z)dz \tag{8.42}$$

and

$$\tilde{b}^k(t, x) := \int_{\mathbb{R}^d} \tilde{b}(t, x - k^{-1}z)J(z)dz, \tag{8.43}$$

for an integer k and $(t, \omega) \in \mathbb{R}_+ \times \Omega$. Let $\tilde{\alpha}^{1,k} := \tilde{a}_-^k \Lambda_-$, $\tilde{\alpha}^{2,k} := \tilde{a}^k \Lambda$ and $\tilde{\beta}^{1,k} := \tilde{b}_-^k \Theta^{\frac{1}{2}}$ and $\tilde{\beta}^{2,k} := \tilde{b}^k \Theta^{\frac{1}{2}}$. The functions $|\tilde{a}^k|$, $|\tilde{b}^k|$, $|(\partial/\partial x)\tilde{a}^k|$, $|(\partial/\partial x)\tilde{b}^k|$ can be estimated by the maximums of $|\tilde{a}|$ and $|\tilde{b}|$ in x. It follows that for every k the functions $\tilde{a}^k$ and $\tilde{b}^k$ satisfy the inequalities (8.35), (8.36), (8.37) and (8.38) in Lemma 8.2.7. Therefore, equation

$$\begin{aligned} x_t^k &= \xi + \int_{0+}^{t} \tilde{a}^{1,k}(s, x_{s-}^k)dA_s + \int_0^{t-} \tilde{a}^{2,k}(s, x_s^k)dA_{s+} \\ &\quad + \int_{0+}^{t} \tilde{b}^{1,k}(s, x_{s-}^k)dM_s + \int_0^{t-} \tilde{b}^{2,k}(s, x_s^k)dM_{s+} \end{aligned} \tag{8.44}$$

admits one and only one solution $(x_t^k)_{t \geq 0}$ for every k. Next, we show that x_t^k converges uniformly on bounded intervals in t as $k \to \infty$.

To do so, let φ be the solution of $d\varphi_t = L_t \varphi_{t-} dV_t^r + L_t \varphi_t dV_{t+}^g$, $\varphi_0 = 1$. Using Equation (8.21) yields

$$\begin{aligned} \varphi_t^{-1}|x_t^k - x_t^l|^2 &= \int_{0+}^{t} \varphi_{s-}^{-1} \left\{ 2(x_{s-}^k - x_{s-}^l)(\tilde{\alpha}^{1,k}(x_{s-}^k) - \tilde{\alpha}^{1,l}(x_{s-}^l)) \right. \\ &\quad + \Delta V_s |\tilde{\alpha}^{1,k}(x_{s-}^k) - \tilde{\alpha}^{1,l}(x_{s-}^l)|^2 + |\tilde{\beta}^{1,k}(x_{s-}^k) - \tilde{\beta}^{1,l}(x_{s-}^l)|^2 \\ &\quad \left. - L_s \left|x_{s-}^k - x_{s-}^l\right|^2 \right\} dV_s^r \\ &\quad + \int_0^{t-} \varphi_s^{-1} \left\{ 2(x_s^k - x_s^l)(\tilde{\alpha}^{2,k}(x_s^k) - \tilde{\alpha}^{2,l}(x_s^l)) \right. \\ &\quad + \Delta V_s |\tilde{\alpha}^{2,k}(x_s^k) - \tilde{\alpha}^{2,l}(x_s^l)|^2 + |\tilde{\beta}^{2,k}(x_s^k) - \tilde{\beta}^{2,l}(x_s^l)|^2 \\ &\quad \left. - L_s \left|x_s^k - x_s^l\right|^2 \right\} dV_s^g + m_t^k. \end{aligned}$$

Using Schwarz inequality and Equation (8.42) we find that the integrand is not greater than

$$\tilde{b}^k(t, x)\theta_j = \int_{\mathbb{R}^d} \tilde{b}_j(t, x - k^{-1}z)J(z)dz,$$

where θ_j and $\tilde{b}_j$ have been defined in the proof of Lemma 8.2.6. So,

$$
\begin{aligned}
I_s^1 = {} & \varphi_s^{-1} \int_{\mathbb{R}^d} \mathrm{J}(z) \left\{ 2 \left((x_{s-}^k - k^{-1}z) - (x_{s-}^l - l^{-1}z) \right) \left(\tilde{\alpha}^1 (x_{s-}^k - k^{-1}z) - \tilde{\alpha}^1 (x_{s-}^l - l^{-1}z) \right) \right. \\
& + \Delta V_s \left| \tilde{\alpha}^1 (x_{s-}^k - k^{-1}z) - \tilde{\alpha}^1 (x_{s-}^l - l^{-1}z) \right|^2 + \left| \tilde{\beta}^1 (x_{s-}^k - k^{-1}z) - \tilde{\beta}^1 (x_{s-}^l - l^{-1}z) \right|^2 \\
& \left. - L_s \left| (x_{s-}^k - k^{-1}z) - (x_{s-}^k - l^{-1}z) \right|^2 \right\} dz \\
& + 2(k^{-1} - l^{-1}) \varphi_s^{-1} \int_{\mathbb{R}^d} \mathrm{J}(z) z \left| \tilde{\alpha}^1 (x_{s-}^k - k^{-1}z) - \tilde{\alpha}^1 (x_{s-}^l - l^{-1}z) \right| dz \\
& + L_s (k^{-1} - l^{-1})^2 \varphi_s^{-1} \int_{\mathbb{R}^d} z^2 \mathrm{J}(z) dz.
\end{aligned}
$$

and

$$
\begin{aligned}
I_s^2 = {} & \varphi_s^{-1} \int_{\mathbb{R}^d} \mathrm{J}(z) \left\{ 2 \left((x_s^k - k^{-1}z) - (x_s^l - l^{-1}z) \right) \left(\tilde{\alpha}^2 (x_s^k - k^{-1}z) - \tilde{\alpha}^2 (x_s^l - l^{-1}z) \right) \right. \\
& + \triangle^+ V_s \left| \tilde{\alpha}^2 (x_s^k - k^{-1}z) - \tilde{\alpha}^2 (x_s^l - l^{-1}z) \right|^2 + \left| \tilde{\beta}^2 (x_s^k - k^{-1}z) - \tilde{\beta}^2 (x_s^l - l^{-1}z) \right|^2 \\
& \left. - L_s \left| (x_s^k - k^{-1}z) - (x_s^k - l^{-1}z) \right|^2 \right\} dz \\
& + 2(k^{-1} - l^{-1}) \varphi_s^{-1} \int_{\mathbb{R}^d} \mathrm{J}(z) z \left| \tilde{\alpha}^2 (x_s^k - k^{-1}z) - \tilde{\alpha}^2 (x_s^l - l^{-1}z) \right| dz \\
& + L_s (k^{-1} - l^{-1})^2 \varphi_s^{-1} \int_{\mathbb{R}^d} z^2 \mathrm{J}(z) dz.
\end{aligned}
$$

Therefore, by inequalities (8.25) and (8.26) we get

$$
\begin{aligned}
I_s^1 &\leq C_1 (k^{-1} - l^{-1}) L_s \varphi_s^{-1} + C_2 (k^{-1} - l^{-1})^2 \varphi_s^{-1} L_s, \\
I_s^2 &\leq C_1 (k^{-1} - l^{-1}) L_s \varphi_s^{-1} + C_2 (k^{-1} - l^{-1})^2 \varphi_s^{-1} L_s,
\end{aligned}
$$

where $C_1 = 4 \int |z| \mathrm{J}(z) dz$, $C_2 = \int z^2 \mathrm{J}(z) dz$. This implies that

$$
\begin{aligned}
\varphi_t^{-1} \left| x_t^k - x_t^l \right|^2 \leq {} & C_1 \left(k^{-1} - l^{-1} \right) \int_{0+}^t L_s \varphi_s^{-1} dV_s^r + C_2 (k^{-1} - l^{-1}) \int_{0+}^t \varphi_s^{-1} L_s dV_s^r \\
& + C_1 \left(k^{-1} - l^{-1} \right) \int_0^{t-} L_s \varphi_s^{-1} dV_{s+}^g + C_2 (k^{-1} - l^{-1}) \int_0^{t-} \varphi_s^{-1} L_s dV_{s+}^g \\
& + m_t^k.
\end{aligned}
$$

From Lemma 8.2.3 we have

$$
\begin{aligned}
\int_{0+}^t L_s \varphi_s^{-1} dV_s^r &= l - \varphi_t^{-1} < 1, \\
\int_0^{t-} L_s \varphi_s^{-1} dV_{s+}^g &= l - \varphi_t^{-1} < 1.
\end{aligned}
$$

Consequently,

$$
\varphi_t^{-1} \left| x_t^k - x_t^l \right|^2 \leq 2C_1 (k^{-1} - l^{-1}) + 2C_2 \left(k^{-1} - l^{-1} \right)^2 + m_t^k.
$$

Hence, we obtain that for any stopping time τ

$$\mathbf{E}\left|x_\tau^k - x_\tau^l\right|^2 \varphi_\tau^{-1} \leq 2C_1(k^{-1} - l^{-1}) + 2C_2\left(k^{-1} - l^{-1}\right)^2. \tag{8.45}$$

Next, define the stopping times $\tau = \tau(k, l) = \inf\left(t : |x_t^k - x_t^l|^2 \varphi_t^{-1} \geq \epsilon\right)$. Since

$$\mathbf{P}\left(\sup_{t \geq 0} |x_t^k - x|^2 \varphi_t^{-1} \geq \epsilon\right) \leq \mathbf{P}\left(\left|x_\tau^k - x\right|^2 \varphi_\tau^{-1}\right| \geq \epsilon\right) \leq \frac{1}{\epsilon}\mathbf{E}\left|x_\tau^k - x_\tau^l\right|^2 \varphi_\tau^{-1},$$

from (8.45) we get $\mathbf{P}\left(\sup_{t \geq 0} |x_t^k - x_t^l|^2 \varphi_t^{-1} \geq \epsilon\right) \to 0$ as $k, l \to \infty$. Therefore, x_t^k converges uniformly in t in probability on every bounded interval. Selecting a convergent subsequence $x_t^{k_i}$ and taking the limit in (8.44) we get the solution. ■

Finally, we put all the lemmas together to complete the proof of Theorem 8.2.1.

Proof. Existence of solution part of Theorem 8.2.1. For every integer n let a^n and b^n denote the functions $\tilde{a}$ and $\tilde{b}$, respectively, defined in Lemma 8.2.6. For every n by Lemma 8.2.8 there exists an adapted *ladlag* process x_t^n, such that

$$\begin{aligned}
x_t^n &= \xi + \int_{0+}^t a^n(s, x_{s-}^n)dA_s^r + \int_0^{t-} a^n(s, x_s^n)dA_{s+}^g \\
&\quad + \int_{0+}^t b^n(s, x_{s-}^n)dM_s^r + \int_0^{t-} b^n(s, x_s^n)dM_{s+}^g.
\end{aligned}$$

Let us define the stopping times in the broad sense $\tau^n = \inf\left(t : |x_t^n| \geq n\right)$ and $\tau^{nm} = \tau^n \wedge \tau^m$. Since $a^n(t, x) = a(t, x)$, $b^n(t, x) = b(t, x)$ for $|x| \leq n$, it follows that the process $x_{t \wedge \tau}^n$, $x_{t \wedge \tau}^m$ satisfies the same equation

$$\begin{aligned}
dz_t &= \mathbf{1}_{(t < \tau^{nm})}a(t, z_{t-})dA_t^r + \mathbf{1}_{(t < \tau^{nm})}a(t, z_t)dA_{t+}^g \tag{8.46} \\
&\quad + \mathbf{1}_{(t < \tau^{nm})}b(t, z_{t-})dM_t^r + \mathbf{1}_{(t < \tau^{nm})}b(t, z_t)dM_{t+}^g \\
z_0 &= \xi.
\end{aligned}$$

The uniqueness of the solution of (8.46) implies that $x_t^n = x_t^m$ on $[0, \tau^{nm}]$ a.s. Consequently, if $n \leq m$ then $\tau^n \leq \tau^m$ a.s., and there exists a stopping time τ in the broad sense such that $\tau = \lim_{n \to \infty} \tau^n$ a.s. So, we can define an adapted *ladlag* process x_t such that $x_t = \lim_{n \to \infty} x_t^n$ a.s. on $[0, \tau]$. From (8.46) it follows that for every n and t,

$$\begin{aligned}
x_{t \wedge \tau^n} &= x_{t \wedge \tau^n}^n = \xi + \int_{0+}^{t \wedge \tau^n} a(s, x_{s-})dA_s^r + \int_0^{t \wedge \tau^n -} a(s, x_s)dA_{s+}^g \tag{8.47} \\
&\quad + \int_{0+}^{t \wedge \tau^n} b(s, x_{s-})dM_s^r + \int_0^{t \wedge \tau^n -} b(s, x_s)dM_{s+}^g
\end{aligned}$$

with probability 1.

Finally we must show that $\tau = \infty$ a.s. Let ϕ_t be the solution of the equation $d\phi_t = K_t(1)\phi_{t-}dV_t^r + K_t(1)\phi_t dV_t^g$, $\phi_0 = 1$, and $\psi_t = \exp(-|\xi|)\phi_t^{-1}$. Using Lemma 8.2.3 and (8.12), (8.13) and (8.45) we get

$$\mathbf{E}\left(\left|x_{\tau^n}^n\right|^2 \psi_{\tau^n} \mathbf{1}_{(\tau^n < \infty)}\right) \leq \mathbf{E}\left(|\xi|^2 \exp\left(-|\xi|\right)\right) = Const.$$

Hence, $\mathbf{E}\psi_{\tau^n}\mathbf{1}_{(\tau^n < \infty)} \leq n^{-2}Const.$ and $\mathbf{E}\psi_{\tau^n}\mathbf{1}_{(\tau^n < \infty)} \to 0$ as $n \to \infty$. Therefore, $\tau^n \to \infty$. ∎

8.2.3   Remarks and applications

We discuss possible applications, motivations and further research in the form of the following examples. The first example shows where the monotonicity condition arises and where the classical Lipschitz condition does not help. The second example forms a fruitful idea on how our technique can be extended to the area of backward stochastic differential equations (BSDEs) which is well-developing now in Stochastic Analysis, Mathematical Finance and Stochastic Control.

Example 1. Consider a Brownian particle in a box experiencing collisions with other particles in the box and the box walls. At every collision event, the particle experiences a sudden change in its velocity. The velocities before collision, in collision and after collision are random and possibly different, but are subject to the law of conservation of momentum. The collisions are assumed to be instantaneous and occur at random times measurable in $\mathbf{F}_+$. Suppose we are given a particle in this box of mass 1. The change in the particle's velocity, v, is given by

$$dv_t = sign(v_t)\mu_t dt + \sigma_t(v)dW_t + \sum \alpha_t^i(v_t) \triangle \mathbf{1}_{(t \geq \tau_i)} + \beta_t^i(v_t) \triangle^+ \mathbf{1}_{(t \leq \tau_i)} \tag{8.48}$$

where W is a Brownian noise force and $sign(v)\mu_t$ is a velocity-direction dependent force field and μ_t is its deterministic amplitude. The term

$$\sum \alpha_i(v) \triangle \mathbf{1}_{(t \geq \tau_i)} + \beta_i(v) \triangle^+ \mathbf{1}_{(t \leq \tau_i)}$$

is the impulse force on the particle as a result of its ith collision with other particles or the wall. Equation (8.48) does not satisfy the Lipchtiz conditions because of the $sign(v)$ function; however, it satisfies the monotonicity and growth conditions under proper conditions placed on the functions σ, α^i and β^i. Therefore, by Theorem 8.2.1 the solution to Equation (8.48) exists and is unique.

Example 2. Backward stochastic differential equations (BSDEs) were introduced by Bismut [97] and studied in a general way by Pardoux and Peng [98]. BSDEs have been extensively used in mathematical finance and stochastic control; see the work of Duffie and Epstein [99, 100] and El Karoui, Peng, Quenez [101]. BSDEs are also connected to a class of partial differential equations (PDEs) which led to solving of PDEs using probabilistic methods [102].

Although the basic theory of BSDEs is well understood, here we introduce a generalization of BSDEs by using optional local martingale driver as follows

$$Y_t = \xi + \int_{[t,T]} f\left(s, X_s, Z_s, Y_s\right) dA_s - \int_{[t,T]} Z_s dM_s \tag{8.49}$$

where A is an increasing continuous process taking values in $\mathbb{R}$, and M is a locally square integrable optional martingale taking values in $\mathbb{R}^k$. ξ is a r.v. in the Euclidean space $\mathbb{R}^d$. $f(t, X_t, Z_t, Y_t)$ is the generator of the BSDE an $\mathbb{R} \times \mathbb{R}^k$ function. X_t is an optional semimartingale and Z_t is the control process. The idea is to find a pair (Z_t, Y_t) of adapted processes that solves (8.49).

According to existence uniqueness Theorem 8.2.1, if we were to identify the functions $a(t, \omega, \cdot) := f(t, X_t, Z_t, \cdot)$ and $b(t, \omega, \cdot) := Z_t$ of Equation (8.16), then Equation (8.49) will have a unique solution for a class of $\{Z_t(\omega)\}$ processes satisfying the integrability condition of Theorem 8.2.1 and for $f(t, X_t, Z_t, \cdot)$ to satisfy the monotonicity and restriction on growth conditions.

An example of BSDE with monotone generator is the optimal consumption and investment problems for an agent with recursive utility of the Epstein-Zin type, studied in [103]. The optional version of the optimal value process is given by

$$Y_t = e^{-\kappa \lambda T} C_T^{1-\gamma} + \int_{[t,T]} \kappa \lambda e^{-\lambda s} C_s^{(1-\frac{1}{\eta})} Y_s^{(1-\frac{1}{\kappa})} ds - \int_{[t,T]} Z_s dM_s, \quad \kappa < 0, \ \gamma, \eta > 1, \ \lambda > 0$$

where C is a consumption process and Z is the investment portfolio.

8.3 Comparison of Solutions of Optional Stochastic Equations

Other equally important results in the study stochastic equations are comparison theorems. Comparison theorems allow us to compare solutions of related stochastic equations. With a comparison theorem, one finds that knowing the structure of the stochastic equation and the set of all possible initial conditions, a stochastic ordering of some sort can be established between processes that are solutions of these stochastic equations. Many have studied comparison of solutions of stochastic equations: Skorokhod [104] proved a comparison theorem for diffusion equations discovering that the solution of these equations must be a nondecreasing function of their drift coefficient. This result was established in another way in [105] under weaker conditions on the diffusion coefficient. In [89] and [106], the comparison theorem was carried over to the case of equations with integrals with respect to continuous martingales. In [107] Galchuk considered equations of a more general form, namely equations containing integrals with respect to continuous martingales and integer-valued random measures where the coefficients of the semimartingale are not Lipschitz but satisfy weaker conditions similar to those of Yamada [105]. Again, the solution is a nondecreasing function of the drift coefficient and in some sense of the jump functions.

Our goal for this section is to study comparison of solutions of stochastic equations driven by optional semimartingales in unusual probability spaces under more general conditions placed on the coefficient of the stochastic equation. To do so, first, we define stochastic equations with respect to components of optional semimartingales, then we prove the comparison theorem. Finally, we give an illustrative example of a possible application of comparison theorem to finance.

Suppose that we are given a complete but *unusual* probability space $(\Omega, \mathcal{F}, \mathbf{F} = (\mathcal{F}_t)_{t \geq 0}, \mathbf{P})$. Let us also introduce $\mathcal{G}(\mathbf{F})$ progressive in addition to $\mathcal{O}(\mathbf{F})$ and $\mathcal{P}(\mathbf{F})$ on $\Omega \times \mathbb{R}_+$. Recall that $\mathcal{G}$ is generated by all progressively measurable processes. We assume that, unless otherwise specified, all processes considered here are $\mathbf{F}$-consistent, and their trajectories have right and left limits but are not necessarily right- or left-continuous.

8.3.1   Comparison theorem

Let there be given an optional semimartingale Z with components: a continuous locally integrable process $a \in \mathcal{A}_{loc}$ with $a_0 = 0$, a continuous martingale $m \in \mathcal{M}_{loc}^c$ with $m_0 = 0$ and integer-valued measures μ^j, p^j for $j = 1$, 2 and η with predictable and optional projections ν^j, λ^j, and θ, respectively.

We shall consider the equations

$$
\begin{aligned}
X_t^i &= X_0^i + f^i(X^i) \cdot a_t + g(X^i) \cdot m_t \\
&\quad + \sum_j U h_j(X^i) * (\mu^j - \nu^j)_t + V h_j^i(X^i) * \mu_t^j + \left(k_j^i(X^i) + l_j^i(X^i) \right) * p_t^j \\
&\quad + \left(r^i(X^i) + w^i(X^i) \right) * \eta_t,
\end{aligned}
\tag{8.50}
$$

where $U = \mathbf{1}_{|u| \leq 1}$ and $V = \mathbf{1}_{|u| > 1}$ and the dependence on the arguments is as follows:

$f^i(X^i) = f^i(\omega, s, X_{s-}^i),$	$g(X^i) = g^i(\omega, s, X_{s-}^i)$
$h_1(X^i) = h_1(\omega, s, u, X_{s-}^i),$	$h_2(X^i) = h_2(\omega, s, u, X_s^i)$
$h_1^i(X^i) = h_1^i(\omega, s, u, X_{s-}^i),$	$h_2^i(X^i) = h_2^i(\omega, s, u, X_s^i)$
$k_1^i(X^i) = k_1^i(\omega, s, u, X_{s-}^i),$	$k_2^i(X^i) = k_2^i(\omega, s, u, X_s^i)$
$l_1^i(X^i) = l_1^i(\omega, s, u, X_{s-}^i),$	$l_2^i(X^i) = l_2^i(\omega, s, u, X_s^i)$
$r^i(X^i) = r^i(\omega, s, u, X_s^i),$	$w^i(X^i) = w^i(\omega, s, u, X_s^i)$

for $i = 1, 2$. In another way to describe the processes X^i for $i = 1, 2$,

$$
\begin{aligned}
X_t^i &= X_0^i + A_t^i(X^i) + M_t(X^i), \\
A_t^i(X^i) &= f^i(X^i) \cdot a_t + \sum_j V h_j^i(X^i) * \mu_t^j + \left(k_j^i(X^i) + l_j^i(X^i) \right) * p_t^j \\
&\quad + \left(r^i(X^i) + w^i(X^i) \right) * \eta_t, \\
M_t(X^i) &= g(X^i) \cdot m_t + \sum_j U h_j(X^i) * (\mu^j - \nu^j)_t
\end{aligned}
$$

the martingale part $M(X^i) \in \mathcal{M}_{loc}$ and finite variation process $A^i(X^i) \in \mathcal{V}$ form the process X^i. It is also assumed that for, $i = 1, 2$, the functions above satisfy these conditions,

(D1) $f^i(\omega, s, x)$ and $g(\omega, s, x)$ are defined on $\Omega \times \mathbb{R}_+ \times \mathbb{R}$
and $\mathcal{P}(\mathbf{F}) \times \mathcal{B}(\mathbb{R})$-measurable,

(D2) $Uh_1(\omega, s, u, x)$ is defined on $\Omega \times \mathbb{R}_+ \times \mathbb{E} \cap (|u| \leq 1) \times \mathbb{R}$
and $\mathcal{P}(\mathbf{F}) \times \mathcal{B}(\mathbb{E} \cap (|u| \leq 1)) \times \mathcal{B}(\mathbb{R})$-measurable,

(D3) $Uh_2(\omega, s, u, x)$ is defined on $\Omega \times \mathbb{R}_+ \times \mathbb{E} \cap (|u| \leq 1) \times \mathbb{R}$
and $\mathcal{O}(\mathbf{F}) \times \mathcal{B}(\mathbb{E} \cap (|u| \leq 1)) \times \mathcal{B}(\mathbb{R})$-measurable,

(D4) $Vh_1^i(\omega, s, u, x)$ is defined on $\Omega \times \mathbb{R}_+ \times \mathbb{E} \cap (|u| > 1) \times \mathbb{R}$
and $\mathcal{G}(\mathbf{F}) \times \mathcal{B}(\mathbb{E} \cap (|u| > 1)) \times \mathcal{B}(\mathbb{R})$-measurable,

(D5) $Vh_2^i(\omega, s, u, x)$ is defined on $\Omega \times \mathbb{R}_+ \times \mathbb{E} \cap (|u| > 1) \times \mathbb{R}$
and $\mathcal{G}(\mathbf{F}) \times \mathcal{B}(\mathbb{E} \cap (|u| > 1)) \times \mathcal{B}(\mathbb{R})$-measurable,

(D6) $k_1^i(\omega, s, u, x)$ is defined on $\Omega \times \mathbb{R}_+ \times \mathbb{E} \times \mathbb{R}$
and $\mathcal{P}(\mathbf{F}) \times \mathcal{B}(\mathbb{E}) \times \mathcal{B}(\mathbb{R})$-measurable such that $k_1^i(X^i) * p^1 \in \mathcal{M}_{loc}^{1,r}(\mathbf{F})$,

(D7) $k_2^i(\omega, s, u, x)$ is defined on $\Omega \times \mathbb{R}_+ \times \mathbb{E} \times \mathbb{R}$
and $\mathcal{O}(\mathbf{F}) \times \mathcal{B}(\mathbb{E}) \times \mathcal{B}(\mathbb{R})$-measurable such that $k_2^i(X^i) * p^2 \in \mathcal{M}_{loc}^{1,g}(\mathbf{F})$,

(D8) $l_1^i(\omega, s, u, x)$ is defined on $\Omega \times \mathbb{R}_+ \times \mathbb{E} \times \mathbb{R}$
and $\mathcal{G}(\mathbf{F}) \times \mathcal{B}(\mathbb{E}) \times \mathcal{B}(\mathbb{R})$-measurable such that $l_1^i(X^i) * p^1 \in \mathcal{V}$,

(D9) $l_2^i(\omega, s, u, x)$ is defined on $\Omega \times \mathbb{R}_+ \times \mathbb{E} \times \mathbb{R}$
and $\mathcal{G}(\mathbf{F}) \times \mathcal{B}(\mathbb{E}) \times \mathcal{B}(\mathbb{R})$-measurable such that $l_2^i(X^i) * p^2 \in \mathcal{V}$,

(D10) $r^i(\omega, s, u, x)$ is defined on $\Omega \times \mathbb{R}_+ \times \mathbb{E} \times \mathbb{R}$
and $\mathcal{O}(\mathbf{F}) \times \mathcal{B}(\mathbb{E}) \times \mathcal{B}(\mathbb{R})$-measurable such that $r^i(X^i) * \eta \in \mathcal{M}_{loc}^{1,g}(\mathbf{F})$,

(D11) $w^i(\omega, s, u, x)$ is defined on $\Omega \times \mathbb{R}_+ \times \mathbb{E} \times \mathbb{R}$
and $\mathcal{G}(\mathbf{F}) \times \mathcal{B}(\mathbb{E}) \times \mathcal{B}(\mathbb{R})$-measurable such that $w^i(X^i) * \eta \in \mathcal{V}$.

Note that the 2 in k^2 is an index whereas in $|k|^2$ it is an exponent.

Let us formulate conditions under which the comparison theorem will be proved:

(A1) $X_0^2 \geq X_0^1$;

(A2) $f^2(s, x) > f^1(s, x)$ for any $s \in \mathbb{R}_+$, $x \in \mathbb{R}$, $f^i(s, x)$ are continuous in (s, x), $i = 1, 2$;

(A3) There exists a non-negative nondecreasing function $\rho(x)$ on $\mathbb{R}_+$ and a $\mathcal{P}(\mathbf{F})$-measurable non-negative function G such that

$$|g(s, x) - g(s, y)| \leq \rho(|x - y|)G(s),$$

$$|G|^2 \cdot \langle m, m \rangle_s < \infty \ \ a.s., \ \ \int_0^\epsilon \rho^{-2}(x)dx = \infty \ \ for \ any \ \ s \in \mathbb{R}_+, \ \epsilon > 0, \ x, y \in \mathbb{R};$$

(A4) There exists a non-negative $\widetilde{\mathcal{P}}(\mathbf{F})$-measurable function H_1 and $\widetilde{\mathcal{O}}(\mathbf{F})$-measurable function H_2 such that

$$\begin{aligned} |h_1(s, u, x) - h_1(s, u, y)| &\leq \rho(|x - y|)H_1(s, u), \ |H_1|^2 * \nu_s^1 < \infty, \\ |h_2(s, u, x) - h_2(s, u, y)| &\leq \rho(|x - y|)H_2(s, u), \ |H_2|^2 * \nu_s^2 < \infty, \end{aligned}$$

a.s., for any $s \in \mathbb{R}_+$, $u \in \mathbb{E}$, $x, y \in \mathbb{R}$;

(A5) For any $s \in \mathbb{R}_+$, $u \in \mathbb{E}$, $x, y \in \mathbb{R}$, $y \geq x$,

$$h_1(s, u, y) \geq h_1(s, u, x), \quad h_2(s, u, y) \geq h_2(s, u, x)$$
$$y + h_1^2(s, u, y)\mathbf{1}_{|u|>1} \geq x + h_1^1(s, u, x)\mathbf{1}_{|u|>1},$$
$$y - h_2^2(s, u, y)\mathbf{1}_{|u|>1} \geq x - h_2^1(s, u, x)\mathbf{1}_{|u|>1},$$
$$y + h_1(s, u, y)\mathbf{1}_{|u|\leq 1} + (k_1^2 + l_1^2)(s, u, y) \geq x + h_1(s, u, x)\mathbf{1}_{|u|\leq 1} + (k_1^1 + l_1^1)(s, u, x),$$
$$y - h_2(s, u, y)\mathbf{1}_{|u|\leq 1} - (k_2^2 + l_2^2)(s, u, y) - \left(r^2 + w^2\right)(s, u, x) \geq$$
$$x - h_2(s, u, x)\mathbf{1}_{|u|\leq 1} - (k_2^1 + l_2^1)(s, u, x) - \left(r^1 + w^1\right)(s, u, x);$$

(A6) The functions $\left(r^i + w^i\right)(s, u, x)$ and $(k_j^i + l_j^i)(s, u, x)$ are continuous in (s, u, x), $i = 1, 2$ and $j = 1, 2$,

$$(k_j^2 + l_j^2)(s, u, x) > (k_j^1 + l_j^1)(s, u, x)$$
$$\left(r^2 + w^2\right)(s, u, x) > \left(r^1 + w^1\right)(s, u, x)$$

for any $s \in \mathbb{R}_+$, $u \in \mathbb{E}$, $x \in \mathbb{R}$;

(A7) For $i = 1, 2$ and $j = 1, 2$,

$$\left|f^i(X^i)\right| \cdot a \in \mathcal{A}_{loc},$$

$$\left|g(X^i)\right|^2 \cdot \langle m, m \rangle \in \mathcal{A}_{loc}, \quad \left|h_j(X^i)\right|^2 * \nu^j \in \mathcal{A}_{loc},$$

$$\left|l_j^i(X^i)\right| * p^j \in \mathcal{A}_{loc}, \quad \left[\left|k_j^i(X^i)\right|^2 * p^j\right]^{1/2} \in \mathcal{A}_{loc},$$

$$\left|r^i(X^i)\right| * \eta \in \mathcal{A}_{loc}, \quad \left[\left|w^i(X^i)\right|^2 * \eta\right]^{1/2} \in \mathcal{A}_{loc},$$

and

$$\mathbf{E}[k_1^i(S, \beta_S^d, X_{S-}^i)|\mathcal{F}_{S-}] = 0$$

a.s., for any predictable stopping time S and

$$\mathbf{E}[k_2^i(T, \beta_T^g, X_T^i)|\mathcal{F}_T] = 0$$

a.s., for any totally inaccessible stopping time T, and

$$\mathbf{E}\left[w(U, \beta_U^g)|\mathcal{F}_U\right] = 0$$

a.s., for any totally inaccessible stopping time in the broad sense U, $i = 1, 2$.

To formulate the next assumption we need to introduce the sequence $\{a_n\}_{n \in \mathbb{N}}$ of positive numbers $a_0 = 1 > a_1 > \cdots$, $\lim_{n \to \infty} a_n = 0$, by the relations

$$\int_{a_{n+1}}^{a_n} \rho^{-2}(x)dx = n + 1, \quad n = 0, 1, \cdots.$$

Now let us write the last assumption: (A8). We assume that there exists a sequence $\{\varepsilon_n\}_{n \in \mathbb{N}}$ of positive numbers such that $\varepsilon_n \leq a_{n-1} - a_n$ for all $n \in \mathbb{N}$ and

$$\frac{1}{n}\left[\frac{\rho(a_{n-1})}{\rho(a_{n-1} - \varepsilon_n)}\right]^2 \to 0, \quad n \to \infty.$$

It is easy to verify that condition (A8) is satisfied by Holder class functions ρ with index $\alpha = 1/2 + \epsilon$, $\epsilon > 0$, and is not satisfied by functions of this class with index $\alpha = 1/2$.

Theorem 8.3.1 Let there exist strong solutions X^i, $i = 1, 2$, of Equation (8.50) and let conditions A1-A8 hold. Then off some set of **P**-measure zero $X_t^2 \geq X_t^1$ for any $t \in \mathbb{R}_+$.

Before proving the theorem, let us perform a useful reduction of the problem.

Lemma 8.3.2 If the comparison theorem is valid for the equations in (8.51),

$$
\begin{aligned}
Y_t^i &= X_0^i + f^i(Y^i) \cdot a_t + g(Y^i) \cdot m_t \\
&\quad + \sum_j U h_j(Y^i) * (\mu^j - v^j)_t + \left(k_j^i(Y^i) + l_j^i(Y^i)\right) * p_t^j \\
&\quad + \left(r^i(Y^i) + w^i(Y^i)\right) * \eta_t
\end{aligned}
\tag{8.51}
$$

with functions X_0^i, f^i, g, h_j, k_j^i, l_j^i, r^i, and w^i satisfying conditions A1-A8, then, it is also valid for Equations (8.50).

Proof. Let $\{\tau_n\}_{n \in \mathbb{N}}$, $\tau_0 = 0$, be a nondecreasing sequence of totally inaccessible stopping times and stopping times in the broad sense, absorbing the jumps of the processes $h_j^i(X^i) * \mu^j$, $i = 1, 2$ and $j = 1, 2$, from Equations (8.50). On $]\tau_0, \tau_1[$ Equations (8.50) and (8.51) coincide. Since $Y^2 \geq Y^1$, on this interval then $X^2 \geq X^1$ on this interval. On the boundary of this interval, from Equations (8.50) we find that at time τ_1,

$$
\begin{aligned}
X_{\tau_1}^2 &= X_{\tau_1-}^2 + \triangle X_{\tau_1}^2 = X_{\tau_1-}^2 + h_1^2\left(\tau_1, \beta_{\tau_1}^1, X_{\tau_1-}^2\right) 1_{|\beta_{\tau_1}^1|>1} \\
&\geq X_{\tau_1-}^1 + h_1^1(\tau_1, \beta_{\tau_1}^1, X_{\tau_1-}^1) 1_{|\beta_{\tau_1}^1|>1} = X_{\tau_1-}^1 + \triangle X_{\tau_1}^1 = X_{\tau_1}^1,
\end{aligned}
$$

and

$$
\begin{aligned}
X_{\tau_0}^2 &= X_{\tau_0+}^2 - \triangle^+ X_{\tau_0}^2 \\
&= X_{\tau_0+}^2 - h_2^2\left(\tau_0, \beta_{\tau_0}^2, X_{\tau_0}^2\right) 1_{|\beta_{\tau_0}^2|>1} \\
&\quad - \left(r^2(\tau_0, \beta_{\tau_0}^2, X_{\tau_0}^2) + w^2(\tau_0, \beta_{\tau_0}^2, X_{\tau_0}^2)\right) 1_{|\beta_{\tau_0}^2|>1} \\
&\geq X_{\tau_0+}^1 - h_2^1(\tau_0, \beta_{\tau_0}^2, X_{\tau_0}^1) 1_{|\beta_{\tau_0}^2|>1} \\
&\quad - \left(r^1(\tau_0, \beta_{\tau_0}^2, X_{\tau_0}^1) + w^1(\tau_0, \beta_{\tau_0}^2, X_{\tau_0}^1)\right) 1_{|\beta_{\tau_0}^2|>1} \\
&= X_{\tau_0+}^1 - \triangle^+ X_{\tau_0}^1 = X_{\tau_0}^1;
\end{aligned}
$$

Therefore, by condition (A5) the comparison theorem holds for (8.50) on $[0, \tau_1]$.

Now let us suppose that the comparison theorem for (8.50) holds on $[0, \tau_n]$, $n \geq 1$ and prove it holds on $]\tau_n, \tau_{n+1}]$. On $]\tau_n, \infty[$ consider the equations ($i = 1, 2$)

$$
\begin{aligned}
Y_t^i &= X_{\tau_n}^i + \int_{\tau_n+}^t f^i(Y_s^i) da_s + \int_{\tau_n+}^t g(Y_s^i) dm_s \\
&\quad + \int_{\tau_n+}^t \int_E U h_1(Y^i) d(\mu^1 - v^1)_s \\
&\quad + \int_{\tau_n+}^{t-} \int_E U h_2(Y_s^i) d(\mu^2 - v^2)_s + \int_{\tau_n+}^t \int_E (k_1^i(Y_s^i) + l_1^i(Y_s^i)) dp^1 \\
&\quad + \int_{\tau_n+}^{t-} \int_E (k_2^i(Y_s^i) + l_2^i(Y_s^i)) dp_s^2 + \int_{\tau_n+}^{t-} \int_E \left(r^i(Y_s^i) + w^i(Y_s^i)\right) d\eta_s
\end{aligned}
\tag{8.52}
$$

Let us transform (8.52) to the form (8.51). For this we make the substitution $t - \tau_n = s$ and set

$$\mathcal{F}_s^{(n)} = \mathcal{F}_{s+\tau_n}, \quad s \in [0, \infty[, \quad a_s^{(n)} = a_{s+\tau_n}, \quad m_s^{(n)} = m_{s+\tau_n},$$

$$\left(\mu^{1(n)} - \nu^{1(n)}\right)(]\rho, \varsigma], \Gamma) = \left(\mu^1 - \nu^1\right)(]\rho + \tau_n, \varsigma + \tau_n], \Gamma),$$

$$\left(\mu^{2(n)} - \nu^{2(n)}\right)([\rho, \varsigma[, \Gamma) = \left(\mu^2 - \nu^2\right)([\rho + \tau_n, \varsigma + \tau_n[, \Gamma),$$

$$p^{1(n)}(]\rho, \varsigma], \Gamma) = p^1(]\rho + \tau_n, \varsigma + \tau_n], \Gamma),$$

$$p^{2(n)}([\rho, \varsigma[, \Gamma) = p^2([\rho + \tau_n, \varsigma + \tau_n[, \Gamma),$$

$$\eta^{(n)}([\rho, \varsigma[, \Gamma) = \eta([\rho + \tau_n, \varsigma + \tau_n[, \Gamma),$$

$$Y_s^{i(n)} = Y_{s+\tau_n}^i, \quad X_s^{i(n)} = X_{s+\tau_n}^i.$$

Introduce further the functions $f^{i(n)}$, $g^{(n)}$, $h_j^{(n)}$, $k_j^{i(n)}$, $l_j^{i(n)}$, $r^{i(n)}$, and $w^{i(n)}$ setting

$$f^{i(n)}(s, x) = f^i(s + \tau_n, x), \quad h_j^{(n)}(s, u, x) = h_j(s + \tau_n, u, x),$$

and proceed analogously for the remaining functions. Equations (8.52) take on the form

$$\begin{aligned}
Y_s^i &= X_0^{i(n)} + f^{i(n)}\left(Y^{i(n)}\right) \cdot a_s^{(n)} + g^{(n)}\left(Y^{i(n)}\right) \cdot m_s^{(n)} \\
&\quad + \sum_j U h_j^{(n)}(Y^{i(n)}) * (\mu^j - \nu^j)_s^{(n)} + (k_j^{i(n)} + l_j^{i(n)})(Y^{i(n)}) * p_s^{j(n)} \\
&\quad + \left(r^{i(n)}(Y^{i(n)}) + w^{i(n)}(Y^{i(n)})\right) * \eta_t^{(n)}
\end{aligned}$$

for $(i = 1, 2)$ and $s \in [0, \infty[$.

These are equations of the form (8.51) with integrands satisfying conditions A1-A8. By the assumption of the lemma, the comparison theorem holds for these equations. Their solutions for $s \in]0, \tau_{n+1} - \tau_n[$ coincide with the solutions of (8.50) on $]\tau_n, \tau_{n+1}[$. Hence the comparison theorem for (8.50) can be extended to the interval $]0, \tau_{n+1}[$. Arguing just as in the case of $[0, \tau_1]$ we extend the comparison theorem for (8.50) to the points τ_n and τ_{n+1}. Then repeating these arguments, we prove the result for Equation (8.50) for all $t \in [0, \infty[$. ∎

Proof. of *Comparison*, (Theorem 8.3.1); by Lemma (8.3.2), it suffices to establish comparison results for (8.51). So it begins; by (A7), all the integrals in (8.51) are defined. Not to resort to an additional localization argument, we shall assume that for $(i = 1, 2)$,

$$\mathbf{E}\left[|G|^2 \cdot \langle m, m \rangle_\infty\right] < \infty, \quad \mathbf{E}\left[|H_j|^2 * \nu_\infty^j\right] < \infty,$$

$$\begin{aligned}
\mathbf{E}\Big[&|f^i(X^i)| \cdot a_\infty + |g(X^i)|^2 \cdot \langle m, m \rangle_\infty \\
&+ \sum_j U |h_j(X^i)|^2 * \nu_\infty^j + |l_j^i(X^i)| * p_\infty^j + \left(|k_j^i(X^i)|^2 * p_\infty^j\right)^{1/2} \\
&+ |r^i(Y^i)| * \eta_\infty + w^i(Y^i) * \eta_\infty\Big] < \infty.
\end{aligned}$$

Let us introduce the sets

$$A = \left\{\omega : X_0^2(\omega) > X_0^1(\omega)\right\}, \quad B = \left\{\omega : X_0^2(\omega) = X_0^1(\omega)\right\}.$$

First we prove the theorem on the set B. Let $Y_0^i = X_0^i 1_B$, $\tilde{f}^i = f^i 1_B$, and similarly define $\tilde{g}$, $\tilde{h}_j$, $\tilde{k}_j^i$, $\tilde{l}_j^i$, $\tilde{r}^i$ and $\tilde{w}^i$. Consider the equations $(i = 1, 2)$,

$$
\begin{aligned}
Y_t^i &= Y_0^i + \tilde{f}^i(Y^i) \cdot a_t + \tilde{g}(Y^i) \cdot m_t \\
&\quad + \sum_j U\tilde{h}_j(Y^i) * (\mu^j - \nu^j)_t + (\tilde{k}_j^i + \tilde{l}_j^i)(Y^i) * p_t^j \\
&\quad + (\tilde{r}^i(Y^i) + \tilde{w}^i(Y^i)) * \eta_t.
\end{aligned}
\tag{8.53}
$$

It is clear that $Y^i = X^i$ on B. Define the quantity T as follows:

$$
\begin{aligned}
T = \ \inf \Big\{ t > 0 : &\ \tilde{f}^1(t, Y_{t-}^1) > \tilde{f}^2(t, Y_{t-}^2) \\
or \ &\ (\tilde{k}_1^1 + \tilde{l}_1^1)(t, \beta_t^1, Y_{t-}^1) > (\tilde{k}_1^2 + \tilde{l}_1^2)(t, \beta_t^1, Y_{t-}^2) \\
or \ &\ (\tilde{k}_2^1 + \tilde{l}_2^1)(t, \beta_t^2, Y_t^1) > (\tilde{k}_2^2 + \tilde{l}_2^2)(t, \beta_t^2, Y_t^2) \\
or \ &\ (\tilde{r}^1 + \tilde{w}^1)(t, \beta_t^2, Y_t^1) > (\tilde{r}^2 + \tilde{w}^2)(t, \beta_t^2, Y_t^2) \Big\}
\end{aligned}
$$

We have $Y_0^2 = Y_0^1$, and, by (A2) and (A6),

$$
\begin{aligned}
\tilde{f}^2(0, Y_0^2) &> \tilde{f}^1(0, Y_0^1), \\
(\tilde{k}_1^2 + \tilde{l}_1^2)(0, \beta_0^1, Y_0^2) &> (\tilde{k}_1^1 + \tilde{l}_1^1)(0, \beta_0^1, Y_0^1), \\
(\tilde{k}_2^2 + \tilde{l}_2^2)(0, \beta_0^2, Y_0^2) &> (\tilde{k}_2^1 + \tilde{l}_2^1)(0, \beta_0^2, Y_0^1), \\
(\tilde{r}^2 + \tilde{w}^2)(0, \beta_0^2, Y_0^2) &> (\tilde{r}^1 + \tilde{w}^1)(0, \beta_0^2, Y_0^1);
\end{aligned}
$$

Since $\beta_t^1 \to \beta_0^1$ and $\beta_t^2 \to \beta_0^2$, $Y_t^i \to Y_0^i$, $t \downarrow 0$, and the functions $\tilde{f}^i(t, x)$, $(\tilde{k}_j^i + \tilde{l}_j^i)(t, u, x)$ and $(\tilde{r}^i + \tilde{w}^i)(t, u, x)$ are continuous in (t, u, x), it follows from what has been said that $T > 0$ a.s. on the set B.

Let $v = t \wedge T$. Set $R = Y^2 - Y^1$,

$$
\begin{aligned}
R_v &= Y_v^2 - Y_v^1 = R_0 + \left(\tilde{f}^2(Y^2) - \tilde{f}^1(Y^1) \right) \cdot a_t + (\tilde{g}(Y^2) - \tilde{g}(Y^1)) \cdot m_t \\
&\quad + \sum_j U \left(\tilde{h}_j(Y^2) - \tilde{h}_j(Y^1) \right) * (\mu^j - \nu^j)_t + \left[(\tilde{k}_j^2 + \tilde{l}_j^2)(Y^2) - (\tilde{k}_j^1 + \tilde{l}_j^1)(Y^1) \right] * p_t^j \\
&\quad + \left((\tilde{r}^2 + \tilde{w}^2)(Y^2) - (\tilde{r}^1 + \tilde{w}^1)(Y^1) \right) * \eta_t
\end{aligned}
$$

and considering the properties of stochastic integrals we have

$$
\begin{aligned}
\mathbf{E} R_v = \ \mathbf{E} \Big[&\left(\tilde{f}^2(Y^2) - \tilde{f}^1(Y^1) \right) \cdot a_v \\
&+ \sum_j \left(\tilde{l}^2(Y^2) - \tilde{l}^1(Y^1) \right) * \lambda_v^j + (\tilde{w}^2(Y^2) - \tilde{w}^1(Y^1)) * \theta_v \Big].
\end{aligned}
\tag{8.54}
$$

Now let $\{\psi_n(x)\}_{n \in \mathbb{N}}$ be a sequence of non-negative continuous functions such that $\mathrm{supp}\psi_n \subseteq (a_n, a_{n-1})$,

$$
\int_{a_n}^{a_{n-1}} \psi_n(x)dx = 1, \quad \psi_n(x) \leq \frac{2}{n}\rho^{-2}(|x|), \quad x \in \mathbb{R},
$$

and the maximum of ψ_n is attained at $a_{n-1} - \epsilon_n$, where the sequence $\{\epsilon_n\}_{n \in \mathbb{N}}$ satisfies A8. Set

$$\varphi_n(x) = \int_0^{|x|} dy \int_0^y \psi_n(u)du, \quad x \in \mathbb{R}, \quad n \in \mathbb{N}.$$

Clearly,

$$\varphi_n \in C^2(R^1), \quad \varphi_n(x) \uparrow |x|, \quad n \to \infty, \quad |\varphi_n'| \le 1,$$

$$\varphi_n''(x) = \psi_n(x) \le \frac{2}{n}\rho^{-2}(|x|), \quad x \in \mathbb{R}. \tag{8.55}$$

Then, by lemma on change of variable formula for the component representation of optional semimartingales,

$$
\begin{aligned}
\varphi_n(R_v) &= \varphi_n(R_0) + \varphi_n'(R)\left(\tilde{f}^2(Y^2) - \tilde{f}^1(Y^1)\right) \cdot a_v + \varphi_n'(R)\left(\tilde{g}(Y^2) - \tilde{g}(Y^1)\right) \cdot m_v \\
&\quad + \frac{1}{2}\varphi_n''(R)\left(|\tilde{g}(Y^2) - \tilde{g}(Y^1)|^2\right) \cdot \langle m, m \rangle_v \\
&\quad + \sum_j U\left[\varphi_n\left(R + \tilde{h}_j(Y^2) - \tilde{h}_j(Y^1)\right) - \varphi_n(R)\right] * (\mu^j - \nu^j)_v \\
&\quad + \sum_j U\left[\varphi_n\left(R + \tilde{h}_j(Y^2) - \tilde{h}_j(Y^1)\right) - \varphi_n(R) + \varphi_n'(R)\left(\tilde{h}_j(Y^2) - \tilde{h}_j(Y^1)\right)\right] * \nu_v^j \\
&\quad + \sum_j \left[\varphi_n\left(R + (\tilde{k}_j^2 + \tilde{l}_j^2)(Y^2) - (\tilde{k}_j^1 + \tilde{l}_j^1)(Y^1)\right) - \varphi_n(R)\right] * p_v^j \\
&\quad + \left[\varphi_n\left(R + (\tilde{r}^2 + \tilde{w}^2)(Y^2) - (\tilde{r}^1 + \tilde{w}^1)(Y^1)\right) - \varphi_n(R)\right] * \eta_v
\end{aligned}
$$

Let

$$
\begin{aligned}
I_1(v) &= \left[\varphi_n'(R)(\tilde{f}^2(Y^2) - \tilde{f}^1(Y^1))\right] \cdot a_v, \\
I_2(v) &= \frac{1}{2}\left[\varphi_n''(R)|\tilde{g}(Y^2) - \tilde{g}(Y^1)|^2\right] \cdot \langle m, m \rangle_v, \\
I_3(v) &= \sum_j U\left[\varphi_n\left(R + \tilde{h}_j(Y^2) - \tilde{h}_j(Y^1)\right) - \varphi_n(R) - \varphi_n'(R)\left(\tilde{h}_j(Y^2) - \tilde{h}_j(Y^1)\right)\right] * \nu_v^j \\
I_4(v) &= \sum_j \left[\varphi_n\left(R + (\tilde{k}_j^2 + \tilde{l}_j^2)(Y^2) - (\tilde{k}_j^1 + \tilde{l}_j^1)(Y^1)\right) - \varphi_n(R)\right] * p_v^j \\
I_5(v) &= \left[\varphi_n\left(R + (\tilde{r}^2 + \tilde{w}^2)(Y^2) - (\tilde{r}^1 + \tilde{w}^1)(Y^1)\right) - \varphi_n(R)\right] * \eta_v.
\end{aligned}
$$

and write

$$
\begin{aligned}
\varphi_n(R_v) &= \left[\varphi_n'(R)(\tilde{g}(Y^2) - \tilde{g}(Y^1))\right] \cdot m_v \\
&\quad + \sum_j U\left[\varphi_n\left(R + \tilde{h}_j(Y^2) - \tilde{h}_j(Y^1)\right) - \varphi_n(R)\right] * (\mu^j - \nu^j)_v \\
&\quad + \sum_{\kappa=1}^5 I_\kappa(v).
\end{aligned}
$$

Taking the expectation we get

$$\mathbf{E}\varphi_n(R_v) = \mathbf{E}\sum_{\kappa=1}^{5} I_\kappa(v). \tag{8.56}$$

Since $\tilde{f}^2(v, Y_{v-}^2) > \tilde{f}^1(v, Y_{v-}^1)$ for $v < T$ and $|\varphi_n'| \leq 1$, we have

$$\mathbf{E}I_1(v) \leq \mathbf{E}\left[\left(f^2(Y^2) - \tilde{f}^1(Y^1)\right) \cdot a_v\right].$$

Further, by (A3) and property (D7), the relations

$$\begin{aligned}
\mathbf{E}|I_2(v)| &\leq \frac{1}{2}\left(\max_{a_n \leq x \leq a_{n-1}}[\varphi_n''^2(|x|)\rho^2(|x|)]\right)\mathbf{E}\left[|G|^2 \cdot \langle m, m\rangle_v\right] \\
&\leq \frac{1}{n}\mathbf{E}|G|^2 \cdot \langle m, m\rangle_\infty \to 0, \ n \to \infty,
\end{aligned}$$

are valid.

Applying Taylor's formula, noting (A4) and (A8) and property (D7) we get

$$\begin{aligned}
|I_3(v)| &\leq \sum_j \frac{1}{2}\left[\left|\varphi_n''\left(R + \alpha(\tilde{h}_j(Y^2) - \tilde{h}_j(Y^1))\right)\right|\left|\tilde{h}_j(Y^2) - \tilde{h}_j(Y^1)\right|^2\right] * \nu_v^j \\
&\leq \sum_j \frac{1}{2}\left[|H_j|^2\rho^2(|R|)|\varphi_n''(R + \alpha(\tilde{h}(Y^2) - \tilde{h}(Y^1)))|\right] * \nu_v^j,
\end{aligned}$$

where $0 \leq \alpha \leq 1$.

Consider here three cases:

(i) if $|R| \in [a_n, a_{n-1}]$, then by (D7) and (A8),

$$|I_3| \leq \sum_j \frac{1}{2}|H_j|^2 * \nu_\infty^j\left(\frac{1}{n}\rho^2(a_{n-1})\rho^{-2}(a_{n-1} - \epsilon_n)\right)$$

(ii) if $|R| < a_n$, then, by (A5) and (A8),

$$\begin{aligned}
|I_3| &\leq \sum_j \frac{1}{2}\rho^2(a_n)\left[|H_j|^2\left|\varphi_n''\left(R + \alpha\left(\tilde{h}(Y^2) - \tilde{h}(Y^1)\right)\right)\right|\right] * \nu_\infty^j \\
&\leq \sum_j \frac{1}{2}|H|^2 * \nu_\infty^j\frac{1}{n}\rho^2(a_{n-1})\rho^{-2}(a_{n-1} - \epsilon_n);
\end{aligned}$$

(iii) if $|R| > a_{n-1}$, then by (A5) and property (D7) we find that $\varphi_n''(R + \alpha(\tilde{h}(Y^2) - \tilde{h}(Y^1))) = 0$ and $I_3 = 0$.

From what has been said, it follows that

$$\mathbf{E}|I_3(v)| \leq \sum_j \frac{1}{2}\mathbf{E}|H_j|^2 * \nu_\infty^j\left(\frac{1}{n}\rho^2(a_{n-1})\rho^{-2}(a_{n-1} - \epsilon_n)\right) \to 0, \quad n \to \infty.$$

Again using Taylor's formula and noting that

$$(\tilde{k}_j^2 + \tilde{l}_j^2)(v, \beta_v^1, Y_{v-}^2) > (\tilde{k}_j^1 + \tilde{l}_j^1)(v, \beta_v^1, Y_{v-}^1)$$

and

$$(\tilde{k}_2^2 + \tilde{l}_2^2)(v, \beta_v^2, Y_v^2) > (\tilde{k}_2^1 + \tilde{l}_2^1)(v, \beta_v^2, Y_v^1)$$

for $v < T$, we have

$$\mathbf{E}I_4(v) = \mathbf{E}\sum_j \left[\varphi_n' \left(R + \alpha \left(\left(\tilde{k}_j^2 + \tilde{l}_j^2\right)(Y^2) - \left(\tilde{k}_j^1 + \tilde{l}_j^1\right)(Y^1) \right) \right) \left(\left(\tilde{k}_j^2 + \tilde{l}_j^2\right)(Y^2) - \left(\tilde{k}_j^1 + \tilde{l}_j^1\right)(Y^1) \right) \right] * p_v^j$$

$$\leq \mathbf{E}\sum_j \left[\left(\tilde{k}_j^2 + \tilde{l}_j^2\right)(Y^2) - \left(\tilde{k}_j^1 + \tilde{l}_j^1\right)(Y^1) \right] * p_v^j.$$

Now noting that $\mathbf{E}\left[\tilde{k}_j^2(Y^2) - \tilde{k}_j^1(Y^1) \right] * p_v^j = 0$, we obtain

$$\mathbf{E}I_4(v) \leq \mathbf{E}\sum_j \left[\tilde{l}_j^2(Y^2) - \tilde{l}_j^1(Y^1) \right] * p_v^j$$

$$= \mathbf{E}\sum_j \left[\tilde{l}_j^2(Y^2) - \tilde{l}_j^1(Y^1) \right] * \lambda_v^j.$$

Applying Taylor's formula one more time and noting that

$$\left(\tilde{r}^2 + \tilde{w}^2\right)(v, \beta_v^2, Y_v^2) > \left(\tilde{r}^1 + \tilde{w}^1\right)(v, \beta_v^2, Y_v^2)$$

for $v < T$, we have

$$\mathbf{E}I_5(v) = \mathbf{E}\left[\varphi_n' \left(R + \alpha \left(\left(\tilde{r}^2 + \tilde{w}^2\right)(Y^2) - \left(\tilde{r}^1 + \tilde{w}^1\right)(Y^1) \right) \right) \left(\left(\tilde{r}^2 + \tilde{w}^2\right)(Y^2) - \left(\tilde{r}^1 + \tilde{w}^1\right)(Y^1) \right) \right] * \eta_v$$

$$\leq \mathbf{E}\left[\left(\tilde{r}^2 + \tilde{w}^2\right)(Y^2) - \left(\tilde{r}^1 + \tilde{w}^1\right)(Y^1) \right] * \eta_v.$$

Now noting that $\mathbf{E}\left[\tilde{r}^2(Y^2) - \tilde{r}^1(Y^1) \right] * \eta_v = 0$, we obtain

$$\mathbf{E}I_5(v) \leq \mathbf{E}\left[\tilde{w}^2(Y^2) - \tilde{w}^1(Y^1) \right] * \eta_v = \mathbf{E}\left[\tilde{w}^2(Y^2) - \tilde{w}^1(Y^1) \right] * \theta_v.$$

by the estimates of $\mathbf{E}I_i$, $f = 1, ..., 5$, and the fact that

$$\mathbf{E}\varphi_n(R) = \mathbf{E}\varphi_n(Y^2 - Y^1) \uparrow \mathbf{E}\left|Y^2 - Y^1\right|$$

as $n \to \infty$, we have from (8.56)

$$\mathbf{E}|Y_v^2 - Y_v^1| \leq \mathbf{E}\left[(\tilde{f}^2(Y^2) - \tilde{f}^1(Y^1)) \cdot a_v + \sum_j (\tilde{l}_j^2(Y^2) - \tilde{l}_j^1(Y^1)) * \lambda_v^j + (\tilde{w}^2(Y^2) - \tilde{w}^1(Y^1)) * \theta_v \right]$$

$$= \mathbf{E}\left(Y_v^2 - Y_v^1\right),$$

where the last equality follows as a result of Equation (8.54). Since the processes Y^i, $i = 1, 2$, are r.l.l.l., it follows from the derived inequality that off some set of **P**-measure zero, $Y_v^2 \geq Y_v^1$ (hence, also $X_v^2 \geq X_v^1$) for $v \leq T$ a.s. on the set B.

Now, let's consider the quantity

$$\varrho = \inf \left(t > T : Y_t^2 < Y_t^1 \right).$$

Let us show that $\varrho = \infty$ a.s. Naturally for $t < \varrho$ or on $[0, \varrho[$ the inequality $Y_t^2 \geq Y_t^1$ is valid a.s., whereas $]\varrho, \infty[$ $Y_t^2 < Y_t^1$ by definition of ϱ. Using (8.53) we obtain, that at time ϱ,

$$Y_\varrho^i = Y_{\varrho-}^i + \tilde{h}_1(\varrho, \beta_\varrho^1, Y_{\varrho-}^i)\mathbf{1}_{|\beta_\varrho^1| \leq 1} + \left(\tilde{k}_1^i + \tilde{l}_1^i\right)(\varrho, \beta_\varrho^1, Y_{\varrho-}^i).$$

And by (A5) we obtain that $Y_\varrho^2 \geq Y_\varrho^1$. Hence, $Y_\varrho^2 \geq Y_\varrho^1$ is true a.s. on $[0, \varrho]$.

Now, let us introduce the sets

$$C = \{Y_\varrho^2 > Y_\varrho^1, \varrho < \infty\}, \quad D = \{Y_\varrho^2 = Y_\varrho^1, \varrho < \infty\}.$$

From the definition of ϱ and the processes Y^i, $i = 1, 2$, it follows that $\mathbf{P}(C) = 0$.

For the set D, we repeat the same arguments that were carried out, above, for B. Moreover, if $\varrho < \infty$, then there is a stopping time S, $\mathbf{P}(S > \varrho, \varrho < \infty) > 0$ such that $Y_t^2 \geq Y_t^1$ for $\varrho < t \leq S$. The latter will contradict the definition of ϱ. Hence, $\varrho = \infty$ a.s.

Now let us prove the theorem for the set A. Consider the quantity $\hat{\varrho} = \inf(t > 0 : X_t^2 < X_t^1)$, where the X^i, $i = 1, 2$ are the solutions of Equations (8.51). For $t < \hat{\varrho}$ we have: $X_t^2 \geq X_t^1$ a.s. on A. Just as above, computing the jumps of the X^i at time ϱ and noting condition (A5) we get: $X_{\hat{\varrho}}^2 \geq X_{\hat{\varrho}}^1$ a.s. on A.

Introduce the sets

$$\hat{C} = \left\{X_{\hat{\varrho}}^2 > X_{\hat{\varrho}}^1, \hat{\varrho} < \infty\right\}, \quad \hat{D} = \left\{X_{\hat{\varrho}}^2 = X_{\hat{\varrho}}^1, \hat{\varrho} < \infty\right\}.$$

From the definition of the stopping time $\hat{\varrho}$ and the right continuity of the processes X^i, it follows that $\mathbf{P}(\hat{C}) = 0$. For the set $\hat{D}$ we repeat the arguments made for the set B. We obtain: $X^2 \geq X^1$ a.s. on A. ∎

8.3.2 Remarks and applications

We give an example showing how the stochastic domination (comparison) theorem can be used.

Example 1. The constant elasticity of variance (CEV) model was proposed by Cox and Ross [108]. It is often used to capture leverage effects and stochasticity of volatility for modeling equities and commodities. Consider a modified version of the CEV model where the stock price is said to satisfy the following integral equation,

$$\begin{aligned}
S_t &= \rho S \cdot A_t + \sigma S^\alpha \cdot M_t, \quad S_0 = s, \\
A_t &= t + V * \mu^1 + V * \mu^2 \\
M_t &= W_t + U * \left(\mu^1 - \nu^1\right)_t + U * \left(\mu^2 - \nu^2\right)_t
\end{aligned} \tag{8.57}$$

where ρ and σ are constants and the martingale M is a jump-diffusion process with left and right jumps. W_t is the Wiener process, $\mu^1 - \nu^1$ is the measure of right jumps and $\mu^2 - \nu^2$ is the measure of left jumps. For $B \in \mathcal{B}(\mathbb{R}_+)$ and $\Gamma \in \mathcal{B}(\mathbb{E})$ the jump measures are defined

$$\begin{aligned}
\mu^1(B \times \Gamma) &\quad : \quad = \#\left\{(t, \triangle L_t^1) \in B \times \Gamma | t > 0 \text{ such that } \triangle L_t^1 \neq 0\right\} \\
\mu^2(B \times \Gamma) &\quad : \quad = \#\left\{(t, \triangle^+ L_t^2) \in B \times \Gamma | t > 0 \text{ such that } \triangle^+ L_t^2 \neq 0\right\}
\end{aligned}$$

where L_t^1 and L_t^2 are independent Poisson with constant intensities γ^1 and γ^2, respectively, and compensators $\nu^1 = \gamma^1 t$ and $\nu^2 = \gamma^2 t$.

Let

$$F(x) = \frac{1}{\sigma} \int_s^x u^{-\alpha} du = \frac{x^{1-\alpha} - s^{1-\alpha}}{\sigma(1-\alpha)},$$

$$F'(x) = \frac{x^{-\alpha}}{\sigma}, \quad F''(x) = \frac{-\alpha x^{-\alpha-1}}{\sigma}.$$

where $0 < \alpha < 1$. Denote $X_t = F(S_t)$ and applying Itô's formula we get

$$
\begin{aligned}
X_t &= \rho S F'(S) \circ A_t + \sigma S^\alpha F'(S) \circ M_t + \frac{\sigma^2}{2} F''(Y) S^{2\alpha} \circ [M,M]_t \\
&= \frac{\rho}{\sigma} S_t^{1-\alpha} \circ A_t - \frac{\alpha\sigma}{2} S^{\alpha-1} \circ [M,M]_t + M_t \\
&= \frac{\rho}{\sigma} \left(\sigma(1-\alpha)X + s^{1-\alpha} \right) \circ A_t - \frac{\alpha\sigma}{2} \left(\sigma(1-\alpha)X + s^{1-\alpha} \right)^{-1} \circ [M,M]_t + M_t
\end{aligned}
$$

where $[M,M]_t = \left(1 + \gamma^1 + \gamma^2\right) t$. With the comparison theorem proved above, we can give an estimate of the process X_t from above by a new process Y_t, satisfying the equation,

$$Y_t = \frac{\rho}{\sigma} \left(\sigma(1-\alpha)Y + s^{1-\alpha} \right) \cdot A_t + M_t, \quad Y_0 = 0$$

which is essentially an Ornstein-Uhlenbeck process with left and right jumps. Applying the comparison theorem to X_t and Y_t yields that, $Y_t \geq X_t = F(S_t)$ a.s., therefore

$$S_t \leq F^{-1}(Y_t) \quad \text{a.s.} \tag{8.58}$$

Now let's consider an increasing function f with an option payoff $f(S_T)$. Assuming zero interest rates, the price of such option is given by $\tilde{\mathbf{E}} f(S_T)$ for an appropriate martingale measure $\tilde{\mathbf{P}}$. Using inequality (8.58), we have that $\mathbf{E} f(S_T) \leq \tilde{\mathbf{E}} f(F^{-1}(Y_T))$ and thus we obtain an estimate for the option price, for which $\tilde{\mathbf{E}} f(F^{-1}(Y_T))$ is easier to compute.

Chapter 9

Optional Financial Markets

In theories of mathematical finance, financial markets are commonly modeled by probability spaces that satisfy the usual conditions and market processes that are r.c.l.l. semimartingales. Here we develop a general model of financial markets based on optional semimartingales on *un*usual probability spaces. Our motivation is to expand the repertoir of mathematical models of financial markets in a way to advance our understanding of financial markets and in turn solve new problems that may arise in the mathematics of stochastic processes.

9.1   Introduction

Let's motivate our proposed approach based on optional semimartingales by revisiting some of the assumptions and concepts of the current theories of mathematical finance concerning the underlying probability space satisfying the usual conditions and processes that are r.c.l.l. semi-martingales.

We start by asking, why must it be that the filteration associated with the present time t, $\mathcal{F}_t$, be set equal to the filteration containing events from the immediate future, $\mathcal{F}_{t+}$, as a model for the flow of market information over time, othewise known as the right continuity of filteration assumption? Moreover, should we accept initial-completion of the information σ-algebras where null sets of the market at end-time are attached to the initial algebra of the market $\mathcal{F}_0$? This condition is also related to the definition of a real-world measure $\mathbf{P}$ which must be defined a priori with the underlying sample space $(\Omega, \mathcal{F})$ on which it acts. Furthermore, market processes are usually assumed to be cadlag for the purpose of having a well-defined stochastic integral by which solutions to stochastic differential and integral equations arising in mathematical financial modeling. These assumptions are artificial and restrictive, and are sometimes not realistic; they are made to reduce the complexity of mathematical modeling of financial systems and to give proofs to certain important results, such as, no-arbitrage theorems.

Recently, a growing number of literature in problems such as portfolio optimization [109] and pricing and hedging [110] under transaction costs as well as optimal stopping [111] demonstrated the application of elements of the theory of optional semimartingales to mathematical finance, thus, attempting to relax some of the usual conditions. For example, Czichowsky et al. [109] showed that the dual optimizer, often used to find optimal trading strategy under transaction costs, is in general a laglad optional strong supermartingale. Kuhn and Teusch [112] proved that an optional process of a non-exploding realized power variation along stopping times is almost surely of ladlag paths. These processes are useful for the analysis of imperfect markets, markets with proportional trading costs, and markets with large traders impacting assets' traded prices and of trading strategies of illiquid stocks.

Moreover, there are problems in mathematical finance that were not treated with the methods of the calculus of optional processes but, possibly, should be. One of these problems is stochastic modeling of energy commodity prices where price spikes occur [113]. An advantage of the application of optional processes to such models is that they can naturally describe spikes in electricity spot prices. Also, optional processes can be applied in multi-asset option pricing, for example, in cases where some underlying assets are left-continuous, while the others are cadlag [114].

In mathematical finance, in the framework of the usual conditions and cadlag semimartingales, two of the most fundamental problems of finance are: the problem of portfolio optimization and hedging and pricing of contingent claims which were studied extensively. In both of these cases, the existence of risk-neutral martingale deflators (measures) is a way to thwart arbitrage (see [115, 116, 117, 118, 119, 120, 121]). Hence, finding martingale deflators became the central goal of arbitrage pricing theory and at large mathematical finance.

In this chapter, we propose a rational framework for financial markets using the calculus of optional processes on *un*usual probability spaces and develop methods and tools for this purpose. The chapter begins by defining a general optional semimartingale market and portfolios. Then, several methods for finding martingale deflators are presented. Hereinafter, we study pricing and hedging in these markets and give several examples including: Black-Scholes with left and right jumps and pricing of European call option, a portfolio of defaultable bond and a stock, an instrument with the option to trade its dividends and debit repayment problem. The results we present here are largely based on the work of Abdelghani and Melnikov [14].

9.2   Market Model

Consider a market on *un*usual probability space $\left(\Omega, \mathcal{F}, \mathbf{F} = (\mathcal{F}_t)_{t \geq 0}, \mathbf{P}\right)$. Let us also introduce the filtration $\mathbf{F}_+ = (\mathcal{F}_{t+})_{t \geq 0}$ and its completion under $\mathbf{P}$, $\mathbf{F}^\star = \left(\mathcal{F}_{t+}^\mathbf{P}\right)_{t \geq 0}$. Our market consists of two types of securities x and X and a portfolio $\pi = (\eta, \xi)$ which is composed of the optional processes η and ξ. η is the volume of the reference asset x, while ξ is the volume of the security X. Suppose $x_t > 0$ and $X_t \geq 0$ for all $t \geq 0$ and write the ratio process $R_t = X_t/x_t$. Then, the value of the portfolio is

$$Y_t = \eta_t + \xi_t R_t. \tag{9.1}$$

$Y \in \mathcal{O}(\mathbf{F})$ is a real-valued optional semimartingale that is r.l.l.l. Furthermore, we restrict the portfolio, π, to be self-financing; that is we must have,

$$Y_t = Y_0 + \xi \circ R_t. \tag{9.2}$$

Reconciling Equations (9.1) and (9.2), we get

$$C_t = \eta_t + R \circ \xi_t + [\xi, R]_t = C_0.$$

where C_t is the consumption process with its initial value C_0. Since the ratio process R is optional semimartingale, then ξ evolves in the space $\mathcal{P}(\mathbf{F}) \times \mathcal{O}(\mathbf{F})$ with the predictable part determining the volume of R^r and the optional part determining the volume of R^g. R^r is the right-continuous part of X/x and R^g is its left-continuous part. Also, η belongs to the space $\mathcal{O}(\mathbf{F})$. Furthermore, for the integral in Equation (9.2) to be well defined ξ must be R-integrable,

$$\int_0^\infty \xi_s^2 d[R, R]_s \in \mathcal{A}_{loc}.$$

In addition to the process Y, let $Y^+ = (Y_{t+})_{t \geq 0}$ which is $\mathcal{O}(\mathbf{F}^\star)$. Y^+ is r.c.l.l. semimartingale on $(\Omega, \mathcal{F}, \mathbf{F}^\star, \mathbf{P})$. The role of Y^+ will be discussed when we discuss no free lunch in optional markets.

Remark 9.2.1 It is also possible to write the market portfolio as $Z_t = \eta_t x_t + \xi_t X_t$ where $Y = Z/x$. However, in this case, the volume of the reference asset, η, is a process that also evolves in $\mathcal{P}(\mathbf{F}) \times \mathcal{O}(\mathbf{F})$. Therefore, a variety of portfolios are possible on *un*usual financial markets.

9.3 Martingale Deflators

Here we present methods for finding local martingale transforms (deflators) for these markets. A local martingale deflator is a strictly positive supermartingale multiplier used to transform the value process of a portfolio to a supermartingale (i.e., a local martingale). It is important for finding fair prices and hedging strategies for contingent claims.

9.3.1 The case of stochastic exponentials

We suppose that the dynamics of securities in our market follow the stochastic exponentials,

$$X_t = X_0 \mathcal{E}_t(H), \quad x_t = x_0 \mathcal{E}_t(h)$$

where x_0 and X_0 are $\mathcal{F}_0$-measurable r.v.s. $h = (h_t)_{t \geq 0}$ and $H = (H_t)_{t \leq 0}$ are optional semimartingales admitting the representations,

$$h_t = h_0 + a_t + m_t, \quad H_t = H_0 + A_t + M_t$$

with respect to (w.r.t.) $\mathbf{P}$. $a = (a_t)_{t \geq 0}$ and $A = (A_t)_{t \geq 0}$ are locally bounded variation processes and predictable. $m = (m_t)_{t \geq 0}$ and $M = (M_t)_{t \geq 0}$ are optional local martingales.

First, we shall study, when the ratio process, R, is a local optional martingale w.r.t. the initial measure $\mathbf{P}$, i.e., when is $R \in \mathcal{M}_{loc}(\mathbf{F}, \mathbf{P})$ for $t \geq 0$? R in exponential form can be written as

$$R_t = \frac{X_t}{x_t} = R_0 \mathcal{E}(H)_t \mathcal{E}^{-1}(h)_t,$$

and using the properties of stochastic exponentials [15] we find

$$
\begin{aligned}
R_t &= R_0 \mathcal{E}(H_t) \mathcal{E}^{-1}(h_t) = R_0 \mathcal{E}(H_t) \mathcal{E}(-h_t^*) \\
&= R_0 \mathcal{E}(H_t - h_t^* - [H, h^*]_t)
\end{aligned}
$$

where

$$
\begin{aligned}
H_t - h_t^* - [H, h^*]_t &= H_t - h_t + \langle h^c, h^c \rangle_t + \sum_{0 < s \leq t} \frac{(\Delta h_s)^2}{1 + \Delta h_s} \\
&\quad + \sum_{0 \leq s < t} \frac{(\Delta^+ h_s)^2}{1 + \Delta^+ h_s} - \langle H^c, h^c \rangle_t - \sum_{0 < s \leq t} \Delta H_s \Delta h_s^* \\
&\quad - \sum_{0 \leq s < t} \Delta^+ H_s \Delta_s^+ h^* \\
&= H_t - h_t + \langle h^c, h^c \rangle_t + \sum_{0 < s \leq t} \frac{(\Delta h_s)^2}{1 + \Delta h_s} \\
&\quad + \sum_{0 \leq s < t} \frac{(\Delta^+ h_s)^2}{1 + \Delta^+ h_s} - \langle H^c, h^c \rangle_t - \sum_{0 < s \leq t} \Delta H_s \frac{\Delta h_s}{1 + \Delta h_s} \\
&\quad - \sum_{0 \leq s < t} \Delta^+ H_s \frac{\Delta^+ h_s}{1 + \Delta^+ h_s} \\
&= H_t - h_t + \langle h^c, h^c - H^c \rangle_t + \sum_{0 < s \leq t} \frac{\Delta h_s (\Delta h_s - \Delta H_s)}{1 + \Delta h_s} \\
&\quad + \sum_{0 \leq s < t} \frac{\Delta^+ h_s (\Delta^+ h_s - \Delta^+ H_s)}{1 + \Delta^+ h_s}.
\end{aligned}
$$

Let

$$J^g = \sum_{0 \le s < t} \frac{\triangle^+ h_s \left(\triangle^+ h_s - \triangle^+ H_s \right)}{1 + \triangle^+ h_s}, \quad J^d = \sum_{0 < s \le t} \frac{\triangle h_s (\triangle h_s - \triangle H_s)}{1 + \triangle h_s}$$

and write

$$\Psi(h_t, H_t) = H_t - h_t + \langle h^c, h^c - H^c \rangle_t + J_t^d + J_t^g.$$

Then, the ratio R satisfies

$$R_t = R_0 + \int_{0+}^t R_{s-} d\Psi_s + \int_0^{t-} R_s d\Psi_{s+}.$$

Considering the properties of stochastic integrals, we get $\Psi(h, H) \in \mathcal{M}_{loc}(\mathbf{P}, \mathbf{F}) \Rightarrow R \in \mathcal{M}_{loc}(\mathbf{P}, \mathbf{F})$ and if $\triangle^+ \Psi \ne -1$ and $\triangle \Psi \ne -1$, then $\Psi(h, H) \in \mathcal{M}_{loc}(\mathbf{P}, \mathbf{F}) \Leftrightarrow R \in \mathcal{M}_{loc}(\mathbf{P}, \mathbf{F})$. Given the decomposition of h and H one can write,

$$\Psi(h, H) = (A - a) + \langle m^c, m^c - M^c \rangle + (M - m) + J^d + J^g \qquad (9.3)$$
$$= (A - a) + \langle m^c, m^c - M^c \rangle + \tilde{J}^d + \tilde{J}^g + (M - m)$$
$$+ \left(J^g - \tilde{J}^g \right) + \left(J^d - \tilde{J}^d \right),$$

where $(M - m) + \left(J^g - \tilde{J}^g \right) + \left(J^d - \tilde{J}^d \right) \in \mathcal{M}_{loc}(\mathbf{P}, \mathbf{F})$. Thus, $\Psi(h, H) \in \mathcal{M}_{loc}(\mathbf{P}, \mathbf{F})$ if and only if

$$(A - a) + \langle m^c, m^c - M^c \rangle + \tilde{J}^d + \tilde{J}^g = 0.$$

If Ψ is a local optional martingale, then R is a local optional martingale and we are done. Otherwise, we have to find a strictly positive transformation $Z \in \mathcal{M}_{loc}(\mathbf{P}, \mathbf{F})$ that will render $ZR \in \mathcal{M}_{loc}(\mathbf{P}, \mathbf{F})$. Z is known as the local martingale transform or deflator. Since our market ratio process R is positive by definition and it would not make any financial sense to search for a local martingale transform Z that leads to $Z_t R_t \le 0$ for some t when $R_t > 0$ for all t. Therefore, we will restrict ourselves to a set of possible local martingale transforms that are strictly positive, i.e., $Z > 0$ a.s. $\mathbf{P}$. For a strictly positive Z, we can define $N \in \mathcal{M}_{loc}(\mathbf{P}, \mathbf{F})$ with $N = \mathcal{L}og(Z) = Z^{-1} \circ Z$ or $Z = \mathcal{E}(N)$. To find N, we have the following theorem to so:

Theorem 9.3.1 Given $R = R_0 \mathcal{E}(\Psi(h, H))$ where $\Psi(h, H)$ as in Equation (9.3) and $Z = \mathcal{E}(N)$ where $Z, N \in \mathcal{M}_{loc}(\mathbf{P}, \mathbf{F})$ and $Z > 0$ then $ZR \in \mathcal{M}_{loc}(\mathbf{P}, \mathbf{F})$ is a local optional martingale if and only if

$$(A - a) + \langle m^c - N^c, m^c - M^c \rangle + \tilde{K}^d + \tilde{K}^g = 0,$$

where $\tilde{K}^d$ and $\tilde{K}^g$ are the compensators of the processes

$$K^d = \sum_{0 < s \le t} \frac{(\triangle h_s - \triangle N_s)(\triangle h_s - \triangle H_s)}{1 + \triangle h_s},$$

$$K^g = \sum_{0 \le s < t} \frac{(\triangle^+ h_s - \triangle^+ N_s)(\triangle^+ h_s - \triangle^+ H_s)}{1 + \triangle^+ h_s}.$$

Proof. Suppose $Z_t = \mathcal{E}(N)_t \in \mathcal{M}_{loc}(\mathbf{P}, \mathbf{F})$, $Z_t > 0$ for all t such that $ZR \in \mathcal{M}_{loc}(\mathbf{P}, \mathbf{F})$ then

$$ZR = R_0 \mathcal{E}(N) \mathcal{E}(\Psi(h, H)) = \mathcal{E}(N + \Psi(h, H) + [N, \Psi(h, H)])$$
$$= R_0 \mathcal{E}(\Psi(h, H, N)),$$

where

$$\Psi(h, H, N) = N_t + H_t - h_t + \langle h^c, h^c - H^c \rangle_t + J_t^d + J_t^g$$
$$+ [N, H] - [N, h] + [N, J^d] + [N, J^g]$$
$$= N_t + H_t - h_t + \langle h^c, h^c - H^c \rangle_t + J_t^d + J_t^g$$
$$+ \langle N^c, H^c \rangle_t + \sum_{0 < s \le t} \triangle N_s \triangle H_s + \sum_{0 \le s < t} \triangle^+ N_s \triangle^+ H_s$$
$$- \langle N^c, h^c \rangle_t - \sum_{0 < s \le t} \triangle N_s \triangle h_s - \sum_{0 \le s < t} \triangle^+ N_s \triangle^+ h_s$$
$$+ \sum_{0 < s \le t} \triangle N_s \frac{\triangle h_s (\triangle h_s - \triangle H_s)}{1 + \triangle h_s}$$
$$+ \sum_{0 \le s < t} \triangle^+ N_s \frac{\triangle^+ h_s (\triangle^+ h_s - \triangle^+ H_s)}{1 + \triangle^+ h_s}$$

hence,

$$\Psi(h, H, N) = N_t + H_t - h_t + \langle h^c - N^c, h^c - H^c \rangle_t$$
$$+ \sum_{0 < s \le t} \frac{\triangle h_s (\triangle h_s - \triangle H_s)}{1 + \triangle h_s} + \triangle N_s (\triangle H_s - \triangle h_s)$$
$$+ \triangle N_s \frac{\triangle h_s (\triangle h_s - \triangle H_s)}{1 + \triangle h_s}$$
$$+ \sum_{0 < s \le t} \frac{\triangle^+ h_s (\triangle^+ h_s - \triangle^+ H_s)}{1 + \triangle^+ h_s} + \triangle^+ N_s (\triangle^+ H_s - \triangle^+ h_s)$$
$$+ \triangle^+ N_s \frac{\triangle^+ h_s (\triangle^+ h_s - \triangle^+ H_s)}{1 + \triangle^+ h_s}.$$

Therefore,

$$\Psi(h, H, N) = N_t + H_t - h_t + \langle h^c - N^c, h^c - H^c \rangle_t$$
$$+ \sum_{0 < s \le t} \frac{(\triangle h_s - \triangle N_s)(\triangle h_s - \triangle H_s)}{1 + \triangle h_s}$$
$$+ \sum_{0 < s \le t} \frac{(\triangle^+ h_s - \triangle^+ N_s)(\triangle^+ h_s - \triangle^+ H_s)}{1 + \triangle^+ h_s}.$$

Let

$$K^d = \sum_{0 < s \le t} \frac{(\triangle h_s - \triangle N_s)(\triangle h_s - \triangle H_s)}{1 + \triangle h_s},$$

$$K^g = \sum_{0 \le s \le t} \frac{(\triangle^+ h_s - \triangle^+ N_s)(\triangle^+ h_s - \triangle^+ H_s)}{1 + \triangle^+ h_s}$$

and write

$$\Psi(h, H, N) = N_t + H_t - h_t + \langle h^c - N^c, h^c - H^c \rangle_t + K^d + K^g.$$

So, if $\Psi(h, H, N) \in \mathcal{M}_{loc}(\mathbf{P}, \mathbf{F})$, then $ZR \in \mathcal{M}_{loc}(\mathbf{P}, \mathbf{F})$. And if $\triangle^+ \Psi(h, H, N) \ne -1$ and $\triangle \Psi(h, H, N) \ne -1$, then $\Psi(h, H, N) \in \mathcal{M}_{loc}(\mathbf{P}, \mathbf{F}) \Leftrightarrow ZR \in \mathcal{M}_{loc}(\mathbf{P}, \mathbf{F})$. Now, let us take into consideration the decomposition of H and h and write

$$\Psi(h, H, N) = (A - a) + (M - m + N) + \langle (m - N)^c, (m - M)^c \rangle + K^d + K^g.$$

So, $\Psi(h, H, N)$ is a local optional martingale under $\mathbf{P}$ if

$$(A - a) + \langle m^c - N^c, m^c - M^c \rangle + \tilde{K}^d + \tilde{K}^g = 0 \qquad (9.4)$$

where $\tilde{K}^d$ and $\tilde{K}^g$ are the compensators of K^d and K^g, respectively. ■

By finding all $N \in \mathcal{M}_{loc}(\mathbf{P}, \mathbf{F})$ such that the above Equation (9.4) is valid and $\mathcal{E}(N) > 0$ we find the set of all appropriate local optional martingale transforms Z such that ZR is a local optional martingale. Note that if Z is a local martingale transform such that ZR is a local martingale, then it is true for all self-financing strategies π.

Theorem 9.3.2 If Z is a local martingale transform of R, that is ZR is a local optional martingale, and π is a self-financing portfolio which is R-integrable, then ZY_t^π is a local optional martingale.

Proof. Z is a local martingale transform of R; therefore $Z > 0$. $\pi = (\eta, \xi)$ is self-financing and R-integrable, then $Y_t^\pi = Y_0 + \xi \circ R_t$ and $Z_t Y_t^\pi$ can be written as,

$$
\begin{aligned}
d(Z_t Y_t^\pi) &= Z_t dY_t^\pi + dZ_t Y_t^\pi + d[Z_t, Y_t^\pi] \\
&= Z_t \xi_t dR_t + dZ_t (\eta_t + \xi_t R_t) + d[Z_t, Y_t^\pi] \\
&= Z_t \xi_t dR_t + dZ_t \xi_t R_t + dZ_t \eta_t + \xi_t d[Z_t, R_t] \\
&= \xi_t (Z_t dR_t + dZ_t R_t + d[Z_t, R_t]) + dZ_t \eta_t \\
&= \xi_t d(Z_t R_t) + dZ_t \eta_t.
\end{aligned}
$$

This leads us to the following result

$$Z_t Y_t^\pi = \xi \circ Z_t R_t + \eta \circ Z_t.$$

$\eta \circ Z_t$ and $\xi \circ Z_t R_t$ are local optional martingales; therefore, their sum $Z_t Y_t^\pi$ is a well-defined local optional martingale. Note, that we have implicitly used the fact that η is bounded, i.e., comes from the fact that π is a self-financing and also that,

$$\int_0^\infty \xi_t^2 d[ZR]_t \in \mathcal{A}_{loc}.$$

■

Remark 9.3.3 On the other hand, if we know that there exist a Z such that ZY^π is a local optional martingale, then what can we say about the portfolio π and the product ZR? It is reasonable to suppose that $Z = \mathcal{E}(N) > 0$, π-self-financing, ξ is R-integrable and η is bounded. In this case, $\xi \circ Z_t R_t = Z_t Y_t^\pi - \eta \circ Z_t$ is a sum of two local optional martingales and therefore a local optional martingale itself, for any optional process ξ, in particular for $\xi = 1$; therefore, ZR is a local optional martingale.

Remark 9.3.4 If $Z \in \mathcal{M}_{loc}(\mathbf{P}, \mathbf{F})$ and $Z > 0$, then one can define $\tilde{\mathbf{P}}_t = \int_\Omega Z_t d\mathbf{P}_t$ is a new measure equivalent to $\mathbf{P}$, i.e., $\tilde{\mathbf{P}} \overset{loc}{\sim} \mathbf{P}$, and $Z_t = \frac{d\tilde{\mathbf{P}}_t}{d\mathbf{P}_t}$.

Now, is there a way to construct N knowing Equation (9.4). We make an educated guess, choosing N as,

$$N = [\alpha_c \ \ \alpha_d \ \ \alpha_g] \circ [m^c \ \ m^d \ \ m^g]^\mathsf{T} + [\beta_c \ \ \beta_d \ \ \beta_g] \circ [M^c \ \ M^d \ \ M^g]^\mathsf{T}, \tag{9.5}$$

where $\alpha_c, \beta_c, \alpha_d, \beta_d \in \mathcal{P}(\mathbf{F})$ and $\alpha_g, \beta_g \in \mathcal{O}(\mathbf{F})$. Note, with α's and β's the local optional martingales (m, M) span a subspace of the local optional martingales space $\mathcal{M}_{loc}(\mathbf{P}, \mathbf{F})$. So if N takes the representation above (9.5), then what is the solution of Equation (9.4)? Substitute for N in (9.4) by Equation (9.5) we get

$$\langle m^c - N^c, m^c - N^c \rangle = \left(1 - 2\alpha_c + \alpha_c^2\right) \langle m^c, m^c \rangle + \beta_c^2 \langle M^c, M^c \rangle,$$

$$K^d = \sum_{0 < s \le t} \frac{\left((1 - \alpha_r) \triangle m_s^r - \beta_r \triangle M_s^r\right)\left(\triangle m_s^r - \triangle M_s^r\right)}{1 + \triangle m_s^r},$$

$$K^g = \sum_{0 < s \le t} \frac{\left((1 - \alpha_g) \triangle^+ m_s^g - \beta_g \triangle^+ M_s^g\right)\left(\triangle^+ m_s^g - \triangle^+ M_s^g\right)}{1 + \triangle^+ m_s^g}.$$

The compensators $\left(\tilde{K}^d, \tilde{K}^g\right)$ of K^d, K^g are not easy to compute in general and will have to be evaluated on a problem-by-problem basis.

9.3.2 The case of stochastic logarithms

Here we consider an alternative approach for finding local martingale deflators, using the methods of stochastic logarithms. What is interesting about this approach is that we don't have to define the process R as a stochastic exponential of the underlying process Ψ. All that is required is that the ratio process R and its predictable version R_- not vanish, except on sets of measures zero.

Consider the following lemmas,

Lemma 9.3.5 Suppose $X = X + A + M$, $x = x + a + m$, $R = X/x$ and $R_- \neq 0$ and $R \neq 0$ a.s. **P**, then $\mathcal{L}og(R)$ is a local optional martingale if and only if

$$\frac{1}{X} \circ A - \frac{1}{x} \circ a + \frac{1}{x^2} [m, m] - \frac{1}{xX} \circ [M, m] = -1.$$

Proof. Consider $\mathcal{L}og(R_t)$; using Galchuk lemma and properties of stochastic logarithm,

$$
\begin{aligned}
\mathcal{L}og(R_t) &= \mathcal{L}og(X_t) + \mathcal{L}og\left(\frac{1}{x_t}\right) + \left[\mathcal{L}og(X), \mathcal{L}og\left(\frac{1}{x}\right)\right]_t \\
&= \mathcal{L}og(X_t) + 1 - \mathcal{L}og(x_t) - \left[x, \frac{1}{x}\right]_t \\
&\quad + \left[\mathcal{L}og(X), 1 - \mathcal{L}og(x) - \left[x, \frac{1}{x}\right]\right]_t \\
&= 1 + \mathcal{L}og(X_t) - \mathcal{L}og(x_t) - \left[x, \frac{1}{x}\right]_t - [\mathcal{L}og(X), \mathcal{L}og(x)]_t \\
&= 1 + \frac{1}{X} \circ X_t - \frac{1}{x} \circ x_t - \left[x, \frac{1}{x}\right]_t - \frac{1}{xX} \circ [X, x]_t.
\end{aligned}
$$

knowing that $x^{-1} = -x^{-2} \circ x + 2/3x^{-3} \circ [x, x]$ then $\mathcal{L}og(R_t) \in \mathcal{M}_{loc}(\mathbf{P}, \mathbf{F})$ if

$$\frac{1}{X} \circ A - \frac{1}{x} \circ a + \frac{1}{x^2} [m, m] - \frac{1}{xX} \circ [M, m] = -1.$$

∎

Lemma 9.3.6 Suppose that R_- and R don't vanish, then $\mathcal{L}og(R) \in \mathcal{M}_{loc}(\mathbf{P}, \mathbf{F}) \Leftrightarrow R \in \mathcal{M}_{loc}(\mathbf{P}, \mathbf{F})$.

Proof. Suppose $R \in \mathcal{M}_{loc}(\mathbf{P}, \mathbf{F})$; since $\mathcal{L}og(R) = R^{-1} \circ R$ and $R^{-1} \circ R$ is a local martingale, then $\mathcal{L}og(R) \in \mathcal{M}_{loc}(\mathbf{P}, \mathbf{F})$. Now suppose that $\mathcal{L}og(R) \in \mathcal{M}_{loc}(\mathbf{P}, \mathbf{F})$, then $R^{-1} \circ R$ is a local martingale and since R_- and R don't vanish, then it must be that $R \in \mathcal{M}_{loc}(\mathbf{P}, \mathbf{F})$. ∎

If R is not a local martingale, then one can find a local martingale deflator $Z > 0$ such that ZR is a local martingale. The following lemma helps with finding local martingale deflators.

Lemma 9.3.7 Let $X = X + A + M$, $x = x + a + m$, $R = X/x$ and suppose that R_- and R don't vanish, then $\mathcal{L}og(ZR)$ is a local optional martingale if and only if

$$1 + \frac{1}{X} \circ A - \frac{1}{x} \circ a + \frac{1}{x^2} \circ [m, m] - \frac{1}{xX} \circ [M, m] + \frac{1}{ZX} \circ [Z, M] - \frac{1}{xZ} \circ [Z, m] = -1;$$

furthermore if $Z = \mathcal{E}(N) > 0$, then

$$\frac{1}{X} \circ A - \frac{1}{x} \circ a + \frac{1}{x^2} \circ [m, m] - \frac{1}{xX} \circ [M, m] + \frac{1}{X} \circ [N, M] - \frac{1}{x} \circ [N, m] = -1.$$

Proof. Consider $\mathcal{L}og\,(ZR)$, using Galchuk lemma and properties of stochastic logarithms,

$$
\begin{aligned}
\mathcal{L}og\,(ZR) &= \mathcal{L}og\,(Z) + \mathcal{L}og\,(R) + [\mathcal{L}og\,(Z)\,,\mathcal{L}og\,(R)] \\
&= \mathcal{L}og\,(Z) + 1 + \mathcal{L}og\,(X) - \mathcal{L}og\,(x) - \left[x, \frac{1}{x}\right] \\
&\quad - [\mathcal{L}og\,(X)\,,\mathcal{L}og\,(x)] \\
&\quad + \left[\mathcal{L}og\,(Z)\,, 1 + \mathcal{L}og\,(X) - \mathcal{L}og\,(x) - \left[x, \frac{1}{x}\right]\right] \\
&\quad - [\mathcal{L}og\,(Z)\,, [\mathcal{L}og\,(X)\,,\mathcal{L}og\,(x)]] \\
&= 1 + \mathcal{L}og\,(Z) + \mathcal{L}og\,(X) - \mathcal{L}og\,(x) - \left[x, \frac{1}{x}\right] \\
&\quad - [\mathcal{L}og\,(X)\,,\mathcal{L}og\,(x)] \\
&\quad + [\mathcal{L}og\,(Z)\,,\mathcal{L}og\,(X)] - [\mathcal{L}og\,(Z)\,,\mathcal{L}og\,(x)] \\
&= 1 + \frac{1}{X} \circ X - \frac{1}{x} \circ x - \left[x, \frac{1}{x}\right] - \frac{1}{xX} \circ [X, x] \\
&\quad + \frac{1}{ZX} \circ [Z, X] - \frac{1}{xZ} \circ [Z, x]
\end{aligned}
$$

then

$$
\frac{1}{X} \circ A - \frac{1}{x} \circ a + \frac{1}{x^2} \circ [m, m] - \frac{1}{xX} \circ [M, m] + \frac{1}{ZX} \circ [Z, M] - \frac{1}{xZ} \circ [Z, m] = -1
$$

and when $Z = \mathcal{E}(N)$ we get

$$
\begin{aligned}
\mathcal{L}og\,(ZR) &= 1 + \mathcal{L}og\,(Z) + \mathcal{L}og\,(X) - \mathcal{L}og\,(x) - \left[x, \frac{1}{x}\right] \\
&\quad - [\mathcal{L}og\,(X)\,,\mathcal{L}og\,(x)] \\
&\quad + [\mathcal{L}og\,(Z)\,,\mathcal{L}og\,(X)] - [\mathcal{L}og\,(Z)\,,\mathcal{L}og\,(x)] \\
&= 1 + N + \frac{1}{X} \circ X - \frac{1}{x} \circ x + \frac{1}{x^2} \circ [m, m] \\
&\quad - \frac{1}{xX} \circ [X, x] + \frac{1}{X} \circ [N, X] - \frac{1}{x} \circ [N, x]
\end{aligned}
$$

then $\mathcal{L}og\,(ZR)$ is a local martingale if

$$
N + \frac{1}{X} \circ A - \frac{1}{x} \circ a + \frac{1}{x^2} \circ [m, m] - \frac{1}{xX} \circ [M, m] + \frac{1}{X} \circ [N, M] - \frac{1}{x} \circ [N, m] = -1
$$

■

Lemma 9.3.8 Suppose that R_- and R don't vanish, then $\mathcal{L}og(ZR) \in \mathcal{M}_{loc}(\mathbf{P}, \mathbf{F}) \Leftrightarrow ZR \in \mathcal{M}_{loc}(\mathbf{P}, \mathbf{F})$.

Proof. Similar to Lemma 9.3.7 ■

9.4   Pricing and Hedging

A contingent claim is an integrable or square-integrable r.v., $\Lambda \in \mathcal{F}$. Λ generates the optional local martingale process $\Lambda_t = \mathbf{E}[\Lambda|\mathcal{F}_t]$ for $t \in [0, T]$, for some final time T. In the market (x, X), for Λ to be priced and for there to exist a hedge portfolio, Λ_t, must admit an integral representation in terms of the ratio process R. If R is a local martingale, then Galchuk [11], theorem 2.3, gives that for any $(\alpha, \beta) \in \mathcal{P}(\mathbf{F}) \times \mathcal{O}(\mathbf{F})$, $(\alpha, \beta) \circ R$ is again a local optional martingale in $(\mathbf{F}, \mathbf{P})$.

So, let $\pi = (\eta, \xi)$ be a general portfolio where $\xi = (\alpha, \beta) \in \mathcal{P}(\mathbf{F}) \times \mathcal{O}(\mathbf{F})$; hence $\pi \in \mathcal{O}(\mathbf{F}) \times \mathcal{P}(\mathbf{F}) \times \mathcal{O}(\mathbf{F})$. Let

$$\mathfrak{R} = \{(\alpha, \beta) \circ R : \xi = (\alpha, \beta) \in \mathcal{P}(\mathbf{F}) \times \mathcal{O}(\mathbf{F}),$$
$$(\eta, \xi) \text{ is self-financing } \eta \in \mathcal{O}(\mathbf{F}) \text{ and } \xi \text{ is } R\text{-integrable}\},$$

be the space of local optional martingales *generated by* the ratio process R. Furthermore, let the space $\mathfrak{A} \subseteq \mathcal{O}(\mathbf{F}) \times \mathcal{P}(\mathbf{F}) \times \mathcal{O}(\mathbf{F})$ be the set of all admissible portfolios such that $(\alpha, \beta) \circ R \in \mathfrak{R}$.

If $\Lambda \in \mathfrak{R}$, then there exists a portfolio $\pi^\Lambda = \left(\eta^\Lambda, \alpha^\Lambda, \beta^\Lambda\right)$ such that

$$\Lambda_t = \Lambda_0 + \int_0^t \xi_t^\Lambda dR_t = \Lambda_0 + \int_{0+}^t \alpha_{t-}^\Lambda dR_t^r + \int_0^{t-} \beta_t^\Lambda dR_{t+}^g, \qquad (9.6)$$
$$\Lambda_t = \eta_t^\Lambda + \xi_t^\Lambda R_t = \eta_t^\Lambda + \alpha_t^\Lambda R_t^r + \beta_t^\Lambda R_t^g, \text{ and}$$
$$C_t = \eta_t + R \circ \xi_t^\Lambda + [\xi^\Lambda, R]_t = C_0.$$

The processes ξ^Λ together with η^Λ form the hedge portfolio for Λ in the market (x, X) and Λ_t is the value of the claim over time $t \in [0, T]$ such that $\Lambda_T = \Lambda$, and the claim price is Λ_0.

Let us give several remarks on optional portfolios.

Remark 9.4.1 (a) Note that the portfolio $(\eta^\Lambda, \alpha^\Lambda, \beta^\Lambda)$ is not unique, so there exist many portfolios such that $\Lambda_T = \Lambda$. This implies that there are many possible initial fare prices for the contingent claim Λ. Hence, *un*usual stochastic markets are *fundamentally incomplete*.

(b) Furthermore, $(\eta^\Lambda, \alpha^\Lambda, \beta^\Lambda)$ is not like traditional hedging portfolios of predictable integrands η^Λ and β^Λ are optional. While traditional predictable portfolios can be approximated by a *single-simple-trading strategy* over the underlying asset ξ^Λ may not be, since, its component α^Λ is predictable and β^Λ is optional, see [122].

(c) Also, $(\eta^\Lambda, \alpha^\Lambda, \beta^\Lambda)$ cannot be traded by an agent given the current structure of financial markets where portfolios are derived knowing $\mathcal{F}_{t-}$. The problem is that components of the price process R, the right-continuous part R^r and the left-continuous part R^g have to be traded differently. Moreover, even if we let $\phi \in \mathcal{O}(\mathbf{F})$ and consider a restricted version of ξ^Λ in which

$\xi^\Lambda = (\alpha^\Lambda, \beta^\Lambda) = (\phi_-, \phi)$ where $\phi_- \in \mathcal{P}(\mathbf{F})$, even in this case, the integrals in Equation (9.11) cannot be approximated by a single simple trading strategy. Indeed, Galchuk [11] optional semi-martingales integration theory showed that α^Λ and β^Λ integrands can only be approximated by two simple functions in $\mathrm{L}^2(\langle R^r \rangle, \langle R^g \rangle)$ or $\mathrm{H}^2([R^r], [R^g])$. The real problem for these portfolios is not α^Λ, because α^Λ is predictable; therefore a possible trading strategy in the classical view of financial markets. But β^Λ is optional and is the source of the "problem" in our classical view of financial systems. Since β^Λ may not be possible, then we are lead to the fact that only a subset of contingent claims in optional semimartingale markets can be hedged by predictable portfolios α which is less than the total set of contingent claims possible in this market. Therefore, once more, we arrive at the conclusion that optional semimartingale markets are *inherently incomplete*.

Remark 9.4.2 However, there is an alternative viewpoint for a portfolio $\pi = (\eta, \xi)$ with consumption, C. The consumption process C is understood to be the dynamic addition, either of, funds, spending, dividends, charity payouts, commission payments, tax payments or debit repayments: that is of course under the usual assumptions, R is r.c.l.l. local martingale and C an increasing/decreasing predictable process or in some cases an optional finite variation process, the value process Y is either a supermartingale or submartingale admitting the representation $Y = Y_0 + \xi \cdot R - C$. In the context of optional semimartingale market (x, X) under the risk neutral measure where the ratio process R is a local optional martingale we can describe the value process by,

$$Y = Y_0 + \alpha \cdot R^r + \beta \circ R^g - C. \tag{9.7}$$

If C is an r.l.l.l. increasing/decreasing strongly predictable process, then Y is an optional super/submartingale by decomposition of optional super/submartingale [10]. If C is a finite variation, then Y is a semimartingale with its decomposition having the same form as Equation (9.7). We can in this case consider $\beta \in \mathcal{O}(\mathbf{F})$ as part of a general consumption plan,

$$D = C - \beta \circ R^g,$$

and consider the following optimization-hedging problem,

$$
\begin{aligned}
u &= \min_{(\alpha, D) \in \mathfrak{A}} \mathbf{E}\left[(\Lambda - \alpha \cdot R_T^r + D_T)^2\right], \\
D_T &= \beta \circ R_T^g - C_T \in \mathcal{O}(\mathbf{F}), \\
\alpha &\in \mathcal{P}(\mathbf{F}),
\end{aligned}
$$

where Λ is the contingent claim we like to hedge, u is its optimal price, (α, D) is the admissible hedge and $\mathfrak{A}$ the set of admissible trade-consumption plans. This optimization problem is interesting but its solution is out of the scope of this work.

Remark 9.4.3 (a) Consider the fact that we live in an increasingly digital world where real economic and financial systems are running on top of digital networks where information is sampled and digitized, transmitted and stored. Information about assets, economic and financial variables are sampled and quantized with varying degrees of resolution where information accuracy costs money.

We believe that financial and economic instruments and variables, independent of the degree of their liquidity, are naturally continuous in time; however, as a result of digital information processing, they are discretized. Thus, market prices are an approximation of underlying processes that are in reality optional r.l.l.l. processes or they may be cadlag.

(b) Finally, let's consider this from the mechanics of actual trading. Real market orders are complex and are not necessarily predictable when executed. Consider conditional orders, for example, limit orders where an order is executed if the price of the assets is below a certain price limit; these orders in conjunction with the order matching process that the market makers undertake between buyers and sellers, the resulting executed orders and prices are not predictable but optional.

Next we consider absence of arbitrage in optional financial markets.

9.5 Absence of Arbitrage

The no-arbitrage arguments culminated in the fundamental theorem of asset pricing: under the *usual conditions*, for a real-valued semimartingale X, there exists a probability measure $\mathbf{Q}$ equivalent to $\mathbf{P}$ under which X is a σ-martingale *if and only if* X does not permit a free lunch with vanishing risk. Given that $K = \{(\phi \cdot X)_\infty : \phi \text{ admissible and } (\phi \cdot X)_\infty = \lim_{t \to \infty}(\phi \cdot X)_t$ exists a.s.$\}$ and $C = \{g \in \mathrm{L}^\infty(\mathbf{P}) : g \leq f \quad \forall f \in K\}$, then X is said to satisfy no free lunch with vanishing risk (NFLVR) if $\bar{C} \cap \mathrm{L}^\infty_+(\mathbf{P}) = \{0\}$ where $\bar{C}$ is the closure of C in the norm topology of $\mathrm{L}^\infty(\mathbf{P})$ [123]. $\mathbf{Q}$ is known as the equivalent local martingale measure (ELMM). However, in some models of financial markets, ELMM may fail to exist. An alternative was developed, an equivalent local martingale deflator (ELMD), which is a strictly positive local martingale that transforms the semimartingale X to a local martingale. It was shown that the existence of strictly positive local martingale deflator is equivalent to the no-arbitrage condition of the first kind (NA1). An $\mathcal{F}_T$ measurable r.v. ζ is called an arbitrage of the first kind if $\mathbf{P}(\zeta \geq 0) = 1$, $\mathbf{P}(\zeta > 0) > 0$ and for any initial wealth $x > 0$ if there exists an admissible ϕ such that $x + \phi \cdot X_T \geq \zeta$ [124]. NA1 is a weaker condition than NFLVR. While extending NFLVR and NA1 and the equivalence relations ELMM and ELMD, respectively, to *un*usual stochastic basis is possible and is of importance, it is definitely out of scope here. However, here we will present an argument to show the viability of financial markets on *un*usual probability spaces.

Again, consider a market on *un*usual probability space $\left(\Omega, \mathcal{F}, \mathbf{F} = (\mathcal{F}_t)_{t \geq 0}, \mathbf{P}\right)$. Introduce the filtration $\mathbf{F}_+ = (\mathcal{F}_{t+})_{t \geq 0}$ and its completion under $\mathbf{P}$, $\tilde{\mathbf{F}} = \left(\mathcal{F}^\mathbf{P}_{t+}\right)_{t \geq 0}$. Also let $Y \in \mathcal{O}(\mathbf{F})$ be a real-valued optional semimartingale that is at least of r.l.l.l. paths and $\tilde{Y} = (Y_{t+})_{t \geq 0}$ is $\mathcal{O}(\tilde{\mathbf{F}})$. $\tilde{Y}$ is r.c.l.l. semimartingale and the stochastic basis $\left(\Omega, \mathcal{F}, \tilde{\mathbf{F}}, \mathbf{P}\right)$ satisfies the usual conditions. Suppose $\tilde{Y}$ satisfied the conditions NFLVR with $\phi \in \mathcal{P}(\tilde{\mathbf{F}})$ admissible portfolio and $(\phi \cdot \tilde{Y})_\infty < \infty$ a.s. then there is ELMM $\mathbf{Q} \sim \mathbf{P}$ such that $\phi \cdot \tilde{Y}$ is a local martingale, i.e., let $\tilde{\mathcal{F}}_s = \mathcal{F}^\mathbf{P}_{s+} \in \tilde{\mathbf{F}}^\star$ then

$\phi \cdot \tilde{Y}_s = \mathbf{E_Q}[\phi \cdot \tilde{Y}_t | \tilde{\mathcal{F}}_s]$. Knowing that $\phi \cdot \tilde{Y}_s$ is a local martingale under $\left(\tilde{\mathbf{F}}, \mathbf{Q}\right)$ where $\phi \in \mathcal{P}(\tilde{\mathbf{F}})$ is admissible, then how do we recover Y? And what is the portfolio ϕ in $(\mathbf{F}, \mathbf{Q})$? Knowing that $\tilde{Y}_t = Y_{t+}$, we are going to employ optional projection of the space $\left(\Omega, \mathcal{F}, \tilde{\mathbf{F}}, \mathbf{Q}\right)$ on $(\Omega, \mathcal{F}, \mathbf{F}, \mathbf{Q})$ to answer these questions. We begin by showing that

$$\phi \circ Y_s = \mathbf{E_Q}[\phi \cdot \tilde{Y}_t | \mathcal{F}_s] \quad \text{a.s. } \mathbf{Q} \tag{9.8}$$

where "$\cdot$" is the stochastic integral with respect to r.c.l.l. semimartingale $\tilde{Y}$ and predictable integrand and "$\circ$" is the stochastic integral with respect to r.l.l.l. optional semimartingales Y with optional integrand. Again, we are going to identify $\mathbf{E_Q}$ with $\mathbf{E}$ for brevity.

Theorem 9.5.1 *Let $\phi \cdot \tilde{Y}_t$ be a $\mathbf{Q}$-local martingale, then $\phi \circ Y_s = \mathbf{E}[\phi \cdot \tilde{Y}_t | \mathcal{F}_s]$ a.s. $\mathbf{Q}$.*

Proof. Let $R_k \uparrow \infty$ be a sequence of stopping times in $\tilde{\mathbf{F}}$ where $\phi \cdot \tilde{Y}_{t \wedge R_k}$ is a martingale for all k and $\mathbf{E}\left[\phi \cdot \tilde{Y}_{t \wedge R_k}\right] < \infty$. Note that $\tilde{Y}_t = Y_{t+}$ and $\triangle^+ Y_t = Y_{t+} - Y_t$, then $\triangle^+ Y_t = Y_{t+} - Y_t = \tilde{Y}_t - Y_t = \tilde{Y}_t - \tilde{Y}_{t-} = \triangle \tilde{Y}_t$. Also, $d\tilde{Y}_t = dY_t + \triangle \tilde{Y}_t = dY_t + \triangle^+ Y_t$ and $\mathbf{E}\left[\phi \cdot \tilde{Y}_t - \phi \cdot \tilde{Y}_s | \mathcal{F}_s\right] = \mathbf{E}\left[\mathbf{E}\left[\phi \cdot \tilde{Y}_t - \phi \cdot \tilde{Y}_s | \tilde{\mathcal{F}}_s\right] | \mathcal{F}_s\right] = 0$ since $\tilde{\mathcal{F}}_s \supseteq \mathcal{F}_s$ for all s. Moreover, if Y_u is evolving in the interval $[s, t)$ then $\tilde{Y}_u$ is evolving in the interval $(s, t]$. Hence,

$$\mathbf{E}\left[\phi \cdot \tilde{Y}_t - \phi \cdot \tilde{Y}_s | \mathcal{F}_s\right] = \mathbf{E}\left[\mathbf{E}\left[\phi \cdot \tilde{Y}_t - \phi \cdot \tilde{Y}_s | \tilde{\mathcal{F}}_s\right] | \mathcal{F}_s\right]$$

$$= \mathbf{E}\left[\int_{(0,t]} \phi_{u-} d\tilde{Y}_u - \int_{(0,s]} \phi_{u-} d\tilde{Y}_u \Big| \mathcal{F}_s\right]$$

$$= \mathbf{E}\left[\int_{(0,s]} \phi_{u-} d\tilde{Y}_u + \int_{(s,t]} \phi_{u-} d\tilde{Y}_u - \int_{(0,s]} \phi_{u-} d\tilde{Y}_u \Big| \mathcal{F}_s\right]$$

$$= \mathbf{E}\left[\int_{(s,t]} \phi_{u-} d\tilde{Y}_u \Big| \mathcal{F}_s\right]$$

$$= \mathbf{E}\left[\int_{(s,t]} \phi_{u-} \left(dY_u + \triangle \tilde{Y}_u\right) \Big| \mathcal{F}_s\right]$$

$$= \mathbf{E}\left[\int_{(s,t]} \phi_{u-} dY_u \Big| \mathcal{F}_s\right] + \mathbf{E}\left[\int_{(s,t]} \phi_{u-} \triangle \tilde{Y}_u \Big| \mathcal{F}_s\right]$$

$$= \mathbf{E}\left[\int_{(s,t]} \phi_{u-} dY_u \Big| \mathcal{F}_s\right] + \mathbf{E}\left[\int_{[s,t)} \phi_v \triangle^+ Y_v \Big| \mathcal{F}_s\right]$$

$$= \mathbf{E}\left[\int_{(s,t]} \phi_{u-}dY_u + \int_{[s,t)} \phi_v \triangle^+ Y_v | \mathcal{F}_s\right]$$

$$= \mathbf{E}\left[\int_{(s,t]} \phi_{u-}dY_u + \int_{[s,t)} \phi_v dY_{v+} | \mathcal{F}_s\right]$$

$$= \mathbf{E}\left[\int_{[s,t]} \phi_u dY_u | \mathcal{F}_s\right] = \mathbf{E}\left[\int_{[0,t]} \phi_u dY_u +\right] \mathbf{E}\left[\int_{[0,s]} \phi_u dY_u | \mathcal{F}_s\right]$$

$$= \mathbf{E}\left[\phi \circ Y_t - \phi \circ Y_s | \mathcal{F}_s\right] = \mathbf{E}\left[\phi \circ Y_t | \mathcal{F}_s\right] - \phi \circ Y_s = 0 \Rightarrow$$

$$\mathbf{E}\left[\phi \circ Y_t | \mathcal{F}_s\right] = \phi \circ Y_s$$

Alternatively, one can use approximation methods of stochastic integrals by sums to arrive to the same result. So, in the limit of a partition Π_n of $[0,T]$ where T is some time horizon and for any $t \in [0,T]$,

$$\phi_t \longleftarrow \phi_0 + \sum \phi_{t_{i-1}} 1_{]t_{i-1}, t_{i+1}]}(t),$$

where we have chosen the intervals, $]t_{i-1}, t_{i+1}]$, for which $\phi_{t_{i-1}} = \phi_{t_i}$. Then,

$$\phi \cdot \tilde{Y} \longleftarrow \sum \phi_{t_{i-1}} \left(\tilde{Y}_{t_{i+1}} - \tilde{Y}_{t_{i-1}}\right) = \sum \phi_{t_{i-1}} \left(\tilde{Y}_{t_{i+1}} - \tilde{Y}_{t_i} + \tilde{Y}_{t_i} - \tilde{Y}_{t_{i-1}}\right)$$

$$= \sum \phi_{t_{i-1}} \left(\tilde{Y}_{t_{i+1}} - \tilde{Y}_{t_i}\right) + \phi_{t_{i-1}} \left(\tilde{Y}_{t_i} - \tilde{Y}_{t_{i-1}}\right) \quad using \ \phi_{t_i} = \phi_{t_{i-1}}$$

$$= \sum \phi_{t_i} \left(\tilde{Y}_{t_{i+1}} - \tilde{Y}_{t_i}\right) + \phi_{t_{i-1}} \left(\tilde{Y}_{t_i} - \tilde{Y}_{t_{i-1}}\right).$$

Note that the stochastic integral "$\cdot$" is defined in the usual sense with respect to $\left(\Omega, \mathcal{F}, \tilde{\mathbf{F}}, \mathbf{Q}\right)$ and is valid under any refinement of the interval $[0,T]$. This allows to use the intervals $]t_{i-1}, t_{i+1}]$ where in the limit the integral is defined. Consider a partition of $[s,t]$ then the projection of the integral $\int_s^t \phi_{u-} d\tilde{Y}_u$ on $\mathcal{F}_s$ is

$$\mathbf{E}\left[\int_s^t \phi_{u-} d\tilde{Y}_u | \mathcal{F}_s\right] = \mathbf{E}\left[\lim \sum \phi_{t_{i-1}} \left(\tilde{Y}_{t_{i+1}} - \tilde{Y}_{t_{i-1}}\right) | \mathcal{F}_s\right]$$

$$= \lim \mathbf{E}\left[\sum \phi_{t_{i-1}} \left(\tilde{Y}_{t_{i+1}} - \tilde{Y}_{t_{i-1}}\right) | \mathcal{F}_s\right], \quad \text{by localization and integrability}$$

$$= \lim \mathbf{E}\left[\sum \phi_{t_{i-1}} \left(\tilde{Y}_{t_{i+1}} - \tilde{Y}_{t_i}\right) + \phi_{t_{i-1}} \left(\tilde{Y}_{t_i} - \tilde{Y}_{t_{i-1}}\right) | \mathcal{F}_s\right]$$

$$= \lim \mathbf{E}\left[\sum \phi_{t_i} \left(\tilde{Y}_{t_{i+1}} - \tilde{Y}_{t_i}\right) + \phi_{t_{i-1}} \left(\tilde{Y}_{t_i} - \tilde{Y}_{t_{i-1}}\right) | \mathcal{F}_s\right], \quad \text{by } \phi_{t_{i-1}} = \phi_{t_i}.$$

By properties of conditional expectations and knowing that for any t_i, $\mathcal{F}_{t_i} \supseteq \mathcal{F}_s$ and ϕ_t is $\mathcal{F}_t$-measurable, we get

$$\mathbf{E}\left[\phi \cdot \tilde{Y}_t - \phi \cdot \tilde{Y}_s | \mathcal{F}_s\right] = \mathbf{E}\left[\int_s^t \phi_{u-} d\tilde{Y}_u | \mathcal{F}_s\right] =$$

$$= \lim \mathbf{E}\left[\sum \phi_{t_i} \left(\tilde{Y}_{t_{i+1}} - \tilde{Y}_{t_i}\right) + \phi_{t_{i-1}} \left(\tilde{Y}_{t_i} - \tilde{Y}_{t_{i-1}}\right) | \mathcal{F}_s\right], \quad \text{by localization and integrability}$$

$$= \mathbf{E}\left[\lim \sum \phi_{t_i} \left(\tilde{Y}_{t_{i+1}} - \tilde{Y}_{t_i}\right) + \phi_{t_{i-1}} \left(\tilde{Y}_{t_i} - \tilde{Y}_{t_{i-1}}\right) | \mathcal{F}_s\right]$$

$$= \mathbf{E}\left[\lim \sum \mathbf{E}\left[\phi_{t_i} \tilde{Y}_{t_{i+1}} | \mathcal{F}_{t_{i+1}}\right] - \mathbf{E}\left[\phi_{t_i} \tilde{Y}_{t_i} | \mathcal{F}_{t_i}\right] + \mathbf{E}\left[\phi_{t_{i-1}} \tilde{Y}_{t_i} | \mathcal{F}_{t_i}\right] - \mathbf{E}\left[\phi_{t_{i-1}} \tilde{Y}_{t_{i-1}} | \mathcal{F}_{t_{i-1}}\right] | \mathcal{F}_s\right]$$

$$= \mathbf{E}\left[\lim \sum \phi_{t_i} \mathbf{E}\left[\tilde{Y}_{t_{i+1}}|\mathcal{F}_{t_{i+1}}\right] - \phi_{t_i}\mathbf{E}\left[\tilde{Y}_{t_i}|\mathcal{F}_{t_i}\right] + \phi_{t_{i-1}}\mathbf{E}\left[\tilde{Y}_{t_i}|\mathcal{F}_{t_i}\right] - \phi_{t_{i-1}}\mathbf{E}\left[\tilde{Y}_{t_{i-1}}|\mathcal{F}_{t_{i-1}}\right]|\mathcal{F}_s\right]$$

by $\tilde{Y}_u = Y_{u+} = Y_u + \triangle^+ Y_u$

$$= \mathbf{E}\left[\lim \sum \left(\begin{array}{c} \phi_{t_{i-1}}\left(Y_{t_i} - Y_{t_{i-1}}\right) + \phi_{t_i}\left(Y_{t_{i+1}} - Y_{t_i}\right) \\ +\phi_{t_{i-1}}\left(\triangle^+ Y_{t_i} - \triangle^+ Y_{t_{i-1}}\right) + \phi_{t_i}\left(\triangle^+ Y_{t_{i+1}} - \triangle^+ Y_{t_i}\right) \end{array} \right)|\mathcal{F}_s\right]$$

$$= \mathbf{E}\left[\lim \sum \left(\begin{array}{c} \phi_{t_{i-1}}\left(Y_{t_i} - Y_{t_{i-1}}\right) + \phi_{t_i}\left(Y_{t_{i+1}} - Y_{t_i}\right) \\ +\phi_{t_{i-1}}\left(\triangle^+ Y_{t_i} - \triangle^+ Y_{t_{i-1}}\right) + \phi_{t_i}\left(\triangle^+ Y_{t_{i+1}} - \triangle^+ Y_{t_i}\right) \end{array} \right)|\mathcal{F}_s\right]$$

$$= \mathbf{E}\left[\int_{s+}^{t} \phi_{u-} dY_u^r + \int_s^{t-} \phi_u dY_{u+}^g |\mathcal{F}_s\right]$$

$$= \mathbf{E}\left[\int_{0+}^{t} \phi_{u-} dY_u^r + \int_0^{t-} \phi_u dY_{u+}^g - \int_{0+}^{s} \phi_{u-} dY_u^r - \int_0^{s-} \phi_u dY_{u+}^g |\mathcal{F}_s\right]$$

$$= \mathbf{E}\left[\int_0^{t} \phi_{u-} d\tilde{Y}_u - \int_{0+}^{s} \phi_{u-} dY_u^r - \int_0^{s-} \phi_u dY_{u+}^g |\mathcal{F}_s\right]$$

$$= \mathbf{E}\left[\phi \cdot \tilde{Y}_t - \phi_- \cdot Y_s^r - \phi \odot Y_{s+}^g |\mathcal{F}_s\right]$$

$$= \mathbf{E}\left[\phi \cdot \tilde{Y}_t - \phi \circ Y_s |\mathcal{F}_s\right] \;=\; \mathbf{E}\left[\phi \cdot \tilde{Y}_t |\mathcal{F}_s\right] - \phi \circ Y_s = 0$$

and

$$= \mathbf{E}\left[\lim \sum \left(\begin{array}{c} \phi_{t_{i-1}}\left(Y_{t_i} - Y_{t_{i-1}}\right) + \phi_{t_i}\left(Y_{t_{i+1}} - Y_{t_i}\right) \\ +\phi_{t_{i-1}}\left(\triangle^+ Y_{t_i} - \triangle^+ Y_{t_{i-1}}\right) + \phi_{t_i}\left(\triangle^+ Y_{t_{i+1}} - \triangle^+ Y_{t_i}\right) \end{array} \right)|\mathcal{F}_s\right]$$

$$= \mathbf{E}\left[\lim \sum \left(\begin{array}{c} \phi_{t_{i-1}}\left(Y_{t_i} - Y_{t_{i-1}}\right) + \phi_{t_i}\left(Y_{t_{i+1}} - Y_{t_i}\right) \\ +\phi_{t_{i-1}}\left(\triangle^+ Y_{t_i} - \triangle^+ Y_{t_{i-1}}\right) + \phi_{t_i}\left(\triangle^+ Y_{t_{i+1}} - \triangle^+ Y_{t_i}\right) \end{array} \right)|\mathcal{F}_s\right]$$

$$= \mathbf{E}\left[\int_{s+}^{t} \phi_{u-} dY_u^r + \int_s^{t-} \phi_u dY_{u+}^g |\mathcal{F}_s\right]$$

$$= \mathbf{E}\left[\int_{0+}^{t} \phi_{u-} dY_u^r + \int_0^{t-} \phi_u dY_{u+}^g - \int_{0+}^{s} \phi_{u-} dY_u^r - \int_0^{s-} \phi_u dY_{u+}^g |\mathcal{F}_s\right]$$

$$= \mathbf{E}\left[\int_0^{t} \phi_{u-} d\tilde{Y}_u - \int_{0+}^{s} \phi_{u-} dY_u^r - \int_0^{s-} \phi_u dY_{u+}^g |\mathcal{F}_s\right]$$

$$= \mathbf{E}\left[\phi \cdot \tilde{Y}_t - \phi_- \cdot Y_s^r - \phi \odot Y_{s+}^g |\mathcal{F}_s\right]$$

$$= \mathbf{E}\left[\phi \cdot \tilde{Y}_t - \phi \circ Y_s |\mathcal{F}_s\right] \;=\; \mathbf{E}\left[\phi \cdot \tilde{Y}_t |\mathcal{F}_s\right] - \phi \circ Y_s = 0$$

Therefore, $\mathbf{E}\left[\phi \cdot \tilde{Y}_t |\mathcal{F}_s\right] = \phi \circ Y_s = \phi \cdot Y_s^r + \phi \odot Y_{s+}^g$ for any $s \le t$. Note that in the limit $\triangle^+ Y_{t_i} - \triangle^+ Y_{t_{i-1}} = \triangle^+ Y_{t_{i+1}} - \triangle^+ Y_{t_i} = 0$. ∎

Lemma 9.5.2 $\phi \circ Y_t$ is a local optional martingale under $\mathbf{Q}$.

    Proof. By the result of the above theorem, for any $u \le s \le t$,

$$\mathbf{E}\left[\phi \circ Y_s |\mathcal{F}_u\right] = \mathbf{E}\left[\mathbf{E}\left[\phi \cdot \tilde{Y}_t |\mathcal{F}_s\right]|\mathcal{F}_u\right] = \mathbf{E}\left[\phi \cdot \tilde{Y}_t |\mathcal{F}_u\right] = \phi \circ Y_u.$$

∎

Remark 9.5.3 By the same method we have established NFLVR in the *unusual* conditions, it is possible to see that if NA1 is satisfied for $\tilde{Y}$ on $\left(\Omega, \mathcal{F}, \tilde{\mathbf{F}}, \mathbf{Q}\right)$, then it is satisfied for Y on $(\Omega, \mathcal{F}, \mathbf{F}, \mathbf{Q})$. Therefore, one can rest assured that if NFLVR or NA1 is satisfied for r.c.l.l. semimartingales on $\left(\Omega, \mathcal{F}, \tilde{\mathbf{F}}, \mathbf{P}\right)$, then they must be satisfied for their optional version by optional projection on $(\Omega, \mathcal{F}, \mathbf{F}, \mathbf{P})$. Hence, optional markets are free of arbitrage. The same is true for optional defaultable markets, as they are a special case of optional markets, where defaultable assets and cash-flows are optional semimartingales (see Chapter 10).

Next we consider some financial examples.

9.6 Examples of Special Cases

Here we present examples of a value process Y^π that is a jump diffusion with a combination of left and right jumps, corresponding to a portfolio π and evolving on unusual probability space. A value process such as Y^π can be the result of either one of the following market structures that we present below.

9.6.1 Ladlag jumps diffusion model

Let us consider the augmented Black-Scholes model with left and right jumps,

$$x_t = x_0 + \int_{0+}^{t} rx_s ds \tag{9.9}$$

$$X_t = X_0 + \int_{0+}^{t} X_{s-} \left(\mu ds + \sigma dW_s + adL_s^r\right) + \int_{0}^{t-} bX_s dL_{s+}^g,$$

where $L_t^r = L_t - \lambda t$, $L_t^g = -\bar{L}_{t-} + \gamma t$, and r, μ, σ, a, and b are constants. W is diffusion term and L and $\bar{L}$ are independent Poisson with constant intensity λ and γ, respectively. Let $\mathcal{F}_t$ be the natural filtration that is neither right- nor left-continuous. Let the initial money market account be x_0 and the initial price be X_0. We can write X as $X_t = X_0 \mathcal{E}(H)$, where $H_t = \mu t + \sigma W_t + a\left(L_t - \lambda t\right) + b\left(\gamma t - \bar{L}_{t-}\right)$, with $H_0 = 0$, and $x_t = x_0 \exp(rt)$ so that $h_t = rt$. In some sense, this model is simpler than Merton jump diffusion model [118] in which Merton assumed that the coefficient a "jump-amplitudes" of the Poisson process is a r.v. having a normal distribution but complicated by the fact that we are adding a left jump Poisson process. In this case the ratio process is

$$R_t = \frac{X_0}{x_0} \exp\left\{ H_t - \frac{1}{2}\langle H^c, H^c \rangle - rt \right\} \prod_{0 \le s < t} \left[(1 + \triangle^+ H_s)\, e^{-\triangle^+ H_s} \right]$$

$$\times \prod_{0 < s \le t} \left[(1 + \triangle H_s)\, e^{-\triangle H_s} \right]$$

$$= X_0 \exp\left\{ \left(\mu - r - \frac{1}{2}\left(\sigma^2 - \lambda a^2 - \gamma b^2 \right) \right) t + \sigma W_t \right\}$$

$$\times \prod_{0 < s \le t} \left[(1 + a\triangle L_t)\, e^{-a\triangle L_t} \right] \prod_{0 \le s < t} \left[\left(1 - b\triangle^+ \bar{L}_{s-} \right) e^{-b\triangle^+ \bar{L}_{s-}} \right],$$

and is not a local optional martingale. So we want $Z = \mathcal{E}(N)$ for which we have to find a local martingale N such that $\Psi(h, H, N)$,

$$\Psi(h, H, N) = N_t + H_t - h_t + \langle h^c - N^c, h^c - H^c \rangle_t$$

$$+ \sum_{0 < s \le t} \frac{(\triangle h_s - \triangle N_s)(\triangle h_s - \triangle H_s)}{1 + \triangle h_s}$$

$$+ \sum_{0 < s \le t} \frac{(\triangle^+ h_s - \triangle^+ N_s)(\triangle^+ h_s - \triangle^+ H_s)}{1 + \triangle^+ h_s}.$$

9.6.1.1   Computing a local martingale deflator

It makes sense to start with the guess $N_t = \varsigma W_t + c(L_t - \lambda t) + d(\gamma t - \bar{L}_{t-})$ an optional local martingale that will render Z an optional scaling factor. If N is as we chose above, then

$$\Psi(h, H, N) = \varsigma W_t + c(L_t - \lambda t) + d(\gamma t - \bar{L}_{t-}) + \mu t + \sigma W_t + a(L_t - \lambda t) + b(\gamma t - \bar{L}_{t-}) - rt$$

$$+ \left\langle [\varsigma W_t + c(L_t - \lambda t) + d(\gamma t - \bar{L}_{t-})]^c, [\sigma W_t + a(L_t - \lambda t) + b(\gamma t - \bar{L}_{t-})]^c \right\rangle$$

$$+ \sum_{0 < s \le t} ac\triangle L_s \triangle L_s + \sum_{0 \le s \le t} bd\triangle_s^+ \bar{L}_{s-} \triangle_s^+ \bar{L}_{s-}$$

$$= (\varsigma + \sigma) W_t + (a + c)(L_t - \lambda t) + (b + d)(\gamma t - \bar{L}_{t-}) + (\mu - r) t + \varsigma \sigma t$$

$$+ ac\lambda t + bd + \gamma t + acL_t + bd\bar{L}_{t-} + ac\lambda t - ac\lambda t + bd\gamma t - bd\gamma t$$

$$= (\varsigma + \sigma) W_t + (a + c)(L_t - \lambda t) + (b + d)(\gamma t - \bar{L}_{t-}) + ac(L_t - \lambda t)$$

$$- bd(\gamma t - \bar{L}_{t-}) + ac\lambda t + bd\gamma t + (\mu - r) t + \varsigma \sigma t + ac\lambda t + bd\gamma t;$$

therefore,

$$\Psi(h, H, N) = (\varsigma + \sigma) W_t + (a + c + ac)(L_t - \lambda t) \tag{9.10}$$

$$+ (b + d - bd)(\gamma t - \bar{L}_{t-})$$

$$+ (\mu - r + \varsigma \sigma + 2ac\lambda + 2bd\gamma) t$$

is local martingale if

$$\mu - r + \varsigma\sigma + 2ac\lambda + 2bd\gamma = 0. \quad (*)$$

So we have to find (ς, c, d) such that the last statement (*) is true, or in other words

$$[\sigma, \quad 2a\lambda, \quad 2b\gamma] \, [\varsigma, \quad c, \quad d]^{\mathsf{T}} = r - \mu.$$

Trying to solve the equation above leads to infinitely many solutions which means that our market, the market of Black-Scholes with left and right jumps is incomplete. Many of these solutions are interesting; for example, one possible solution is $(\varsigma, c, d) = (\sigma, a, b)/|(\sigma, a, b)|^2$. Another interesting solution is to let $d = 0$ which leads to right-continuous local martingale measure. Yet another solution is one which will eliminate the effects of jumps on drift that is by letting $d = -1/b\gamma$ and $c = 1/a\lambda$, in this case $\varsigma = (r - \mu)/\sigma$. Now that we have found a local martingale measure, we are going to use this knowledge to price a European call option in this market.

9.6.1.2 Pricing of a European call option

A European call option is a contingent claim. Generally, a contingent claim is a r.v. $Y \in \mathcal{F}_T$ at some time T. Y generates the optional martingale process $Y_t = \mathbf{E}[Y|\mathcal{F}_t]$ for $t \in [0, T]$. In the optional Black-Scholes market (x, X) Equation (9.9), we are going to assume that Y_t is a solution of the integral

$$Y_t = Y_0 + \int_{0+}^{t} \alpha_{t-} dR_t^r + \int_{0}^{t-} \beta_t dR_{t+}^g, \qquad (9.11)$$

where $R = X/x$. Recall that Y also satisfies the portfolio equation $Y_t = \eta_t + \xi_t R_t$ and $\xi_t R_t = \alpha_t R_t^r + \xi_t R_t^g$ where $\xi = (\alpha_-, \beta) \in \mathcal{P}(\mathbf{F}) \times \mathcal{O}(\mathbf{F})$. Under the risk neutral measure $\mathbf{Q} \sim \mathbf{P}$ where $\mathbf{Q}_t = \int_{\Omega} Z_t d\mathbf{P}_t$ where Z is the martingale transform of R is strictly larger than 0. $\mathbf{Q}$ exists if $Z > 0$. In the context of our example of the Black-Scholes market, this fact is warranted as we have established above. Furthermore, when pricing a contingent claim, we are only concerned with those contingent claims that can be written as in Equation (9.11).

Now let's turn our attention to the problem at hand, pricing and hedging a European contingent claim. Let's consider pricing of a European call option whose value at maturity is $C_T = (X_T - \mathsf{K})^+$ where K is the strike price and T is the maturity date. In a way, we can think of C_T as the value of a portfolio at time T. The normalized value of this portfolio or the option is

$$\tilde{C}_T = \frac{C_T}{x_T} = \left(\frac{X_T}{x_T} - \frac{\mathsf{K}}{x_T} \right)^+,$$

where $x_T = e^{rT}$ is the discounting factor. At time t and under the risk neutral measure $\tilde{C}_T$ is a local optional martingale whose value is given by

$$\tilde{C}_t = \mathbf{E}_{\mathbf{Q}} \left[\frac{C_T}{x_T} | \mathcal{F}_t \right] = \mathbf{E}_{\mathbf{Q}} \left[(R_T - e^{-rT}\mathsf{K})^+ | \mathcal{F}_t \right]. \quad (*)$$

where R_T is the ratio process. Since $\tilde{C}_t = C_t/e^{rt}$ then we can write the above equation (*) as

$$C_t = e^{rt}\mathbf{E_Q}\left[\left(R_T - e^{-rT}\mathsf{K}\right)^+ |\mathcal{F}_t\right]$$
$$= e^{rt}\mathbf{E_Q}\left[\left(R_T - e^{-rT}\mathsf{K}\right)\mathbf{1}_{(R_T \geq e^{-rT}\mathsf{K})}|\mathcal{F}_t\right].$$

The value of the portfolio at time $t = 0$ which is the option price at the time of its offering is

$$C_0 = \mathbf{E_Q}\left[\frac{C_T}{x_T}\right] = \mathbf{E_Q}\left[\left(R_T - e^{-rT}\mathsf{K}\right)^+\right].$$

To compute the value C_t of the option, we must first choose $\mathbf{Q}$, an appropriate local martingale measure. Previously, we have shown that there exist infinitely many choices. Here we are going to choose a $\mathbf{Q}$ in a way which makes our calculations of the price simpler. However, in practice the option seller would want to choose $\mathbf{Q}$ which maximizes the price of the option, while the buyer would want to choose one to minimize the price of the option. To choose $\mathbf{Q}$ we must choose the parameters that make $\Psi(h, H, N)$ a martingale. According to Equation (9.10) we are going to choose

$$\left\{c = \frac{-a}{1+a}, \quad d = \frac{-b}{1-b}, \quad \varsigma = \frac{r - \mu + 2\lambda\frac{a^2}{1+a} + 2\gamma\frac{b^2}{1+b}}{\sigma}.\right\}$$

This choice leads to the normalized price R under $\mathbf{Q}$ to be a function of just the Wiener process W,

$$R_t = R_0\mathcal{E}_t\left(\Psi(h, H, N)\right) = R_0\mathcal{E}\left((\varsigma + \sigma)W_t\right)$$
$$= X_0 \exp\left((\varsigma + \sigma)W_t - \frac{1}{2}(\varsigma + \sigma)^2 t\right), \quad x_0 = 1.$$

Hence, the option value at time t is given by

$$C_t = e^{rt}\mathbf{E_Q}\left[\left(R_T - e^{-rT}\mathsf{K}\right)\mathbf{1}_{(R_T \geq e^{-rT}\mathsf{K})}|\mathcal{F}_t\right]$$
$$= e^{rt}\mathbf{E_Q}\left[\left(R_T - e^{-rT}\mathsf{K}\right)\mathbf{1}_{(R_T \geq e^{-rT}\mathsf{K})}|\mathcal{F}_t\right]$$
$$= e^{rt}\mathbf{E_Q}\left[R_T\mathbf{1}_{(R_T \geq e^{-rT}\mathsf{K})}|\mathcal{F}_t\right] - e^{-r(T-t)}\mathsf{K}\mathbf{Q}\left(R_T \geq e^{-rT}\mathsf{K}|\mathcal{F}_t\right)$$

and its price at the time of its initial offering, $t = 0$, is

$$C_0 = \mathbf{E_Q}\left[R_T\mathbf{1}_{(R_T \geq e^{-rT}\mathsf{K})}\right] - e^{-rT}\mathsf{K}\mathbf{Q}\left(R_T \geq e^{-rT}\mathsf{K}\right).$$

Now let's compute the price C_0. We will start by computing

$$\mathbf{Q}\left(R_T > e^{-rT}\mathsf{K}\right) = \mathbf{Q}\left(\exp\left((\varsigma + \sigma)W_T - \frac{1}{2}(\varsigma + \sigma)^2 T\right) > \frac{e^{-rT}\mathsf{K}}{X_0}\right)$$
$$= \mathbf{Q}\left(W_T \geq \frac{1}{(\varsigma + \sigma)}\left[-rT - \ln\left[\frac{X_0}{\mathsf{K}}\right] + \frac{1}{2}(\varsigma + \sigma)^2 T\right]\right)$$
$$= \mathbf{Q}\left(Z < \frac{1}{(\varsigma + \sigma)\sqrt{T}}\left[\ln\left[\frac{X_0}{\mathsf{K}}\right] + \left(r - \frac{1}{2}(\varsigma + \sigma)^2\right)T\right]\right)$$
$$= \Phi\left(\tilde{\mathsf{K}}\right).$$

where Z here is a standard normal r.v. and

$$\tilde{K} = \frac{1}{(\varsigma+\sigma)\sqrt{T}}\left[\ln\left[\frac{X_0}{K}\right] + (r - \frac{1}{2}(\varsigma+\sigma)^2)T\right].$$

The other part of the price is the expectation

$$\mathbf{E_Q}\left[R_T \mathbf{1}_{(R_T \geq e^{-rT}K)}\right] = \frac{X_0}{\sqrt{2\pi}}\int_{-\infty}^{\tilde{K}} \exp -\frac{1}{2}\left(z^2 - 2(\varsigma+\sigma)\sqrt{T}z + (\varsigma+\sigma)^2 T\right)dz$$

$$= \frac{X_0}{\sqrt{2\pi}}\int_{-\infty}^{\tilde{K}} \exp -\frac{1}{2}\left(z + (\varsigma+\sigma)\sqrt{T}\right)^2 dz$$

$$= X_0\Phi\left(\tilde{K} + (\varsigma+\sigma)\sqrt{T}\right).$$

Therefore, the price of the option is going to be given by

$$C_0 = X_0\Phi\left(\tilde{K} + (\varsigma+\sigma)\sqrt{T}\right) - e^{-rT}K\Phi\left(\tilde{K}\right).$$

Note that this formula has the same form as the regular Black-Scholes pricing formula for a European call option. However, the volatility σ has been changed by ς "the effective volatility" that is a result of the left and right jumps. Now that we have computed the price of the option, how do we go about finding the evolution of the price through time?

The evolution of the price of a European call option can also be derived in the following way

$$C_t = e^{rt}\mathbf{E_Q}\left[(R_T - e^{-rT}K)\mathbf{1}_{(R_T \geq e^{-rT}K)}|\mathcal{F}_t\right]$$

$$= e^{rt}\mathbf{E_Q}\left[R_T\mathbf{1}_{(R_T \geq e^{-rT}K)}|\mathcal{F}_t\right] - e^{-r(T-t)}K\mathbf{Q}\left(R_T \geq e^{-rT}K|\mathcal{F}_t\right)$$

Note that $R_T \geq e^{-rT}K$ if and only if

$$W_T \geq \frac{1}{(\varsigma+\sigma)}\left[-\ln\left[\frac{X_0}{K}\right] - \left[r - \frac{1}{2}(\varsigma+\sigma)^2\right]T\right]$$

which is also true if and only if

$$W_T - W_t \geq \frac{1}{(\varsigma+\sigma)}\left[-\ln\left[\frac{X_0}{K}\right] - \left[r - \frac{1}{2}(\varsigma+\sigma)^2\right](T-t)\right]$$

by using a simple time change from T by $T-t$ and knowing that W_{T-t} has the same distribution as $W_T - W_t$. Replacing the inequality in $\mathbf{Q}\left(R_T \geq e^{-rT}K|\mathcal{F}_t\right)$ by the inequality involving $W_T - W_t$; therefore,

$$\mathbf{Q}\left(R_T \geq e^{-rT}K|\mathcal{F}_t\right)$$

$$\iff \mathbf{Q}\left(W_T \geq \frac{1}{(\varsigma+\sigma)}\left[-rT - \ln\left[\frac{X_0}{K}\right] + \frac{1}{2}(\varsigma+\sigma)^2 T\right]|\mathcal{F}_t\right)$$

$$\iff \mathbf{Q}\left(W_T - W_t \geq \frac{-1}{(\varsigma+\sigma)}\left[\ln\left[\frac{X_0}{K}\right] - \left[r - \frac{1}{2}(\varsigma+\sigma)^2\right](T-t)\right]|\mathcal{F}_t\right)$$

$$= \mathbf{Q}\left(W_T - W_t \geq \frac{-1}{(\varsigma+\sigma)}\left[\ln\left[\frac{X_0}{K}\right] - \left[r - \frac{1}{2}(\varsigma+\sigma)^2\right](T-t)\right]\right)$$

$$= \Phi\left(\tilde{K}_t\right)$$

where

$$\tilde{K}_t = \frac{1}{(\varsigma+\sigma)\sqrt{T-t}}\left[\ln\left[\frac{X_0}{K}\right] + \left(r - \frac{1}{2}(\varsigma+\sigma)^2\right)(T-t)\right].$$

For the other term

$$\mathbf{E_Q}\left[R_T \mathbf{1}_{(R_T \geq e^{-rT}K)}|\mathcal{F}_t\right]$$

$$= \mathbf{E_Q}\left[R_t \exp\left((\varsigma+\sigma)(W_T - W_t) - \frac{1}{2}(\varsigma+\sigma)^2(T-t)\right)\mathbf{1}_{(R_T \geq e^{-rT}K)}|\mathcal{F}_t\right]$$

$$= R_t \mathbf{E_Q}\left[\exp\left((\varsigma+\sigma)(W_T - W_t) - \frac{1}{2}(\varsigma+\sigma)^2(T-t)\right)\right.$$

$$\left. \mathbf{1}_{\left(W_T - W_t \geq \frac{-1}{(\varsigma+\sigma)}\left[\ln\left[\frac{X_0}{K}\right] - \left[r - \frac{1}{2}(\varsigma+\sigma)^2\right](T-t)\right]\right)}|\mathcal{F}_t\right]$$

$$- R_t \mathbf{E_Q}\left[\exp\left((\varsigma+\sigma)(W_T - W_t) - \frac{1}{2}(\varsigma+\sigma)^2(T-t)\right)\right.$$

$$\left. \mathbf{1}_{\left(W_T - W_t \geq \frac{-1}{(\varsigma+\sigma)}\left[\ln\left[\frac{X_0}{K}\right] - \left[r - \frac{1}{2}(\varsigma+\sigma)^2\right](T-t)\right]\right)}\right]$$

$$= R_t \mathbf{E_Q}\left[\exp\left((\varsigma+\sigma)\sqrt{T-t}Z - \frac{1}{2}(\varsigma+\sigma)^2(T-t)\right)\right.$$

$$\left. \mathbf{1}_{\left(\sqrt{T-t}Z \geq \frac{-1}{(\varsigma+\sigma)}\left[\ln\left[\frac{X_0}{K}\right] - \left[r - \frac{1}{2}(\varsigma+\sigma)^2\right](T-t)\right]\right)}\right]$$

$$= R_t \Phi\left(\tilde{K}_t + (\varsigma+\sigma)\sqrt{T-t}\right).$$

So the price of the option evolves according to

$$C_t = e^{rt}R_t\Phi\left(\tilde{K}_t + (\varsigma+\sigma)\sqrt{T-t}\right) + e^{-r(T-t)}K\Phi\left(\tilde{K}_t\right).$$

9.6.1.3 Hedging of a European call option

Since $\tilde{C}_t$ is a local martingale under $\mathbf{Q}$, then by martingale representation theorem we can write

$$C_t = e^{-r(T-t)}\mathbf{E_Q}\left[C_T\right] + e^{rt}\int_0^t \beta_s dR_s \text{ a.s. } \mathbf{Q}.$$

This representation of C_t will help us determine the replicating (hedging) portfolio by computing

$$\beta_t = \frac{d\left[\tilde{C}, R\right]_t}{d\left[R, R\right]_t}.$$

Note that

$$\left[\tilde{C}, R\right]_t = \left[R_t\Phi\left(\tilde{K}_t + (\varsigma+\sigma)\sqrt{T-t}\right) + e^{-r(T-t)}K\Phi\left(\tilde{K}_t\right), R_t\right]$$

$$= \left[R_t\Phi\left(\tilde{K}_t + (\varsigma+\sigma)\sqrt{T-t}\right), R_t\right] + e^{-r(T-t)}K\left[\Phi\left(\tilde{K}_t\right), R_t\right]$$

$$= \Phi\left(\tilde{K}_t + (\varsigma+\sigma)\sqrt{T-t}\right)[R, R]_t$$

where $\left[\Phi\left(\tilde{\mathsf{K}}_t\right), R_t\right] = 0$. Therefore, the hedging strategy is

$$\beta_t = \frac{d\left[\tilde{C}, R\right]_t}{d\left[R, R\right]_t} = \frac{\Phi\left(\tilde{\mathsf{K}}_t + (\varsigma + \sigma)\sqrt{T - t}\right)d\left[R, R\right]_t}{d\left[R, R\right]_t}$$

$$= \Phi\left(\tilde{\mathsf{K}}_t + (\varsigma + \sigma)\sqrt{T - t}\right)$$

and $\eta_t = e^{-rT}\mathsf{K}\Phi\left(\tilde{\mathsf{K}}_t\right).$

Remark 9.6.1 Note that in general hedging of contingent claims in optional semimartingale markets leads to portfolios that are not predictable. However, in this example, we have found predictable hedging portfolios. So how is this possible? It turns out that, in some special cases one can choose a risk neutral measure that absorbs at least all the left jumps in the market and renders the market processes right-continuous or even in some cases continuous like what we have shown in the example above.

9.6.2 Basket of stocks

Here we present an example of a market of optional semimartingales where it is possible to trade in the usual sense. Furthermore, we discuss portfolio structure in this market. Consider the market: a money market account x and two assets X^1 and X^2 evolving according to

$$x_t = 1$$

$$X_t^1 = X_0^1 + \int_{0+}^{t} X_{s-}^1\left(\mu ds + \sigma dW_s + adL_s^r\right),$$

$$X_t^2 = X_0^2 + \int_0^{t-} bX_s^2 dL_{s+}^g,$$

where $L_t^r = L_t - \lambda t$, $L_t^g = -\bar{L}_{t-} + \gamma t$, and μ, σ, a, and b are constants. W is diffusion term and L and $\bar{L}$ are independent Poisson processes with constant intensities λ and γ, respectively. The initial prices are X_0^1 and X_0^2. In this case, one can write a portfolio of this market as

$$Y_t = Y_0 + \alpha \cdot X_t^1 + \beta \odot X_t^2,$$

$$= \eta_t + \alpha_t X_t^1 + \beta_t X_t^2.$$

In this case, (η, α, β) can be traded independently since X_t^1 and X_t^2 are two different assets that are available in this market. Also, each trading strategy can be approximated by a simple trading strategy.

The portfolio value process Y is an optional semimartingale defined on the filtration generated by the pair X^1 and X^2 which is not necessarily right-continuous. One notices here that even if the individual processes comprising the market are either right- or left-continuous, the market information as a whole is not necessarily right- or left-continuous. Optional semimartingales in *un*usual probability spaces provide a way for dealing with complicated market structure such as the examples we have presented above.

9.6.3   Defaultable bond and a stock

Consider a market composed of a money market account x and assets X evolving according to $x_t = x_0 \mathcal{E}(h)_t$ and $X_t = X_0 \mathcal{E}(H)_t$ where

$$
\begin{aligned}
h_t &= rt + bL_t^g, \ h_0 = 0, \\
H_t &= \mu t + \sigma W_t + aL_t^d, \ H_0 = 0.
\end{aligned}
$$

$L_t^d = L_t - \lambda t$, $L_t^g = -\bar{L}_{t-} + \gamma t$, and r, μ, σ, a, and b are constants. W is diffusion term and L and $\bar{L}$ are Poisson with constant intensity λ and γ, respectively. Let $\mathcal{F}_t$ be the natural filtration that is neither right- or left-continuous. Let the initial money market account be x_0 and the initial price be X_0.

In this example we have modeled the money market account value by a left-continuous process. A similar model was given by Duffie and Singleton [125] for bonds that can experience defaults. We believe that the model we present above is a better description of a portfolio of stocks and bonds than r.c.l.l. processes on usual probability space.

In this case the ratio process is $R_t = \frac{X_0}{x_0} \mathcal{E}(H_t - h_t^* - [H, h^*]_t)$. We want to find $Z = \mathcal{E}(N)$ such that ZR is a local martingale. In Section 9.3 we have shown that associated with the product $ZR = \frac{X_0}{x_0} \mathcal{E}(\Psi(h, H, N))$ is the process $\Psi(h, H, N)$. To compute a reasonable form for $\Psi(h, H, N)$,

$$
\begin{aligned}
\Psi(h, H, N) = N_t &+ H_t - h_t + \langle h^c - N^c, h^c - H^c \rangle_t \\
&+ \sum_{0 < s \leq t} \frac{(\triangle h_s - \triangle N_s)(\triangle h_s - \triangle H_s)}{1 + \triangle h_s} \\
&+ \sum_{0 < s \leq t} \frac{(\triangle^+ h_s - \triangle^+ N_s)(\triangle^+ h_s - \triangle^+ H_s)}{1 + \triangle^+ h_s},
\end{aligned}
$$

it makes sense to suppose that $N_t = \varsigma W_t + cL_t^d + \theta L_t^g$ is an optional local martingale for which Z

is an optional local martingale deflator. In this case $\Psi(h, H, N)$ is

$$\Psi(h, H, N) = \left[\varsigma W_t + cL_t^d + \theta L_t^g\right] + \left[\mu t + \sigma W_t + aL_t^d\right] - \left[rt + bL_t^g\right]$$
$$+ \left\langle \left[rt + bL_t^g\right]^c - \left[\varsigma W_t + cL_t^d + \theta L_t^g\right]^c, \left[rt + bL_t^g\right]^c - \left[\mu t + \sigma W_t + aL_t^d\right]^c\right\rangle$$
$$+ \sum_{0 < s \le t} ac \left(\triangle L_s^d\right)^2 + \sum_{0 \le s < t} \frac{b(b - \theta)\left(\triangle^+ L_s^g\right)^2}{1 + b\triangle^+ L_s^g}$$
$$= (\mu - r + \varsigma\sigma)t + (\varsigma + \sigma)W_t + (c + a)L_t^d + (\theta - b)L_t^g + acL_t$$
$$+ (b - \theta)L_t^g + b(\theta - b)[L_t^g, L_t^g]$$
$$= (\mu - r + \varsigma\sigma)t + (\varsigma + \sigma)W_t + (c + a)L_t^d + acL_t + b(\theta - b)\bar{L}_{t-}.$$

W_t and L_t^d are martingales. $\Psi(h, H, N)$ is a martingale if and only if $\mu - r + \varsigma\sigma + ca\lambda + \theta b - b^2\gamma = 0$. The solution of this equation leads to infinitely many solutions, which means the market is incomplete. Many of these solutions are interesting; for example, one possible solution is to let $\theta = 0$ which leads to right-continuous local martingale deflator. Yet another solution is one which will eliminate the effects of jumps on drift that is by letting $\theta = -1/b$ and $c = 1/a\lambda$, in this case $\varsigma = (r - \mu + b^2\gamma)/\sigma$.

Chapter 10

Defaultable Markets on Unusual Space

Here we introduce by far the most important application of optional process on *un*susual probability space to defaultable markets.

10.1 Introduction

Defaultable markets are markets with the possibility of default events occurring. Research on default risk is concerned with modeling of default time, defaultable claims and recovery rules and with pricing and hedging strategies. The two main approaches to modeling default are the structural and the reduced form models. In structural models the value of the firm determines if a default event occurs. This approach was founded by [115] and [126, 127] and extended by many, see [108] [128, 129, 130, 131]. The advantage of structural models is that one sees how the corporate conditions affect default. However, most often the firm's value is not in itself a tradable asset but the result of an accounting of all of the firm's assets and liabilities where some are visible to the market and others are not. Hence, the parameters of structural models are difficult to estimate in practice. Moreover, in structural models where a boundary condition is used to demarcate the onset of default, it is difficult to define what the boundary condition should be.

In contrast to structural models, in reduced-form models the firm's value plays only an auxiliary role and defaults arrive as a total surprise to all counterparties. Therefore, default is the result of exogenous factors. In this case, the random time of default is defined as a totally inaccessible stopping time on an enlarged filtration that encompasses all market information, assets and defaults. To compute pricing rules and hedging portfolios, a fundamental problem in reduced-form approach is the computation of the conditional probability of default given available market information. To compute the conditional probability of default, several approaches arose depending on whether or not the information about default-free market is available. If default-free market information is not given, then one must suppose the existence of an intensity process associated with the probability of default to be able to price a defaultable claim [125]. This approach is known as the intensity based approach. The main problem with this approach is that the pricing rule is hard to compute. On the other hand, in hazard process approach the reference filtration of the default-free market is enlarged by the progressive knowledge of default events. It is more convenient to derive probability of default in this framework; however, one needs to assume knowledge of the default-free market information. Much of reduced-form pricing theory has been elaborated by many mathematicians, see [125, 132, 133, 134, 135, 136, 137, 138] and [139, 140, 141] for a review.

The basic mathematics of structural and reduced-form models is given here as a review and to compare these models to models we will present in this paper. Let us consider a usual probability space $\left(\Omega, \mathcal{G}, \mathbf{G} = (\mathcal{G}_t)_{t \geq 0}, \mathbf{P}\right)$ where the market evolves: asset, X, and default, τ. Associated with default time is the default process is $H_t = \mathbf{1}_{(\tau \leq t)}$. Let $\mathcal{F}_t$ be the default-free σ-algebra generated by X_t and $\mathcal{H}_t$ be the σ-algebra generated by H_t with $\mathbf{F} = (\mathcal{F}_t)_{t \geq 0}$ and $\mathbf{H} = (\mathcal{H}_t)_{t \geq 0}$. Both $\mathcal{F}_t$ and $\mathcal{H}_t$ are sub-σ-algebras of $\mathcal{G}_t$ for all t. In structural form models, the process X is a determinant factor of τ. In other words, the filtrations $\mathcal{F}_t$ and $\mathcal{H}_t$ may have common elements for any time t. And, in structural models where default time τ is predictable, then $\mathcal{H}_t \subseteq \mathcal{F}_t$. On the other hand, in reduced-form models default is not measurable in the reference-filtration $\mathbf{F}$, i.e., $(\tau \leq t) \notin \mathcal{F}_t$ for any time t. A way to analyze reduced-form models is to consider τ to be an inaccessible stopping time with respect to the market filtration $\mathbf{G}$. With this consideration of τ one can proceed to compute the conditional probability of default $\mathbf{P}\left(\tau \leq t | \mathcal{G}_t\right)$ given $\mathcal{G}_t$ using intensity-based techniques. If one is able to define $\mathbf{F}$ the default-free filtration, then a new filtration can be constructed by joining the filtration $\mathbf{H} = (\mathcal{H}_t)_{t \geq 0}$ generated by $H_t = \mathbf{1}_{(\tau \leq t)}$ to $\mathbf{F}$, i.e., $\tilde{\mathcal{G}}_t = \sigma\left(\mathcal{F}_t \vee \mathcal{H}_t\right) \subseteq \mathcal{G}_t$ and $\tilde{\mathbf{G}} = \left(\tilde{\mathcal{G}}_t\right)_{t \geq 0}$, then compute $\mathbf{P}\left(\tau \leq t | \tilde{\mathcal{G}}_t\right)$ using hazard process based techniques. But this enlargement of the filtration $\mathbf{F}$ by $\mathbf{H}$ leads to several problems in trying to establish a pricing theory based on a no-arbitrage principle. Enlargement of filtration changes the properties of martingales and semimartingales. For example, if X is a martingale under $\mathbf{F}$, it might not be a martingale under $\tilde{\mathbf{G}}$ or $\mathbf{G}$. To deal with these effects and to establish existence of local martingale measures for pricing, one has to invoke two invariance principles known as the $\mathbf{H}$ and $\mathbf{H}'$ hypotheses. $\mathbf{H}$ hypothesis states that every local martingale in the smaller filtration $\mathcal{F}$ is a local martingale in the larger filtration $\tilde{\mathcal{G}}$, whereas the $\mathbf{H}'$ hypothesis states that a semimartingale under the smaller filtration $\mathcal{F}$ remains a semimartingale under the larger filtration $\tilde{\mathcal{G}}$.

While there are still many open and interesting problems to consider in the classical approaches to credit risk, we choose to take another approach to the problem. Our approach is based on the calculus of processes on *un*usual probability spaces. Even though, we are going to use a *non*standard calculus to approach the problem of default, it is possible to view our models as extension or enrichment of reduced-form models by methods from the calculus of processes on *un*usual probability spaces. So, our goal with this paper is to develop initial results and present applications of the calculus of optional processes to defaultable markets.

However, before we delve into our approach, let us *motivate* it by a few examples where we take a closer look at the mechanics of default and how it affects the value of an asset X. Let us place ourselves on the usual probability space $\left(\Omega, \mathcal{F}, \mathbf{F} = (\mathcal{F}_t)_{t \geq 0}, \mathbf{P}\right)$ where the market evolves. Also, suppose the asset, X, remains in existence after default while its value changes. Now, let us *fix* an instance of time t. If default time is predictable or a stopping time in $\mathcal{F}_t$ then $(\tau \leq t) \in \mathcal{F}_t$ – information about default is incorporated in $\mathcal{F}_t$. On the other hand, if $(\tau \leq t) \notin \mathcal{F}_t$, so that $H_t = \mathbf{1}_{(\tau \leq t)}$ is not $\mathcal{F}_t$ measurable, then the default time τ is a random time that is the result of external factors. However, *after default takes place*, the surprising information about it gets incorporated in *future-values* of the asset, X. If X_t is r.c.l.l. and $\mathcal{F}_t = \mathcal{F}_{t+}$ then obviously $(\tau \leq t) \notin \mathcal{F}_{t+}$ and, *loosely speaking*, we may want to say $(\tau \leq t) \in \mathcal{F}_s$ for any $s > t$ such that $\mathcal{F}_s \supset \mathcal{F}_{t+}$ to mean that the information about default time τ at t is part of a *future-filtration* of the market. In reduced-form modeling, to be *precise*, an enlarged filtration $\mathbf{F} \vee \mathbf{H}$ is constructed, such that, the view that information about default is incorporated in future values of the asset is subsumed in the definition of the enlarged market filtration $\mathbf{F} \vee \mathbf{H}$. However, we are going to show in this paper that with the use of the calculus of processes on *un*usual probability spaces, one needs not consider a filtration enlargement to incorporate default information in the dynamic changes of the value of the asset.

Let us consider default from a different viewpoint. Suppose that the event $A \in \mathcal{F}_\infty$ such that $\mathbf{P}(A) > 0$. Ideally, $\mathcal{F}_\infty$ contains all events including defaults, for example, for the default time τ, $\sigma(\tau) \subseteq \mathcal{F}_\infty$. The expected value $a_t = \mathbf{E}[\mathbf{1}_A|\mathcal{F}_t]$ at some *fixed* time t is implicitly affected by a future default event τ. For example, consider for some *fixed* time u such that $(2t < u < \infty)$ and $\tau = u$ where it is possible that $A \cap (\tau = u) \neq \emptyset$. However, at time t, $\mathcal{F}_t$ contains absolutely no information about the default event $(\tau = u)$ but the natural filtration $\mathbf{F}^a$ generated by the process, a, is not default-free. Essentially, in a defaultable market processes that are affected by default at some point over their lifetime, their current value is "implicitly" affected by default eventhough current information does not announce the existence of future default events. Therefore, the construction of a defaultable market filtration $\mathbf{G}$ from a default-free filtration $\mathbf{F}$ and a filtration associated with the default process $\mathbf{H}$ is hard to realize in real defaultable markets. This issue can be circumvented by considering defaultable markets on *un*usual probability spaces.

Let us consider another example. Suppose the asset X that evolves according to, $X_t^{-1} dX_t = \mu dt + \sigma dW_t$, where W is a Wiener process and μ and σ are unknown constants to be estimated from market-traded values of X. In this market, even though default events might be future events beyond the current market time t are implicit in the *psyche of the market*, e.g., corporate bonds are riskier than US/Canadian government bonds, and can weakly influence the current traded values of X. As a result, the estimated values of μ and σ from the traded values of X are inadvertently contaminated by default events. Therefore, in a defaultable market assets and defaults are stitched together and are difficult to segregate.

To avoid difficulties of previous approaches: the reconstruction of *enlarged market filtrations*, the requiring of *invariance* properties to be satisfied as a way to get decomposition results, the *initial completion* of filtration by all null sets from the final filtration, and the requiring that the *filtration be right continuous*, without all these constructs, we propose a different and a more *natural* approach based on the stochastic calculus of optional processes on *un*usual probability spaces. The chapter is organized as follows: Section 10.2 presents the essence of our approach to defaultable markets on *un*usual stochastic basis. Sections 10.3 deals with defaultable claims and cash-flows. Section 10.4 is devoted to the conditional probability of default. In Section 10.5, we consider valuation of defaultable contingent claims, discuss no-arbitrage conditions in defaultable markets and give illustrative examples.

10.2   Optional Default

To illustrate our approach, we begin with an example of a simple market that experiences default. Even though simple, the ideas we present here are general and can be carried over to other more delicate market structures. Let us consider *now* the *un*usual stochastic basis $\left(\Omega, \mathcal{F}, \mathbf{F} = (\mathcal{F}_t)_{t\geq 0}, \mathbf{P}\right)$ and an asset whose value is given by the process Y on this space. Assume that the asset final value is η realized at the end of time is a strictly positive random variable measurable with respect to $\mathcal{F}_\infty$. Let $\mathbf{F} = (\mathcal{F}_t)_{t\geq 0}$ represent the history of the market, the good, the bad and defaults. However bad, we assume that the asset never vanquishes, in other words, $Y > 0$ for all time. All is well with the market except for a single default event $\tau \in \mathcal{T}(\mathbf{F}_+)$ that happens at some time, $t \geq 0$. Notice that we have defined *default time*, $\tau \in \mathcal{T}(\mathbf{F}_+)$, as an *inaccessible stopping time* in the *broad sense*. Furthermore, we assume that the value of the asset before default τ follows X and immediately after default evolves according to x. Both x and X are optional semimartingales adapted to $\mathbf{F}$; however, X is observable only up to default and x is only observable after default. Also, we define the default process $H = \mathbf{1}_{(\tau < \cdot)}$ as a left-continuous optional finite variation process with respect to $\mathbf{F}$. Then, Y is given by

$$Y_t = (1 - H_t)X_t + H_t x_t. \tag{10.1}$$

In the integral representation of Y, we have,

$$
\begin{aligned}
Y_t &= X_t - \int_{0+}^{t} H_{s-} d\left(X_s - x_s\right) - \int_{0}^{t-} H_s d\left(X_{s+} - x_{s+}\right) \\
&\quad + \int_{0+}^{t} \left(X_{s-} - x_{s-}\right) dH_s + \int_{0}^{t-} \left(X_s - x_s\right) dH_{s+} \\
&\quad + \int_{0+}^{t} \triangle H_s d\left(X_s - x_s\right) + \int_{0}^{t-} \triangle^+ H_s d\left(X_{s+} - x_{s+}\right).
\end{aligned}
$$

But since H has a single left jump but otherwise continuous, then $\triangle H = 0$ and Y reduces to,

$$
\begin{aligned}
Y_t &= X_t - \int_{0+}^{t} H_{s-} d\left(X_s - x_s\right) - \int_{0}^{t-} H_s d\left(X_{s+} - x_{s+}\right) \\
&\quad + \int_{0+}^{t} \left(X_{s-} - x_{s-}\right) H_s ds + \int_{0}^{t-} \left(X_s - x_s\right) H_{s+} ds + \left(X_{\tau+} - x_{\tau+}\right) H_t.
\end{aligned}
$$

To highlight the effects of default τ on the value of Y, which will manifest itself as a left jump on the value of Y at time of default, we are going to assume that x and X are continuous, hence

$$Y_t = X_t - H \circ \left(X_t - x_t\right) + \left(X_{\tau+} - x_{\tau+}\right) H_t + \int_{\tau}^{t} \left(X_s - x_s\right) ds.$$

To summarize, the basic tenants to our approach are: (1) The market evolves in the *un*usual probability space $\left(\Omega, \mathcal{F}, \mathbf{F} = (\mathcal{F}_t)_{t\geq 0}, \mathbf{P}\right)$; (2) $\mathbf{F}_+$ is the smallest right-continuous enlargement of

F; (3) the value of an asset or a defaultable cash-flow is an **F** optional semimartingale; (4) finally, but most importantly is that default time, τ, is defined as τ is a totally inaccessible stopping time in the broad sense, i.e., $\tau \in \mathcal{T}(\mathbf{F}_+)$, and the associated default process is $H_t = \mathbf{1}_{(\tau < t)}$ which is an optional left-continuous process adapted to $\mathcal{F}_t$. The definition of default time as an inaccessible stopping time in the broad sense makes default events come as a surprise to the market given up to time t information, $\mathcal{F}_t$. However, as market evolves to $\mathcal{F}_{t+}$ the surprising information about default has been incorporated in $\mathcal{F}_{t+}$. In this way $\mathcal{F}_{t+}$ in *un*usual probability space models of defaultable market takes the place of filtration enlargement in reduced-form models. Next, we consider defaultable cash-flow pricing and hedging in the context of *un*usual stochastic basis.

10.3   Defaultable Cash-Flow

Again, let the market evolve in the *un*usual probability space $\left(\Omega, \mathcal{F}, \mathbf{F} = (\mathcal{F}_t)_{t \geq 0}, \mathbf{P}\right)$. $\mathbf{P}$ is the real-world probability, as opposed to the *local optional martingale measure* $\mathbf{Q}$, which we *assume* that at least one exists in this market or *choose one* if many exist [142]. Also, we have $\mathbf{F}_+$ the *smallest right-continuous enlargement* of $\mathbf{F}$. The filtration $\mathbf{F}$ supports the following objects. The claim, Λ, is the payoff received by the owner of the claim at maturity time, T, if there was *no* default prior to or *at* T. The process, A, with $A_0 = 0$, is the promised dividends if there was *no* default prior to or *at* T. The recovery claim, ρ, represents the recovery payoff received at T, if default occurs prior to the claims maturity date T. The recovery process, R, specifies the recovery payoff at time of default, if it occurs prior to T. Finally, default time, τ, is a totally inaccessible stopping time in the broad sense, i.e., $\tau \in \mathcal{T}(\mathbf{F}_+)$. Also, we define the associated default process by, $H_t = \mathbf{1}_{(\tau < t)}$ which is an optional and left-continuous process with respect to $\mathcal{F}_t$.

Furthermore, we assume that the processes X, R, and A are progressively measurable with respect to the filtration $\mathbf{F}$. And, the random variable Λ is $\mathcal{F}_T$-measurable and ρ is at least $\mathcal{F}_{T+}$-measurable. Also, we assume without mentioning that all random objects introduced above are *at least* r.l.l.l. and satisfy suitable integrability conditions that are needed for evaluating integrals, stochastic or otherwise. This brings us to the recovery rules. If default occurs after time T, then the promised claim Λ is paid in full at time T. Otherwise, default time $\tau \leq T$ and depending on the agreed upon recovery rules either the amount R_τ is paid at the time of default τ, or the amount ρ is paid at the maturity date T. Therefore, in its more general setting we consider simultaneously both kinds of recovery payoff and thus define a defaultable claim formally as a quintuple, $DCT = (A, \Lambda, \rho, R, H)$. Notice that the date, T, the information structures, $\mathbf{F}$ and $\mathbf{F}_+$, and the real-world probability, $\mathbf{P}$, are intrinsic components of the definition of a defaultable claim (see [141] for the usual case). For $DCT = (A, \Lambda, \rho, R, H)$ the dividend process is defined as:

Definition 10.3.1 The *dividend process, D,* of a defaultable claim $DCT = (A, \Lambda, \rho, R, H)$ equals

$$D_t = \tilde{X}\mathbf{1}_{(t \geq T)} + (1 - H) \circ A_t + R \circ H_t, \tag{10.2}$$

where $\tilde{X} = [\Lambda(1 - H_T) + \rho H_T]$. The process D is optional and **F**-measurable.

Lemma 10.3.2 The process D is finite variation over finite time segments including $[0, T]$.

Proof. By using Galchuk-Ito lemma and properties of the components of D. ∎

The risk neutral value of a defaultable claim in this market is the discounted value of the dividend process D.

Definition 10.3.3 The *ex-dividend price process* $X(\cdot, T)$ of a defaultable claim $DCT = (A, \Lambda, \rho, R, H)$ which settles at time T is given by

$$X(t, T) = B_t \mathbf{E_Q} \left(B^{-1} \circ D_T - B^{-1} \circ D_t | \mathcal{F}_t \right) \tag{10.3}$$

$$= B_t \mathbf{E_Q} \left(\int_{t+}^{T} B_{u-}^{-1} dD_u + \int_{t}^{T-} B_u^{-1} dD_{u+} | \mathcal{F}_t \right), \quad \forall t \in [0, T],$$

where B is the discounting process, an optional strictly positive r.l.l.l. semimartingale.

Under $\mathbf{Q}$, the ex-dividend process, $B_t^{-1} X(t, T)$, is a local optional martingale, $B^{-1} X(\cdot, T) \in \mathcal{M}_{loc}(\mathbf{F}, \mathbf{Q})$, i.e., $B_s^{-1} X(s, T) = \mathbf{E_Q} \left[B_t^{-1} X(t, T) | \mathcal{F}_s \right]$ for all $s \leq t$. Expression (10.3) is referred to as the risk-neutral valuation formula of a defaultable claim, see [83, 125] for its definition in usual probability spaces. For brevity, write $X_t = X(t, T)$ and combine (10.2) with (10.3), and, knowing that H is left-continuous we obtain

$$X_t = B_t \mathbf{E_Q} \left(\begin{array}{c} B_T^{-1} \tilde{X} + \int_{t+}^{T} B_{u-}^{-1} (1 - H_{u-}) dA_u \\ + \int_{t}^{T-} B_u^{-1} (1 - H_u) dA_{u+} + \int_{t}^{T-} B_u^{-1} R_u dH_{u+} \end{array} \middle| \mathcal{F}_t \right).$$

Remark 10.3.4 r.c.l.l Cash-Flow. We have considered a general model of defaultable cash-flow, namely that the components of the cash-flow are optional r.l.l. processes affected by default. However, we should not think that we are limited by this generalization of defaultable cash-flow. We can consider that the underlying components of defaultable cash-flow are r.c.l.l. processes affected by default, which induces a left-continuous optional jump on the resultant defaultable cash-flow, and the theory of defaultable markets on optional space remains the same.

From now on, we are going to work in the deflated probability space $\left(\Omega, \mathcal{F}, \mathbf{F} = (\mathcal{F}_t)_{t \geq 0}, \mathbf{Q} \right)$ and let the expectation operator $\mathbf{E}$ to mean $\mathbf{E_Q}$ – expected values under $\mathbf{Q}$. Next we provide a justification of Definition (10.3.3).

10.3.1 Portfolio with default

Consider a portfolio of 3 primary securities: $S^2 = B$ the value process of a money market account, S^1 a default-free non-dividend-paying assets, and S^0 a dividend paying asset, D with $D_0 = 0$, with the possibility of default. Introduce the discounted price processes $\tilde{S}^i$ by setting $\tilde{S}_t^i = S_t^i / S_t^2$. The market life-span is the time interval $[0, \infty]$ and $\phi = (\phi^0, \phi^1, \phi^2)$ is an **F**-optional self-financing trading strategy on (S^0, S^1, S^2). It is straightforward to generalize the 3 assets portfolio to any number of assets.

To begin with, let us examine a simple trading strategy: suppose that at time $t = 0$ we purchase one unit of the 0th asset at the initial price S_0^0 and hold it until time T, then invest all the proceeds from dividends in the money market account. More specifically, we consider a buy-and-hold strategy $\phi = (1, 0, \phi^2)$. The associated wealth process U equals

$$U_t = S_t^0 + \phi_t^2 B_t, \quad \forall t \geq 0, \tag{10.4}$$

with initial wealth $U_0 = S_0^0 + \phi_0^2 B_0$. Since, ϕ, is self-financing then,

$$U_t - U_0 = S_t^0 - S_0^0 + \phi^2 \circ B_t + D_t \tag{10.5}$$

and $B \circ \phi_t^2 + [\phi^2, B] = 0$, where D is the dividend paid by S^0. Now, let's divide $(U_t - S_t^0)$ by B, and use the product rule,

$$\begin{aligned} B_t^{-1}(U_t - S_t^0) &= B^{-1} \circ (U_t - S_t^0) + (U_t - S_t^0) \circ B_t^{-1} + [B^{-1}, (U - S^0)]_t \\ &= B^{-1} \circ (U_0 - S_0^0 + \phi^2 \circ B_t + D_t) + (\phi_t^2 B_t) \circ B_t^{-1} + [B^{-1}, \phi_t^2 B_t]_t \\ &= \phi^2 B^{-1} \circ B_t + B^{-1} \circ D_t + \phi^2 B \circ B_t^{-1} + [B^{-1}, \phi^2 B]_t, \end{aligned}$$

where we have used Equations 10.4 and 10.5. But since $B^{-1} \circ B + B \circ B^{-1} = B^{-2} \circ [B, B]$,

$$\begin{aligned} [B^{-1}, \phi^2 B] &= [B^{-1}, \phi^2 \circ B + B \circ \phi^2 + [\phi^2, B]] = \phi^2 \circ [B^{-1}, B] \\ &= \phi^2 \circ [-B^{-2} \circ B + B^{-3} \circ [B, B], B] = -B^{-2} \phi^2 \circ [B, B], \end{aligned}$$

then we are able to find the simple relation

$$B_t^{-1}(U_t - S_t^0) = \phi^2 B_t^{-1} \circ B_t + \phi^2 B \circ B_t^{-1} - \phi^2 B^{-2} \circ [B, B]_t + B^{-1} \circ D_t = B^{-1} \circ D_t.$$

Considering the difference between the value of a defaultable portfolio at time t, i.e., U_t, and its value at maturity T, U_T, we arrive at the following difference equation,

$$B_T^{-1}(U_T - S_T^0) - B_t^{-1}(U_t - S_t^0) = B^{-1} \circ D_T - B^{-1} \circ D_t,$$
$$B_T^{-1} U_T - B_t^{-1} U_t = B_T^{-1} S_T^0 - B_t^{-1} S_t^0 + B^{-1} \circ D_T - B^{-1} \circ D_t. \tag{10.6}$$

Now we are ready to derive the risk-neutral valuation formula for the ex-dividend price S_t^0. To this end, we have assumed that our model admits a local optional martingale measure **Q** equivalent to **P** such that the discounted wealth process $B^{-1} U^\phi$ of any admissible self-financing trading strategy ϕ follows local optional martingales under **Q** with respect to the filtration **F**. Moreover, we make an assumption that the market value at time t of the 0th security comes exclusively from the future dividends stream; this means that we have to postulate that $S_T^0 = 0$. This postulate makes sense, because the value of a defaultable cash-flow at maturity which has paid all its value in dividends is essentially nothing. We shall refer to S^0 as the ex-dividend price of the 0th asset – the defaultable claim. Given that,

$$\mathbf{E}\left(B_T^{-1}U_T - B_t^{-1}U_t|\mathcal{F}_t\right) = 0, \ \mathbf{E}\left(B_t^{-1}S_t^0|\mathcal{F}_t\right) = B_t^{-1}S_t^0, \ S_T^0 = 0,$$

and Equation (10.6), we arrive at the definition of the value of defaultable claim,

$$
\begin{aligned}
\mathbf{E}\left(B_T^{-1}U_T - B_t^{-1}U_t|\mathcal{F}_t\right) &= \mathbf{E}\left(B_T^{-1}S_T^0 - B_t^{-1}S_t^0 + B^{-1} \circ D_T - B^{-1} \circ D_t|\mathcal{F}_t\right) = 0 \Rightarrow \\
B_t^{-1}S_t^0 &= \mathbf{E}\left(B^{-1} \circ D_T - B^{-1} \circ D_t|\mathcal{F}_t\right).
\end{aligned}
$$

Hence, $B_t^{-1}S_t^0$ is an **F** local optional martingale under **Q** and

$$B_t^{-1}S_t^0 = \mathbf{E}\left(B^{-1} \circ D_T - B^{-1} \circ D_t|\mathcal{F}_t\right). \tag{10.7}$$

Let us now examine trading with a general self-financing trading strategy $\phi = (\phi^0, \phi^1, \phi^2)$. The associated wealth process is $U_t(\phi) = \sum_{i=0}^2 \phi_t^i S_t^i$. Since ϕ is self-financing then it must be that, $U_t(\phi) = U_0(\phi) + G_t(\phi)$ for every $t \geq 0$, where the gains process $G(\phi)$ is,

$$G_t(\phi) := \phi^0 \circ D_t + \sum_{i=0}^2 \phi^i \circ S_t^i = \phi^0 \circ \left(S_t^0 + D_t\right) + \phi^1 \circ S_t^1 + \phi^2 \circ S_t^2.$$

As before, $S_t^2 = B$ the value of our money market account, $S_t^0 = X(t,T)$ is the dividend, D, paying asset and S^1 is the default-free, non-dividend paying instrument. The term "$\phi^0 \circ \left(S_t^0 + D_t\right)$" of the gain process is the gain acquired as a result of trading ϕ^0 of the 0th asset having current value S_t^0 and paid dividend D_t.

Theorem 10.3.5 For any self-financing trading strategy, ϕ, and $\tilde{S}^i = B^{-1}S^i$, $i = 0, 1, 2$ are $(\mathbf{F}, \mathbf{Q})$ local optional martingales and the discounted wealth process $B_t^{-1}U_t(\phi)$ follows a local optional martingale under $(\mathbf{F}, \mathbf{Q})$.

Proof. Given $\tilde{S}^i = B^{-1}S^i = S^i \circ B^{-1} + B^{-1} \circ S^i + [B^{-1}, S^i]$, then, by product rule we get,

$$
\begin{aligned}
B_t^{-1}U_t(\phi) &= B_t^{-1} \circ U_t(\phi) + U_t(\phi) \circ B_t^{-1} + \left[B^{-1}, U(\phi)\right]_t \\
&= B^{-1}\phi^0 \circ D_t + \sum_{i=0}^2 B^{-1}\phi^i \circ S_t^i + \sum_{i=0}^2 \left(\phi^i S^i\right) \circ B_t^{-1} + \left[B^{-1}, \sum_{i=0}^2 \phi^i S_t^i\right] \\
&= B^{-1}\phi^0 \circ D_t + \sum_{i=0}^2 B^{-1}\phi^i \circ S_t^i + \sum_{i=0}^2 \phi^i S^i \circ B_t^{-1} + \sum_{i=0}^2 \phi^i \circ \left[B^{-1}, S^i\right]_t \\
&= \phi^0 B^{-1} \circ D_t + \sum_{i=0}^2 \phi^i B^{-1} \circ S_t^i + \phi^i S^i \circ B_t^{-1} + \phi^i \circ \left[B^{-1}, S^i\right]_t \\
&= \phi^0 B^{-1} \circ D_t + \sum_{i=0}^2 \phi^i \circ B_t^{-1}S_t^i = \phi^0 \circ \hat{S}_t^0 + \sum_{i=1}^2 \phi^i \circ \tilde{S}_t^i
\end{aligned}
$$

where the process $\hat{S}^0$ is given by $\hat{S}^0_t := \tilde{S}^0_t + B^{-1} \circ D_t$. The process $\hat{S}^0_t$ is the discounted cumulative dividend price at time t of the defaultable dividend D. The processes $\tilde{S}^i$, $i = 0, 1, 2$ are local optional martingales in $(\mathbf{F}, \mathbf{Q})$. To finalize this proof, observe that using Equation 10.7 the process $\hat{S}^0$ satisfies $\hat{S}^0_t = \mathbf{E}\left(B^{-1} \circ D_T | \mathcal{F}_t\right)$, and thus follows a local martingale under $\mathbf{Q}$. Hence $B^{-1}_t U_t(\phi)$ is a local optional martingale in $(\mathbf{F}, \mathbf{Q})$. ∎

10.4 Probability of Default

The essential component of the defaultable cash-flows, D, Equation (10.2) is the default process $H_u = \mathbf{1}_{(\tau < u)}$ where $u \in [0, \infty]$ is a time horizon. The ex-dividend price process Equation (10.3) relies implicitly on the conditional expected value of $\mathbf{1}_{(\tau < u)}$ given all known up to time t market information $\mathcal{F}_t$. So, our goal for this section is to understand the conditional probability of default $F(t, u) = \mathbf{E}[H_u | \mathcal{F}_t]$, also known as the *hazard process*, and its properties. Associated with the hazard process F is the *survival process* $G(t, u) := 1 - F(t, u)$, which we will discuss some of its properties too. This section is expository, where we will simply list a few interesting related results about F and G. We begin by the lemma:

Lemma 10.4.1 The process H_t is a submartingale.

Proof. Established by conditioning on $\mathcal{F}_s$ for any $s \le t$. ∎

Lemma 10.4.2 F and G are positive. And, for a fixed time horizon both are optional martingales. However, for a fixed time t and variable time horizon, F is a submartingale and G a supermartingale.

Proof. F and G are positive and one can write $F(t, u)$,

$$
\begin{aligned}
F(t, u) &= \mathbf{E}[H_u | \mathcal{F}_t] = \mathbf{Q}(\tau < u | \mathcal{F}_t) = \mathbf{1}_{(u>t)} \mathbf{E}\left[H_t + \mathbf{1}_{(t \le \tau < u)} | \mathcal{F}_t\right] + \mathbf{1}_{(u \le t)} H_u \\
&= \mathbf{1}_{(u>t)}\left(H_t + \mathbf{E}[(1 - H_t) H_u | \mathcal{F}_t]\right) + \mathbf{1}_{(u \le t)} H_u.
\end{aligned}
$$

For a fixed time horizon u and any $0 \le s \le t < u$, $F(t, u) = \mathbf{E}[\mathbf{E}[H_u | \mathcal{F}_t] | \mathcal{F}_s] = \mathbf{E}[H_u | \mathcal{F}_s] = F(s, u)$ is a positive optional martingale. Also, for $0 \le u \le s \le t$, $F(t, u) = F(s, u) = H_u$ a constant. But, for $0 \le s \le u \le t$, $F(t, u) = \mathbf{E}[\mathbf{E}[H_u | \mathcal{F}_t] | \mathcal{F}_s] = \mathbf{E}[H_u | \mathcal{F}_s] = F(s, u)$, again, an optional martingale. Essentially for any fixed u, $F(s, u)$ is an optional martingale. The same is true for G. For a fixed, t, and any time horizons, $0 \le v \le u$ where u and $v \in [0, \infty]$ then $(\tau < v) \subseteq (\tau < u)$ and

$$
F(t, u) = \mathbf{E}[H_u | \mathcal{F}_t] = \mathbf{E}[H_v | \mathcal{F}_t] + \mathbf{E}\left[\mathbf{1}_{(v \le \tau < u)} | \mathcal{F}_t\right] = F(t, v) + \mathbf{E}\left[\mathbf{1}_{(v \le \tau < u)} | \mathcal{F}_t\right] \ge F(t, v).
$$

Hence, $F(t, u)$ is increasing along its time horizon. Moreover, for a fixed t and any $v \leq u$,

$$
\begin{aligned}
\mathbf{E}(F(t, u)|\mathcal{F}_v) &= \mathbf{E}(\mathbf{E}(H_u|\mathcal{F}_t)|\mathcal{F}_v) \\
&= \mathbf{1}_{(v \geq t)}\mathbf{E}(\mathbf{E}(H_v + \mathbf{1}_{(v \leq \tau < u)}|\mathcal{F}_t)|\mathcal{F}_v) + \mathbf{1}_{(v < t)}\mathbf{E}(\mathbf{E}(H_t + \mathbf{1}_{(t \leq \tau < u)}|\mathcal{F}_t)|\mathcal{F}_v) \\
&= \mathbf{1}_{(v \geq t)}\mathbf{E}(H_v + \mathbf{1}_{(v \leq \tau < u)}|\mathcal{F}_t) + \mathbf{1}_{(v < t)}\mathbf{E}(H_t + \mathbf{1}_{(t \leq \tau < u)}|\mathcal{F}_v) \\
&\geq \mathbf{1}_{(v \geq t)}F(t, v) + \mathbf{1}_{(v < t)}F(v, t).
\end{aligned}
$$

Therefore, if $v \geq t$ then $F(t, u)$ is a submartingale on the other hand if $v < t$ then we re-label v as t and t as v to get $\mathbf{E}(F(v, u)|\mathcal{F}_t) \geq F(v, t)$ which is also a submartingale. Hence F is a submartingale in either case. It follows that $G = 1 - F$ is a supermartingale. ∎

On a related note, [54] showed that for the process $\xi \mathbf{1}_{(\tau < t)}$, where ξ is $\mathcal{F}_{\tau+}$-measurable and integrable r.v. and τ is a stopping time in the broad sense, can be decomposed to an optional martingale that is $\mathcal{F}_t$-measurable and an $\mathcal{F}_t$-measurable continuous finite variation process. Therefore,

Corollary 10.4.3 The default process H_t has the following decomposition, $H_t = \mathbf{1}_{(\tau < t)} = M_t + \mu_t$, where $M \in \mathcal{M}(\mathbf{F}, \mathbf{Q})$ is an optional martingale and μ is the **F**-compensator of H which is **F**-measurable continuous finite variation process.

Proof. A consequence of theorem 3.5 in [54]. ∎

Using the above corollary, let's consider evaluating $\mathbf{E}[H_T|\mathcal{F}_t]$. For a time horizon $T > t$,

Corollary 10.4.4 $F(t, T) = H_t + \mathbf{E}[\mu_T - \mu_t|\mathcal{F}_t]$ and $\mathbf{Q}(t \leq \tau < T|\mathcal{F}_t) = \mathbf{E}[\mu_T|\mathcal{F}_t] - \mu_t$.

Proof. $F(t, T) = \mathbf{E}[\mathbf{1}_{(\tau < T)}|\mathcal{F}_t]$; therefore,

$$
F(t, T) = \mathbf{E}[\mathbf{1}_{(\tau < T)}|\mathcal{F}_t] = \mathbf{E}[M_T + \mu_T|\mathcal{F}_t] = M_t + \mathbf{E}[\mu_T|\mathcal{F}_t] = H_t + \mathbf{E}[\mu_T - \mu_t|\mathcal{F}_t].
$$

Consequently one can write,

$$
\begin{aligned}
\mathbf{Q}(t \leq \tau < T|\mathcal{F}_t) &= [(1 - H_t)H_T|\mathcal{F}_t] = \mathbf{E}[H_T - H_t|\mathcal{F}_t] = \mathbf{E}[H_T|\mathcal{F}_t] - H_t \\
&= \mathbf{E}[\mu_T - \mu_t|\mathcal{F}_t] = \mathbf{E}[\mu_T|\mathcal{F}_t] - \mu_t.
\end{aligned}
$$

∎

Let us now understand the relation between the process $\mathbf{1}_{(\tau \leq t)}$, for a stopping time in the broad sense τ, and the default process $H_t = \mathbf{1}_{(\tau < t)}$ as regards the filtration **F**.

Remark 10.4.5 Recall that under the usual conditions, $\mathbf{1}_{(\tau \leq t)}$ is known as the default process; however, in the *unusual* conditions, it is replaced with the $\mathcal{O}(\mathbf{F})$ process $H_t = \mathbf{1}_{(\tau < t)}$. This choice of $\mathbf{1}_{(\tau < t)}$ as the default process was motivated by the fact that the known market information up to time t is $\mathcal{F}_t$ on which $\mathbf{1}_{(\tau < t)}$ is adapted and optional while $\mathbf{1}_{(\tau \leq t)}$ is not. Furthermore, the forward derivative $\triangle^+ H_t = \mathbf{1}_{(\tau = t)}$ marks the occurrence of default, which is in $\mathcal{O}(\mathbf{F}_+)$.

Proposition 10.4.6 Let τ be a stopping time in the broad sense and $\bar{H}_t = H_{t+} = \mathbf{1}_{(\tau \leq t)}$, where $(\tau \leq t) = (\tau < t+)$, is r.c.l.l. and $\mathcal{F}_{t+}$ measurable or $\mathcal{O}(\mathbf{F}_+)$. Also, $\bar{H}_{t-} = H_t$ or $(\tau \leq t-) = (\tau < t)$. $\bar{H}_{t-}$ and H_t are of right-limits and left-continuous paths and are $\mathcal{O}(\mathbf{F})$. Moreover, $(\tau \leq t) = (\tau < t) \cup (\tau = t)$ or that $\bar{H}_t = H_t + \mathbf{1}_{(\tau = t)}$.

Proof. For every $\omega \in (\tau \leq t)$, $\tau(\omega) \leq t < t+$ so $(\tau \leq t) \subseteq (\tau < t+)$. On the other hand, $(\tau(\omega) < t+) = \cap_{\epsilon \geq 0} (\tau < t + \epsilon) \subseteq (\tau \leq t)$. So, $H_{t+} = \mathbf{1}_{(\tau < t+)} = \mathbf{1}_{(\tau \leq t)}$ or that $(\tau \leq t) = (\tau < t+)$. By definition $\mathbf{1}_{(\tau \leq t)}$ is r.c.l.l. and $\mathcal{O}(\mathbf{F}_+)$. Furthermore, it can be easily seen that the process $\mathbf{1}_{(\tau \leq t)}$ can be decomposed as, $\mathbf{1}_{\tau \in [0,t]} = \mathbf{1}_{\tau \in [0,t)} + \mathbf{1}_{(\tau = t)}$. Also, it is obvious that $\bar{H}_{t-} = \mathbf{1}_{(\tau \leq t-)} = \mathbf{1}_{(\tau < t)} = H_t$. Finally, for τ a stopping time in the broad sense, by definition, $(\tau \leq t) \in \mathcal{F}_{t+}$ for any time t. Moreover, since $(\tau \leq t - 1/n) \in \mathcal{F}_{(t-1/n)+} \subseteq \mathcal{F}_t$ for all n then it must be that $(\tau < t) \in \mathcal{F}_t$. ∎

Let's study the properties of the process $F(t, t+) = \mathbf{E}(H_{t+}|\mathcal{F}_t) = \mathbf{Q}(\tau \leq t|\mathcal{F}_t)$. Set $\bar{F}_t := \mathbf{E}(H_{t+}|\mathcal{F}_t)$ and consider the following lemma:

Lemma 10.4.7 The ex-hazard process $\bar{F}$ can be decomposed to the default process H and the jump hazard process δ. Both $\bar{F}$ and δ are $\mathcal{O}(\mathbf{F}_+)$, whereas H is $\mathcal{O}(\mathbf{F})$.

Proof. The optional projection $\mathbf{E}(H_{t+}|\mathcal{F}_t)$ can be reduced to,

$$\mathbf{E}(H_{t+}|\mathcal{F}_t) = \mathbf{E}(\mathbf{1}_{(\tau \leq t)}|\mathcal{F}_t) = \mathbf{E}(\mathbf{1}_{(\tau < t)} + \mathbf{1}_{(\tau = t)}|\mathcal{F}_t) = H_t + \mathbf{E}(\mathbf{1}_{(\tau = t)}) = H_t + \delta_t,$$

where δ_t can be derived by formula, $\delta_t = \lim_{h \to 0} \frac{1}{h} \mathbf{Q}(t \leq \tau < t + h|\mathcal{F}_t)$. ∎

Corollary 10.4.8 $F(t, T+) = F(t, T) + \mathbf{E}(\delta_T|\mathcal{F}_t) = M_t + \mathbf{E}(\mu_T + \delta_T|\mathcal{F}_t)$.

Proof. The process $F(t, T+) = \mathbf{E}(H_{T+}|\mathcal{F}_t)$ is evaluated as,

$$F(t, T+) = \mathbf{E}(H_T|\mathcal{F}_t) + \mathbf{E}(\mathbf{1}_{\tau = T}|\mathcal{F}_t) = H_t + \mathbf{E}[\mu_T - \mu_t|\mathcal{F}_t] + \mathbf{E}(\delta_T|\mathcal{F}_t) = F(t, T) + \mathbf{E}(\delta_T|\mathcal{F}_t)$$

or $F(t, T+) = H_t - \mu_t + \mathbf{E}(\mu_T + \delta_T|\mathcal{F}_t) = M_t + \mathbf{E}(\mu_T + \delta_T|\mathcal{F}_t)$. ∎

10.5 Valuation of Defaultable Cash-Flow and Examples

Our next goal is to establish a convenient representation of the value of a defaultable cash-flow. The *ex-dividend* value of defaultable cash-flow is, $B_t^{-1} X(t, T) = \mathbf{1}_{(\tau \geq t)} \mathbf{E}(B^{-1} \circ D_T - B^{-1} \circ D_t|\mathcal{F}_t)$

and in terms of $D_t = (\Lambda (1 - H_T) + \rho H_T) \mathbf{1}_{(t \geq T)} + (1 - H) \circ A_t + R \circ H_t$ is

$$
\begin{aligned}
B_t^{-1} X_t &= \mathbf{E}\left[B_T^{-1} (\Lambda (1 - H_T) + \rho H_T) | \mathcal{F}_t \right] \\
&\quad + \mathbf{E}\left[\int_{t+}^{T} B_{u-}^{-1}(1 - H_{u-})dA_u + \int_{t}^{T-} B_u^{-1}(1 - H_u)dA_{u+} | \mathcal{F}_t \right] \\
&\quad + \mathbf{E}\left[\int_{t}^{T-} B_u^{-1} R_u dH_{u+} | \mathcal{F}_t \right].
\end{aligned}
$$

We begin valuation of $X(t,T)$ with the value of $\mathbf{E}\left[B_T^{-1} \Lambda (1 - H_T) | \mathcal{F}_t \right]$. Let $\lambda_T = B_T^{-1}\Lambda$ and $\lambda_t = \mathbf{E}(\lambda_T | \mathcal{F}_t)$ is a martingale. λ_t can be thought of as the default-free contingent claim price at time t. Let $G(t,u) = \mathbf{E}(1 - H_u | \mathcal{F}_t)$. G can be thought of as the survival process, whereas $F(t,u) = \mathbf{E}(H_u | \mathcal{F}_t)$ is the associated default process on the interval $[t, u]$.

Lemma 10.5.1 The value of $\tilde{\Lambda}_t = \mathbf{E}\left[B_T^{-1}\Lambda (1 - H_T) | \mathcal{F}_t \right]$ at time t is given by

$$
\tilde{\Lambda}_t = \mathbf{E}\left(B_T^{-1}\Lambda | \mathcal{F}_t \right) (1 - H_t) + \mathbf{E}\left[\int_{t}^{T-} \left(\lambda_u + \triangle^+ \lambda_u \right) dG_{u+} | \mathcal{F}_t \right]. \tag{10.8}
$$

where $\lambda_u = \mathbf{E}\left[B_T^{-1}\Lambda | \mathcal{F}_u \right]$ and $G_{u+} = G(u, u+) = \mathbf{E}(1 - H_{u+} | \mathcal{F}_u)$.

Proof. Using the product rule on $\lambda_T (1 - H_T)$ we find that,

$$
\begin{aligned}
\lambda_T (1 - H_T) &= \lambda_0 (1 - H_0) + \lambda \circ (1 - H_T) + G \circ \lambda_T + [\lambda, (1 - H)]_T \\
&= \lambda_0 (1 - H_0) + \lambda \circ (1 - H_t) + G \circ \lambda_t + [\lambda, (1 - H)]_t \\
&\quad + \int_{t+}^{T} \lambda_{u-} d(1 - H)_u + \int_{t}^{T-} \lambda_u d(1 - H)_{u+} \\
&\quad + \int_{t+}^{T} (1 - H)_{u-} \, d\lambda_u + \int_{t}^{T-} (1 - H)_u \, d\lambda_{u+} + \int_{t}^{T-} \triangle^+ \lambda_u d(1 - H)_{u+} \\
&= \lambda_t (1 - H)_t + \int_{t}^{T} \lambda_u d(1 - H)_u + \int_{t}^{T} (1 - H)_u \, d\lambda_u + \int_{t}^{T-} \triangle^+ \lambda_u d(1 - H)_{u+}.
\end{aligned}
$$

Since H is a left-continuous finite variation process, the quadratic variation,

$$
[\lambda, H]_t = \sum_{0 \leq u < t} \triangle^+ \lambda_u \, \triangle^+ H_{u+} = \int_{0}^{t-} \triangle^+ H_u d\lambda_{u+} = \int_{0}^{t-} \triangle^+ \lambda_u dH_{u+}.
$$

We have chosen the definition $[\lambda, H] = \triangle^+ \lambda \odot H$. Thus, the conditional expected value of $\lambda_T (1 - H_T)$ is $\mathbf{E}(\lambda_T (1 - H_T) | \mathcal{F}_t) = \mathbf{E}(B_T^{-1}\Lambda (1 - H_T) | \mathcal{F}_t)$ and

$$
\begin{aligned}
\mathbf{E}(\lambda_T (1 - H_T) | \mathcal{F}_t) &= \mathbf{E}\left[\begin{array}{c} \lambda_t (1 - H_t) + \int_{t}^{T} (1 - H_u) \, d\lambda_u + \int_{t}^{T-} \lambda_u d(1 - H_{u+}) \\ + \int_{t}^{T-} \triangle^+ \lambda_u d(1 - H_{u+}) \end{array} \middle| \mathcal{F}_t \right] \\
&= \lambda_t (1 - H_t) + \mathbf{E}\left[\int_{t}^{T-} \left(\lambda_u + \triangle^+ \lambda_u \right) dG_{u+} | \mathcal{F}_t \right],
\end{aligned}
$$

where we have used the fact that $\mathbf{E}\left[\int_t^T G_u d\lambda_u | \mathcal{F}_t\right] = 0$ since λ is a local martingale, $G_{u+} = G(u, u+)$ and $\mathbf{E}\left(\lambda_t\left(1 - H_t\right) | \mathcal{F}_t\right) = \lambda_t\left(1 - H_t\right)$. So, we arrive at the result,

$$
\begin{aligned}
\mathbf{E}\left[B_T^{-1}\Lambda\left(1 - H_T\right) | \mathcal{F}_t\right] &= \mathbf{E}\left(B_T^{-1}\Lambda | \mathcal{F}_t\right)\left(1 - H_t\right) \\
&\quad + \mathbf{E}\left[\int_t^{T-}\left[\mathbf{E}\left(B_T^{-1}\Lambda | \mathcal{F}_u\right) + \triangle^+\mathbf{E}\left(B_T^{-1}\Lambda | \mathcal{F}_u\right)\right] dG_{u+} | \mathcal{F}_t\right].
\end{aligned}
$$

∎

Remark 10.5.2 We have replaced the integral over $1 - H$ with an integral over the process G as a result of the following statement,

$$
\begin{aligned}
\mathbf{E}\left[\int_t^{T-}\alpha_u dH_{u+} | \mathcal{F}_t\right] &\leftarrow \sum_{i=0}^n \mathbf{E}\left[\alpha_{t_i}\mathbf{E}\left(H_{t_{i+1}\wedge T} - H_{t_i\wedge T} | \mathcal{F}_{t_i}\right) | \mathcal{F}_t\right] \\
&= \sum_{i=0}^n \mathbf{E}\left[\alpha_{t_i}F_{t_{i+1}\wedge T} - F_{t_i\wedge T} | \mathcal{F}_t\right] \\
&= \mathbf{E}\left[\left(\sum_{i=0}^n \alpha_{t_i}F_{t_{i+1}\wedge T} - F_{t_i\wedge T}\right) | \mathcal{F}_t\right],
\end{aligned}
$$

where $F_{u+} = F(u, u+) = \mathbf{E}\left(H_{u+} | \mathcal{F}_u\right)$.

Remark 10.5.3 How do we evaluate dG_{u+}? Here is how: $G(u, u+) = \mathbf{E}\left(1 - H_{u+} | \mathcal{F}_u\right)$ and $dG_{u+} = \triangle^+ G_u = G(u, u+) - G(u, u)$. Therefore,

$$
dG_{u+} = \mathbf{E}\left(1 - H_{u+} | \mathcal{F}_u\right) - \mathbf{E}\left(1 - H_u | \mathcal{F}_u\right) = \mathbf{E}\left(H_u - H_{u+} | \mathcal{F}_u\right),
$$

where $H_{u+} = \mathbf{1}_{(\tau \leq u)} = H_u + \mathbf{1}_{(\tau=u)}$. Hence $dG_{u+} = -\mathbf{E}\left(\mathbf{1}_{(\tau=u)} | \mathcal{F}_u\right) = -\mathbf{Q}(\tau = u | \mathcal{F}_u)$. Note that the event $(\tau = u)$ is $\mathcal{F}_{u+}$ measurable so dG_{u+} is the projection of $(\tau = u)$ on $\mathcal{F}_u$.

For the second term, $\mathbf{E}\left(B_T^{-1}\rho H_T | \mathcal{F}_t\right)$: Let $\varrho_T = B_T^{-1}\rho$ thus $\varrho_t = \mathbf{E}\left(\varrho_T | \mathcal{F}_t\right)$ is a local martingale. And, we have the following lemma:

Lemma 10.5.4 The value $\tilde{\rho}_t = \mathbf{E}\left(\varrho_T H_T | \mathcal{F}_t\right)$ is given by

$$
\tilde{\rho}_t = \varrho_t H_t - \mathbf{E}\left[\int_t^{T-}\left(\varrho_u + \triangle^+\varrho_u\right) dG_{u+} | \mathcal{F}_t\right]
$$

where $\varrho_u = \mathbf{E}\left(B_T^{-1}\rho | \mathcal{F}_u\right)$.

Proof. Applying a similar argument to the one used in the above lemma, we arrive at the result

$$
\begin{aligned}
\tilde{\rho}_t &= \mathbf{E}\left(\varrho_T H_T | \mathcal{F}_t\right) = \mathbf{E}\left[\varrho_t H_t + \int_t^T H_u d\varrho_u + \int_t^T \varrho_u dH_u + \int_t^{T-} \triangle^+ \varrho_u dH_{u+} | \mathcal{F}_t\right] \\
&= \varrho_t H_t + \mathbf{E}\left[\int_t^{T-} \left(\varrho_u + \triangle^+ \varrho_u\right) dF_{u+} | \mathcal{F}_t\right] \\
&= \varrho_t H_t - \mathbf{E}\left[\int_t^{T-} \left(\varrho_u + \triangle^+ \varrho_u\right) dG_{u+} | \mathcal{F}_t\right].
\end{aligned}
$$

where $\mathbf{E}\left[\int_t^T H_u d\varrho_u | \mathcal{F}_t\right] = 0$ since ϱ is a local martingale. $\blacksquare$

As for the value of the defaultable cash-flow recovery-stream we have the following lemma:

Lemma 10.5.5 The value of the defaultable cash-flow recovery stream is

$$
\mathbf{E}\left[\int_t^{T-} B_u^{-1} R_u dH_{u+} | \mathcal{F}_t\right] = -\mathbf{E}\left[\int_t^{T-} B_u^{-1} R_u dG_{u+} | \mathcal{F}_t\right]
$$

Proof. Using the sum approximation of the left optional integral, the ex-dividend value of defaultable claim is derived as follows

$$
\begin{aligned}
\mathbf{E}\left[\int_t^{T-} B_u^{-1} R_u dH_{u+} | \mathcal{F}_t\right] &= \mathbf{E}\left[\int_t^{T-} \mathbf{E}\left[B_u^{-1} R_u dH_{u+} | \mathcal{F}_u\right] | \mathcal{F}_t\right] \\
&= \mathbf{E}\left[\int_t^{T-} B_u^{-1} R_u dF_{u+} | \mathcal{F}_t\right] \\
&= -\mathbf{E}\left[\int_t^{T-} B_u^{-1} R_u dG_{u+} | \mathcal{F}_t\right].
\end{aligned}
$$

Note that $B_u^{-1} R_u$ is $\mathcal{F}_u$ measurable. $\blacksquare$

Finally we look at the process $\mathbf{E}\left[B^{-1}(1-H) \circ A_T - B^{-1}(1-H) \circ A_t | \mathcal{F}_t\right]$. Let $\hat{A}_t = B^{-1} \circ A_t$ then

$$
\begin{aligned}
\mathbf{E}\left[B^{-1}(1-H) \circ A_T - B^{-1}(1-H) \circ A_t | \mathcal{F}_t\right] &= \mathbf{E}\left[(1-H) \circ \hat{A}_T - (1-H) \circ \hat{A}_t | \mathcal{F}_t\right] \\
&= \mathbf{E}\left[\int_t^T (1-H_u) d\hat{A}_u | \mathcal{F}_t\right].
\end{aligned}
$$

Now, let's apply the product rule to $(1-H_t)\hat{A}_t$ from $[t, T]$,

$$
(1-H_T)\hat{A}_T = (1-H_t)\hat{A}_t + \int_t^T \hat{A} d(1-H)_u + \int_t^T (1-H) d\hat{A}_u + \int_t^T \triangle^+ \hat{A}_u d(1-H)_{u+}.
$$

Using the above equation we can write

$$\int_t^T (1-H)d\hat{A}_u = (1-H_T)\hat{A}_T - (1-H_t)\hat{A}_t - \int_t^{T-} \hat{A}_u d(1-H)_u - \int_t^{T-} \triangle^+ \hat{A}_u d(1-H)_{u+}.$$

Without loss of generality, we are going to assume that $A_T = 0$ because if there is any dividend payment made at time T, it is going to be part of the claim Λ paid at maturity time T. Hence the equation above reduces to,

$$\int_t^T (1-H)d\hat{A}_u = -(1-H_t)\hat{A}_t - \int_t^{T-} \hat{A}_u d(1-H)_{u+} - \int_t^{T-} \triangle^+ \hat{A}_u d(1-H)_{u+}. \qquad (10.9)$$

We are going to use this equation to prove the next lemma.

Lemma 10.5.6 The value of dividend payout stream is

$$\mathbf{E}\left[B^{-1}(1-H) \circ A_T - B^{-1}(1-H) \circ A_t | \mathcal{F}_t\right] = -(1-H_t)\left(B^{-1} \circ A_t\right)$$
$$-\mathbf{E}\left[\int_t^{T-} \left((B^{-1} \circ A_u) + \triangle^+ \left(B^{-1} \circ A_u\right)\right) dG_{u+} | \mathcal{F}_t\right].$$

Proof. Using Equation (10.9), the sum approximation of the left optional integral and the fact that $(1-H_t)\left(B^{-1} \circ A_t\right)$ is $\mathcal{F}_t$ measurable we get,

$$\mathbf{E}\left[B^{-1}(1-H) \circ A_T - B^{-1}(1-H) \circ A_t | \mathcal{F}_t\right] = \mathbf{E}\left[\int_t^T (1-H)d\hat{A}_u | \mathcal{F}_t\right]$$

$$= \mathbf{E}\left[-(1-H_t)\hat{A}_t - \int_t^{T-} \hat{A}_u d(1-H)_{u+} | \mathcal{F}_t\right]$$
$$-\mathbf{E}\left[\int_t^{T-} \triangle^+ \hat{A}_u d(1-H)_{u+} | \mathcal{F}_t\right]$$
$$= -(1-H_t)\hat{A}_t$$
$$-\mathbf{E}\left[\int_t^{T-} \left(\hat{A}_u + \triangle^+ \hat{A}_u\right) d(1-H)_{u+} | \mathcal{F}_t\right]$$
$$= -(1-H_t)\left(B^{-1} \circ A_t\right)$$
$$-\mathbf{E}\left[\int_t^{T-} \left((B^{-1} \circ A_u) + \triangle^+ \left(B^{-1} \circ A_u\right)\right) dG_{u+} | \mathcal{F}_t\right].$$

where $\hat{A} = B^{-1} \circ A$. ∎

Consequently, we arrive at the following theorem for the ex-dividend price process:

Theorem 10.5.7 The *ex-dividend price process* $X(\cdot, T)$ that settles at time T is

$$
\begin{aligned}
B_t^{-1} X(t, T) &= \lambda_t (1 - H_t) + \mathbf{E}\left[\int_t^{T-} \left(\lambda_u + \triangle^+ \lambda_u \right) dG_{u+} | \mathcal{F}_t \right] \\
&\quad + \varrho_t H_t - \mathbf{E}\left[\int_t^{T-} \left(\varrho_u + \triangle^+ \varrho_u \right) dG_{u+} | \mathcal{F}_t \right] \\
&\quad - (1 - H_t) \hat{A}_t - \mathbf{E}\left[\int_t^{T-} \left(\hat{A}_u + \triangle^+ \hat{A}_u \right) dG_{u+} | \mathcal{F}_t \right] \\
&\quad - \mathbf{E}\left[\int_t^{T-} \hat{R}_u dG_{u+} | \mathcal{F}_t \right] \\
&= \left(\lambda_t - \hat{A}_t \right) (1 - H_t) + \varrho_t H_t \\
&\quad + \mathbf{E}\left[\int_t^{T-} \left(\lambda_u - \varrho_u - \hat{A}_u - \hat{R}_u + \triangle^+ \left(\lambda_u - \varrho_u - \hat{A}_u \right) \right) dG_{u+} | \mathcal{F}_t \right]
\end{aligned}
$$

where $\lambda_t = \mathbf{E}\left(B_T^{-1} \Lambda | \mathcal{F}_t \right)$, $\varrho_t = \mathbf{E}\left(B_T^{-1} \rho | \mathcal{F}_t \right)$, $\hat{A}_t = B^{-1} \circ A_t$ and $\hat{R}_t = B_t^{-1} R_t$.

Proof. Follows from the lemmas above. ∎

In the special case where components of the dividends process are r.c.l.l. processes, we get the following corollary.

Corollary 10.5.8 Suppose we are given a defaultable cash-flow A, R and B are r.c.l.l. Then, the *ex-dividend price process* $X(\cdot, T)$ that settles at time T is,

$$
\begin{aligned}
X(t, T) &= B_t \left(\lambda_t - \hat{A}_t \right) (1 - H_t) + B_t \varrho_t H_t \\
&\quad + B_t \mathbf{E}\left[\int_t^{T-} \left(\lambda_u - \varrho_u - \hat{A}_u - \hat{R}_u + \triangle^+ (\lambda_u - \varrho_u) \right) dG_{u+} | \mathcal{F}_t \right]
\end{aligned}
$$

Furthermore, if $\lambda_t = \mathbf{E}\left(B_T^{-1} \Lambda | \mathcal{F}_t \right)$ and $\varrho_t = \mathbf{E}\left(B_T^{-1} \rho | \mathcal{F}_t \right)$ have r.c.l.l. modifications, then $\triangle^+ \mathbf{E}\left(B_T^{-1} \Lambda | \mathcal{F}_t \right) = 0$ and $\triangle^+ \mathbf{E}\left(B_T^{-1} \rho | \mathcal{F}_t \right) = 0$ for all t and

$$
\begin{aligned}
X(t, T) &= B_t \left(\lambda_t - \hat{A}_t \right) (1 - H_t) + B_t \varrho_t H_t \\
&\quad + B_t \mathbf{E}\left[\int_t^{T-} \left(\lambda_u - \varrho_u - \hat{A}_u - \hat{R}_u \right) dG_{u+} | \mathcal{F}_t \right]
\end{aligned}
$$

Proof. Follows from the definition of the ex-dividend price process. ∎

We were able to establish a convenient representation of the value of a defaultable claim in terms of the probability of default. Now we provide some examples.

Example 10.5.9 Zero-Coupon Defaultable Bond. The price of a zero-coupon bond with face-value \$1 at maturity date T is $B_t \mathbf{E}\left(B_T^{-1}|\mathcal{F}_t\right)$ at time t. On the other hand, the price of a zero-coupon bond that may experience default is $B_t \mathbf{E}\left(B_T^{-1}\mathbf{1}_{(\tau \geq T)}|\mathcal{F}_t\right)$. Lemma 10.5.1 tells us how to compute the price $B_t \mathbf{E}\left(B_T^{-1}\mathbf{1}_{(\tau \geq T)}|\mathcal{F}_t\right)$ at time t. Let $\lambda_T = B_T^{-1}$ thus $\lambda_t = \mathbf{E}\left(\lambda_T|\mathcal{F}_t\right)$ and

$$
\begin{aligned}
B_t^{-1}X(t,T) &= \mathbf{E}\left(\lambda_T\left(1 - H_T\right)|\mathcal{F}_t\right) \\
&= \lambda_t\left(1 - H_t\right) + \mathbf{E}\left[\int_t^{T-}\left(\lambda_u + \triangle^+\lambda_u\right)dG_{u+}|\mathcal{F}_t\right] \\
&= \lambda_t\mathbf{1}_{(\tau \geq t)} + \mathbf{E}\left[\int_t^{T-}\left(\lambda_u + \triangle^+\lambda_u\right)dG_{u+}|\mathcal{F}_t\right].
\end{aligned}
$$

Suppose that $B_t = e^{rt}$, hence $B(t,T) = e^{-r(T-t)} = B_t B_T^{-1}$, with a constant interest rate r and that the survival process admits a constant intensity γ such that

$$
\begin{aligned}
dG_{u+} &= \mathbf{E}\left(H_u - H_{u+}|\mathcal{F}_u\right) = -\mathbf{E}\left(\mathbf{1}_{(\tau=u)}|\mathcal{F}_u\right) \\
&= -\delta\left(u - \tau\right)\gamma e^{-\gamma u}du,
\end{aligned}
$$

where $\delta\left(u - \tau\right)$ is Dirac delta function at a *particular value* of default time τ. Then,

$$
\begin{aligned}
X(t,T) &= \mathbf{1}_{(\tau \geq t)}e^{-r(T-t)} - e^{-r(T-t)}\mathbf{E}\left[\int_t^T \delta\left(u - \tau\right)\gamma e^{-\gamma u}du|\mathcal{F}_t\right] \\
&= e^{-r(T-t)}\left[\mathbf{1}_{(\tau \geq t)} - \mathbf{1}_{(t \leq \tau < T)}e^{-\gamma t}\left(1 - e^{-\gamma(T-t)}\right)\right].
\end{aligned}
$$

At $t = 0$, $X(0,T) = e^{-rT}\left[1 - \mathbf{1}_{(0 \leq \tau < T)}\left(1 - e^{-\gamma T}\right)\right]$ and at $t = T$, $X(T,T) = \mathbf{1}_{(\tau \geq T)}$. If $\tau < T$ then

$$
X(0,T) = e^{-(r+\gamma)T}. \tag{10.10}
$$

Therefore, for a bond of value \$1 at maturity date T, default decreases the present value of the bond by a factor $e^{-\gamma T}$.

In practice, it may be of interest to know how to transform a defaultable process, Y, to a local optional martingale with a strictly positive local optional martingale deflator, Z. Here, we will demonstrate with a simple example a method to derive Z.

Example 10.5.10 Consider $Y = X + H(X - x)$ as in Equation 10.1. But suppose that X and x are continuous local martingales and that $[H,X] = [H,x] = 0$ (see Remark 10.5.11). We are going to find a strictly positive local optional martingale deflator Z for Y. Given that, $H = M + \mu$ and knowing that the sum of local optional martingales, is a local optional martingale and that the products of Z with continuous local martingales X and $(X - x)$ are local optional martingales then it suffices to consider finding Z such that ZH is a local optional martingale. To do so, we choose $Z = \mathcal{E}(\theta \circ M) > 0$ and find the appropriate conditions that θ must satisfy so that ZH is a local martingale. We do this as follows,

$$\mathcal{E}(\theta \circ M) H = H\mathcal{E}(\theta \circ M) \theta \circ M + \mathcal{E}(\theta \circ M) \circ M + \mathcal{E}(\theta \circ M) \circ \mu + \theta\mathcal{E}(\theta \circ M) \circ [M, M] \Rightarrow$$

$$\mathcal{E}(\theta \circ M) \circ \mu + \theta\mathcal{E}(\theta \circ M) \circ [M, M] = 0 \Rightarrow$$

$$\theta = -\frac{d\mu}{d[M, M]}.$$

With $\theta = -\frac{d\mu}{d[M,M]}$ and $(1 + \theta \bigtriangleup M)(1 + \theta \bigtriangleup^+ M) > 0$, one can easily see that the process ZY is a local optional martingale.

Remark 10.5.11 Implicit in the definition HY is the fact that X and x are the default-free evolution of the components of the asset Y over its life and H is the default process. This allows us to suppose that H and (X, x) are orthogonal. In reality, X and x can't be observed in isolation of H. In defaultable markets we observe the process $X + H(X - x)$. The above calculation of local martingale deflator assumes knowledge of x, X and H and that they are related in a particular way. This is done to prove that local optional martingale deflators can be constructed for defaultable asset Y.

Chapter 11

Filtering of Optional Semimartingales

This chapter is devoted to the study of the filtering theory of semimartingales. First we will review results on the filtering theory of cadlag semimartingales under the usual conditions. In the next part of the chapter, we show how to apply uniform Doob-Meyer decomposition to the *filtering problem* of optional ladlag semimartingales without the usual conditions.

Up to now, in filtering theory of cadlag semimartingales under the usual conditions only martingale representations were exploited for the construction of optimal filters. Here we show how uniform Doob-Meyer decomposition (UDMD) of optional supermartingale can be used instead of the traditionally used martingale representations to obtain optimal filters for optional processes.

Finally, in Section 11.4, we apply UDMD and filtering of optional supermartingales to the problem of pricing and hedging of contingent claims in *incomplete* markets on *un*usual probability space where the flow of information is also *incomplete*. Other theoretical findings are supported by relevant examples as well.

11.1   The Filtering Problem

Let $(X, Y) = (X_t, Y_t)_{t \geq 0}$ be a partially observable stochastic process where Y is the unobservable component and X is the observed process. The filtering problem is understood as constructing for each $t \geq 0$ an optimal estimate of Y_t by observations of X_t.

On the *usual* stochastic basis, Hadjiev [77] was the first to study the filtering of semimartingales, Y, by observations of a point process, X, by means of dual projection and martingale representation theorems. Using the same approach, Vetrov [79] investigated the filtering problem when both X and Y are semimartingales under some conditions. The most important condition among the required conditions is that the probability measure $\mathbf{P}^X$ restricted to the filtration generated by X (i.e., $\mathbf{F}^X$) is unique. It turns out that the uniqueness of $\mathbf{P}^X$ is equivalent to the filtration $\mathbf{F}^X$ being right-continuous and complete (see [2, 51]).

11.2   The Usual Case of Optimal Filtering

Here we summarize results pertaining to filtering theory of cadlag semimartingale on usual probability space. We will begin by reviewing some definitions and results under the usual conditions about cadlag semimartingales relevant to the filtering problems. The reader can find much of this material in chapters 1-8 in [51].

11.2.1   Auxiliary results

Let $\mathfrak{C}$ and $\mathfrak{D}$ denote the spaces of real-valued functions on $[0, \infty]$ that are continuous, respectively r.c.l.l. Clearly $\mathfrak{C} \subset \mathfrak{D}$. An optional σ-algebra $\mathcal{O}(\mathbf{F})$ coincides with the smallest σ-algebra of random sets $\mathfrak{D}(\mathbf{F})$, generated by $\mathbf{F}$-adapted processes taking values in $\mathfrak{D}$.

Definition 11.2.1 A predictable σ-algebra $\mathcal{P}(\mathbf{F})$ coincides with the smallest σ-algebra of random sets $\mathfrak{C}(\mathbf{F})$, generated by $\mathbf{F}$-adapted processes taking values in $\mathfrak{C}$.

Remark 11.2.2 Even in the absence of right-continuity of the flow $\mathbf{F}$ we have, $\mathcal{P}(\mathbf{F}) = \mathcal{P}(\mathbf{F}_-) = \mathcal{P}(\mathbf{F}_+)$.

Proposition 11.2.3 For a locally square integrable martingale $M \in \mathcal{M}^2_{loc}$ with $M_0 = 0$ we have the following inequalities

$$\mathbf{P}\left(M_T^* \geq c\right) \leq \frac{1}{c^2}\mathbf{E}\left[M, M\right]_T = \mathbf{E}\left\langle M, M \right\rangle_T \tag{11.1}$$

for a constant $c > 0$, and for $\tau = \inf(t : |M_t| > c) \wedge T$ we also have

$$\mathbf{P}\left(M_\tau^* \geq c\right) \leq \frac{1}{c^2}\mathbf{E}M_\tau^2; \tag{11.2}$$

These inequalities are related by the fact that $\mathbf{E}M_\tau^2 = \mathbf{E}\left[M, M\right]_\tau$ and $[M, M]_\tau \leq [M, M]_T$.

Lemma 11.2.4 For a concave function f and a measurable process X such that $\mathbf{E}|X_\tau| < \infty$ and $\mathbf{E}|f(X_\tau)| < \infty$ for any $\tau \in \mathcal{T}$, then by Jensen inequality we have that the predictable projection inequality $\ ^p f(X) \geq f(\ ^p X)$.

Theorem 11.2.5 For any r.c.l.l. process X with $\mathbf{E}\sup_{t\geq 0}|X_t| < \infty$, there is an r.c.l.l. modification of its optional projection ${}^o X$.

Theorem 11.2.6 If X is a martingale, then there exists a random variable X_∞ such that

$$X_t \longrightarrow X_\infty, \; t \longrightarrow \infty \text{ a.s.}$$
$$\mathbf{E}|X_t - X_\infty| \longrightarrow 0, \; t \longrightarrow \infty$$
$$X_t = \mathbf{E}[X_\infty|\mathcal{F}_t] \; \mathbf{P}\text{-a.s., } t \geq 0.$$

and for each $\sigma, \tau \in \mathcal{T}$ we have $X_{\tau \wedge \sigma} = \mathbf{E}[X_\sigma|\mathcal{F}_\tau]$ $\mathbf{P}$-a.s. And, for an integrable random variable Y then there exists one and only one (up to $\mathbf{P}$-indistinguishability) martingale X such that ($\mathbf{P}$-a.s.) for each $t \geq 0$ $X_t = \mathbf{E}[Y|\mathcal{F}_t]$. Furthermore, if $(\tau_n)_{n\geq 1}$ is a sequence of increasing Markov times, then $\lim_n X_{\tau_n} = \mathbf{E}[X_{\lim_n \tau_n}|\sigma(\cup_n \mathcal{F}_{\tau_n})]$ $\mathbf{P}$-a.s. and if τ then $X_{\tau-} = \mathbf{E}[X_\tau|\mathcal{F}_{\tau-}]$.

Definition 11.2.7 To characterize the class of uniformly integrable martingales within the class of local martingales: A process X is of the Dirichlet (D) class if the family of random variables $\{X_\tau : \tau < \infty\}$ is uniformly integrable, i.e.,

$$\sup_{\tau \in \mathcal{T}} \mathbf{E}|X_\tau|\mathbf{1}(|X_\tau| > N) \longrightarrow 0, \quad N \longrightarrow \infty.$$

A local martingale is a uniformly integrable martingale if and only if it is of the Dirichlet class. Recall that uniformly integrable process means

$$\sup_{t\geq 0} \mathbf{E}|X_\tau|\mathbf{1}(|X_t| > N) \longrightarrow 0, \quad N \longrightarrow \infty.$$

Proposition 11.2.8 Let X be r.c.l.l. $\mathbf{F}$ measurable process and its limit $\lim X_t (= X_\infty)$ exist $\mathbf{P}$-a.s. Then, it is necessary and sufficient for X to be a uniformly integrable martingale that $\mathbf{E}X_0 < \infty$ and $\mathbf{E}X_\tau = \mathbf{E}X_0$ for each stopping time τ.

Proposition 11.2.9 Let X be local martingale, $X \geq 0$ and $T = \inf(t : X_t = 0)$ ($\inf \emptyset = \infty$), then the process $X\mathbf{1}_{[\![T,\infty[\![}$ is indistinguishable from zero.

Lemma 11.2.10 If A is a predictable process of bounded variation over finite interval, then a localizing sequence (T_n) of Markov times can be found such that $\mathbf{Var}(A)_{T_n} \leq n$.

Lemma 11.2.11 Let X and Y be two non-negative measurable processes and suppose that for each Markov time τ, $\mathbf{E}[X_\tau 1_{\tau<\infty}] = \mathbf{E}[Y_\tau 1_{\tau<\infty}]$ then, for each $A \in \mathcal{V}^+$ and $\tau \in \mathcal{T}$,

$$\mathbf{E} \int_0^\tau X_s dA_s = \mathbf{E} \int_0^\tau Y_s dA_s,$$

Theorem 11.2.12 Let A be a locally integrable variation increasing process, then there is one and only one predictable process $\tilde{A}$ that is predictable of locally integrable variation called the dual predictable projection or compensator of A such that $A - \tilde{A}$ is a local martingale. Furthermore if A is predictable, then $A = \tilde{A}$, whereas if A is locally integrable variation, then $^p(\triangle A) = \left(\triangle \tilde{A}\right)$ and for any stopping time τ, $A^\tau = \left(\tilde{A}\right)^\tau$.

Proposition 11.2.13 If $\tilde{A}$ is the compensator of A, then $\mathbf{Var}\tilde{A} \leq \widetilde{\mathbf{Var}A}$.

Proposition 11.2.14 For a non-negative process X and $A \in \mathcal{V}^+$ then $\mathbf{E}[X \circ A_\infty] = \mathbf{E}[\,^o X \circ A_\infty]$ and A is predictable, then $\mathbf{E}[X \circ A_\infty] = \mathbf{E}[\,^p X \circ A_\infty]$.

Proposition 11.2.15 For each local martingale M and N the predictable projection $^p(\triangle M) = 0$ and $\mathbf{Var}\langle M, N \rangle \leq \frac{1}{2}(\langle M \rangle + \langle N \rangle)$.

Theorem 11.2.16 For a uniformly integrable martingale M and Markov time T with $\mathbf{E}|M|^p < \infty$ for $p > 0$ then $\mathbf{E}\,(M_T^*)^p \leq \left(\frac{p}{p-1}\right)^p \mathbf{E}|M|_T^p$ and if M is a local martingale with zero intial value and $|M|^p$ is locally integralbe process then $\mathbf{E}\,(M_T^*)^p \leq \left(\frac{p}{p-1}\right)^p \mathbf{E}\widetilde{|M|_T^p}$.

Theorem 11.2.17 The Lenglart-Rebolledo inequality tells us that X and Y are non-negative processes from the class D with zero initial values but also Y is $\mathcal{V}^+$. Let Y dominate X in the sense that for each stopping time τ, $\mathbf{E}[X_\tau] \leq \mathbf{E}[Y_\tau]$. Then, for each Markov time T and any $a > 0$, $b > 0$,

$$\mathbf{P}\left(\sup_{t \leq T} X_t \geq a\right) \leq \frac{1}{a}\mathbf{E}\left[Y_T \wedge \left(b + \sup_{t \leq T} \triangle Y_t\right)\right] + \mathbf{P}(Y_T \geq b).$$

if, in addition, Y is predictable then

$$\mathbf{P}\left(\sup_{t \leq T} X_t \geq a\right) \leq \frac{1}{a}\mathbf{E}\left[Y_T \wedge b\right] + \mathbf{P}(Y_T \geq b).$$

Theorem 11.2.18 The Burkholder-Gandy inequalities; let T be a Markov time $(T \leq \infty)$, M a local martingale with zero initial value and $p > 1$. There exist universal constants c_p and C_p, independent of T and M, such that

$$c_p \mathbf{E}[M, M]_T^{p/2} \leq \mathbf{E} (M_T^*)^p \leq C_p \mathbf{E}[M, M]_T^{p/2}.$$

Theorem 11.2.19 Let M and N be locally square integrable martingales, then there exists a predictable function H such that $\langle M, N \rangle = h \circ \langle M \rangle$ and if there is also another function $h\prime$ satisfying the same equality then $(h - h')^2 \circ \langle M \rangle$ is negligible. Also, $\langle N \rangle - h^2 \circ \langle M \rangle \in \mathcal{V}^+$.

Proposition 11.2.20 Let X be an optional stochastic process such that ($\mathbf{P}$-a.s.) the set $\Gamma = \{t : X_t \neq 0\}$ is an at most countable subset. Let

$$V_t(X) = \left(\sum_{s \leq t} X_s^2 \right)^{1/2}, \quad L_t(X) = \sum_{s \leq t} \frac{X_s^2}{1 + |X_s|}$$

then $V(X) \leq \frac{1}{4} + 2L(X)$.

Proposition 11.2.21 For square integrable martingales M and N and H a locally square integrable integrand of M then $\langle H \circ M, N \rangle = H \circ \langle M, N \rangle$.

Proposition 11.2.22 For a positive local martingale M and $T = \inf(t : M_{t-} = 0)$ then $M 1_{[\![T, \infty[\![}$ is indistinguishable from M.

Proposition 11.2.23 For any local martingale M and $f = x^2/(1 + |x|)$ we have $\langle M^c \rangle + \sum f (\triangle M)$ an increasing locally integrable variation process.

Definition 11.2.24 The Banach space H^p, $p \geq 1$ is a space of martingales for which $||X||_{\mathrm{H}^p} < \infty$ where $||X||_{\mathrm{H}^p} = ||X_\infty^*||_p$ and $||\cdot||$ is the norm in $\mathrm{L}^p = \mathrm{L}^p (\Omega, \mathcal{F}_\infty, \mathbf{P})$. In case $p = 1$ we write simply H instead of H^1. For $p = 2$, in particular, the following inequalities take place

$$||X_\infty||_2 \leq ||X_\infty^*||_2 \leq 2||X_\infty||_2 \tag{11.3}$$

the first inequality is evident, and the second one is a consequence of Doob's inequality.

Proposition 11.2.25 Let us recall the definition of integer valued measures – an integer valued measure μ is called optional (predictable), if for any non-negative $\tilde{\mathcal{O}}$-measurable (respectively $\tilde{\mathcal{P}}$-measurable) function X the random process

$$X * \mu_t = \int X(\omega, s, x)\mu(\omega, ds, dx)$$

is optional (respectively predictable). Further, to each random measure μ and probability measure $\mathbf{P}$ one can relate the Doleans measure

$$\mathsf{M}_\mu^{\mathbf{P}}(d\omega, dt, dx) = \mathbf{P}(d\omega)\mu(\omega, dt, x)$$

and $\mathsf{M}_\mu^{\mathbf{P}}(X) = \mathbf{E}X * \mu$. Let $f = f(x)$ be a concave function, let $\mu \in \tilde{\mathcal{V}}_{\mathcal{P}}^+$ and let X be a $\tilde{\mathcal{F}}$-measurable function such that $|X|\mu \in \tilde{\mathcal{V}}_{\mathcal{P}}^+$ and $|f(X)|\mu \in \tilde{\mathcal{V}}_{\mathcal{P}}^+$ then the conditional mathematical expectation $\mathsf{M}_\mu^{\mathbf{P}}(\cdot|\tilde{\mathcal{P}})$ Jensen's inequality

$$\mathsf{M}_\mu^{\mathbf{P}}(f(X)|\tilde{\mathcal{P}}) \geq f(\mathsf{M}_\mu^{\mathbf{P}}(X|\tilde{\mathcal{P}})) \quad (\mathsf{M}_\mu^{\mathbf{P}}\text{-a.s.})$$

takes place.

Proposition 11.2.26 Let $\mu \in \tilde{\mathcal{V}}_{\mathcal{P}}^+$, and let the processes $X = X(\omega, t)$ and $X^* = X^*(\omega, t) = \sup_{s \leq t} |X(\omega, s)|$ be $\mathcal{F} \times \mathcal{B}(\mathbb{R}_+)$-measurable functions, then the conditional mathematical expectation $\mathsf{M}_\mu^{\mathbf{P}}(X|\tilde{\mathcal{P}})$ is defined provided $\mathbf{E}X^*(\omega, t) < \infty$ for each $t \in \mathbb{R}_+$.

Theorem 11.2.27 Let μ be an integer-valued random measure ($\mu \in \tilde{\mathcal{V}}_{\mathcal{P}}^+$), and v its compensator ($a_t = v(\{t\} \times E)$, $a \leq 1$). For $M \in \mathcal{M}_{loc}$ the conditional mathematical expectations $\mathsf{M}_p^{\mathbf{P}}(\triangle M|\mathcal{P})$ and $\mathsf{M}_\mu^{\mathbf{P}}(\triangle M|\tilde{\mathcal{P}})$ are formed. We have

$$\mathsf{M}_p^{\mathbf{P}}(\triangle M|\mathcal{P}) = -\frac{\mathbf{1}(0 < a < 1)}{1 - a}\mathsf{M}_\mu^{\mathbf{P}}\widehat{(\triangle M}|\tilde{\mathcal{P}}) \quad (\mathsf{M}_p^{\mathbf{P}}\text{-a.s.})$$

where $\frac{0}{0} = 0$ and

$$\mathsf{M}_p^{\mathbf{P}}(\triangle M|\mathcal{P}) = -\frac{\mathbf{1}(0 < a < 1)}{1 - a}\mathsf{M}_\mu^{\mathbf{P}}\widehat{(\triangle M}|\tilde{\mathcal{P}}) \quad (\{0 < a < 1\}; \; \mathsf{M}_p^{\mathbf{P}}\text{-a.s.})$$

Proposition 11.2.28 Let an integer-valued random measure $\mu \in \tilde{\mathcal{V}}_{\mathcal{P}}^+$, let v be its compensator with $a \leq 1$ and

$$\alpha_t = \mathbf{1}(a_t = 1)(1 - \mu(\{t\} \times E)) = \mathbf{1}(a_t = 1)(v(\{t\} \times E) - \mu(\{t\} \times E)),$$

then the process $\alpha = (\alpha_t)_{t \geq 0}$ is negligible.

Proposition 11.2.29 Let an integer-valued random measure $\mu \in \tilde{\mathcal{V}}_\mathcal{P}^+$, let v be its compensator with a ≤ 1 and $M \in \mathcal{M}_{loc}$, then

$$\mathbf{1}(a_t = 1)M_\mu^\mathbf{P}\widehat{(\triangle M|\tilde{\mathcal{P}})} = 0$$

$(M_p^\mathbf{P}$-a.s.$)$.

Lemma 11.2.30 Let N be a counting process with the deterministic compensator A (independent of ω), then N is a process with independent increments.

Lemma 11.2.31 Let μ be an integer-valued random measure, $\mu \in \tilde{\mathcal{V}}^+ \cap \tilde{\mathcal{P}}$, v its compensator and $U = U(\omega, t, x)$ a $\tilde{\mathcal{P}}$-measurable function such that $|U| * v \in \mathcal{A}_{loc}^+$. Then, we have $|U| * \mu \in \mathcal{A}_{loc}^+$ and the integrals

$$U * (\mu - v)_t, U * \mu_t - U * v_t, t \geq 0, \tag{11.4}$$

are well defined and the process $U * (\mu - v) = (U * (\mu - v)_t)_{t \geq 0}$ is a local martingale. Hence, $\mu - v$ is a martingale measure. For a large supply of functions U of the stochastic integral $U * (\mu - v)$ with respect to a martingale measure $\mu - v$ is a process of the class $\mathcal{M}_{loc}^d$.

Definition 11.2.32 Let μ be an integer-valued random measure and v its compensator with $a_t(\omega) = v(\omega; \{t\} \times E) \leq 1$. Besides, let p be an integer-valued random measure on $(\mathbb{R}_+, \mathcal{B}(\mathbb{R}_+))$ with

$$p(\omega; \Gamma) = \sum_{s \in \Gamma} \mathbf{1}(a_s(\omega) > 0)(1 - \mu(\omega; \{s\} \times E)), \quad \Gamma \in \mathcal{B}(\mathbb{R}_+),$$

which has the compensator q with

$$q(\omega; \Gamma) = \sum_{s \in \Gamma} \mathbf{1}(a_s(\omega) > 0)(1 - a_s(\omega))).$$

Proposition 11.2.33 Let $U = U(\omega, t, x)$ be a $\tilde{\mathcal{P}}$-measurable function such that for each Markov time T

$$\mathbf{1}(T < \infty) \int_E |U(\omega, T, x)| v(\omega; \{T\}, dx) < \infty \quad (\mathbf{P}\text{-a. s.}). \tag{11.5}$$

for such a function U we define

$$\hat{U}(\omega, t) = \int_E U(\omega, t, x) v(\omega; \{t\}, dx). \tag{11.6}$$

Define the increasing processes $G(U)$ and $G^i(U)$, $i = 1, 2$, by setting

$$G(U) = \frac{(U - \hat{U})^2}{1 + |U - \hat{U}|} * v + \frac{\hat{U}^2}{1 + |\hat{U}|} * q, \tag{11.7}$$

$$G^i(U) = |U - \hat{U}|^i * v + |\hat{U}|^i * q. \tag{11.8}$$

We say that a $\tilde{\mathcal{P}}$-measurable function $U \in \mathcal{G}_{loc}$ if $G(U) \in \mathcal{A}^+_{loc}$ and $U \in \mathcal{G}^i_{loc}$ if $G^i(U) \in \mathcal{A}^+_{loc}$, $i = 1, 2$. Evidently,

$$\mathcal{G}^i_{loc} \subseteq \mathcal{G}_{loc}, \; i = 1, 2. \tag{11.9}$$

If $U \in \mathcal{G}^i_{loc}$ and c is a constant, then $cU \in \mathcal{G}^i_{loc}$; if $U_1, U_2 \in \mathcal{G}^i_{loc}$, then $U_1 + U_2 \in \mathcal{G}^i_{loc}$.

Lemma 11.2.34 (1) if $U \in \mathcal{G}_{loc}$ and c is a constant, then $cU \in \mathcal{G}_{loc}$.
(2) If $U_1, U_2 \in \mathcal{G}_{loc}$, then $U_1 + U_2 \in \mathcal{G}_{loc}$.
(3) If $U \in \mathcal{G}_{loc}$, then functions $U_1 \in \mathcal{G}^1_{loc}$, and $U_2 \in \mathcal{G}^2_{loc}$ can be found such that $U = U_1 + U_2$.
(4) If $U \in \mathcal{G}^2_{loc}$, then

$$G^2_t(U) = U^2 * v_t - \sum_{0 < s \leq l} \hat{U}^2(s).$$

Theorem 11.2.35 Let $U \in \mathcal{G}_{loc}$ then the stochastic integral $U * (\mu - v)$ possesses the following properties:
(1) $U * (\mu - v) \in \mathcal{M}^d_{loc,0}$;
(2) $\triangle(U * (\mu - v))_t = \int_E U(t, x)\mu(\{t\}, dx) - \hat{U}(t)$;
(3) $[U * (\mu - v), U * (\mu - v)] = (U - \hat{U})^2 * \mu + \hat{U}^2 * p$;
(4) $(U * (\mu - v))^T = U 1_{[\![0,T]\!]} * (\mu - v) \; \forall T \in \mathcal{T}$;
(5) $U \in \mathcal{G}^1_{loc} \Leftrightarrow \mathbf{Var}(v * (\mu - v)) \in \mathcal{A}^+_{loc}$ and in this case

$$U * (\mu - v) = (U - \hat{U}) * \mu - \hat{U} * p - (U - \hat{U}) * v + \hat{U} * q;$$

(6) $U \in \mathcal{G}^2_{loc} \Leftrightarrow U * (\mu - v) \in \mathcal{M}^{2,d}_{loc}$ and in this case

$$\langle U * (\mu - v) \rangle = G^2(U).$$

Theorem 11.2.36 Let μ be an integer-valued random measure ($\mu \in \tilde{\mathcal{V}}^+ \cap \tilde{\mathcal{P}}$), v its compensator ($a_t = v(\{t\} \times E)$) and $M \in \mathcal{M}_{loc}$. Then given $f(x) = x^2 (1 + |x|)^{-1}$:
(1) $f(\mathsf{M}^{\mathbf{P}}_\mu(\triangle M | \tilde{\mathcal{P}})) * v \in \mathcal{A}^+_{loc}$,

$$\sum_s \mathbf{1}(a_s > 0)(f(\mathsf{M}^{\mathbf{P}}_p(\triangle M | \mathcal{P})(s))(1 - a_s) \in \mathcal{A}^+_{loc}$$

(2) $U = \mathsf{M}^{\mathbf{P}}_\mu(\triangle M | \tilde{\mathcal{P}}) - \mathsf{M}^{\mathbf{P}}_p(\triangle M | \mathcal{P}) \mathbf{1}(0 < a < 1) \in \mathcal{G}_{loc}$;
(3) if $|\triangle M| \leq c$ and if $N = U * (\mu - v)$ with a function U defmed in (2) then $|\triangle N| \leq 4c$;
(4) if $M = U' * (\mu - v)$, $U' \in \mathcal{G}_{loc}$ and if $N = U * (\mu - v)$ with U defined in (2), then the processes M and N are indistinguishable.

Proposition 11.2.37 Let μ be an integer-valued random measure and v its compensator, let U be a $\tilde{\mathcal{P}}$-measurable function such that $|U|*v \in \mathcal{A}_{loc}^+$ then $U \in \mathcal{G}_{loc}^1$, and the stochastic integral $U*(\mu - v)$ and the process $U*\mu - U*v$ are indistinguishable.

Proposition 11.2.38 Let $U \in \mathcal{G}_{loc}$ and $M = U*(\mu - v)$ then $\mathsf{M}_\mu^\mathbf{P}(\triangle M|\tilde{\mathcal{P}}) = U - 0$. $\mathsf{M}_\mu^\mathbf{P}$-a.s.

Proposition 11.2.39 Let μ be an integer-valued random measure, v its compensator and $U^1, U^2 \in \mathcal{G}_{loc}$, $U^1 = U^2$ ($\{a < 1\}$; $\mathsf{M}_\mu^\mathbf{P}$-a.s.) then the stochastic processes $U^1*(\mu - v)$ and $U^2*(\mu - v)$ are indistinguishable.

Theorem 11.2.40 Let $X \in \mathcal{S}(\mathbf{F}, \mathbf{P})$ be a semimartingale with predictable characteristic $T = (B, C, v)$ and $\tilde{\mathbf{P}} <<_{loc} \mathbf{P}$. Then, $X \in \mathcal{S}(\mathbf{F}, \tilde{\mathbf{P}})$ and its predictable characteristic $T = (\tilde{B}, \tilde{C}, \tilde{v})$ with respect to the measure $\tilde{\mathbf{P}}$ is given by following formulas

$$\tilde{B} = B + \beta \circ C + \mathbf{1}_{|x| \leq 1} x(Y - 1)*v, \tag{11.10}$$

$$\tilde{C} = C, \tag{11.11}$$

$$\tilde{v} = Yv. \tag{11.12}$$

Let Z be the local density process of a measure $\tilde{\mathbf{P}}$ with respect to $\mathbf{P}$ then Y and β are given by

$$Y = \mathbf{1}(Z_- > 0) Z_-^{-1} M_\mu^\mathbf{P}(Z|\tilde{\mathcal{P}}) \tag{11.13}$$

$$\beta = \mathbf{1}(Z_- > 0) Z_-^{-1} \frac{d\langle X^c, Z^c \rangle}{d\langle X^c \rangle}. \tag{11.14}$$

where

$$x\mathbf{1}_{|x| \leq 1}(Y - 1)*v \in \mathcal{A}_{loc}^+ \left(\tilde{\mathcal{P}}\right), \quad \beta^2 \circ C \in \mathcal{A}_{loc}^+ \left(\tilde{\mathcal{P}}\right)$$

Let the flow of σ-algebras $\mathbf{G} = (\mathcal{G}_t)_{t \geq 0}$ be given and satisfying the usual conditions and let

$$\mathcal{F}_{t+}^X \subseteq \mathcal{G}_t \subseteq \mathcal{F}_t, \ t \geq 0. \tag{11.15}$$

Theorem 11.2.41 Under the assumption 11.15 we have

$$X \in \mathcal{S}(\mathbf{F}, \mathbf{P}) \Rightarrow X \in \mathcal{S}(\mathbf{G}, \mathbf{P}). \tag{11.16}$$

and

$$C \overset{\mathbf{P}}{=} \overline{C} \tag{11.17}$$

where $\overline{C}$ is the quadratic characteristic of the continuous local martingale $\overline{X}^c$ in the canonical representation of X relative to $(\mathbf{G}, \mathbf{P})$.

The above theorem tells us that the semimartingale property is preserved under reduction of flow of σ-algebras. Next, we consider a particular cases:

Theorem 11.2.42 Let $X = \mathcal{S}(\mathbf{F}, \mathbf{P})$ with

$$X_t = B_t + M_t, \ t \geq 0, \tag{11.18}$$

where $B = (B_t)_{t \geq 0} \in \mathcal{A} \cap \mathcal{P}(\mathbf{F})$ and $M = (M_t)_{t \geq 0} \in \mathcal{M}(\mathbf{F}, \mathbf{P})$, $M_0 = 0$ with $\mathbf{E} \sup_{t \geq 0} |M_t| < \infty$. Then, $X \in \mathcal{S}(\mathbf{G}, \mathbf{P})$.

According to Definition 11.2.24, the space $\mathrm{H}(\mathbf{F}, \mathbf{P})$ is the space of local martingales M with the norm $\mathbf{E} \sup_{t \geq 0} |M_t| < \infty$. To prove this theorem we need the following two lemmas in which the optional projections of relative to $(\mathbf{G}, \mathbf{P})$ are denoted by $\pi(Y) = (\pi_t(Y))_{t \geq 0}$.

Lemma 11.2.43 Let $A \in \mathcal{A}^+ \cap \mathcal{P}(\mathbf{F})$. Then $\pi(A)$ presents a submartingale of the class D and a semimartingale relative to $(\mathbf{G}, \mathbf{P})$.

Proof. A modification $\pi(A) \in \mathfrak{D} \cap \mathbf{G}$ can be chosen, and in accordance with optional projection, as $t > s$ we have (**P**-a.s.)

$$\mathbf{E}(\pi_t(A)|\mathcal{G}_s) = \mathbf{E}(\mathbf{E}(A_t|\mathcal{G}_t)|\mathcal{G}_s) = \mathbf{E}(A_t|\mathcal{G}_s) \geq \mathbf{E}(A_s|\mathcal{G}_s) = \pi_s(A).$$

Hence, $\pi(A)$ is a submartingale. Next, for each stopping time τ (relative to $\mathbf{G}$) we have $\pi_t(A) \leq \mathbf{E}(A_\infty|\mathcal{G}_t)$. Therefore, by Definition 11.2.7 the family $\{\pi_t(A), \tau \in \mathcal{T}(\mathbf{G})\}$ is uniformly integrable, i.e., $\pi(A)$ belongs to class D. Hence, by the Doob-Meyer theorem, the submartingale $\pi(A)$ is a semimartingale. ∎

Lemma 11.2.44 Let $M = (M_t)_{t \geq 0} \in \mathcal{M}(\mathbf{F}, \mathbf{P})$ with $M_0 = 0$. Then $\pi(M) \in (\mathbf{G}, \mathbf{P})$.

Proof. Using 11.2.5 one can choose $\pi(M) \in \mathfrak{D} \cap \mathbf{G}$. Next, by the definition of the optional projection as $t > s$ we have (**P**-a.s.)

$$
\begin{aligned}
\mathbf{E}(\pi_t(M)|\mathcal{G}_s) &= \mathbf{E}(\mathbf{E}(M_t|\mathcal{G}_t)|\mathcal{G}_s) = \mathbf{E}(M_t|\mathcal{G}_s) \\
&= \mathbf{E}(\mathbf{E}(M_t|\mathcal{F}_s)|\mathcal{G}_s) = \mathbf{E}(M_s|\mathcal{G}_s) = \pi_s(M),
\end{aligned}
$$

i.e., $\pi(M)$ is a martingale relative to $(\mathbf{G}, \mathbf{P})$. Finally,

$$\pi_t(M) = \mathbf{E}(\mathbf{E}M_\infty|\mathcal{F}_t)|\mathcal{G}_t) = \mathbf{E}(M_\infty|\mathcal{G}_t) \ (\mathbf{P}\text{-a.s.}).$$

Hence $\pi(M)$ is a uniformly integrable martingale (Theorem 11.2.6). ∎

Remark 11.2.45 The assertion of the lemma remains is (with $\mathcal{M}_{loc}(\mathbf{G})$ replaced by $\mathcal{M}(\mathbf{G})$) for $M \in \mathcal{M}_{1oc}(\mathbf{F}, \mathbf{P})$ if there exists a localizing sequence $(\tau_n)_{n \geq 1}$ such that $M^{\tau_n} \in \mathrm{H}(F, P)$, $\tau^n \in \mathcal{T}(\mathbf{G})$, $n \geq 1$.

Proof. Continue the proof of Theorem 11.2.42. Let $B^i = (B^i_t)_{t \geq 0} \in \mathcal{A}^+ \cap \mathcal{P}(\mathbf{F})$, $i = 1, 2$ be processes such that $B_t = B^1_t - B^2_t$ and $\mathbf{Var}(B) = B^1 + B^2$. Then,

$$X = B^1 - B^2 + M. \tag{11.19}$$

Since $\mathcal{F}^X_{t+} \subseteq \mathcal{G}_t$, $t \geq 0$, we have $\pi(X) = X$. Consequently,

$$X = \pi(B^1) - \pi(B^2) + \pi(M), \tag{11.20}$$

and the desired assertion follows by Lemmas 11.2.43 and 11.2.44. ∎

11.2.2 Martingales' integral representation

Let $(\Omega, \mathcal{F}, \mathbf{F} = (\mathcal{F}_t)_{t \geq 0}, \mathbf{P})$ be a stochastic basis, $\mathcal{F} = \mathcal{F}_\infty$, $X \in \mathcal{S}(\mathbf{F}, \mathbf{P})$ and let $T = (B, C, v)$ be the triplet of its predictable characteristics. If h is a $\mathcal{P}$-measurable function and H a $\tilde{\mathcal{P}} = \mathcal{P} \times \mathcal{B}(\mathbb{R}_0)$-measurable function such that

$$h \in \mathrm{L}^2_{loc}(\mathbf{F}, \langle X^c \rangle), \quad H \in \mathcal{G}_{loc},$$

where $\langle X^c \rangle = C$ and $\mathcal{G}_{loc}$ is related to the jump measure of X, then with a semimartingale X one may associate a local martingale M ($M \in \mathcal{M}_{loc}(\mathbf{F}, \mathbf{P})$), which presents the sum of a $\mathcal{F}_0$-measurable and integrable random variable M_0 and the stochastic integrals $h \cdot X^c$ and $H * (\mu - v)$ with respect to the continuous component X^c and to the martingale measure $\mu - v$ (μ is the jump measure of X), i.e.,

$$M_t = M_0 + h \cdot X^c_t + H * (\mu - v)_t. \tag{11.21}$$

In the present subsection the following converse problem is considered. Let $M \in \mathcal{M}_{loc}(\mathbf{F}, \mathbf{P})$ and $X \in \mathcal{S}(\mathbf{F}, \mathbf{P})$. It is asked in which case the representation (11.21) for M takes place. In case when (11.21) is valid, we say that a local martingale M admits the integral representation (with respect to the continuous martingale component X^c and the martingale measure $\mu - v$ of a semimartingale X).

In order to formulate conditions ensuring the integral representation for $M \in \mathcal{M}_{loc}(\mathbf{F}, \mathbf{P})$, we define the family

$$\mathcal{L}_{\mathbf{P}} = \{\tilde{\mathbf{P}} : \tilde{\mathbf{P}} << \mathbf{P}, \ \tilde{\mathbf{P}}_0 = \mathbf{P}_0, \ \tilde{T} \overset{\mathbf{P}}{=} T\} \tag{11.22}$$

which presents the family of probability measures possessing the following properties: if $\tilde{\mathbf{P}} \in \mathcal{L}_{\mathbf{P}}$ then $\tilde{\mathbf{P}} << \mathbf{P}$, the restriction of the measure $\tilde{\mathbf{P}}$ to the σ-algebra $\mathcal{F}_0$ coincides with the restriction of $\mathbf{P}$ to $\mathcal{F}_0$ ($\tilde{\mathbf{P}}_0 = \mathbf{P}_0$), and the triplet $\tilde{T} = (\tilde{B}, \tilde{C}, \tilde{v})$ of predictable characteristics of a $(\mathbf{F}, \tilde{\mathbf{P}})$-semimartingale X ($X \in \mathcal{S}(\mathbf{F}, \tilde{\mathbf{P}})$ by Theorem 11.2.40) is $\mathbf{P}$-indistinguishable from the triplet $T = (B, C, v)$.

Theorem 11.2.46 The following conditions are equivalent:
(a) each local martingale $M \in \mathcal{M}_{loc}(\mathbf{F}, \mathbf{P})$ admits the integral representation,
(b) $\mathcal{L}_{\mathbf{P}} = \{\mathbf{P}\}$, i.e., the set $\mathcal{L}_{\mathbf{P}}$ consists of one point $\mathbf{P}$ only.

The proof of this theorem relies on a number of auxiliary results, formulated below as lemmas.

Lemma 11.2.47 Let $M \in \mathcal{M}_{loc}(\mathbf{F}, \mathbf{P})$ admit the integral representation (11.21). Then in this representation one can take

$$h = \frac{d\langle M^c, X^c \rangle}{d\langle X^c \rangle}, \quad H = \mathsf{M}_\mu^{\mathbf{P}}(\triangle M | \tilde{\mathcal{P}}) - \mathbf{1}\,(0 < \mathrm{a} < 1)\, \mathsf{M}_p^{\mathbf{P}}(\triangle M | \mathcal{P}).$$

Proof. From (11.21) it follows that $M^c = h \circ X^c$, hence $\langle M^c, X^c \rangle = h \circ \langle X^c \rangle$ (see Proposition 11.2.21), i.e., for h the desired representation holds. Since by Proposition 11.2.28 we have

$$H * (\mu - v) = (H - \hat{H}\mathbf{1}(\mathrm{a} = 1)) * (\mu - v),$$

instead of H, involved in (11.21), we can take $h - \hat{H}\mathbf{1}(\mathrm{a} = 1)$. By Proposition 11.2.38 we have $\mathsf{M}_\mu^{\mathbf{P}}(\triangle M | \tilde{\mathcal{P}}) = H - \hat{H}$, and hence

$$\widehat{\mathsf{M}_\mu^{\mathbf{P}}(\triangle M | \tilde{\mathcal{P}})} = \hat{H}\,(1 - \mathrm{a})\,.$$

By taking into consideration the definition of $\mathsf{M}_p^{\mathbf{P}}(\triangle M | \mathcal{P})$ (cf. Theorem 11.2.27), this gives

$$\begin{aligned}
\hat{H} &= \frac{\mathbf{1}\,(0 < \mathrm{a} < 1)}{1 - \mathrm{a}} \widehat{\mathsf{M}_\mu^{\mathbf{P}}(\triangle M | \tilde{\mathcal{P}})} + \hat{H}\mathbf{1}(\mathrm{a} = 1) \\
&= -\mathbf{1}\,(0 < \mathrm{a} < 1)\, \mathsf{M}_p^{\mathbf{P}}(\triangle M | \mathcal{P}) + \hat{H}\mathbf{1}(\mathrm{a} = 1).
\end{aligned}$$

Consequently,

$$\hat{H}\mathbf{1}(\mathrm{a} < 1) = -\mathbf{1}(0 < \mathrm{a} < 1)\mathsf{M}_p^{\mathbf{P}}(\triangle M | \mathcal{P})$$

and

$$\begin{aligned}
H - \hat{H}\mathbf{1}(\mathrm{a} &= 1) = \mathsf{M}_\mu^{\mathbf{P}}(\triangle M | \tilde{\mathcal{P}}) + \hat{H}\mathbf{1}(\mathrm{a} < 1) \\
&= \mathsf{M}_\mu^{\mathbf{P}}(\triangle M | \tilde{\mathcal{P}}) - \mathbf{1}(0 < \mathrm{a} < 1)\mathsf{M}_p^{\mathbf{P}}(\triangle M | \mathcal{P}).
\end{aligned}$$

∎

Corollary 11.2.48 If for $M \in \mathcal{M}_{loc}(\mathbf{F}, \mathbf{P})$ the integral representation takes place, then

$$\langle M^c \rangle = \left(\frac{d\langle M^c, X^c \rangle}{d\langle X^c \rangle} \right)^2 \circ C, \tag{11.23}$$

$$\begin{aligned}
\mathbf{1}_{(\tau \leq \infty)} \triangle M_\tau = \mathbf{1}_{(\tau < \infty)} \Big[&\mathsf{M}_\mu^{\mathbf{P}}(\triangle M | \tilde{\mathcal{P}})(\tau, \triangle X_\tau)\mathbf{1}(\triangle X_\tau \neq 0) \\
&- \mathbf{1}(0 < \mathrm{a} < 1)\mathsf{M}_p^{\mathbf{P}}(\triangle M | \mathcal{P})(\tau)\mathbf{1}(\triangle X_\tau = 0) \Big], \quad \forall \tau \in \mathcal{T}
\end{aligned} \tag{11.24}$$

Lemma 11.2.49 If for each $M \in \mathcal{M}_{loc}(\mathbf{F}, \mathbf{P})$ such that

$$\mathbf{E}\langle M^c\rangle_\infty + \sum_{s \geq 0} f(\triangle M_s)) < \infty \qquad (11.25)$$

with

$$f(x) = \frac{x^2}{1 + |x|}$$

the integral representation takes place, then it takes place for each $M \in \mathcal{M}_{loc}(\mathbf{F}, \mathbf{P})$ as well.

Proof. According to Proposition 11.2.23 for each $M \in \mathcal{M}_{loc}(\mathbf{F}, \mathbf{P})$ we have

$$\langle M^c\rangle + \sum_s f(\triangle M_s) \in \mathcal{A}_{loc}^+.$$

Therefore, for each $M \in \mathcal{M}_{loc}(\mathbf{F}, \mathbf{P})$ there exists a localizing sequence $(\tau_n)_{n \geq 1}$ such that

$$\mathbf{E}\left[\langle M^c\rangle_{\tau_n} + \sum_{\tau_n \geq s > 0} f(\triangle M_s))\right] < \infty.$$

By assumption M^{τ_n} admits the integral representation. Besides,

$$\frac{d\langle (M^{\tau_n})^c, X^c\rangle}{d\langle X^c\rangle} = \mathbf{1}_{[0,\tau_n]}\frac{d\langle M^c, X^c\rangle}{d\langle X^c\rangle} = \mathbf{1}_{[0,\tau_n]}h,$$

$$\begin{aligned}
\mathsf{M}_\mu^P(\triangle M^{\tau_n}|\tilde{\mathcal{P}}) - \mathbf{1}(0 \quad < \quad \mathrm{a} < 1)\mathsf{M}_p^P(\triangle M^{\tau_n}|\mathcal{P}) \\
&= \mathbf{1}_{[0,\tau_n]}(\mathsf{M}_\mu^\mathbf{P}(\triangle M|\tilde{\mathcal{P}}) - \mathbf{1}(0 < \mathrm{a} < 1)\mathsf{M}_p^\mathbf{P}(\triangle M|\mathcal{P})) \\
&= \mathbf{1}_{[0,\tau_n]}H.
\end{aligned}$$

Since $h \in \mathrm{L}_{loc}^2(\mathbf{F}, \langle X^c\rangle)$ (Theorem 11.2.19) and $H \in \mathcal{G}_{loc}$ (Lemma 11.2.36), the process $M' \in \mathcal{M}_{loc}(\mathbf{F}, \mathbf{P})$ is defined with

$$M_t = M_0 + h \cdot X_t^c + H * (\mu - v)_t.$$

Obviously, M^{τ_n} and $(M')^{\tau_n}$ are $\mathbf{P}$-indistinguishable processes for every $n \geq 1$. Therefore M and M' are $\mathbf{P}$-indistinguishable processes and hence M admits the integral representation. ∎

Lemma 11.2.50 If for any $M \in \mathcal{M}_{loc}(\mathbf{F}, \mathbf{P})$, satisfying Condition (11.25) and the inequality $|\triangle M| \leq const$, the integral representation takes place, then it takes place for any $M \in \mathcal{M}_{loc}(\mathbf{F}, \mathbf{P})$ as well.

Proof. By Lemma 11.2.49 it suffices to establish the existence of the integral representation for $M \in \mathcal{M}_{loc}(\mathbf{F}, \mathbf{P})$ with the property (11.25). Therefore, we will assume henceforth that for M the property (11.25) holds. For $k \geq 1$ set

$$A_t^k = \sum_{0 < s \leq t} \triangle M_s \mathbf{1}(|\triangle M_s| > k).$$

Then
$$\mathbf{EVar}(A^k)_\infty < \infty. \tag{11.26}$$

In fact
$$\mathbf{Var}(A^k)_\infty = \sum_{s>0} |\triangle M_s| \mathbf{1}(|\triangle M_s| > k) \le \frac{k}{1+k} \sum_{s>0} f(\triangle M_s), \tag{11.27}$$

and consequently (11.26) follows from (11.25).

Due to (11.26) there exists, by Theorem 11.2.12, the compensator $\tilde{A}_k$ of the process $A^k = (A^k_t)_{t\ge0}$. Besides (cf. Proposition 11.2.13)

$$\mathbf{EVar}(\tilde{A}^k)_\infty \le \widetilde{\mathbf{EVar}(A^k)}_\infty = \mathbf{EVar}(A^k)_\infty < \infty.$$

Consequently
$$M^k = A^k - \tilde{A}^k \in \mathcal{M}(\mathbf{F}, \mathbf{P}) \cap \mathcal{A}(\mathbf{F}, \mathbf{P}).$$

Consider the process $M - M^k$. Clearly, $M - M^k \in \mathcal{M}_{loc}(\mathbf{F}, \mathbf{P})$. Besides $\triangle(M - M^k) = \triangle M \mathbf{1}(|\triangle M| \le k) + \triangle(\tilde{A}^k)$. By 11.2.12 we have $\triangle(\tilde{A}^k) = {}^p(\triangle A^k)$, while by Theorem 11.2.15 we have ${}^p\triangle M = 0$. Therefore

$$\triangle(\tilde{A}^k) = -{}^p\left(\triangle M \mathbf{1}(|\triangle M| \le k)\right)$$

and consequently

$$\triangle(M - M^k) = \triangle M \mathbf{1}(|\triangle M| \le k) - {}^p\left(\triangle M \mathbf{1}(|\triangle M| \le k)\right). \tag{11.28}$$

By the representation (11.28) and the definition of the predictable projection we get

$$|\triangle(M - M^k)| \le 2k. \tag{11.29}$$

Let us show that $M - M^k$ possesses the property (11.25). Since $(M - M^k)^c = M^c$, it suffices to verify that

$$\mathbf{E} \sum_{s>0} f(\triangle(M - M^k)_s) < \infty.$$

It has been established in the course of proving Lemma 11.2.34 that $f(x + y) \le 2[f(x) + f(y)]$. Therefore, by (11.28) we get

$$f(\triangle(M - M^k)) \le 2[f(\triangle M \mathbf{1}(|\triangle M| \le k)) + f({}^p(\triangle M \mathbf{1}(|\triangle M| \le k)))].$$

Further, by 11.2.4
$$f\left({}^p\left(\triangle M \mathbf{1}(|A| \le k)\right)\right) \le {}^p f(\triangle M \mathbf{1}(|\triangle M| \le k)).$$

Denote
$$B^k = \sum_s f(\triangle M_s \mathbf{1}(|\triangle M_s| \le k)).$$

Since $f(\triangle M \mathbf{1}(|\triangle M| \le k)) \le f(\triangle M)$, by (11.25) we have $\mathbf{E}B^k_\infty < \infty$, and the process B^k has the compensator $\triangle\tilde{B}^k$ (Theorem 11.2.12). Besides, by Theorem 11.2.12 we have $\triangle\tilde{B}^k = {}^p(\triangle B^k) =$

$^p f(\triangle M \mathbf{1}(|\triangle M| \le k))$. In view of these properties

$$\mathbf{E}\sum_{s>0} f(\triangle(M-M^k)_s) \le 2\mathbf{E}\left[\sum_{s>0} f(\triangle M_s \mathbf{1}(|\triangle M_s| \le k)) + \sum_{s>0}\triangle\left(\tilde{B}^k\right)_s\right]$$

$$\le 2\mathbf{E}\left(B^k_\infty + \tilde{B}^k_\infty\right) = 4\mathbf{E}B^k_\infty$$

$$= 4\mathbf{E}\sum_{s>0} f(\triangle M_s \mathbf{1}(|\triangle M_s| \le k))$$

$$\le 4\mathbf{E}\sum_{s>0} f(\triangle M_s) < \infty.$$

Thus $M - M^k \in \mathcal{M}_{loc}(\mathbf{F}, \mathbf{P})$ possesses the property (11.25) and satisfies the inequality $|\triangle(M - M^k)| \le const$. By assumption for this local martingale the integral representation

$$(M - M^k)_t = M_0 + h \circ X^c_t + H^k * (\mu - v)_t$$

takes place with

$$h = \frac{d\langle M^c, X^c\rangle}{d\langle X^c\rangle}$$

(since $(M - M^k)^c = M^c$) and

$$H^k = \mathsf{M}^{\mathbf{P}}_\mu(\triangle(M - M^k)|\tilde{\mathcal{P}}) - \mathbf{1}(0 < \mathrm{a} < 1)\mathsf{M}^{\mathbf{P}}_p(\triangle(M - M^k)|\mathcal{P}),$$

by Lemma 11.2.47.

Let $M' \in \mathcal{M}_{loc}(\mathbf{F}, \mathbf{P})$ be defined by the integral representation

$$M_t = M_0 + h \circ X^c_t + H * (\mu - v)_t \tag{11.30}$$

with

$$h = \frac{d\langle M^c, X^c\rangle}{d\langle X^c\rangle}$$

and

$$H = \mathsf{M}^{\mathbf{P}}_\mu(\triangle M|\tilde{\mathcal{P}}) - \mathbf{1}(0 < \mathrm{a} < 1)\mathsf{M}^{\mathbf{P}}_p(\triangle M|\mathcal{P}).$$

Let us show that the processes M md M' are $\mathbf{P}$-indistinguishable. We have

$$M - M' = M^k + (M - M^k) - M' = M^k + H^k * (\mu - v) - H * (\mu - v)$$

$$= A^k - \tilde{A}^k - U^k * (\mu - v)$$

with

$$U^k = H - H^k = \mathsf{M}^{\mathbf{P}}_\mu(\triangle M^k|\tilde{\mathcal{P}}) - \mathbf{1}(0 < \mathrm{a} < 1)\mathsf{M}^{\mathbf{P}}_p(\triangle M^k|\mathcal{P}).$$

From this it follows that (cf. Theorem 11.2.27)

$$\mathbf{Var}(M - M')_\infty \le \mathbf{Var}(A^k)_\infty + \mathbf{Var}\left(\tilde{A}^k\right)_\infty + |\mathsf{M}^{\mathbf{P}}_\mu(\triangle M^k|\tilde{\mathcal{P}})| * (\mu + v)_\infty \tag{11.31}$$

$$+ \sum_{s>0} \mathbf{1}(\mathrm{a}_s > 0)|\mathsf{M}^{\mathbf{P}}_p(\triangle M^k|\mathcal{P})|[(1 - \mu(\{s\} \times \mathbb{R}_0)) + (1 - \mathrm{a}_s)].$$

Let us establish now the inequality

$$\mathbf{EVar}(M - M')_\infty \le 10\mathbf{EVar}(A^k)_\infty. \tag{11.32}$$

By 11.2.13

$$\mathbf{EVar}\left(\tilde{A}^k\right)_\infty \le \widetilde{\mathbf{EVar}\,(A^k)}_\infty = \mathbf{EVar}\left(A^k\right)_\infty. \tag{11.33}$$

This implies, in particular,

$$\mathbf{E}\sum_{s>0}|\triangle M_s^k| \le \mathbf{E}(\mathbf{Var}(A^k)_\infty + \mathbf{Var}(\tilde{A}^k)_\infty) \le 2\mathbf{EVar}(A^k)_\infty. \tag{11.34}$$

Therefore, due to Jensen's inequality for $\mathsf{M}_\mu^{\mathbf{P}}(\cdot|\tilde{\mathcal{P}})$ (11.2.25), we get

$$
\begin{aligned}
\mathbf{E}|\mathsf{M}_\mu^{\mathbf{P}}(\triangle M^k|\tilde{\mathcal{P}})| * (\mu + v)_\infty &= 2\mathbf{E}\mathsf{M}_\mu^{\mathbf{P}}(\triangle M^k|\tilde{\mathcal{P}})| * \mu_\infty \le 2\mathbf{E}\mathsf{M}_\mu^{\mathbf{P}}(|\triangle M^k||\tilde{\mathcal{P}}) * \mu_\infty \quad (11.35)\\
&= 2\mathbf{E}|\triangle M^k| * \mu_\infty \le 2\mathbf{E}\sum_{s>0}|\triangle\phi_s|\\
&\le 4\mathbf{EVar}(A^k)_\infty
\end{aligned}
$$

and analogously

$$
\begin{aligned}
\mathbf{E}\sum_{s>0}\mathbf{1}(a_s > 0)|\mathsf{M}_p^{\mathbf{P}}(\triangle M^k|\mathcal{P})|[(1 - \mu(\{s\} \times \mathbb{R}_0)) + (1 - a_s)] &\quad (11.36)\\
= 2\mathbf{E}|\mathsf{M}_p^{\mathbf{P}}(\triangle M^k|\mathcal{P})| * p_\infty \le 2\mathbf{E}\mathsf{M}_p^{\mathbf{P}}(|\triangle M^k||\mathcal{P}) * p_\infty&\\
= 2\mathbf{E}|\triangle M^k| * p_\infty \le 2\mathbf{E}\sum_{s>0}|\triangle M_s^k| \le 4\mathbf{EVar}(A^k)_\infty.&
\end{aligned}
$$

The desired inequality (11.32) follows from (11.31), (11.33), (11.35) and (11.36).

To prove $\mathbf{P}$-indistinguishability of the processes M and M', it suffices, by the obtained inequality (11.32), to show that

$$\lim_k \mathbf{EVar}(A^k)_\infty = 0. \tag{11.37}$$

In virtue of (11.27) as $k \ge 1$ we have

$$
\begin{aligned}
\mathbf{Var}(A^k)_\infty &\le \mathbf{1}\left(M_\infty^* > \frac{k}{2}\right)\sum_{s>0}|\triangle M_s|\mathbf{1}(|\triangle M_s| > 1)\\
&\le \frac{1}{2}\mathbf{1}\left(M_\infty^* > \frac{k}{2}\right)\sum_{s>0}f(\triangle M_s).
\end{aligned}
$$

This shows that (11.37) will take place if $M \in \mathrm{H}(\mathbf{F}, \mathbf{P})$. For this in turn it suffices to show that $M^c \in \mathrm{H}(\mathbf{F}, \mathbf{P})$ and $M^d \in \mathrm{H}(\mathbf{F}, \mathbf{P})$.

Let us show that the following implication takes place:

$$\mathbf{E}\left(\langle M^c\rangle_\infty + \sum_{s>0}f(\triangle M_s)\right) < \infty \implies \left\{\begin{array}{l} M^c \in \mathrm{H}(\mathbf{F}, \mathbf{P}),\\ M^d \in \mathrm{H}(\mathbf{F}, \mathbf{P}). \end{array}\right. \tag{11.38}$$

In fact, by Doob's inequality (Theorem 11.2.16) and (11.25) we have

$$\mathbf{E}((M^c)^*_\infty)^2 \leq 4\mathbf{E}\langle M^c\rangle_\infty < \infty,$$

i.e., $M^c \in \mathrm{H}(\mathbf{F},\mathbf{P})$.

Next, by the inequalities (11.25) and (11.2.20) in Chapter 2 we have

$$\mathbf{E}[M^d, M^d]^{1/2}_\infty \leq \frac{1}{4} + 2\mathbf{E}\sum_{s>0} f(\triangle M_s) < \infty.$$

By Davis inequality (Theorem 11.2) this gives

$$\mathbf{E}(M^d)^*_\infty \leq C\mathbf{E}[M^d, M^d]^{1/2}_\infty < \infty$$

i.e., $M^d \in \mathrm{H}(\mathbf{F},\mathbf{P})$. Thus Lemma 11.2.50 is proved. ∎

Next, we go back to prove Theorem 11.2.46.

Proof. Theorem 11.2.46. (a)⇒(b). Let $\tilde{\mathbf{P}} \in \mathcal{L}_\mathbf{P}$ and $\tilde{\mathbf{P}} \neq \mathbf{P}$. By the definition of $\mathcal{L}_\mathbf{P}$ we have $\tilde{\mathbf{P}} << \mathbf{P}$. Let Z be the density process of the measure $\tilde{\mathbf{P}}$ with respect to $\mathbf{P}$. Since $Z \in \mathcal{M}(\mathbf{F},\mathbf{P})$, then Z admits the integral representation

$$Z_t = Z_0 + h \circ X^c_t + H * (\mu - v)_t.$$

By the equality $\tilde{\mathbf{P}}_0 = \mathbf{P}_0$ the random variable $Z_0 = 1$. By Theorem 11.2.40 the process X belongs to $\mathcal{S}(\mathbf{F},\tilde{\mathbf{P}})$ and possesses the triplet $\tilde{T} = (\tilde{B}, \tilde{C}, \tilde{v})$ with

$$\tilde{B} = B + \beta \circ C + x\mathbf{1}_{(|x|\leq 1)}(Y-1)*v, \tag{11.39}$$
$$\tilde{C} = C,$$
$$\tilde{v} = Yv,$$

where

$$Y = \mathbf{1}(Z_- > 0)Z^{-1}_-\mathsf{M}^\mathbf{P}_\mu(Z|\tilde{\mathcal{P}}), \quad \beta = \mathbf{1}_{(Z_->0)}Z^{-1}_-\frac{d\langle X^c, Z^c\rangle}{d\langle X^c\rangle}$$

Denote by $\mathsf{M}^\mathbf{P}_C$ the measure with $\mathsf{M}^\mathbf{P}_C(d\omega, dt, dx) = \mathbf{P}(d\omega)dC_t(\omega)$. Since by the definition of the class $\mathcal{L}_\mathbf{P}$ we have $\tilde{T} \overset{\mathbf{P}}{=} T$, then $Y = 1$ ($\mathsf{M}^\mathbf{P}_v$-a.s.) and $\beta = 0$ ($\mathsf{M}^\mathbf{P}_C$-a.s.). Set $\tau = \inf(t : Z_t < 1/2)$. From the definition of Y and β it follows that on the set $[\![0,\tau]\!]$ we have

$$\mathsf{M}^\mathbf{P}_\mu(\triangle Z|\tilde{\mathcal{P}}) = 0 \quad (\mathsf{M}^\mathbf{P}_v\text{-a.s.})$$

and

$$\frac{d\langle X^c, Z^c\rangle}{d\langle X^c\rangle} = 0 \quad (\mathsf{M}^\mathbf{P}_C\text{-a.s.}).$$

Next (cf. Theorem 11.2.27),

$$(1-a)\mathbf{1}_{[\![0,\tau]\!]}\mathsf{M}^\mathbf{P}_p(\triangle Z|\mathcal{P}) = -\mathbf{1}_{[\![0,\tau]\!]}\mathbf{1}(0 < a < 1)\mathsf{M}^\mathbf{P}_\mu\widehat{(\triangle Z|\tilde{\mathcal{P}})} = 0.$$

Since by Lemma 11.2.47 in the integral representation for Z we have

$$h = \frac{d\langle X^c, Z^c \rangle}{d\langle X^c \rangle}, \quad H = \mathsf{M}_\mu^{\mathbf{P}}(\triangle Z | \tilde{\mathcal{P}}) - \mathbf{1}(0 < \mathrm{a} < 1) \mathsf{M}_p^{\mathbf{P}}(\triangle Z | \mathcal{P}),$$

then $Z^\tau = Z_0 = 1$. This gives, in particular, $\mathbf{P}(\tau = \infty) = 1$, i.e., $Z_\infty = 1$ ($\mathbf{P}$-a.s.).

Therefore $\tilde{\mathbf{P}} = \mathbf{P}$. (a)$\Rightarrow$(b). By Lemma 11.2.50 it suffices to consider local martingales M possessing Conditions (11.25) and $|\triangle M| \leq c$. Observe that in accordance with (11.38) we have $M \in \mathrm{H}(\mathbf{F}, \mathbf{P})$.

Set

$$h = \frac{d\langle M^c, X^c \rangle}{d\langle X^c \rangle}, \quad H = \mathsf{M}_\mu^{\mathbf{P}}(\triangle M | \tilde{\mathcal{P}}) - \mathbf{1}(0 < \mathrm{a} < 1) \mathsf{M}_p^{\mathbf{P}}(\triangle M | \mathcal{P}).$$

By Theorem 11.2.19 we have $h^2 \circ C \in \mathcal{A}_{loc}^+$ and by Theorem 11.2.36 we have $H \in \mathcal{G}_{loc}$, consequently $M' \in \mathcal{M}_{loc}(\mathbf{F}, \mathbf{P})$ is defined with

$$M_t = M_0 + h \circ X_t^c + H * (\mu - v)_t.$$

Let us show that by Condition (11.25) for M the local martingale $M' \in \mathrm{H}(\mathbf{F}, \mathbf{P})$. It suffices to show, by (11.38), that

$$\mathbf{E}\left[\langle M' \rangle_\infty + \sum_{s>0} f(\triangle M_s) \right] < \infty \tag{11.40}$$

By Theorem 11.2.19 we have $\langle M'^c \rangle = h^2 \circ C_\infty \leq \langle M^c \rangle_\infty$, and hence

$$\mathbf{E}\langle M'^c \rangle_\infty < \infty. \tag{11.41}$$

Since by Proposition 11.2.28 we have

$$H * (\mu - v) = \left[H - \hat{H}\mathbf{1}\,(\mathrm{a} = 1) \right] * (\mu - v),$$

then from the definition of M' it follows that

$$\triangle M'_s = \left[H(s, \triangle X_S) - \hat{H}(s) \right] \mathbf{1}(\triangle X_s \neq 0) - \hat{H}(s)\mathbf{1}(\mathrm{a}_s < 1)\mathbf{1}(\triangle X_s = 0). \tag{11.42}$$

Therefore, by the inequality $f(x + y) \leq 2[f(x) + f(y)]$ we get

$$\sum_{s>0} f(\triangle M'_s) \leq 2f(H - \hat{H}) * \mu_\infty + 2f(\hat{H}\mathbf{1}(\mathrm{a} < 1)) * p_\infty. \tag{11.43}$$

But by the definition of H and by Problem 11.2.29 we have

$$H - \hat{H} = \mathsf{M}_\mu^{\mathbf{P}}(\triangle M | \tilde{\mathcal{P}}), \quad \hat{H}\mathbf{1}(\mathrm{a} < 1) + -\mathbf{1}(0 < \mathrm{a} < 1)\mathsf{M}_p^{\mathbf{P}}(\triangle M | \mathcal{P}). \tag{11.44}$$

In view of Jensen's inequality, by this we get

$$\begin{aligned}
\mathbf{E}\sum_{s>0} f(\triangle M'_s) & \leq 2\mathbf{E}f(\mathsf{M}_\mu^{\mathbf{P}}(\triangle M | \tilde{\mathcal{P}})) * \mu_\infty + 2\mathbf{E}f(\mathsf{M}_p^{\mathbf{P}}(\triangle M | \mathcal{P})) * p_\infty \tag{11.45} \\[6pt]
& \leq 2\mathbf{E}\mathsf{M}_\mu^{\mathbf{P}}(f(\triangle M) | \tilde{\mathcal{P}}) * \mu_\infty + 2\mathbf{E}\mathsf{M}_p^{\mathbf{P}}(f(\triangle M) | \mathcal{P}) * p_\infty \\[6pt]
& = 2\mathbf{E}f(\triangle M) * \mu_\infty + 2\mathbf{E}f(\triangle M) * p_\infty \leq 4\mathbf{E}\sum_{s>0} f(\triangle M_s) < \infty.
\end{aligned}$$

Thus, (11.40) follows from (11.41) and (11.45), and hence $M' \in H(\mathbf{F}, \mathbf{P})$. Then, $N = M - M' \in H(\mathbf{F}, \mathbf{P})$ and the process N possesses, as it will be shown below, the following properties:

$$|\triangle N| \leq |\triangle M| + |\triangle M'| \leq 5c, \tag{11.46}$$

$$\langle N^c, X^c \rangle = 0, \tag{11.47}$$

$$\mathsf{M}_\mu^\mathbf{P}(\triangle N | \tilde{\mathcal{P}}) = 0 \quad (\mathsf{M}_\mu^\mathbf{P} \text{ - a.s.}). \tag{11.48}$$

In fact, the property (11.46) takes place, since $|\triangle M| \leq c$ and by Theorem 11.2.36 we have $|\triangle M'| \leq 4c$. The property (11.47) holds, since $\langle M^c, X^c \rangle = \langle M'^c, X^c \rangle$. Finally, the property (11.48) follows from (11.42) and (11.44).

We assume $N_\infty^* \leq a$, making use of a localizing sequence $\tau_n = \inf(t : |N_t| \geq n)$, $n \geq 1$ if necessary.

Denote

$$Z = 1 + \frac{N}{2a}.$$

Then $Z \in H(\mathbf{F}, \mathbf{P})$ and $Z \geq \frac{1}{2}$. Define the probability measure $\tilde{\mathbf{P}}$ with $d\tilde{\mathbf{P}} = Z_\infty d\mathbf{P}$. We will show that $\tilde{\mathbf{P}} \in \mathcal{L}_\mathbf{P}$.

By construction $\tilde{\mathbf{P}} << \mathbf{P}$. By the property $Z_0 = 1$ we have $\tilde{\mathbf{P}}_0 = \mathbf{P}_0$. Next, by Theorem 11.2.40 we have $X \in \mathcal{S}(\mathbf{F}, \tilde{\mathbf{P}})$. From (11.47) and (11.48) it follows that $\langle Z^c, X^c \rangle = 0$ and $\mathsf{M}_\mu^\mathbf{P}(\triangle Z | \tilde{\mathcal{P}}) = 0$.

Therefore, the local absolutely continuous change of measures (see 11.39), leads here to the triplet transformation with $\beta = 0$ and $Y = 1$. Consequently, by Theorem 11.2.40 and the remark to it the triplet $\tilde{T}$ of the semimartingale $X \in \mathcal{S}(\mathbf{F}, \mathbf{P})$ is $\mathbf{P}$-indistinguishable from the triplet T.

Thus $\tilde{\mathbf{P}} \in \mathcal{L}_\mathbf{P}$. But, by definition $\mathcal{L}_\mathbf{P} = \{\mathbf{P}\}$. Hence $\tilde{\mathbf{P}} = \mathbf{P}$ and the local density process Z is $\mathbf{P}$-indistinguishable from the process identically equal to one, i.e., the process N is $\mathbf{P}$-negligible and the processes M and M' are $\mathbf{P}$-indistinguishable. ∎

Let $X \in \mathcal{S}(\mathbf{F}, \mathbf{P})$ and define the σ-algebras,

$$\mathcal{F}_t^X = \sigma(\{X_s, 0 \leq s \leq t\} \vee \mathcal{N}), \quad \mathcal{F}_{t+}^X = \bigcap_{\epsilon > 0} \mathcal{F}_{t+\epsilon}^X,$$

where $\mathcal{N}$ is the system of null sets of $\mathcal{F}$ of $\mathbf{P}$-measure zero. Set

$$\mathcal{F}_\infty^X = \sigma \left(\bigcup_{t \geq 0} \mathcal{F}_t^X \right).$$

The restriction of a measure $\mathbf{P}$ on the σ-algebras $\mathcal{F}_\infty^X$, $\mathcal{F}_{0+}^X$ and $\mathcal{F}_0^X$ is denoted by $\mathbf{P}^X$, $\mathbf{P}_{0+}^X$ and $\mathbf{P}_0^X$, respectively. By Theorem 11.2.41 we have

$$X \in \mathcal{S}(\mathbf{F}_+^X, \mathbf{P}^X).$$

Therefore, arises a question about what is the integral representation for any local martingale $M \in \mathcal{M}_{loc}(\mathbf{F}_+^X, \mathbf{P}^X)$? In the present special case one can strengthen the results of Theorem 11.2.46 to some extent.

Denote by T^X the triplet of predictable characteristics of $(\mathbf{F}_+, \mathbf{P}^X)$-semimartingale X. If $\tilde{\mathbf{P}}^X$ is a probability measure on $(\Omega, \mathcal{F}^X_\infty)$, absolutely continuous with respect to a measure $\mathbf{P}^X$ ($\tilde{\mathbf{P}}^X \ll \mathbf{P}^X$), then by Theorem 11.2.40 we have

$$X \in \mathcal{S}(\mathbf{F}^X_+, \tilde{\mathbf{P}}^X).$$

The triplet of predictable characteristics of this semimartingale is denoted by $\tilde{T}^X$. Define two families of probability measures on $(\Omega, \mathcal{F}^X_\infty)$ (cf. (11.22)):

$$\mathcal{L}_{\mathbf{P}^X_\oplus} = \left\{ \tilde{\mathbf{P}}^X : \tilde{\mathbf{P}}^X \ll \mathbf{P}^X, \ \tilde{\mathbf{P}}^X_{0+} = \mathbf{P}^X_{0+}, \ \tilde{T}^X \overset{\mathbf{P}^X}{=} T^X \right\},$$

$$\mathcal{L}_{\mathbf{P}^X} = \left\{ \tilde{\mathbf{P}}^X : \tilde{\mathbf{P}}^X \ll \mathbf{P}^X, \ \tilde{\mathbf{P}}^X_0 = \mathbf{P}^X_0, \ \tilde{T}^X \overset{\mathbf{P}^X}{=} T^X \right\},$$

where $\mathbf{P}^X_{0+}$ and $\mathbf{P}^X_0$ are the restrictions of the measures $\mathbf{P}^X$ to $\mathcal{F}^X_{0+}$ and $\mathcal{F}^X_0$. Clearly, $\mathcal{L}_{\mathbf{P}^X} \supseteq \mathcal{L}_{\mathbf{P}^X_\oplus}$. Thus, we are led to the following theorem.

Theorem 11.2.51 The following conditions are equivalent:
  (a) each local martingale $M \in \mathcal{M}_{loc}(\mathbf{F}^X_+, \mathbf{P}^X)$ admits the integral representation and $\mathbf{F}^X_+ = \mathbf{F}^X$;
  (b) $\mathcal{L}_{\mathbf{P}^X} = \mathcal{L}_{\mathbf{P}^X_\oplus} = \{\mathbf{P}^X\}$.

Proof. (a)$\Rightarrow$(b): by Theorem 11.2.46 and the equality $\mathbf{F}_+ = \mathbf{F}$ we have

$$\mathcal{L}_{\mathbf{P}^X} = \mathcal{L}_{\mathbf{P}^X_\oplus} = \{\mathbf{P}\}.$$

(b)$\Rightarrow$(a): by Theorem 11.2.46 and the condition $\mathcal{L}_{\mathbf{P}^X_\oplus} = \{\mathbf{P}\}$ for each $M \in \mathcal{M}_{loc}(\mathbf{F}^X_+, \mathbf{P}^X)$ the integral representation takes place.

Let us show now that $\mathbf{F}^X_+ = \mathbf{F}^X$. To this end, we establish first that $\mathcal{F}^X_{0+} = \mathcal{F}^X_0$, and then we show that $\mathcal{F}^X_{t+} = \mathcal{F}^X_t$ for each $t > 0$. Let $\Gamma \in \mathcal{F}^X_{0+}$. Define the random variable

$$Z_\infty = 1 + \frac{1}{2}(1_\Gamma - \mathbf{P}^X(\Gamma | \mathcal{F}^X_0)).$$

Clearly,

$$\frac{1}{2} \leq Z_\infty \leq \frac{3}{2}, \quad \int_\Omega Z_\infty d\mathbf{P}^X = 1$$

and for each set $A \in \mathcal{F}^X_0$ we have

$$\int_A Z_\infty d\mathbf{P}^X = \mathbf{P}^X(A).$$

Therefore, the probability measure $\tilde{\mathbf{P}}^X$ on $(\Omega, \mathcal{F}^X_\infty)$ with $d\tilde{\mathbf{P}}^X = Z_\infty d\mathbf{P}^X$ is equivalent to the measure $\mathbf{P}^X$ ($\tilde{\mathbf{P}}^X \sim \mathbf{P}^X$), and $\tilde{\mathbf{P}}^X_0 = \mathbf{P}^X_0$. Let $Z = (Z_t)_{t \geq 0}$ be the density process of the measure $\tilde{\mathbf{P}}^X$ with respect to $\mathbf{P}^X$ (and the family $\mathbf{F}^X_+$). Clearly, $Z_t = Z_\infty$ ($\mathbf{P}^X$-a.s.). This means that for the martingale Z the continuous martingale component Z^c and the jump process $\triangle Z$ are negligible. By Theorem 11.2.40 this gives

$$\tilde{T}^X \overset{\mathbf{P}^X}{=} T^X.$$

Hence
$$\tilde{\mathbf{P}}^X \in \mathcal{L}_{\mathbf{P}_\oplus^X}.$$

By definition
$$\mathcal{L}_{\mathbf{P}^X} = \mathcal{L}_{\mathbf{P}_\oplus^X} = \{\mathbf{P}^X\},$$

i.e., $\tilde{\mathbf{P}}^X = \mathbf{P}^X$, and consequently $Z_\infty = 1$. Therefore, $\mathbf{1}_\Gamma = \mathbf{P}^X(\Gamma|\mathcal{F}_0^X)$, i.e., $\Gamma \in \mathcal{F}_0^X$ (recall that the σ-algebra $\mathcal{F}_0^X$ is completed by sets from $\mathcal{F}$ of zero probability $\mathbf{P}$). As the set Γ is arbitrary, we have $\mathcal{F}_{0+}^X = \mathcal{F}_0^X$.

Let $t > 0$ and $\Gamma \in \mathcal{F}_{t+}^X$. Consider a process $M \in \mathcal{M}_{loc}(\mathbf{F}_+^X, \mathbf{P}^X)$ that is the optional projection of $\mathbf{1}_\Gamma$ relative to $(\mathbf{F}_+^X, \mathbf{P}^X)$. Clearly, $M_t = \mathbf{1}_\Gamma$. On the other hand M admits the integral representation, in particular
$$\mathbf{1}_\Gamma = \mathbf{P}^X(\Gamma|\mathcal{F}_0^X) + h \circ X_t^c + H * (\mu - v)_t,$$

where X^c is the continuous $(\mathbf{F}_+, \mathbf{P}^X)$-martingale component of $X \in \mathcal{S}(\mathbf{F}_+, \mathbf{P}^X)$ and v the compensator of the jump measure of the process X relative to $(\mathbf{F}_+, \mathbf{P}^X)$.

Let $s > t$. The random variables M_s are $\mathcal{F}_t^X$-measurable. Hence M_t is a $\mathcal{F}_t^X$-measurable random variable. Therefore, it suffices to show that $\triangle M_t$ is a $\mathcal{F}_t^X$-measurable random variable. From the integral representation it follows that
$$\triangle M_t = H(t, \triangle X_t)\mathbf{1}(\triangle X_t \neq 0) - \hat{H}(t).$$

The functions $H(t, x)$ and $\hat{H}(t)$ are measurable with respect to the σ-algebras $\mathcal{P}(\mathbf{F}_+) \times \mathcal{B}(\mathbb{R}_0)$ and $\mathcal{P}(\mathbf{F}^X)$. Since $\mathcal{P}(\mathbf{F}_+^X) = \mathcal{P}(\mathbf{F}^X)$ (11.2.2), for each fixed t the variables $H(t, x)$ and $\hat{H}(t)$ are $\mathcal{F}_t^X \times \mathcal{B}(\mathbb{R}_0)$-and $\mathcal{F}_t^X$-measurable functions. Since $\triangle X_t$ is $\mathcal{F}_t^X$-measurable, this shows that $\triangle M_t$ is a $\mathcal{F}_t^X$-measurable random variable.

Let $X \in \mathcal{S}(\mathbf{F}, \mathbf{P})$ and $\mathbf{Q}$ be a probability measure on $(\Omega, \mathbf{F})$, locally absolutely continuous with respect to a measure $\mathbf{P}$ ($\mathbf{Q} <<_{loc} \mathbf{P}$). By Theorem 11.2.40 we have $X \in \mathcal{S}(\mathbf{F}, \mathbf{Q})$. Besides, the triplets $T = (B, C, v)$ and $T^\mathbf{Q} = (B^\mathbf{Q}, C^\mathbf{Q}, v^\mathbf{Q})$ of the corresponding semimartingales are related by the following relations
$$\begin{aligned}
B^\mathbf{Q} &= B + \beta \circ C + x\mathbf{1}(|x| \leq 1)(Y - 1) * v, \qquad (11.49)\\
C^\mathbf{Q} &= C,\\
v^\mathbf{Q} &= Yv,
\end{aligned}$$

that are understood in the sense of indistinguishability of the corresponding left-hand and right-hand parts relative to the measures $\mathbf{Q}$ and $\mathbf{P}$ on the set $[\![0, \tau[\![$, where $\tau = \inf(t : Z_t = 0)$ and Z is the local density process of the measure $\mathbf{Q}$ with respect to $\mathbf{P}$. Moreover, on the set $[\![0, \tau[\![$
$$\beta = Z_-^{-1}\frac{d\langle Z^c, X^c \rangle}{d\langle X^c \rangle}, \quad Y = Z_-^{-1}\mathsf{M}_\mu^\mathbf{P}(\triangle Z|\tilde{\mathcal{P}}) + 1.$$

We will consider the relations (11.49) from the point of view of defining the functions β and Y by the triplets T and $T^\mathbf{Q}$.

Suppose $\mathcal{L}_\mathbf{P} = \{\mathbf{P}\}$. Then by Theorem 11.2.46 any local martingale $M \in \mathcal{M}_{loc}(\mathbf{F}, \mathbf{P})$ admits the integral representation. Since $Z \in \mathcal{M}_{loc}(\mathbf{F}, \mathbf{P})$, for Z the integral representation
$$Z_t = Z_0 + h \circ X_t^c + H * (\mu - v)_t$$

takes place where

$$Z_0 = \frac{d\mathbf{Q}_0}{d\mathbf{P}_0}$$

($\mathbf{Q}_0$, $\mathbf{P}_0$ are the restrictions of the measures $\mathbf{Q}$ and $\mathbf{P}$ to the σ-algebra X_0),

$$h = \frac{d\langle Z^c, X^c\rangle}{d\langle X^c\rangle}, \quad H = \mathsf{M}^{\mathbf{P}}_\mu(\triangle Z|\tilde{\mathcal{P}}) - \mathbf{1}(0 < a < 1)\mathsf{M}^{\mathbf{P}}_p(\triangle Z|\mathcal{P})$$

(cf. Lemma 11.2.47) with ($\{a < 1\}$: $\mathsf{M}^{\mathbf{P}}_{\mu-}$-a.s.)

$$\mathsf{M}^{\mathbf{P}}_p = -\frac{\mathbf{1}(0 < a < 1)}{1 - a}\widehat{\mathsf{M}^{\mathbf{P}}_\mu(\triangle Z|\tilde{\mathcal{P}})}.$$

Therefore, on the set $[\![0, \tau[\![$

$$h = Z_-\beta, \quad H = Z_-\left(Y - 1 + \frac{\mathbf{1}(0 < a < 1)}{1 - a}(\hat{Y} - a)\right)$$

(cf. 11.2.39). Let

$$\tau_n = \inf\left(t : Z_t < \frac{1}{n}\right), \quad n \geq 1.$$

On the set $[\![0, \tau_n[\![$ we have $Z_- \geq \frac{1}{n}$. Therefore, the process $M^n \in \mathcal{M}_{loc}(\mathbf{F}, \mathbf{P})$ is defined with

$$M^n = \mathbf{1}_{[\![0, \tau_n]\!]}\beta \circ X^c + \mathbf{1}_{[\![0, \tau_n]\!]}\left(Y - 1 + \frac{\mathbf{1}(0 < a < 1)}{1 - a}(\hat{Y} - a)\right) * (\mu - v) \tag{11.50}$$

■

The following theorem is frequently utilized in various applications, in particular, in statistics of stochastic processes, describes the structure of the densities of probability measures related to two semimartingales.

Theorem 11.2.52 Let $X \in \mathcal{S}(\mathbf{F}, \mathbf{P})$ and $\mathcal{L}_{\mathbf{P}} = \{\mathbf{P}\}$. If $\mathbf{Q} <<_{loc} \mathbf{P}$ and Z is the local density process of a measure $\mathbf{Q}$ with respect to $\mathbf{P}$, then (in the sense of $\mathbf{P}$-indistinguishability)

$$Z^{\tau_n}_t = Z_0\mathcal{E}_t(M^n), \quad t \geq 0, \quad n \geq 1,$$

where $\mathcal{E}(M^n)$ is the stochastic exponential ([51] Ch. 2, §4),

$$\mathcal{E}_t(M^n) = \exp\left(M^n_t - \frac{1}{2}\langle M^{n,c}\rangle_t\right)\prod_{0 < s \leq t}(1 + \triangle M^n_s)\exp(-\triangle M^n_s). \tag{11.51}$$

Proof. From the definition of h, H and the process M^n it follows that

$$\begin{aligned} Z^{\tau_n}_t &= 1 + Z_-\beta \circ X^c_{t \wedge \tau_n} + Z_-\left(Y - 1 + \frac{\mathbf{1}(0 < a < 1)}{1 - a}(\hat{Y} - a)\right) * (\mu - v)_{t \wedge \tau_n} \quad (11.52) \\ &= 1 + Z^{\tau_n}_- \circ M^n_t \end{aligned}$$

i.e., Z^{τ_n} is defined by Doleans equation which has the unique solution that gives the desired representation. ∎

The representation for Z^{τ_n}, obtained in Theorem 11.2.52, can be used for checking the property $\mathcal{L}_{\mathbf{P}} = \{\mathbf{P}\}$.

Theorem 11.2.53 Let $X \in \mathcal{S}(\mathbf{F}, \mathbf{P})$ and $\mathbf{Q} <<_{loc} \mathbf{P}$. Then

$$\mathcal{L}_{\mathbf{P}} = \{\mathbf{P}\} \Rightarrow \mathcal{L}_{\mathbf{Q}} = \{\mathbf{Q}\}. \tag{11.53}$$

Proof. Let $\tilde{\mathbf{Q}} \in \mathcal{L}_{\mathbf{Q}}$ and $\tilde{\mathbf{Q}} \neq \mathbf{Q}$. Since $\tilde{\mathbf{Q}} << \mathbf{Q}$ and $\mathbf{Q} <<_{loc} \mathbf{P}$, then $\tilde{\mathbf{Q}} <<_{loc} \mathbf{P}$. Let Z and $\tilde{Z}$ be the density processes of the measures $\mathbf{Q}$ and $\tilde{\mathbf{Q}}$ with respect to $\mathbf{P}$ on τ_n and $\tilde{\tau}_n$ which are defined by the relations:

$$\tau_n = \inf\left(t : Z_t < \frac{1}{n}\right), \quad \tilde{\tau}_n = \inf\left(t : \tilde{Z}_t < \frac{1}{n}\right).$$

Set

$$\sigma_n = \tau_n \wedge \tilde{\tau}_n.$$

Since $\tilde{\mathbf{Q}} \in \mathcal{L}_{\mathbf{Q}}$, the triplets $T^{\mathbf{Q}}$ and $T^{\tilde{\mathbf{Q}}}$, corresponding to $X \in \mathcal{S}(\mathbf{F}, \mathbf{Q})$ and $X \in \mathcal{S}(\mathbf{F}, \tilde{\mathbf{Q}})$ are $\mathbf{Q}$-indistinguishable. From this it is not hard to deduce, by using formulas of type (11.49), that the functions β, $\tilde{\beta}$ and Y, $\tilde{Y}$ can be chosen in such a way that

$$\beta = \tilde{\beta}([\![0, \sigma_n[\![; \ \mathsf{M}_C^{\mathbf{P}}\text{-a.s.}),$$

with

$$\mathsf{M}_C^{\mathbf{P}}(d\omega, dt) = \mathbf{P}(d\omega) dC_t(\omega)),$$

and

$$Y = \tilde{Y} \ ([\![0, \sigma_n[\![; \ \mathsf{M}_\mu^{\mathbf{P}}\text{-a.s.}).$$

This implies the coincidence on the set $[\![0, \sigma_n]\!]$ of the local martingales M^n and $\tilde{\mathsf{M}}^n$, involved in the representation of type (11.52). Since $\mathbf{Q}_0 = \tilde{\mathbf{Q}}_0$, the equality $Z_0 = \tilde{Z}_0$ holds and from Theorem 11.2.52 it follows that the processes Z^{σ_n} and $\tilde{Z}^{\sigma_n}$ are $\mathbf{P}$-indistinguishable and $\tau_n = \tilde{\tau}_n$, $n \geq 1$, i.e., the processes Z and $\tilde{Z}$ are $\mathbf{P}$-indistinguishable on $\cup_{n \geq 1}[\![0, \tau_n]\!]$.

Let $\tau = \inf\left(t; Z_t \wedge \tilde{Z}_t = 0\right)$. Since

$$[\![0, \tau[\![\subseteq \cup_{n \geq 1}[\![0, \tau_n]\!],$$

the processes Z and $\tilde{Z}$ are $\mathbf{P}$-indistinguishable on $[\![0, \tau[\![$.

Let us show that $Z1_{[\![\tau, \infty[\![}$, and $\tilde{Z}1_{[\![\tau, \infty[\![}$ are $\mathbf{P}$-indistinguishable processes. The set $\{t < \infty\}$ is representable $\mathbf{P}$-a.s. as the sum of sets $\{Z_{\tau-} = \tilde{Z}_{\tau-} = 0\}$ and $\{Z_{\tau-} = \tilde{Z}_{\tau-} > 0\}$. By Proposition 11.2.22 we have

$$Z1_{[\![\tau, \infty[\![} = \tilde{Z}1_{[\![\tau, \infty[\![} \quad (\{Z_{\tau-} = \tilde{Z}_{\tau-} = 0\}) \quad \mathbf{P}\text{-a.s.}$$

If

$$\omega \in \{Z_{\tau(\infty)-}(\omega) = Z_{\tau(\omega)-}(\omega) > 0\},$$

then for a given ω we have

$$\cup_{n \geq 1} [\![0, \tau_n(\omega)]\!] = [\![0, \tau(\omega)]\!], \quad Z_{\tau(\omega)}(\omega) = \tilde{Z}_{\tau(\omega)}(\omega) = 0,$$

and this gives

$$Z 1_{[\![\tau, \infty[\![} = \tilde{Z} 1_{[\![\tau, \infty[\![}} \quad (\{Z_{\tau-} = \tilde{Z}_{\tau-} > 0\}) \quad \textbf{P}\text{-a.s.}$$

by 11.2.9.

Thus Z and $\tilde{Z}$ are $\textbf{P}$-indistinguishable processes. This means $\textbf{Q}_t = \tilde{\textbf{Q}}_t$, $t \geq 0$, where $\textbf{Q}_t$, $\tilde{\textbf{Q}}_t$ are the restrictions of the measures $\textbf{Q}$, $\tilde{\textbf{Q}}$ to the σ-algebra $\mathcal{F}_t$. This implies $\textbf{Q} = \tilde{\textbf{Q}}$. $\blacksquare$

11.2.3   Filtering of *cadlag* semimartingales

Let $(X, Y) = (X_t, Y_t)_{t \geq 0}$ be a partially observable stochastic process where Y is an unobservable component, while X is a process that is observed. The problem of filtering out the process Y by means of the process X is understood as constructing for each $t \geq 0$ an "optimal" (in one or another sense) estimate of variables Y_t by observations on X_t, $s \leq t$. Based on previous results, in the present section the problem will be studied of filtering out a semimartingale Y by means of a semimartingale X.

Thus, let $(\Omega, \mathcal{F}, \textbf{F}, \textbf{P})$ be a stochastic basis and $Y \in \mathcal{S}p(\textbf{F}, \textbf{P})$, $X \in \mathcal{S}(\textbf{F}, \textbf{P})$. It is useful to define the process X by its canonical representation

$$X_t = X_0 + B_t + X_t^c + \int_0^t \int_{|x| \leq 1} x(d\mu - d\nu) + \int_0^t \int_{|x| > 1} x d\mu \tag{11.54}$$

Let $T = (B, C, \nu)$ be the triplet of predictable characteristics of the process X. Assume that the process Y is a special semimartingale

$$Y_t = Y_0 + M_t + A_t \tag{11.55}$$

with $M \in \mathcal{M}_{loc,0}$ and $A \in \mathcal{A}_{loc} \cap \mathcal{P}(\textbf{F})$. Assume

$$\textbf{E} |Y_0| < \infty, \quad \textbf{EVar}(A)_\infty < \infty \tag{11.56}$$
$$M \in \mathcal{H}(\textbf{F})$$

and denote by $\textbf{F}_+^X = (\mathcal{F}_{t+}^X)_{t \geq 0}$ the family of σ-algebras $\mathcal{F}_{t+}^X = \bigcap_{\varepsilon > 0} \mathcal{F}_{t+\varepsilon}^X$ where

$$\mathcal{F}_t^X = \sigma \{X_s, s \leq t\} \vee \mathcal{N}$$

and $\mathcal{N}$ is the family of sets from $\mathcal{F}$ of $\textbf{P}$-measure zero.

Denote by $\pi(Y) = (\pi_t(Y))_{t \geq 0}$ the optional projection of Y with respect to $\textbf{F}_+^X$. Analogously to proving Theorem 11.2.42, it can be established that $\pi(Y) \in \mathcal{S}p(\textbf{F}_+^X, \textbf{P})$, i.e.,

$$\pi_t(Y) = \pi_0(Y) + A_t^X + M_t^X \tag{11.57}$$

with $A_t^X \in \mathcal{A}_{loc} \cap \mathcal{P}(\textbf{F}_+^X)$ and $M \in \mathcal{M}_{loc}(\textbf{F}_+^X)$.

Prove that the representation (11.57) is valid. If instead of (11.56) the conditions

$$\mathbf{E}\,|Y_0| \;<\; \infty, \; \mathbf{E}\,(\mathbf{Var}(A)_\infty)^2 < \infty \tag{11.58}$$
$$M \;\in\; \mathcal{H}^2(\mathbf{F})$$

are fullfilled, then evidently $\mathbf{E}\sup_{t\geq 0} Y_t^2 < \infty$ and for each $\tau \in T\left(\mathbf{F}_+^X\right)$

$$\mathbf{E}\pi_t^2\,(Y) = \mathbf{E}\left(\mathbf{E}\left(Y_\tau | \mathcal{F}_\tau^X\right)\right)^2 \leq \mathbf{E}\left(Y_\tau^2\right) < \infty.$$

Therefore, in this case the process $\pi\,(Y)$ may be interpreted as the filtration estimate of the process Y by observations on X, optimal in the mean square sense. More specifically, $\pi_t\,(Y)$ is the optimal in the mean square sense estimate of Y by observations on $\cap_{\varepsilon > 0}\{X_s, 0 \leq s \leq t + \varepsilon\}$. From the point of view of applications, it is more natural to assume the flow of observations $\mathbf{F}^X$ is given instead of $\mathbf{F}_+^X$.

If $\mathbf{F}_+^X = \mathbf{F}^X$, then clearly these cases become equivalent. It will be shown below that in fact under certain additional conditions the flows $\mathbf{F}^X$ and $\mathbf{F}_+^X$ coincide. Also, we will clarify the structure of the processes A^X and M^X (see (11.57)), determining the filtration estimate $\pi_t\,(Y)$.

We begin with formulating the conditions under which the procedure determining the process $\pi\,(Y)$ will be worked out.

Group A conditions:

A1: $\mathbf{E}Y_0^2 < \infty$,

A2: $\mathbf{E}\,(\mathbf{Var}(A)_t)^2 < \infty, \; t \geq 0$,

A3: $\mathbf{E}M_t^2 < \infty, \; t \geq 0$.

By Theorem 11.2.41 $X \in \mathcal{S}(\mathbf{F}_+^X, \mathbf{P})$. Therefore $X \in \mathcal{S}(\mathbf{F}_+^X, \mathbf{P}^X)$ where $\mathbf{P}^X$ is the restriction of the measure $\mathbf{P}$ to the σ-algebra $\mathcal{F}_\infty^X = \sigma\left(\cup_{t\geq 0}\mathcal{F}_t^X\right)$. The triplet of a semimartingale $X \in \mathcal{S}(\mathbf{F}_+^X, \mathbf{P})$ is denoted by $T^X = \left(B^X, C^X, v^X\right)$. Assume $\mathbf{a}_t^X = v\left(\{t\} \times \mathbb{R}_0\right)$. If $\tilde{\mathbf{P}}^X$ is the probability measure on $\left(\Omega, \mathcal{F}_\infty^X\right)$ and $\tilde{\mathbf{P}}^X << \mathbf{P}^X$, then by Theorem 11.2.40 we have $X \in \mathcal{S}(\mathbf{F}_+^X, \tilde{\mathbf{P}}^X)$ (the triplet of this semimartingale is denoted by $\tilde{T}^X$). Define the family of probability measures on $\left(\Omega, \mathcal{F}_\infty^X\right)$:

$$\mathcal{L}_{\mathbf{P}^X} = \left\{\tilde{\mathbf{P}}^X : \tilde{\mathbf{P}}^X << \mathbf{P}^X, \tilde{\mathbf{P}}_0^X = \mathbf{P}_0^X, \tilde{T}^X = T^X\right\}$$

Group B conditions:

(B1) $\mathcal{L}_{\mathbf{P}^X} = \left\{\mathbf{P}^X\right\}$, i.e., $\mathcal{L}_{\mathbf{P}^X}$ consists of a single point $\mathbf{P}^X$,

(B2) $\mathbf{1}_{|x|\leq 1}x * (v - v^X)^c \in \mathcal{A}_{loc}$,

(B3) there exists a $\mathcal{P}(\mathbf{F})$-measurable function $\phi = \phi(\omega, t)$ such that

$$\phi \circ C = B^c - (B^X)^c - \mathbf{1}_{|x|\leq 1}x * \left(v - v^X\right)^c,$$

(B4) $\mathbf{E}\,|Y_-\phi| \circ C_t < \infty, t \geq 0$.

Furthermore, the optional and predictable projections with respect to $\mathbf{F}_+^X$ are denoted by $\pi(\cdot)$ and $^p\pi(\cdot)$, respectively. Also, suppose that $\tilde{\mathcal{P}}\left(\mathbf{F}_+^X\right) = \mathcal{P}\left(\mathbf{F}_+^X\right) \times \mathcal{B}(\mathbb{R}_0)$.

Theorem 11.2.54 (1) If Condition (B1) is fullfilled, then $\mathbf{F}_+ = \mathbf{F}^X$.

(2) If Conditions (A) and (B1) are fullfilled, then $Y \in \mathcal{S}(\mathbf{F}^X, \mathbf{P})$ and the processes A^X and M^X, involved in the semimartingale decomposition (11.57), possess the following property:

$$\mathbf{E}\left[\mathbf{Var}\left(A^X\right)_t\right]^2 < \infty, \quad \mathbf{E}\left[M_t^X\right]^2 < \infty, \quad t \geq 0.$$

Besides M^X admits the integral representation

$$M_t^X = h \cdot \overline{X}_t^c + H * \left(\mu - v^X\right)_t \tag{11.59}$$

where $\overline{X}^c$ is the continuous martingale component of a $(\mathbf{F}_+^X, \mathbf{P})$-semimartingale X and h and H are $\mathcal{P}(\mathbf{F}_+^X)$-and $\tilde{\mathcal{P}}(\mathbf{F}_+^X)$-measurable functions, respectively.

(3) If conditions (A) and (B) are fullfilled, then in the integral representation (11.59)

$$h = {}^p\pi\left(\frac{d\langle M^c, X^c\rangle}{d\langle X^c\rangle} + Y_-\phi\right), \tag{11.60}$$

$$H = U + \frac{1(0 < a^X < 1)}{1 - a^X}\hat{U} \tag{11.61}$$

with

$$\hat{U}(t) = \int_{\mathbb{R}_0} U(t, x)v^X(\{t\}, dx),$$

$$U = M_\mu^{\mathbf{P}}(\triangle M|\tilde{\mathcal{P}}\left(\mathbf{F}^X\right)) + \left[M_\mu^{\mathbf{P}}\left(Y_-|\tilde{\mathcal{P}}\left(\mathbf{F}^X\right)\right) - \pi_-(Y)\right] + M_\mu^{\mathbf{P}}\left(\triangle A|\tilde{\mathcal{P}}\left(\mathbf{F}^X\right)\right) - \triangle A^X].$$

Proof. (1) The equality $\mathbf{F}_+ = \mathbf{F}^X$ holds by [51] Theorem 11.2.51 and Condition (B1).

(2) Instead of (A) suppose the stronger conditions (11.58) are fullfilled. Then, obviously $\pi(Y) \in \mathcal{S}p(\mathbf{F}_+, \mathbf{P})$ and the decomposition (11.57) holds. Let us show that

$$\mathbf{E}\left(\mathbf{Var}(A^X)_\infty\right)^2 < \infty, \quad M^X \in \mathbf{H}^2\left(\mathbf{F}_+^X\right). \tag{11.62}$$

Let A^i, $i = 1, 2$, be increasing processes involved in the decomposition of A: $A = A^1 - A^2$, $\mathbf{Var}(A) = A^1 + A^2$. Then by the second inequality in (11.58) we have $\mathbf{E}(A_\infty^i)^2 < \infty$, $i = 1, 2$. According to [51], Lemma 11.2.43 $\pi(A^i)$ are submartingales of the class D

$$\pi(A^i) = A^{X,i} + M^{X,i} \tag{11.63}$$

with

$$A^{X,i} \in \mathcal{A}^+ \cap \mathcal{P}\left(\mathbf{F}_+^X\right)$$

and

$$M^{X,i} \in \mathcal{M}\left(\mathbf{F}_+^X\right), \quad i = 1, 2.$$

Let us show that by the second inequality in (11.58)

$$\mathbf{E}\left(A_\infty^{X,i}\right)^2 < \infty, \quad M^{X,i} \in \mathbf{H}^2\left(\mathbf{F}_+^X\right), \quad i = 1, 2. \tag{11.64}$$

Using Ito's formula we get

$$\begin{aligned}
(\pi(A^i))^2 &= 2\pi_-(A^i) \cdot A^{X,i} + 2\left(\pi_-(A^i) + \triangle A^{X,i}\right) \cdot M^{X,i} \\
&\quad + \sum_s \left(\triangle A_s^{X,i}\right)^2 + \left[M^{X,i}, M^{X,i}\right].
\end{aligned} \tag{11.65}$$

Let $(\tau_n)_{n\geq 1}$ be a localizing sequence for

$$\left(\pi_-(A^i) + \triangle A^{X,i}\right) \cdot M^{X,i} \in \mathcal{M}_{loc}\left(\mathbf{F}_+^X\right).$$

Then (11.65) entails the inequality

$$\mathbf{E}[M^{X,i}, M^{X,i}]_{\tau_n} \leq \mathbf{E}(\pi_{\tau_n}(A^i))^2 \leq \mathbf{E}(A_{\tau_n}^i)^2, \quad n \geq 1,$$

and hence

$$\mathbf{E}[M^{X,i}, M^{X,i}]_\infty \leq \mathbf{E}(A_\infty^i)^2 < \infty.$$

Therefore, by the Burkholder-Gandy inequality (Theorem 11.2.18) the second property in (11.64) holds. But then

$$\mathbf{E}\left(A_\infty^{X,i}\right)^2 \leq 2\mathbf{E}\left[\left(\pi_\infty(A^i)\right)^2 + \left(M_\infty^{X,i}\right)^2\right] \leq 2\mathbf{E}[(A_\infty^i)^2 + (M_\infty^{X,i})^2],$$

i.e., the first property in (11.64) holds.

On proving Theorem 11.2.42 it has been established that

$$A^X = A^{X,1} - A^{X,2}, \quad M^X = M^{X,1} - M^{X,2} + \pi(M) \tag{11.66}$$

The first property in (11.62) holds by (11.64), since $\mathbf{Var}(A^X) \leq A^{X,1} + A^{X,2}$.

The second property in (11.62) holds also by (11.64), since by Lemma 11.2.44 we have

$$\pi(M) \in \mathcal{M}(\mathbf{F}_+)$$

and by Doob's inequality (Theorem 11.2.16)

$$\mathbf{E}\sup_{t\geq 0} \pi^2(M) \leq 4\mathbf{E}(\pi_\infty(M))^2 \leq 4\mathbf{E}M_\infty^2 < \infty.$$

If exclusively Condition (A) is fullfilled, then consider semimartingales Y^n, $n \geq 1$ with $Y_t^n = \mathbf{1}_{[\![0,n]\!]} \circ Y_t$. For each $n \geq 1$

$$Y_t^n = Y_0 + A_t^n + M_t^n,$$

with $A^n = \mathbf{1}_{[\![0,n]\!]} \circ A$ and $M^n = \mathbf{1}_{[\![0,n]\!]} \circ M$ which satisfy the conditions in (11.58). Therefore $\pi(Y^n)$ admits the semimartingale decomposition

$$\pi_t(Y^n) = \pi_0(Y) + A_t^{n,X} + M_t^{n,X}, \quad n \geq 1$$

with

$$A^{n,X} \in \mathcal{A} \cap \mathcal{P}(\mathbf{F}_+^X), \mathbf{E}(\mathbf{Var}(A^{n,X})_\infty)^2 < \infty \text{ and } M^{n,X} \in \mathrm{H}^2\left(\mathbf{F}_+^X\right).$$

The definition of Y^n yields

$$\pi_{t\wedge n}(Y^{n+1}) = \pi_{t\wedge n}(Y^n) = \pi_t(Y^n).$$

Therefore, the equalities

$$A_{t\wedge n}^{n+1,X} + M_{t\wedge n}^{n+1,X} = A_t^{n,X} + M_t^{n,X}$$

hold. As the semimartingale decomposition for special martingales is unique, we have

$$A_{t\wedge n}^{n+1,X} = A_t^{n,X}, \quad M_{t\wedge n}^{n+1,X} = M_t^{n,X}, \quad n \geq 1.$$

This fact allows one to define the processes A^X and M^X,

$$A_t^X = A_t^{n,1} + \sum_{n\geq 1}(A_t^{n+1,X} - A_t^{n,X}),$$

$$M_t^X = M_t^{n,1} + \sum_{n\geq 1}(M_t^{n+1,X} - M_t^{n,X})$$

with the properties indicated in the semimartingale decomposition (11.57), such that

$$\mathbf{E}(\mathbf{Var}(A^X)_t)^2 < \infty, \quad \mathbf{E}(M_t^X)^2 < \infty, \ t > 0.$$

The integral representation (11.59) for M^X takes place by Theorem 11.2.51 and Condition (B_1).

(3) Let us establish that in the integral representation (11.59) for M^X the functions h and H are defined by the formulas (11.60) and (11.61). Examining the proof of assertion (2) of the theorem, we see that it suffices to proof the formulas (11.60) and (11.61) under the stronger condition (11.58) instead of condition (A) and under the condition

$$\mathbf{E}|Y_-\phi| \circ C_\infty < \infty \tag{11.67}$$

instead of condition (B4).

By Lemma 11.2.47 (see also Theorem 11.2.27)

$$h = \frac{d\langle M^{X,c}, \overline{X}^c\rangle}{d\langle \overline{X}^c\rangle} \tag{11.68}$$

$$H = \mathsf{M}_\mu^{\mathbf{P}}(\triangle M^X|\tilde{\mathcal{P}}\,(\mathbf{F}_+^X)) + \frac{\mathbf{1}\,(0 < \mathrm{a}^X < 1)}{1 - \mathrm{a}^X}\mathsf{M}_\mu^{\mathbf{P}}(\widehat{\triangle M^X}|\tilde{\mathcal{P}}\,(\mathbf{F}_+^X))$$

where $\overline{X}^c$ is the continuous martingale component of a process X, belonging to $\mathcal{S}(\mathbf{F}_+, \mathbf{P})$. ∎

To calculate the right-hand side in the equalities (11.68), we need two auxiliary results formulated as the following lemmas:

Lemma 11.2.55 Let Conditions (B2) and (B3) be fullfilled. Then

$$\overline{X}^c = X^c + \phi \circ C.$$

Proof. A process $X \in \mathcal{S}(\mathbf{F}_+^X, \mathbf{P})$ has the canonical representation (cf. (11.54))

$$X_t = X_0 + B_t^X + \overline{X}_t^c + \int_0^t \int_{|x| \leq 1} x d(\mu - v^X) + \int_0^t \int_{|x| > 1} x d\mu. \tag{11.69}$$

By (11.54) and (11.69), the representation for $\phi \circ C$ (see (B_3)) and the fact that

$$\triangle B_t = \int_{|x| \leq 1} x v(\{t\}, dx), \quad \triangle B_t^X = \int_{|x| \leq 1} x v^X(\{t\}, dx)$$

we get

$$\overline{X}^c - X^c - \phi \circ C = \mathbf{1}_{|x| \leq 1} x * (\mu - v) - \mathbf{1}_{|x| \leq 1} x * (\mu - v^X) + \mathbf{1}_{|x| \leq 1} x * (\mu - v^X). \tag{11.70}$$

Let us show that the right-hand side of the Equation (11.70) determines a negligible process. To this end define the process

$$M^\epsilon = \mathbf{1}_{|x| \leq \epsilon} x * (\mu - v)$$

with $0 < \epsilon < 1$. Clearly,

$$M^\epsilon \in \mathcal{M}_{loc}^2(F)$$

and (see Lemma 11.2.34 and Theorem 11.2.35)

$$\langle M^\epsilon \rangle = \mathbf{1}_{(|x| \leq \epsilon)} x^2 * v - \sum_s \left(\int_{|x| \leq \epsilon} x v(\{s\}, dx) \right)^2 \leq \mathbf{1}_{(|x| \leq \epsilon)} x^2 * v. \tag{11.71}$$

Since $\mathbf{E}(M_\tau^\epsilon)^2 \leq \mathbf{E}\langle M^\epsilon \rangle_\tau$ for each stopping time τ relative to $\mathbf{F}_+$ (11.2.3), by (11.71) we have

$$\mathbf{E}(M^\epsilon)^2 \leq \mathbf{E}\mathbf{1}_{|x| \leq \epsilon} x^2 * v_\mathbf{r}.$$

Therefore, by the Lenglart-Reboedo inequality (Theorem 11.2.17)

$$\mathbf{P}\left(\sup_{s \leq t} |M_s| \geq a \right) \leq \frac{b}{a^2} + \mathbf{P}\left(\mathbf{1}_{|x| \leq \epsilon} x^2 * v_t \geq b \right) \tag{11.72}$$

for each $a > 0$, $b > 0$ and $t > 0$. For each $t > 0$ we have $\mathbf{1}_{|x| \leq 1} x^2 * v_t < \infty$ ($\mathbf{P}$-a.s.). Consequently, $\mathbf{1}_{|x| \leq \epsilon} x^2 * v_t \to 0$ ($\mathbf{P}$-a.s.) as $\epsilon \to 0$. In view of this, taking the limit $\lim_{b \to 0} \lim_{\epsilon \to 0}$ in (11.72), we get

$$\lim_{\epsilon \to 0} \mathbf{P}\left(\sup_{s \leq t} |M_s^\epsilon| \geq a \right) = 0 \tag{11.73}$$

for each $a > 0$ and $t > 0$. It is established analogously that

$$\lim_{\epsilon \to 0} \mathbf{P}\left(\sup_{s \leq t} |h\delta_s^\epsilon| \geq a \right) = 0 \tag{11.74}$$

with $M^{X,\epsilon} = \mathbf{1}_{|x| \leq \epsilon} x * (\mu - v^X)$. Besides, by Assumption (B_2)

$$\mathbf{1}_{|x| \leq 1} x * (v - v^X) \in C_{loc}$$

and consequently (**P**-a.s.),

$$\mathbf{Var}(1_{|x|\leq\epsilon}x*(v-v^X))_t\to 0,\ \epsilon\to 0, \tag{11.75}$$

for each $t>0$. Observe now that by Proposition 11.2.37 as $0<\epsilon<1$

$$\mathbf{1}(\epsilon<|x|\leq 1)x*(\mu-v)=\mathbf{1}(\epsilon<|x|\leq 1)x*\mu-\mathbf{1}(\epsilon<|x|\leq 1)x*v.$$

The analogous representation takes place for $\mathbf{1}(\epsilon<|x|\leq 1)x*(\mu-v^X)$ too. Therefore, the process

$$\mathbf{1}(\epsilon<|x|\leq 1)x*(\mu-v)-\mathbf{1}(\epsilon<|x|\leq 1)x*(\mu-v^X)+\mathbf{1}(\epsilon<|x|\leq 1)x*(v-v^X),$$

is negligible. By taking this and equality (11.70) into consideration, for each $t>0$ we get

$$\sup_{s\leq t}\left|\overline{X}_s^c-X_s^c-\phi\circ C\right|\leq\sup_{s\leq t}|M_s^\epsilon|+\sup_{s\leq t}|M_s^{X,\epsilon}|+\mathbf{Var}(\mathbf{1}(|x|\leq\epsilon)x*(v-v^X))_t.$$

From this and (11.73)-(11.75), it follows that the process $\overline{X}^c-X^c-\phi\circ C$ is negligible. ∎

Lemma 11.2.56 Let conditions (A) be fullfilled. Then,

$$\mathsf{M}_\mu^\mathbf{P}(\pi(Y)|\tilde{\mathcal{P}}(\mathbf{F}_+^X))=\mathsf{M}_\mu^\mathbf{P}(Y|\tilde{\mathcal{P}}(\mathbf{F}_+^X))\quad\mathsf{M}_\mu^P\text{-a.s.}$$

Proof. By (A)

$$\mathbf{E}\sup_{s\leq t}|Y_s|\leq\mathbf{E}\left(|Y_0|+\mathbf{Var}(A)_t+\sup_{s\leq t}|M_s|\right)<\infty,$$

since according to the Cauchy-Bunyakovsky-Doob inequalities (Theorem 11.2.16)

$$\left(\mathbf{E}\sup_{s\leq t}|M_s|\right)^2\leq\mathbf{E}\sup_{s\leq t}(M_s)^2\leq 4\mathbf{E}M_t^2.$$

Analogously, using the representation (11.57) for $\pi(Y)$, assertion (2) of the theorem proved already and the inequality $\mathbf{E}|\pi_0(Y)|\leq\mathbf{E}|Y_0|$, we get

$$\mathbf{E}\sup_{s\leq t}|\pi_s(Y)|<\infty.$$

Therefore, in accordance with 11.2.26 there exist conditional mathematical expectations $\mathsf{M}_\mu^\mathbf{P}(Y|\tilde{\mathcal{P}}(\mathbf{F}_+^X))$ and $\mathsf{M}_\mu^\mathbf{P}(\pi(Y)|\tilde{\mathcal{P}}(\mathbf{F}_+^X))$.

Hence, it suffices to show that

$$\mathsf{M}_\mu^\mathbf{P}(Y^+|\tilde{\mathcal{P}}(\mathbf{F}_+^X))=\mathsf{M}_\mu^\mathbf{P}(\pi(Y^+)|\tilde{\mathcal{P}}(\mathbf{F}_+^X)) \tag{11.76}$$

and

$$\mathsf{M}_\mu^\mathbf{P}(Y^-|\tilde{\mathcal{P}}(\mathbf{F}_+^X))=\mathsf{M}_\mu^\mathbf{P}(\pi(Y^-)|\tilde{\mathcal{P}}(\mathbf{F}_+^X)) \tag{11.77}$$

with $Y^+=Y\vee 0$ and $Y^-=-(Y\wedge 0)$. Let Z be a $\tilde{\mathcal{P}}(\mathbf{F}_+^X)$-measurable non-negative function such that $Z*\mu_\infty\leq const$.

Denote $B = Z * \mu$. Clearly $B \in \mathcal{A}^+$ According to Proposition 11.2.14

$$\mathbf{E}Y^+ Z * \mu_\infty = \mathbf{E}Y^+ \circ B_\infty = \mathbf{E}\pi(Y^+) \circ B_\infty = \mathbf{E}\pi(Y^+) Z * \mu_\infty.$$

This easily gives the equality

$$\mathbf{E}Y^+ Z * \mu_\infty = \mathbf{E}\pi(Y^+) Z * \mu_\infty$$

for each non-negative $\tilde{\mathcal{P}}(\mathbf{F}_+)$-measurable function Z. Consequently,

$$\mathbf{E}\mathbf{M}_\mu^\mathbf{P}(Y^+ | \tilde{\mathcal{P}}(\mathbf{F}_+^X)) Z * \mu_\infty = \mathbf{E}\mathbf{M}_\mu^\mathbf{P}(\pi(Y^+) | \tilde{\mathcal{P}}(\mathbf{F}_+^X)) Z * \mu_\infty$$

and the equality (11.76) takes place. The equality (11.77) is proved analogously. We will turn now to detemining the right-hand sides of the equalities (11.68) that define h and H, assuming Conditions (11.58) and (11.67) hold.

Determination of h. Since by Theorem 11.2.41 $\langle \overline{X}^c \rangle = C$, by (11.68) we have $\langle M^{X,c}, \overline{X}^c \rangle = h \circ C$. On the other hand,

$$N = M^{X,c}\overline{X}^c - \langle M^{X,c}, \overline{X}^c \rangle \in \mathcal{M}_{loc}(\mathbf{F}_+^X).$$

Therefore, the process $M^{X,c}\overline{X}^c \in \mathcal{S}p(\mathbf{F}_+^X, \mathbf{P})$ has the semimartingale decomposition

$$M^{X,c}\overline{X}^c = \langle M^{X,c}, \overline{X}^c \rangle + N.$$

As this decomposition is unique, the process $\langle M^{X,c}\overline{X}^c \rangle$ may be defined as a certain process $L = (L_t)_{t \geq 0} \in \mathcal{V} \cap \mathcal{P}(\mathbf{F}_+)$ such that

$$M^{X,c}\overline{X}^c - L \in \mathcal{M}_{loc}(\mathbf{F}_+^X). \tag{11.78}$$

It has been established in the course of proving assertion (2) of the theorem that under Condition (11.58) we have $M^X \in \mathrm{H}^2(\mathbf{F}_+)$ (see (11.62)). In the subsequent considerations we will assume for convenience

$$(\overline{X}^c)_\infty^* + C_\infty + \mathbf{Var}(A^X)_\infty \leq const, \tag{11.79}$$

passing to localizing sequences relative to $\mathbf{F}_+^X$ if necessary (see Lemma 11.2.10).

Under the assumption (11.79) we have $\overline{X}^c \in \mathcal{M}^2(\mathbf{F}_+^X)$. Therefore

$$\mathbf{EVar}\left(\left\langle M^{X,c}, \overline{X}^c \right\rangle\right)_\infty < \infty,$$

(11.2.15). Now, to verify (11.78) it suffices to show, by 11.2.8, that for each $\tau \in \mathcal{T}(\mathbf{F}_+)$

$$\mathbf{E}M_\tau^{X,c}\overline{X}_\tau^c = \mathbf{E}L_\tau. \tag{11.80}$$

Next, since the martingales $M^X - M^{X,c}$ and $\overline{X}^c$ are strongly orthogonal, instead of (11.80) it suffices to verify the equality

$$\mathbf{E}M_\tau^X \overline{X}_\tau^c = \mathbf{E}L_\tau. \tag{11.81}$$

By the semimartingale decomposition (11.57) for $\pi(Y)$ we have

$$\mathbf{E}M_\tau^X \overline{X}_\tau^c = \mathbf{E}(\pi_\tau(Y) - \pi_0(Y) - A_\tau^X)\overline{X}_\tau^c = \mathbf{E}(Y_\tau - A_\tau^X)\overline{X}_\tau^c.$$

Under the assumption (11.58) we have

$$\mathbf{EVar}(A^X)_\infty < \infty$$

(see 11.62). Therefore,

$$\mathbf{E} A_\tau^X \overline{X}_\tau^c = \mathbf{E} \overline{X}^c \circ A_\tau^X$$

(Theorem 11.2.11). Thus,

$$\mathbf{E} M_\tau^{X,c} \overline{X}_\tau^c = \mathbf{E}(Y_\tau \overline{X}_\tau^c - \overline{X}^c \circ A_\tau^X). \tag{11.82}$$

Observe now that if $(\sigma_n)_{n\geq 1}$ is a sequence of Markov times relative to $\mathcal{F}$ such that $\sigma_n \uparrow \infty$, $n \to \infty$, then under the assumptions (11.58) and (11.79) made, we get, by the Lebesgue dominated convergence theorem, that

$$\mathbf{E} Y_\tau \overline{X}_\tau^c = \lim_n \mathbf{E} Y_{\tau_n} \overline{X}_{\tau_n}^c, \quad \tau_n = \tau \wedge \sigma_n. \tag{11.83}$$

Set

$$\sigma_n = \inf \left(t : \sup_{s\leq t} |X_s^c| + |\phi|_s C_t \geq n \right)$$

and apply the decomposition (11.55) for Y and Lemma 11.2.55. Then

$$\mathbf{E} Y_{\tau_n} \overline{X}_{\tau_n}^c = \mathbf{E} \alpha_0 + A_{\tau_n} + M_{\tau_n})(X_{\tau_n}^c + \phi \circ C_{\tau_n}),$$

and by Theorem 11.2.11 and the definition of the mutual quadratic characteristic for square integrable martingales, we get

$$\mathbf{E} Y_{\tau_n} \overline{X}_{\tau_n}^c = \mathbf{E}(X^c \circ A_{\tau_n} + \langle M^c, X^c \rangle_{\tau_n} + (Y_0 \phi) \circ C_{\tau_n} + A_{\tau_n}(\phi C_{\tau_n}) + M_- \phi C_{\tau_n}). \tag{11.84}$$

Next, by Ito's formula we get

$$A_{\tau_n}(\phi \circ C_{\tau_n}) = A_- \phi \circ C_{\tau_n} + (\phi \circ C) \circ A_{\tau_n} \tag{11.85}$$

From (11.84) and (11.85) it follows that

$$\mathbf{E} Y_{\tau_n} \overline{X}_{\tau_n}^c = \mathbf{E}[(X^c + \phi \circ C) \circ A + \langle M^c, X^c \rangle_{\tau_n} + \phi(Y_0 + A_- + M_-) \circ C_{\tau_n}]$$

which, in view of the representation

$$\langle M^c, X^c \rangle = \frac{d\langle M^c, X^c \rangle}{d\langle X^c \rangle} \circ C$$

(Theorem 11.2.19), yields

$$\mathbf{E} Y_{\tau_n} \overline{X}_{\tau_n}^c = \mathbf{E}[\overline{X}^c \circ A_{\tau_n} + \frac{d\langle M^c, X^c \rangle}{d\langle X^c \rangle} \circ C_{\tau_n} + \phi Y_- \circ C_{\tau_n}].$$

By taking the limit (Lebesgue's dominated convergence theorem) we arrive at the equality

$$\mathbf{E} Y_\tau \overline{X}_\tau^c = \mathbf{E}\left[\overline{X}^c \circ A_{\tau_n} + \left(\frac{d\langle M^c, X^c \rangle}{d\langle X^c \rangle} + \phi Y_- \right) \circ C_\tau \right], \tag{11.86}$$

since by (11.79) we have $(\overline{X}^c)_\infty^* \leq const$, by (11.67) we have $\mathbf{E}|\phi Y_-| \circ C_\infty < \infty$ and finally, by the Cauchy-Bunyakovsky inequality, Theorem 11.2.19, the inequality (11.79) (according to which

$C_\infty \leq const$) and the condition $M \in H^2(\mathbf{F})$ (see (11.58)) we have

$$\mathbf{E}\left|\frac{d\langle M^c, X^c\rangle}{d\langle X^c\rangle}\right| \circ C_\infty \leq \left(\mathbf{E}C_\infty \left(\frac{d\langle M^c, X^c\rangle}{d\langle X^c\rangle}\right)^2 \circ C_\infty\right)^{1/2}$$

$$\leq (\mathbf{E}C_\infty \langle M^c\rangle_\infty)^{1/2}$$

$$\leq const \times (\mathbf{E}\langle M\rangle_\infty)^{1/2} < \infty.$$

Thus, (11.82) and (11.86) entail

$$\mathbf{E}M_\tau^{X,c}\overline{X}_\tau^c = \mathbf{E}\left(\frac{d\langle M^c, X^c\rangle}{d\langle X^c\rangle} + \phi Y_-\right) \circ C_\tau + \mathbf{E}[\overline{X}^c \circ A_\tau - \overline{X}^c \circ A_\tau^X].$$

Let A^i, $i = 1, 2$, be increasing processes involved in the decomposition of A: $A = A^1 - A^2$, let $\mathbf{Var}(A) = A^1 + A^2$, and let $A^{X,i}$, $i = 1, 2$, be increasing processes involved in the semimartingale decomposition of $\pi(A^i)$ (see 11.63). From (11.63) (see also 11.64) it follows that

$$\mathbf{E}A_\tau^i = \mathbf{E}A_\tau^{X,i} \quad i = 1, 2,$$

for each $\tau \in \mathcal{T}(\mathbf{F}_+)$. It is not hard to deduce from this that

$$\mathbf{E}g \circ A_\infty^i = \mathbf{E}g \circ A_\infty^{X,i}, \quad i = 1, 2, \tag{11.87}$$

for any non-negative bounded function $g = g(\omega, t)$ measurable relative to $\mathcal{P}_\zeta(\mathbf{F}_+)$, that is the σ-algebra in $\mathbb{R}_+ \times \Omega$ generated by sets of type $A \times \{0\}$ ($A \in \mathcal{F}_{0+}^X$) and of type $]\sigma, \tau]$ with $\sigma, \tau \in \mathcal{T}(\mathbf{F}_+^X)$. It has been established in the course of proving Theorem 11.2.1 that $\mathcal{P}_\zeta = \mathcal{P}$, i.e., $\mathcal{P}_\zeta(\mathbf{F}_+^X) = \mathcal{P}(\mathbf{F}_+^X)$. Hence the equality (11.87) holds for each bounded and $\mathcal{P}(\mathbf{F}_+^X)$-measurable function g.

But then by the equality $A^X = A^{X,1} - A^{X,2}$ (see (11.66)) we have

$$\mathbf{E}\overline{X}^c \circ A_\tau = \mathbf{E}\overline{X}^c \circ A_\tau^X,$$

and hence

$$\mathbf{E}M_\tau^{X,c}\overline{X}_\tau^c = \mathbf{E}\left(\frac{d\langle M^c, X^c\rangle}{d\langle X^c\rangle} + \phi Y\right)_- \circ C_\tau.$$

Therefore, by 11.2.14,

$$\mathbf{E}M_\tau^{X,c}\overline{X}_\tau^c = \mathbf{E}\,{}^p\pi\left(\frac{d\langle M^c, X^c\rangle}{d\langle X^c\rangle} + \phi Y_-\right) \circ C_\tau.$$

Consequently,

$$L = {}^p\pi\left(\frac{d\langle M^c, X^c\rangle}{d\langle X^c\rangle} + \phi Y_-\right) \circ C,$$

and the representation (11.70) for h holds.

Next we work on the determination of H. By the representation (11.68) for H it suffices to calculate $\mathsf{M}_\mu^\mathbf{P}(\triangle M^X | \tilde{\mathcal{P}}(\mathbf{F}_+^X))$. (Conditional expectation $\mathsf{M}_\mu^\mathbf{P}(\cdot | \tilde{\mathcal{P}}(\mathbf{F}_+^X))$ calculated below exists by Proposition 11.2.26 and Theorem 11.2.27.)

By (11.57)

$$\triangle M^X = \pi(Y) - \pi_-(Y) - \triangle A^X$$

Therefore, taking into account Lemma 11.2.56 we get

$$
\begin{aligned}
\mathsf{M}^{\mathbf{P}}_{\mu}(M^X|\tilde{\mathcal{P}}(\mathbf{F}^X_+)) &= \mathsf{M}^{\mathbf{P}}_{\mu}(\pi(Y)|\tilde{\mathcal{P}}((\mathbf{F}^X_+)) - \pi_-(Y) - \triangle A^X \\
&= \mathsf{M}^{\mathbf{P}}_{\mu}(Y|\tilde{\mathcal{P}}(\mathbf{F}^X_+)) - \pi_-(Y) - \triangle A^X
\end{aligned}
$$

The representation (11.55) for Y yields

$$
Y = Y_- + \triangle A + \triangle M.
$$

Consequently

$$
\mathsf{M}^{P}_{\mu}(\triangle M^X|\tilde{\mathcal{P}}(\mathbf{F}^X_+)) = [\mathsf{M}^{P}_{\mu}(Y_-|\tilde{\mathcal{P}}(\mathbf{F}^X_+)) - \pi_-(Y)] + [\mathsf{M}^{P}_{\mu}(\triangle A|\tilde{\mathcal{P}}(\mathbf{F}^X_+)) - \triangle A^X] + \mathsf{M}^{P}_{\mu}(\triangle M|\tilde{\mathcal{P}}(\mathbf{F}^X_+)) \,,
$$

and hence $\mathsf{M}^{P}_{\mu}(\triangle M^X|\tilde{\mathcal{P}}(\mathbf{F}^X_+))$ coincides with the function U, involved in the definition of H by the formula (11.61). ∎

11.3   The *Un*usual case of Optimal Filtering

Here we are going to extend filtering theory to the spaces of optional processes on *un*usual probability spaces by the use of uniform Doob-Meyer decomposition. In doing so, we are able to relax the assumptions of right-continuity of the filtration $\mathbf{F}^X$ and the requirement that $\mathbf{P}^X$ is unique. We are going to do this in two ways. In the first way, we are going to start anew, with an *un*usual probability space and develop a theory for the filtering of optional local supermartingales using UDMD. In the second approach, we will begin with the usual filtering theory of special semimartingale but allow for the filtration $\mathbf{F}^X$ not to be right-continuous and employ the optional projection on *un*usual stochastic basis in conjunction with UDMD to develop a filtering theory for optional local supermartingale. We begin with the first method, then develop the second one.

11.3.1   Filtering on *un*usual stochastic basis

Let X be an $\mathbb{R}^d$ valued r.l.l.l. optional semimartingale on the *un*usual stochastic basis $(\Omega, \mathcal{F}, \mathbf{F} = (\mathcal{F}_t)_{t\geq 0}, \mathbf{P})$. Let Y be an $\mathbb{R}^d$ valued r.l.l.l. optional semimartingale on the same space. The processes X and Y are related; $(X, Y) = (X_t, Y_t)_{t\geq 0}$ is a partially observable stochastic process where Y is the unobservable component and X is the observed process. Let $\mathbf{F}^X = (\mathcal{F}^X_t)_{t\geq 0}$ be the filtration generated by X which is not necessarily right continuous nor complete, obviously $\mathbf{F}^X \subseteq \mathbf{F}$. However, $\mathcal{F}^X := \mathcal{F}^X_\infty \vee \mathcal{N}$ is complete and $\mathbf{P}^X$ is the restriction of the measure $\mathbf{P}$ on $\mathcal{F}^X$. Assume that X remains an optional semimartingale under $(\Omega, \mathcal{F}^X, \mathbf{F}^X, \mathbf{P}^X)$.

Lemma 11.3.1 For the partially observable system of optional semimartingales $(X_t, Y_t)_{t \geq 0}$, let $ZX \in \mathcal{M}_{1oc}(\mathbf{F}^X, \mathbf{P}^X)$ be a local optional martingale for any strictly positive local optional martingale Z with $Z_0 = 1$, $Z \in \mathbb{Z}(X)$. There are, corresponding to any Z the optional local martingale measures $\mathbf{Q} \in \mathbb{Q}(X)$ such that $\mathbf{Q}_t \sim \mathbf{P}_t^X \ll \mathbf{P}_{t+}^X$ for all $t \geq 0$ where $\mathbf{P}_t^X$ is the restriction of $\mathbf{P}^X$ on $\mathcal{F}_t^X$ and $\mathbf{P}_{t+}^X$ is the restriction of $\mathbf{P}^X$ on $\mathcal{F}_{t+}^X$. If ZY is a local optional supermartingale for every $Z \in \mathbb{Z}(X)$, then the optimal estimate of Y given $\mathcal{F}_t^X$ is the optional projection $\pi_t(Y) := \mathbf{E}\left[Z_t Y_t | \mathcal{F}_t^X\right]$ satisfying the following stochastic equation,

$$\pi_t(Y) = \pi_0(Y) + \phi \circ X - C, \quad \forall \mathbf{Q} \ a.s.$$

where ϕ is $\mathcal{P}(\mathbf{F}^X) \times \mathcal{O}(\mathbf{F}^X)$ is a unique X integrable process and C is an optional process. ϕ is given by

$$\phi_t = \mathbf{E_Q}\left[\frac{d \langle Y^c, X^c \rangle_t}{d \langle X^c, X^c \rangle_t} | \mathcal{F}_{t-}^X\right] + \mathbf{E_Q}\left[\frac{\triangle Y_t^d - \triangle C_t^d}{\triangle X_t^d} | \mathcal{F}_{t-}^X\right]$$
$$+ \mathbf{E_Q}\left[\frac{\triangle^+ Y_t^g - \triangle^+ C_t^g}{\triangle^+ X_t^g} | \mathcal{F}_t^X\right]$$

and $-C_t = \pi_t(Y) - \pi_0(Y) - \phi \circ X$. Moreover, given that $Y = Y_0 + M - A$ then $C_t = \mathbf{E_Q}\left[A_t | \mathcal{F}_t^X\right]$ and $\mathbf{E_Q}\left[M_t | \mathcal{F}_t^X\right] = \phi \circ X_t$.

Proof. Let $\mathbb{Z}(X)$ be the set of all strictly positive local optional martingales Z with $Z_0 = 1$ such that ZX is a local optional martingale, i.e., ZX belongs to $\mathcal{M}_{1oc}(\mathbf{F}^X, \mathbf{P}^X)$. Denote by $\mathbb{Q}(X)$ the set of all probability measures $\mathbf{Q}$ such that $\mathbf{Q} \ll \mathbf{P}_+^X$ and $\mathbf{Q} \sim \mathbf{P}^X$ where X is a local optional martingale with respect to $\mathbf{Q}$. Assume $\mathbb{Q}(X) \neq \emptyset$ or that $\mathbb{Z}(X)$ is not empty. We assume that the unobserved process Y is an r.l.l.l. local supermartingale with respect to any $\mathbf{Q} \in \mathbb{Q}(X)$. Consequently, the optional project of Y on $\mathcal{F}^X$ under $\mathbf{Q}$ is $\pi_t(Y) := \mathbf{E_Q}\left[Y_t | \mathcal{F}_t^X\right]$ is a local optional supermartingale that is $\mathcal{F}_t^X$ measurable. By uniform Doob-Meyer decomposition of Theorem 7.5.1 we arrive at

$$\pi_t(Y) = \pi_0(Y) + \phi \circ X - C.$$

Let us identify $\hat{Y} := \pi_t(Y)$. Using the definition of the quadratic variation of optional processes we get

$$\left[\hat{Y}, X\right] = \phi_t \circ [X, X]_t - [C, X]_t,$$
$$\Rightarrow \phi_t \circ [X, X]_t = \left[\hat{Y} - C, X\right].$$

Consider each part of the quadratic variation $\left[\hat{Y}, X\right]$, continuous, discrete right-continuous and discrete left-continuous independently:

$$\left\langle \hat{Y}^c, X^c \right\rangle_t = \phi^c \cdot \langle X^c, X^c \rangle_t, \quad \text{since } \langle C^c, X^c \rangle = 0,$$
$$\left[\hat{Y}^d, X^d\right] = \phi^d \cdot \left[X^d, X^d\right]_t - \left[C^d, X^d\right]_t,$$
$$\left[\hat{Y}^g, X^g\right] = \phi^g \circ [X^g, X^g]_t - [C^g, X^g]_t.$$

Therefore,

$$\phi_t^c = \frac{d\left\langle \hat{Y}^c, X^c \right\rangle_t}{d\left\langle X^c, X^c \right\rangle_t}, \quad \phi_t^d = \frac{d\left[\hat{Y}^d - C^d, X^d\right]_t}{d\left[X^d, X^d\right]_t},$$

$$\phi_t^g = \frac{d\left[\hat{Y}^g - C^g, X^g\right]_t}{d\left[X^g, X^g\right]_t} \Rightarrow$$

$$\phi_t^c = \mathbf{E_Q}\left[\frac{d\left\langle Y^c, X^c \right\rangle_t}{d\left\langle X^c, X^c \right\rangle_t}\Big|\mathcal{F}_{t-}^X\right], \quad \phi_t^d = \mathbf{E_Q}\left[\frac{\triangle Y_t^d - \triangle C_t^d}{\triangle X_t^d}\Big|\mathcal{F}_{t-}^X\right],$$

$$\phi_t^g = \mathbf{E_Q}\left[\frac{\triangle^+ Y_t^g - \triangle^+ C_t^g}{\triangle^+ X_t^g}\Big|\mathcal{F}_t^X\right]. \tag{11.88}$$

Since $Y = Y_0 + M - A$ is $\mathbf{Q}$ local optional supermartingale, then it must be that $\mathbf{E_Q}\left[M_t|\mathcal{F}_t^X\right] = \phi \circ X_t$ and $\mathbf{E_Q}\left[A_t|\mathcal{F}_t^X\right] = C_t$. ∎

Remark 11.3.2 The filtering equation developed in the above theorem assumes that the dominant measure $\mathbf{P}^X$ is known or that its restriction $\mathbf{P}_{t+}^X$ to $\mathcal{F}_{t+}^X$ can be constructed from observations of X_t. However, the measure $\mathbf{P}$ is not necessarily known even though we have assumed it is defined and complete. A consequence of this is that the equations we have developed here are different from the ones developed by Vetrov [79, 143], Lipster and Shiryaev [51, 78] where they have assumed that $\mathbf{P}$ is known.

Remark 11.3.3 In Theorem 11.3.1 we have made the stronger assumption that ZY is local optional supermartingale for every $Z \in \mathbb{Z}(X)$. However, we only needed $\pi_t(Y) := \mathbf{E}\left[Z_t Y_t|\mathcal{F}_t^X\right]$ to be a local optional supermartingale.

11.3.2   Filtering on mixed stochastic basis

Consider the usual probability space $(\Omega, \mathcal{F}, \mathbf{F} = (\mathcal{F}_t)_{t\geq 0}, \mathbf{P})$. Let X and Y be $\mathbb{R}^d$ valued r.c.l.l. semimartingales but Y is special. The processes X and Y are related; $(X, Y) = (X_t, Y_t)_{t\geq 0}$ is a partially observable stochastic process where Y is the unobservable component and X is the observed process. X is a semimartingle, i.e., $X \in \mathcal{S}(\mathbf{F}, \mathbf{P})$. Then $X \in \mathcal{S}(\mathbf{F}_+^X, \mathbf{P}^X)$ where $\mathbf{P}^X$ is the restriction of the measure $\mathbf{P}$ to the σ-algebra $\mathcal{F}_\infty^X = \sigma\left(\cup_{t\geq 0}\mathcal{F}_t^X\right)$ and $\mathbf{F}_+^X = (\mathcal{F}_{t+}^X)_{t\geq 0}$ the family of σ-algebras $\mathcal{F}_{t+}^X = \bigcap_{\varepsilon > 0}\mathcal{F}_{t+\varepsilon}^X$ where $\mathcal{F}_t^X = \sigma\{X_s, s \leq t\} \vee \mathcal{N}$. If $\tilde{\mathbf{P}}^X$ is any probability measure on $(\Omega, \mathcal{F}_\infty^X)$ such that $\tilde{\mathbf{P}}^X \ll \mathbf{P}^X$, then $X \in \mathcal{S}(\mathbf{F}_+^X, \tilde{\mathbf{P}}^X)$. If there is more than one probability measure $\{\tilde{\mathbf{P}}^X\}$, then $\mathbf{F}^X$ is not right-continuous. Moreover, X is a different semimartingale under the different measure $\tilde{\mathbf{P}}^X$.

At this point, we are going to construct the probability space $(\Omega, \mathcal{F}^X, \mathbf{F}^X, \mathbf{P}^X)$ which is *un*usual. That is $\mathbf{F}^X$ may not be right-continuous or complete. Let $\tilde{\mathbb{P}}^X = \left\{ \tilde{\mathbf{P}}^X : \tilde{\mathbf{P}}^X \ll \mathbf{P}^X \right\}$, then $X \in \mathcal{S}(\mathbf{F}_+^X, \tilde{\mathbf{P}}^X)$ is an r.c.l.l. semimartingale and the projection of X on $\mathbf{F}^X$ is an optional semimartingale in $\mathcal{S}(\mathbf{F}^X, \tilde{\mathbf{P}}^X)$. Also, for every $\tilde{\mathbf{P}}^X$ there exists a Doob-Meyer decomposition of the optional semimartingale X.

Consider that a subset of $\tilde{\mathbb{P}}^X$ exist, such that $\tilde{\mathbf{P}}^X > 0$ and under which X is a local optional martingale. It is the set of equivalent optional local martingale measures,

$$\mathbb{Q}^X := \left\{ \mathbf{Q}^X : \mathbf{Q}^X \sim \tilde{\mathbf{P}}^X \right\} \subseteq \tilde{\mathbb{P}}^X$$

or

$$\mathbb{Q}^X := \left\{ \begin{array}{c} \mathbf{Q}^X : \mathbf{Q}^X > 0, \ \mathbf{Q}^X \ll \mathbf{P}^X, \\ \text{and } X \text{ is local optional martingale under } \mathbf{Q}^X \end{array} \right\}.$$

Furthermore, assume that the projection of Y on $\mathbf{F}^X$, $\pi_t(Y) := \mathbf{E}_\mathbf{Q}\left[Y_t | \mathcal{F}_t^X \right]$, is a local optional supermartingale that is $\mathcal{F}_t^X$-measurable. Hence, by optional uniform Doob-Meyer decomposition we get

$$\pi_t(Y) = \pi_0(Y) + \phi \circ X - C.$$

This brings us to our next corollary.

Corollary 11.3.4 Given the above exposition, the filtering equation of $\pi_t(Y)$ is

$$\pi_t(Y) = \pi_0(Y) + \phi \circ X - C, \quad \forall \mathbf{Q} \ a.s.$$

where ϕ is $\mathcal{P}(\mathbf{F}^X) \times \mathcal{O}(\mathbf{F}^X)$ is unique, X integrable process and C an optional process. ϕ is given by

$$\begin{aligned} \phi_t &= \mathbf{E}_\mathbf{Q}\left[\frac{d\langle Y^c, X^c \rangle_t}{d\langle X^c, X^c \rangle_t} | \mathcal{F}_{t-}^X \right] + \mathbf{E}_\mathbf{Q}\left[\frac{\triangle Y_t^d - \triangle C_t^d}{\triangle X_t^d} | \mathcal{F}_{t-}^X \right] \\ &\quad + \mathbf{E}_\mathbf{Q}\left[\frac{\triangle^+ Y_t^g - \triangle^+ C_t^g}{\triangle^+ X_t^g} | \mathcal{F}_t^X \right] \end{aligned} \qquad (11.89)$$

and $-C_t = \pi_t(Y) - \pi_0(Y) - \phi \circ X$. Since $Y = Y_0 + M - A$ is a $\mathbf{P}$ semimartingale, then $C_t = \mathbf{E}_\mathbf{Q}\left[A_t | \mathcal{F}_t^X \right]$ and $\mathbf{E}_\mathbf{Q}\left[M_t | \mathcal{F}_t^X \right] = \phi \circ X_t$.

Proof. Similar to Lemma 11.3.1. ∎

Remark 11.3.5 In Lemma 11.3.1 and Corollary 11.3.4, we have assumed that in determining the integrand ϕ of the observation process X, that C is known. However, Equations (11.88) and (11.89) can be thought of as equations relating the processes ϕ with C, X and Y. It is important to note that in Foellmer and Kabanov [144] the integrator in optional decomposition was determined by an optimization procedure where it is identified as the optimal Lagrange multiplier. As a consequence of determining the optimal Lagrange multiplier, the optional decreasing process was determined. In Lemma 11.3.1 and Corollary 11.3.4, C can be determined as a consequence of the optimization procedure.

11.4   Filtering in Finance

Filtering theory has a long history of applications to finance, economics and engineering (see e.g. [81, 145, 146, 147, 148, 149, 150]). However, some of the most fundamental connections between filtering and finance are in studying the problems of: insider trading [151, 152, 153], hedging and pricing under incomplete information and asymmetric information existing between market agents [154, 155, 156, 157, 158].

Here we study a particular model of financial markets based on the calculus of optional semimartingales on a *mixed probability spaces*, i.e., usual and *un*usual, where some elements of this market may not be observed. In other words, we are going to study an optional semimartingale market under incomplete information. Examples of these markets are the pricing and hedging of derivatives of non-tradable assets that are however correlated with tradable ones, and derivatives of assets that are illiquid such as over the counter traded assets or private funds and dark pools (see e.g. [159, 160, 161, 162, 163]). In any one of these cases the *market-value* of the underlying assets are not observed directly or can't be hedged with.

Let $\left(\Omega, \mathcal{F}, \mathbf{F} = (\mathcal{F}_t)_{t \geq 0}, \mathbf{P}\right)$, $t \in [0, \infty)$ be the *usual* stochastic basis and that the financial market stays on this space. Let $Y_t = Y_0 \mathcal{E}_t(y)$ be the unobserved asset where the stochastic exponential $\mathcal{E}_t(y)$ is given by

$$\mathcal{E}_t(y) = \exp\left\{ y_t - y_0 - \frac{1}{2} \langle y^c, y^c \rangle \right\} \prod_{0 < s \leq t} \left[(1 + \triangle y_s) e^{-\triangle y_s} \right]$$

and $X_t = X_0 \mathcal{E}_t(x)$ be the observed asset that is *related to* Y. Both X and Y are positive for all $t \geq 0$. The initial values X_0 and Y_0 are $\mathcal{F}_0$-measurable random variables. $x = (x_t)_{t \geq 0}$ and $y = (y_t)_{t \geq 0}$ are r.c.l.l. semimartingales admitting the representation, $x_t = x_0 + a_t + m_t$ and $y_t = y_0 + b_t + n_t$ with respect to $\mathbf{P}$. $a = (a_t)_{t \geq 0}$ and $b = (b_t)_{t \geq 0}$ are locally finite variation processes. $m = (m_t)_{t \geq 0}$ and $n = (n_t)_{t \geq 0}$ are local martingales.

Remark 11.4.1 Under the usual conditions, the stochastic exponential technique in the context of semimartingale markets was developed in Melnikov et al. [121]. This technique was extended to *un*usual spaces by Abdelghani and Melnikov [14].

A financial derivative based on Y is a function of Y. Let h be positive and continuous function and consider the financial derivative $H_t = h(Y_t)$. Suppose Y is unknown; however, the process X, which is related to Y, is observed. We are going to consider the pricing rule given by the projection of $h(Y_t)$ on our knowledge, $\mathcal{F}_t^X$, under the risk neutral measure $\mathbf{Q}$,

$$\pi_t(H) := \mathbf{E_Q}\left[H_t | \mathcal{F}_t^X \right].$$

Note that the filtration $\mathbf{F}^X = \left(\mathcal{F}_t^X\right)_{t \geq 0}$ may not be right-continuous. Furthermore, the measure $\mathbf{P}$ may not be accessible. In this case, we have to construct an *unusual* probability space $(\Omega, \mathcal{F}^X, \mathbf{F}^X, \mathbf{P}^X)$. In accordance with the second approach of Section 11.3.2, we suppose the set of risk-neutral measures

$$\mathbb{Q}^X := \left\{ \begin{array}{c} \mathbf{Q} : \mathbf{Q} > 0, \ \mathbf{Q} \ll \mathbf{P}^X, \\ \text{and } X \text{ is local optional martingale under } \mathbf{Q} \end{array} \right\}$$

is not empty. It was shown in [14] that one can find a non-empty set of positive local (optional) martingale measures, respectively, for the stochastic exponential process $X_t = X_0 \mathcal{E}_t(x)$ under some conditions on x. Therefore, by Corollary 11.3.4 we can price and hedge H by

$$\pi_t(H) = \pi_0(H) + \phi \circ X - C, \quad \forall \, \mathbf{Q} \ a.s.$$

where ϕ is given by

$$\begin{aligned}
\phi_t &= \mathbf{E}_{\mathbf{Q}} \left[\frac{d \langle H^c, X^c \rangle_t}{d \langle X^c, X^c \rangle_t} | \mathcal{F}_{t-}^X \right] + \mathbf{E}_{\mathbf{Q}} \left[\frac{\triangle H_t^d - \triangle C_t^d}{\triangle X_t^d} | \mathcal{F}_{t-}^X \right] \\
&\quad + \mathbf{E}_{\mathbf{Q}} \left[\frac{\triangle^+ H_t^g - \triangle^+ C_t^g}{\triangle^+ X_t^g} | \mathcal{F}_t^X \right].
\end{aligned}$$

Remark 11.4.2 In a market of derivatives of non-tradable assets, such as temperature derivatives, it is possible to measure temperature but its market value is unknown. Temperature market value can be estimated by observations through proxies – other tradable assets that are correlated with temperature, such as, the price of gas. Even though temperature is an observed variable, its market value is not known. Therefore, temperature derivatives can only be evaluated by optional projection on observed related assets.

Remark 11.4.3 In OTC markets, the traded value of the asset is the observed market price which may or may not contain information about the value of the asset traded over the counter. Therefore, a contingent claim written on an illiquid asset is subject to asymmetric information where the writer could have access to information in the OTC market while the buyer doesn't, resulting in illiquidity premium.

Example 11.4.4 Suppose the unobserved asset is $Y_t = Y_0 \mathcal{E} \left(\mu t + \sigma \tilde{W}_t \right)$, where μ and σ are constants, evolving on the usual probability space $\left(\Omega, \tilde{\mathcal{F}}, \tilde{\mathbf{F}} = \left(\tilde{\mathcal{F}}_t \right)_{t \geq 0}, \mathbf{P} \right)$. On the other hand, the observed asset follows the stochastic equation

$$X_t = X_0 \mathcal{E} \left(\alpha t + \eta W_t + \gamma L_t \right),$$

where α, η and γ are constants. $L_t = N_{t-} - \lambda t$ is compensated left-continuous Poisson process $(N_t)_{t \geq 0}$ which is $\mathcal{F}_{t+}$-measurable with constant intensity λ. Also, $\triangle^+ L > -\frac{1}{\gamma}$ since $X > 0$. X

is evolving on the *unusual* but *observable* probability space $\left(\Omega, \mathcal{F}^X, \mathbf{F}^X = \left(\mathcal{F}_t^X\right)_{t \geq 0}, \mathbf{P}\right)$, where $\mathcal{F}_t^X := \sigma\left(X_s, s \leq t\right)$ is not right-continuous as a result of X dependence on L. X and Y are correlated by the Wiener processes W and $\tilde{W}$ with $\left\langle W, \tilde{W} \right\rangle = qt$, where q is the covariation constant, and by the time trends αt and μt. The overall probability space $\left(\Omega, \mathcal{G}, \mathbf{G} = \left(\mathcal{G}_t\right)_{t \geq 0}, \mathbf{P}\right)$ generated by $\left(\tilde{W}, W, N\right)$ may be constructed in the usual way. Even though the overall probability space may be a usual probability space, the observed space is *unusual*. To compute a local martingale deflator for X, it makes sense to start with the guess $Z_t = \mathcal{E}\left(aW_t + bL_t\right)$ an optional positive local martingale such that ZX is an optional local martingale:

$$
\begin{aligned}
ZX &= X_0 \mathcal{E}\left(aW_t + bL_t\right) \mathcal{E}\left(\alpha t + \eta W_t + \gamma L_t\right) \\
&= X_0 \mathcal{E}\left(aW_t + bL_t + \alpha t + \eta W_t + \gamma L_t + a\eta t + b\gamma L - \lambda b\gamma t\right) \\
&= X_0 \mathcal{E}\left((\alpha + a\eta - \lambda b\gamma)t + (a+\eta)W_t + (b+\gamma + b\gamma)L_t\right);
\end{aligned}
$$

Hence, for ZX to be a local optional martingale, we must have $\alpha + a\eta - \lambda\gamma b = 0$ or that $(\eta, -\lambda\gamma)(a, b)^\top = -\alpha$ which results in many solutions. Let $H = Y_T$ be a contingent claim (e.g., a future), at time T that we like to price and hedge knowing $\mathcal{F}_t^X$. To do so, we must first find a Z such that ZY is a local optional supermartingale:

$$
\begin{aligned}
ZY &= Y_0 \mathcal{E}\left(\mu t + \sigma \tilde{W}_t\right) \mathcal{E}\left(aW_t + bL_t\right) \\
&= Y_0 \mathcal{E}\left((\mu + \sigma q a)t + \sigma \tilde{W}_t + aW_t + bL_t\right) \\
&= Y_0 \mathcal{E}\left((\mu + \sigma q a)t\right) \mathcal{E}\left(\sigma \tilde{W}_t + aW_t + bL_t\right) \\
&= Y_0 \exp\left((\mu + \sigma q a)t\right) \mathcal{E}\left(\sigma \tilde{W}_t + aW_t + bL_t\right),
\end{aligned}
$$

where $\mathcal{E}\left(\sigma \tilde{W}_t + aW_t + bL_t\right)$ is a local optional martingale, but also a supermartingale by Fatou lemma, i.e., $\mathbf{E}\left[Z_{t \wedge \tau} Y_{t \wedge \tau}\right] \leq \mathbf{E}\left[Y_0 \exp\left((\mu + \sigma q a)(t \wedge \tau)\right)\right]$ for any stopping time τ. Because $Z > 0$ and $ZY > 0$, then it must be that $\mu + \sigma q a \geq 0$ and $\triangle^+ L > -\frac{1}{b}$. But since $\triangle^+ L > -\frac{1}{\gamma}$ then we must at least choose $b \leq \gamma$. Putting all these conditions together we arrive at the set of equations

$$
(\eta, -\lambda\gamma)(a, b)^\top = -\alpha, \quad a \geq -\frac{\mu}{\sigma q}, \quad b \leq \gamma.
$$

The pricing formula is

$$
\begin{aligned}
\mathbf{E}\left[Z_T Y_T | \mathcal{F}_t^X\right] &= \\
&= Y_0 \exp\left((\mu + \sigma q a)T\right) \mathbf{E}\left[\mathcal{E}\left(\sigma \tilde{W}_T + aW_T + bL_T\right) | \mathcal{F}_t^X\right].
\end{aligned}
$$

Considering the sup over all Z,

$$
\begin{aligned}
\sup_{Z \in \mathbb{Z}(X)} \mathbf{E}\left[Z_T Y_T | \mathcal{F}_t^X\right] &= \\
&= Y_0 \sup_{(a,b)} \left\{\exp\left((\mu + \sigma q a)T\right) \mathbf{E}\left[\mathcal{E}\left(\sigma \tilde{W}_T + aW_T + bL_T\right) | \mathcal{F}_t^X\right]\right\},
\end{aligned}
$$

such that $(\eta, -\lambda\gamma)(a, b)^\top = -\alpha, \quad a \geq -\frac{\mu}{\sigma}, \quad b \leq \gamma.$

Using a geometric view of the equations above, one arrives at the solution $b = \gamma$ and $a = \eta^{-1}\left(\lambda\gamma^2 - \alpha\right)$ that gives us the maximum supermartingale

$$\sup_{Z\in\mathbb{Z}(X)} \mathbf{E}\left[Z_T Y_T | \mathcal{F}_t^X\right] = Y_0 \exp\left(\left(\mu + \eta^{-1}\sigma q\left(\lambda\gamma^2 - \alpha\right)\right)T\right) \times$$

$$\mathbf{E}\left[\mathcal{E}\left(\sigma\tilde{W}_T + \eta^{-1}\left(\lambda\gamma^2 - \alpha\right)W_T + \gamma L_T\right)|\mathcal{F}_t^X\right]$$

$$= Y_0 \exp\left(\left(\mu + \eta^{-1}\sigma q\left(\lambda\gamma^2 - \alpha\right)\right)T\right) \times$$

$$\mathcal{E}\left(\left(\sigma\sqrt{q} + \eta^{-1}\left(\lambda\gamma^2 - \alpha\right)\right)W_t + \gamma L_t\right)$$

Consequently, the maximizing local optional martingale is $\tilde{Z}_t = \mathcal{E}\left(\eta^{-1}\left(\lambda\gamma^2 - \alpha\right)W_t + \gamma L_t\right)$ for which we write $\tilde{H}_t = \mathbf{E}\left[\tilde{Z}_T H | \mathcal{F}_t^X\right]$ as $\hat{H}_t = \hat{H}_0 + \phi \circ X_t - C_t$ by UDMD under $\tilde{\mathbf{Q}} = \tilde{Z} \circ \mathbf{P}$. In our particular market framework, we have that the supremum value of the contingent claim is

$$\sup_{Z\in\mathbb{Z}(X)} \mathbf{E}\left[Z_T Y_T | \mathcal{F}_t^X\right] = \phi \circ \tilde{Z}_t X_t - C_t \quad \text{or that}$$

$$
\begin{aligned}
\hat{H}_t &= \mathbf{E}\left[\tilde{Z}_T Y_T | \mathcal{F}_t^X\right] = \mathbf{E}\left[\tilde{Z}_T \mathbf{E}\left[Y_T | \mathcal{F}_T^X\right] | \mathcal{F}_t^X\right] \\
&= \mathbf{E}\left[\tilde{Z}_T \pi_T\left(Y\right) | \mathcal{F}_t^X\right] = \mathbf{E}\left[\tilde{Z}_T \hat{Y}_T | \mathcal{F}_t^X\right] \\
&= \hat{H}_0 + \phi \circ \tilde{Z}_t X_t - C_t.
\end{aligned}
$$

where ϕ is chosen as the minimum value such that $\hat{H}_t - \phi \circ \tilde{Z}_t X_t \leq \hat{H}_0$ for all t; thereby the process C is the maximum difference.

The UDMD for optional supermartingales on *un*usual probability spaces is a natural and important generalization of the classical DMD and optional decomposition. We expect that this decomposition will play the same role played by optional decomposition in cadlag markets, but for financial markets evolving on *un*usual spaces (for example, see Abdelghani and Melnikov [14]). Clearly, we have demonstrated how such a decomposition can be used to build optimal filters in filtering of optional supermartingale, further generalizing the theory of optimal filtering of semimartingales.

Bibliography

[1] T. R. Fleming and D. P. Harrington. *Counting processes and survival analysis*. John Wiley and Sons, New York, 2011.

[2] Gopinath Kallianpur. *Stochastic filtering theory*, volume 13. Springer Science & Business Media, 2013.

[3] Jean-François Mertens. Processus stochastiques généraux et surmartingales. *Z. Wahrsch. Verw. Geb*, 22:45–68, 1972.

[4] Claude Dellacherie and Paul-André Meyer. Un nouveau théoreme de projection et de section. In *Séminaire de Probabilités IX Université de Strasbourg*, pages 239–245. Springer, Berlin, Heidelberg, 1975.

[5] Joseph Leo Doob. Stochastic process measurability conditions. *Ann. Inst. Fourier*, 25:163–167, 1975.

[6] Dominique Lépingle. Sur la représentation des sauts des martingales. *Séminaire de Probabilités de Strasbourg*, 11:418–434, 1977.

[7] J. Horowitz. Optional supermartingales and the andersen-jessen theorem. *Z. Wahrscheinlichkeitstheorie und verw Gebiete*, 43(263-272):3, 1978.

[8] E. Lenglart. Tribus de meyer et theorie des processus. *Lecture Notes in Mathematics*, 784:500–546, 1980.

[9] I. Galchuk, Leonid. On the existence of optional modifications for martingales. *Theory Prob. Appl.*, 22:572–573, 1977.

[10] I. Galchuk, Leonid. Decomposition of optional supermartingales. *Math. Ussr Sbornik*, 43(2):145–158, 1982.

[11] I. Galchuk, Leonid. Stochastic integrals with respect to optional semimartingales and random measures. *Theory of probability and its applications XXIX*, 1:93–108, 1985.

[12] K. V. Gasparyan. Stochastic equations with respect to optional semimartingales. *Izvestiya Vysshikh Uchebnykh Zavedenii*, 12:57–60, 1985.

[13] C. Kühn and M. Stroh. Optimal portfolios of a small investor in a limit order market: a shadow price approach. *Mathematics and Financial Economics*, 3(45-72):2, 2010.

[14] Mohamed N. Abdelghani and Alexander V. Melnikov. Financial markets in the context of the general theory of optional processes. In *Mathematical and Computational Approaches in Advancing Modern Science and Engineering*, pages 519–528. Springer, 2016.

[15] Mohamed N. Abdelghani and Alexander V. Melnikov. On linear stochastic equations of optional semimartingales and their applications. *Statistics & Probability Letters*, 125:207–214, 2017.

[16] Mohamed N. Abdelghani and Alexander V. Melnikov. Optional defaultable markets. *Risks*, 5(4):56, 2017.

[17] Mohamed N. Abdelghani and Alexander V. Melnikov. Existence and uniqueness of stochastic equations of optional semimartingales under monotonicity condition. *Stochastics*, 92(1):67–89, 2019.

[18] Mohamed N. Abdelghani and Alexander V. Melnikov. Optional decomposition of optional supermartingales and applications to filtering and finance. *Stochastics*, 91(6):797–816, 2019.

[19] Claude Dellacherie and Paul-André Meyer. *Probabilities and potential. Chapters I-IV*. North-Holland Publishing Co., Amsterdam, 1978.

[20] Claude Dellacherie and Paul-André Meyer. *Probabilities and potential, B: Theory of Martingales*. North-Holland Publishing Co., Amsterdam, 1982.

[21] Nicolas Bourbaki. *Elements of mathematics: General topology*. Springer, 1995.

[22] Maurice Sion. On capacitability and measurability. In *Annales de l'institut Fourier*, volume 13, pages 83–98, 1963.

[23] Bernard Maisonneuve. Topologies du type de Skorohod. *Séminaire de probabilités de Strasbourg*, 6:113–117, 1972.

[24] Kai Lai Chung and Joseph Leo Doob. Fields, optionality and measurability. *American Journal of Mathematics*, 87(2):397–424, 1965.

[25] Joseph Leo Doob. *Stochastic processes*. J. Wiley & Sons London, Chapman & Hall, New York, 1953.

[26] Kiyosi Itô. The canonical modification of stochastic processes. *Journal of the Mathematical Society of Japan*, 20(1-2):130–150, 1968.

[27] Kai Lai Chung. On the fundamental hypotheses of hunt processes. In *Chance And Choice: Memorabilia*, pages 211–220. World Scientific, 2004.

[28] Claude Dellacherie. *Capacités et processus stochastiques*. Number 67 in Ergebnisse der Mathematik und ihrer Grenzgebiete. Springer-Verlag, Berlin, Heidelberg, 1972.

[29] Frank B Knight. Existence of small oscillations at zeros of brownian motion. *Séminaire de probabilités de Strasbourg*, 8:134–149, 1974.

[30] Claude Dellacherie. Deux remarques sur la separabilite optionnelle. In *Séminaire de Probabilités XI*, volume 11, pages 47–50, Berlin, Heidelberg, 1977. Springer Berlin Heidelberg.

[31] Jean Ville. Etude critique de la notion de collectif. *Bull. Amer. Math. Soc*, 45(11):824, 1939.

[32] James Laurie Snell. Applications of martingale system theorems. *Transactions of the American Mathematical Society*, 73(2):293–312, 1952.

[33] Donald L Burkholder. Martingale transforms. *The Annals of Mathematical Statistics*, 37(6):1494–1504, 1966.

[34] Elias M Stein. *Singular integrals and differentiability properties of functions*, volume 2. Princeton University Press, 1970.

[35] Jacques Neveu. Bases mathématiques du calcul des probabilités. 1970.

[36] David Blackwell, Lester E Dubins, et al. A converse to the dominated convergence theorem. *Illinois Journal of Mathematics*, 7(3):508–514, 1963.

[37] Richard Gundy. On the class L Log L, martingales, and singular integrals. *Studia Mathematica*, 1(33):109–118, 1969.

[38] C. S. Chou. Sur les martingales de la classe L Log L. *CRAS Paris*, 277:751–752, 1973.

[39] Lester E Dubins et al. A note on upcrossings of semimartingales. *The Annals of Mathematical Statistics*, 37(3):728–728, 1966.

[40] Lester E Dubins and David A Freedman. A sharper form of the Borel-Cantelli lemma and the strong law. *The Annals of Mathematical Statistics*, 36(3):800–807, 1965.

[41] Jacques Azéma. Théorie générale des processus et retournement du temps. In *Annales scientifiques de l'École Normale Supérieure*, volume 6, pages 459–519, 1973.

[42] Philippe Morando. Mesures aléatoires. *Séminaire de probabilités de Strasbourg*, 3:190–229, 1969.

[43] J. Horowitz. Une remarque sur les bimesures. *Sem. Probe XI*, 581:59–64, Sem. Probe XI, LN.

[44] Jean-François Mertens. Théorie des processus stochastiques généraux applications aux surmartingales. *Probability Theory and Related Fields*, 22(1):45–68, 1972.

[45] H Kunita. Théorie du filtrage. *Cours de 3e Cycle à l'Université Paris VI (1974-75)*.

[46] Claude Dellacherie. *Sur les Théorèmes Fondamentaux de la Théorie Générale des Processus*, pages 75–84. Springer, Berlin, Heidelberg, 2002.

[47] Jean-Michel Bismut. Temps d'arrêt optimal, quasi-temps d'arrêt et retournement du temps. *The Annals of Probability*, pages 933–964, 1979.

[48] Marie Anne Maingueneau. Temps d'arrêt optimaux et théorie générale. In *Séminaire de Probabilités XII*, pages 457–467. Springer, 1978.

[49] Jean-François Mertens. Strongly supermedian functions and optimal stopping. *Probability Theory and Related Fields*, 26(2):119–139, 1973.

[50] JM Bismut and B Skalli. Le probleme général d'arrêt optimal. *CR Acad. Sci. Paris, Sér. AB*, 283(6), 1976.

[51] Robert Liptser and Albert N. Shiryaev. *Theory of Martingales*, volume 49. Springer Science & Business Media, 2012.

[52] Michel Emery. Stabilité des solutions des équations différentielles stochastiques application aux intégrales multiplicatives stochastiques. *Probability Theory and Related Fields*, 41(3):241–262, 1978.

[53] Jean-Michel Bismut. Régularité et continuité des processus. *Zeitschrift für Wahrscheinlichkeitstheorie und Verwandte Gebiete*, 44(3):261–268, 1978.

[54] I. Galchuk, Leonid. Optional martingales. *Mathematics of the ussr-sbornik*, 40(4):435, 1981.

[55] Paul A. Meyer. *Probability and Potentials*. Blaisdell Publ. Co., Waltham, Mass, 1966.

[56] Claude Dellacherie. Sur la régularisation des surmartingales. In *Séminaire de Probabilités XI*, pages 362–364. Springer, 1977.

[57] J. Jacod. Multivariate point processes: predictable projection, radon-nykodym derivatives, representation of martingales. *Z. Wahrscheinlichkeitstheorie und verw. Gebiete*, 31(3):235 253, 1975.

[58] Albert Benveniste. Separabilite optionnelle, d'apres Doob. In *Séminaire de Probabilités X Université de Strasbourg*, pages 521–531. Springer, 1976.

[59] P. A. Meyer. Un cours sur les integrales stochastiques. *Lecture Notes in Mathematics*, 511:245–400, 1976.

[60] C. Doleans-Dade and P. A. Meyer. Integrals stochastiques par rapport aux martingales locales. *Sem. Prob., Lecture Notes in Math.*, 4(124), 1970.

[61] Chou Ching-sung. Le processus des sauts d'une martingale locale. In *Séminaire de Probabilités XI*, pages 356–361. Springer, 1977.

[62] Paul-André Meyer. *Un cours sur les intégrales stochastique, sem. Probabilites, X (Univ. Strasbourg), lecture notes in math.* Springer-Verlag, 1977.

[63] J. Jacod. Calcul stochastique problemes de martingales. *Lecture Notes in Mathematics*, 714, 1979.

[64] Jacques Neveu. Martingales à temps discret. 1972.

[65] Ronald K Getoor. On the construction of kernels. In *Séminaire de Probabilités IX Université de Strasbourg*, pages 443–463. Springer, 1975.

[66] Yurii M. Kabanov, Robert Sh. Liptser, and Albert N. Shiryaev. Absolute continuity and singularity of locally absolutely continuous probability distributions. i. *Matematicheskii Sbornik*, 149(3):364–415, 1978.

[67] Ju M Kabanov, R Š Lipcer, and AN Širjaev. Absolute continuity and singularity of locally absolutely continuous probability distributions. ii. *Mathematics of the USSR-Sbornik*, 36(1):31, 1980.

[68] I. Galchuk, Leonid. On the change of variables formula. *Matem. Zametki*, 26(4):633–641, 1979.

[69] N. El Karoui and M. C. Quenez. Dynamic programming and pricing of contingent claims in an incomplete market. *SIAM Journal on Control and optimization*, 33(1):27–66, 1995.

[70] D. O. Kramkov. Optional decomposition of supermartingales and hedging in incomplete security markets. *Probability Theory and Related Fields*, 105:459–479, 1996.

[71] Hans Föllmer and Yuri M. Kabanov. Optional decomposition and Lagrange multipliers. *Finance and Stochastics*, 2(1):69–81, 1997.

[72] Christophe Stricker and Jia-An Yan. Some remarks on the optional decomposition theorem. In *Séminaire de Probabilités XXXII*, pages 56–66. Springer, 1998.

[73] Saul Jacka. A simple proof of Kramkov's result on uniform supermartingale decompositions. *Stochastics An International Journal of Probability and Stochastic Processes*, 84(5-6):599–602, 2012.

[74] Ioannis Karatzas, Constantinos Kardaras, et al. Optional decomposition for continuous semimartingales under arbitrary filtrations. *Electronic Communications in Probability*, 20, 2015.

[75] H. Kushner. Nonlinear filtering: The exact dynamical equations satisfied by the conditional mode. *IEEE Transactions on Automatic Control*, 12(3):262–267, 1967.

[76] Moshe Zakai. On the optimal filtering of diffusion processes. *Zeitschrift für Wahrscheinlichkeitstheorie und verwandte Gebiete*, 11(3):230–243, 1969.

[77] D. I. Hadjiev. On the filtering of semimartingales in case of observations of point processes. *Theory of Probability and Its Applications*, 23(1):169–178, 1978.

[78] R. S. Lipster and A. N. Shiryaev. *Statistics of Random Processes*. Springer-Verlag, New York-Heidelberg, 1974.

[79] L. G. Vetrov. On filtering, interpolation and extrapolation of semimartingales. *Theory of Probability & Its Applications*, 27(1):24–36, 1982.

[80] A. H. Jazwinski. *Stochastic processes and filtering theory*. Courier Corporation, 2007.

[81] R. Bhar. *Stochastic filtering with applications in finance*. World Scientific, 2010.

[82] S. Cohen and R. J. Elliott. *Stochastic calculus and applications*. Birkhäuser, 2015.

[83] R. J. Elliott and P. E. Kopp. *Mathematics of financial markets*. Springer-Verlag, Berlin, New York, 1999.

[84] M. Metivier. A stochastic Gronwall lemma and its application to a stability theorem for stochastic differential equations. *C.R. Acad. Sci. Paris, Ser. A.*, 289, 1979.

[85] A. V. Melnikov. Gronwall lemma and stochastic equations for components of semimartingale. *Matematicheskie Zametki*, 32(3):411–423, 1982.

[86] E. N. Stewart and T. G. Kurtz. *Markov Processes: Characterization and Convergence, Probability and Mathematical Statistics*. John Wiley & Sons Inc., New York, 1986.

[87] K. G. Thomas and P. Protter. Weak limit theorems for stochastic integrals and stochastic differential equations. *Ann. Probab.*, 19(3):1035–1070, 1991.

[88] I. Galchuk, Leonid. Strong convergence of the solution of a stochastic integral equation with respect to components of a semimartingale. *Mat. Zametki*, 35:299–306, 1984.

[89] A. V. Melnikov. Stochastic differential equations: singularity of coefficients, regression models and stochastic approximation. *Russian Mathematical Surveys*, 51(5):43–136, 1996.

[90] P. Protter. *Stochastic Integration and Differential Equations*. Stochastic Modelling and Applied Probability. Springer Berlin Heidelberg, 2013.

[91] K. Ito. On stochastic differential equations. *Mem. Amer. Math. Soc.*, 4:1–51, 1951.

[92] I. I. Gihman and A. V. Skorohod. *Stochastic Differential Equations*. Ergebnisse der Mathematik und ihrer Grenzgebiete. 2. Folge. Springer Berlin Heidelberg, 2014.

[93] Norihiko Kazamaki. On the existence of solutions of martingale integral equations. *Tohoku Mathematical Journal, Second Series*, 24(3):463–468, 1972.

[94] Philip E. Protter. On the existence, uniqueness, convergence and explosions of solutions of systems of stochastic integral equations. *The Annals of Probability*, pages 243–261, 1977.

[95] I. Galchuk, Leonid. On the existence and uniqueness of solutions for stochastic equations with respect to martingales and random measures. *Proc. II Vilnius Conf. On Prob. Theory*, 1:88–91, 1977.

[96] I. Gyongy and N. Krylov. On stochastic equations with respect to semimartingales i. *Stochastics*, 4:1–21, 1980.

[97] Jean-Michel Bismut. Conjugate convex functions in optimal stochastic control. *Journal of Mathematical Analysis and Applications*, 44(2):384–404, 1973.

[98] Etienne Pardoux and Shige Peng. Adapted solution of a backward stochastic differential equation. *Systems & Control Letters*, 14(1):55–61, 1990.

[99] Darrell Duffie and Larry G Epstein. Stochastic differential utility. *Econometrica: Journal of the Econometric Society*, pages 353–394, 1992.

[100] Darrell Duffie and Larry G Epstein. Asset pricing with stochastic differential utility. *The Review of Financial Studies*, 5(3):411–436, 1992.

[101] Nicole El Karoui, Shige Peng, and Marie Claire Quenez. Backward stochastic differential equations in finance. *Mathematical Finance*, 7(1):1–71, 1997.

[102] Etienne Pardoux and Shige Peng. Backward stochastic differential equations and quasilinear parabolic partial differential equations. In *Stochastic partial differential equations and their applications*, pages 200–217. Springer, 1992.

[103] Hao Xing. Consumption–investment optimization with epstein–zin utility in incomplete markets. *Finance and Stochastics*, 21(1):227–262, 2017.

[104] A. V. Skorokhod. *Investigations in the Theory of Random Processes*. Kiev, KGU, 1961.

[105] T. Yamada. On comparison theorem for solutions of stochastic differential equations and its applications. *Journal of Mathematics Kyoto University*, 13:497–512, 1973.

[106] A. Uppman. *Deux Applications des Semimartingales Exponentielles aux Equations Differentielles Stochastiques*. C. R. Acad. Sci., Paris, 1980.

[107] I. Galchuk, Leonid. A comparison theorem for stochastic equations with integral with respect to martingales and random measures. *Theory Probability and Applications*, 27(3):450–460, 1982.

[108] John C Cox and Stephen A Ross. The valuation of options for alternative stochastic processes. *Journal of Financial Economics*, 3(1-2):145–166, 1976.

[109] Christoph Czichowsky, Walter Schachermayer, et al. Duality theory for portfolio optimisation under transaction costs. *The Annals of Applied Probability*, 26(3):1888–1941, 2016.

[110] Jean-Francois Chassagneux and Bruno Bouchard. Representation of continuous linear forms on the set of ladlag processes and the hedging of american claims under proportional costs. *Electronic Journal of Probability*, 14:612–632, 2009.

[111] Miryana Grigorova, Peter Imkeller, Elias Offen, Youssef Ouknine, Marie-Claire Quenez, et al. Reflected BSDEs when the obstacle is not right-continuous and optimal stopping. *The Annals of Applied Probability*, 27(5):3153–3188, 2017.

[112] Christoph Kühn and Marc Teusch. Optional processes with non-exploding realized power variation along stopping times are làglàd. *Electronic Communications in Probability*, 16:1–8, 2011.

[113] Shijie Deng. *Stochastic models of energy commodity prices and their applications: Mean-reversion with jumps and spikes*. University of California Energy Institute Berkeley, 2000.

[114] Carol Alexander and Aanand Venkatramanan. Analytic approximations for multi-asset option pricing. *Mathematical Finance: An International Journal of Mathematics, Statistics and Financial Economics*, 22(4):667–689, 2012.

[115] Fischer Black and Myron Scholes. The pricing of options and corporate liabilities. *Journal of Political Economy*, 81(3):637–654, 1973.

[116] J. M. Harrison and D. M. Kreps. Martingales and arbitrage in multiperiod securities markets. *J. Economic Theory*, 20(3):381–408, 1979.

[117] J. M. Harrison and S. R. Pliska. Martingales and stochastic integrals in the theory of continuous trading. *Stochastic Process Appl.*, 11:215–260, 1981.

[118] R. Merton. *Continuous Time Finance*. Blackwall, Cambridge MA, Oxford UK, 1990.

[119] A. N. Shiryaev, Yu. M. Kabanov, D. O. Kramkov, and A. V. Melnikov. Towards the theory of pricing options of both European and American types I discrete time. *Theory Probab. Appl.,*, 39:14–60, 1994.

[120] A. N. Shiryaev, Yu. M. Kabanov, D. O. Kramkov, and A. V. Melnikov. Towards the theory of pricing options of both European and American types II continuous time. *Theory Probab. Appl.*, 39:61–102, 1994.

[121] A. V. Melnikov, S. N. Volkov, and M. L. Nechaev. *Mathematics of Financial Obligations*, volume 212. American Mathematical Society, Providence, Rhode Island, 2002.

[122] Christoph Kühn and Maximilian Stroh. A note on stochastic integration with respect to optional semimartingales. *Electronic Communications in Probability*, 14:192–201, 2009.

[123] Freddy Delbaen and Walter Schachermayer. *The Mathematics of Arbitrage*. Springer Science & Business Media, 2006.

[124] Constantinos Kardaras. Market viability via absence of arbitrage of the first kind. *Finance and stochastics*, 16(4):651–667, 2012.

[125] D. Duffie and K. Singleton. Modeling term structures of defaultable bonds. *Rev. Finan. Stud.*, 12:687–720, 1999.

[126] R. C. Merton. On the pricing of corporate debt: The risk structure of interest rates. *J. Finance*, 29:449–470, 1974.

[127] R. C. Merton. Option pricing when underlying stock returns are discontinuous. *J. Finan. Econom.*, 3:125–144, 1976.

[128] R. Geske. The valuation of compound options. *J. Finan. Econom.*, 7:63–81, 1979.

[129] R. Geske and H. E. Johnson. The valuation of corporate liabilities as compound options: A correction. *J. Finan. Quant. Anal.*, 19:231–232, 1984.

[130] H. E. Leland. Corporate debt value, bond covenants, and optimal capital structure. *J. Finance*, 49:1213–1252, 1994.

[131] F. A. Longstaff and E. S. Schwartz. A simple approach to valuing risky fixed and floating rate debt. *J. Finance*, 50:789–819, 1995.

[132] Philippe Artzner and Freddy Delbaen. Default risk insurance and incomplete markets 1. *Mathematical Finance*, 5(3):187–195, 1995.

[133] D. Duffie and K. Singleton. An econometric model of the term structure of interest-rate swap yields. *J. Finance*, 52:1287–1321, 1997.

[134] D. Lando. On cox processes and credit-risky securities. *Rev. Derivatives Res.*, 2:99–120, 1998.

[135] L. Schlogl. An exposition of intensity-based models of securities and derivatives with default risk. 1998. Working Paper.

[136] D. Wong. A unifying credit model. 1998. Working Paper.

[137] R. J. Elliott, M. Jeanblanc, and M. Yor. On models of default risk. *Math. Finance*, 10:179–195, 2000.

[138] Alain Bélanger, Steven Shreve, and Dennis Wong. A unified model for credit derivatives. *Mathematical Finance*, 2001.

[139] N. El Karoui, M. Jeanblanc, and Y. Jiao. What happens after a default: the conditional density approach. *Stochastic processes and their applications,*, 120(7):1011–1032, 2010.

[140] M. Jeanblanc. Enlargements of filtrations. 2010.

[141] Tomasz R Bielecki and Marek Rutkowski. *Credit risk: modeling, valuation and hedging*. Springer Science & Business Media, 2013.

[142] Jean Jacod and Philip Protter. Risk-neutral compatibility with option prices. *Finance and Stochastics*, 14(2):285–315, 2010.

[143] L. G. Vetrov. On linearization of stochastic differential equations of optimal nonlinear filtering. *Theory of Probability & Its Applications*, 25(2):393–401, 1981.

[144] H. Foellmer and Yu. M. Kabanov. Optional decomposition and Lagrange multipliers. *Finance and Stochastics*, 2(1):69–81, 1997.

[145] W. J. Runggaldier. Estimation via stochastic filtering in financial market models. *Contemporary mathematics*, 351:309–318, 2004.

[146] P. Date and K. Ponomareva. Linear and non-linear filtering in mathematical finance: a review. *IMA Journal of Management Mathematics*, 22(3):195–211, 2011.

[147] D. Brigo and B. Hanzon. On some filtering problems arising in mathematical finance. *Insurance: Mathematics and Economics*, 22(1):53–64, 1998.

[148] P. Christoffersen, C. Dorion, K. Jacobs, and L. Karoui. Nonlinear kalman filtering in affine term structure models. *Management Science*, 60(9):2248–2268, 2014.

[149] C. Wells. *The Kalman filter in finance*, volume 32. Springer Science & Business Media, 2013.

[150] A. V. Borisov, B. M. Miller, and K. V. Semenikhin. Filtering of the Markov jump process given the observations of multivariate point process. *Automation and Remote Control*, 76(2):219–240, Feb 2015.

[151] P. Vitale. Insider trading in sequential auction markets with risk-aversion and time-discounting. *The European Journal of Finance*, pages 1–13, 2016.

[152] A. Danilova. Stock market insider trading in continuous time with imperfect dynamic information. *Stochastics An International Journal of Probability and Stochastics Processes*, 82(1):111–131, 2010.

[153] S. Luo and Q. Zhang. Dynamic insider trading. *AMS IP Studies in Advanced Mathematics*, 26:93–104, 2002.

[154] J. B. Detemple. Asset pricing in a production economy with incomplete information. *The Journal of Finance*, 41(2):383–391, 1986.

[155] R. C. Merton. A simple model of capital market equilibrium with incomplete information. *The Journal of Finance*, 42(3):483–510, 1987.

[156] T. Berrada. Incomplete information, heterogeneity, and asset pricing. *Journal of Financial Econometrics*, 4(1):136–160, 2006.

[157] X. Guo, R. A. Jarrow, and Y. Zeng. Credit risk models with incomplete information. *Mathematics of Operations Research*, 34(2):320–332, 2009.

[158] B. De Meyer. Price dynamics on a stock market with asymmetric information. *Games and economic behavior*, 69(1):42–71, 2010.

[159] H. Geman and M. Leonardi. Alternative approaches to weather derivatives pricing. *Managerial Finance*, 31(6):46–72, 2005.

[160] A. Ang, A. A. Shtauber, and P. C. Tetlock. Asset pricing in the dark: The cross-section of otc stocks. *Review of Financial Studies*, 26(12):2985–3028, 2013.

[161] D. Duffie. *Dark markets: Asset pricing and information transmission in over-the-counter markets*. Princeton University Press, 2012.

[162] H. Mi and S. Zhang. Dynamic valuation of options on non-traded assets and trading strategies. *Journal of Systems Science and Complexity*, 26(6):991–1001, 2013.

[163] Mohamed N. Abdelghani and Alexander V. Melnikov. On macrohedging problem in semi-martingale markets. *Frontiers in Applied Mathematics and Statistics*, 1:3, 2015.

Index